Automotive Technician Training: Theory

Automotive Technician Training is the definitive student textbook for automotive engineering. It covers all the theory and technology sections that students need to learn in order to pass levels 1, 2 and 3 automotive courses. It is recommended by the Institute of the Motor Industry and is ideal for courses and exams run by other awarding bodies. This revised edition overhauls the coverage of general skills and advanced diagnostic techniques. It also includes a new chapter about electric and hybrid vehicles and advanced driver-assistance systems, along with new online learning activities.

Unlike current textbooks on the market, this takes a blended-learning approach, using interactive features that make learning more enjoyable and effective. It is ideal to use on its own but when linked with IMI eLearning online resources, it provides a comprehensive package that includes activities, video footage, assessments and further reading. Information and activities are set out in sequence to meet teacher and learner needs, as well as qualification requirements.

Tom Denton is the leading automotive author in the UK with a teaching career spanning lecturer to head of automotive engineering in a large college. He has published over forty automotive textbooks since 1995 and all are bestsellers.

Hayley Pells is an experienced MOT tester, owner/operator of award winning, Avia Sports Cars Ltd. She is a versatile automotive aftermarket writer, regularly published since 2017, an authoritative public speaker and consultant specialising in digital literacy, blended learning and electric vehicle technology.

Automotive Technician Training: Theory

Second Edition

Tom Denton and Hayley Pells

Routledge
Taylor & Francis Group

LONDON AND NEW YORK

Second edition published 2022
by Routledge
2 Park Square, Milton Park, Abingdon, Oxon, OX14 4RN

and by Routledge
605 Third Avenue, New York, NY 10158

Routledge is an imprint of the Taylor & Francis Group, an informa business

First edition published by Routledge 2014

British Library Cataloguing-in-Publication Data
A catalogue record for this book is available from the British Library

Library of Congress Cataloging-in-Publication Data
A catalog record for this book has been requested

ISBN: 978-1-032-00233-0 (hbk)
ISBN: 978-1-032-00220-0 (pbk)
ISBN: 978-1-003-17323-6 (ebk)

DOI:10.1201/9781003173236

Typeset in Univers LT
by Apex CoVantage, LLC

Printed and bound by CPI Group (UK) Ltd, Croydon, CR0 4YY

Contents

Contents

Contents

Preface

This second edition of *Automotive Technician Training* (ATT) is aimed at students who are taking a course in automotive technology or are interested in learning for themselves about this fascinating industry. Like the previous edition it is an ideal companion to the IMI eLearning but we have also included more activities and the book now works as a standalone package too.

I am delighted to be partnering on this book with my friend and colleague Hayley Pells, a multi award winning owner of a highly respected service and repair business, Avia Autos, in Bridgend, South Wales.

Tom

Comments, suggestions and feedback are always welcome at my website: www.tomdenton.org. On this site, you will also find lots of **free** online resources to help with your studies. Several other books are available too:

▶ *Automobile Mechanical and Electrical Systems*
▶ *Automobile Electrical and Electronic Systems*
▶ *Advanced Automotive Fault Diagnosis*
▶ *Electric and Hybrid Vehicles*
▶ *Alternative Fuel Vehicles*
▶ *Automated Driving and Driver Assistance Systems*

We never stop learning, so I hope you find automotive technology as interesting as I still do.

It is a privilege to work with my friend and inspiration Tom Denton, a highly respected technical author and experienced automotive engineer. Working within the automotive industry has proved to me how the skills learned at the beginning of a career can be transferred to numerous aspects of work and personal life.

Hayley

I run my own workshop which you can see at https://aviaautos.com. Here we communicate with our motorists as well as our business clients; we also offer insight and information about our environment and how we continue to learn and engage with further education as a team.

I wish you every success with your journey, and hope this learning enables you to increase your opportunity to enjoy our sector.

Acknowledgements

Over the years many people have helped in the production of my books. I am therefore very grateful to the following companies who provided information and/or permission to reproduce photographs and/or diagrams:

AA
AC Delco
ACEA
Alpine Audio Systems
Audi
Autologic Data Systems
BMW UK
Bosch
Brembo brakes
C&K Components
Citroën UK
Clarion Car Audio
Continental
CU-ICAR
Dana
Delphi Media
Eberspaecher
First Sensor AG
Fluke Instruments UK
Flybrid systems
Ford Motor Company
Freescale Electronics
General Motors
GenRad
Google (Waymo)
haloIPT (Qualcomm)
Hella
HEVT
Honda

Hyundai
Institute of the Motor Industry (IMI)
Jaguar Cars
Kavlico
Ledder
Loctite
Lucas UK
LucasVarity
Mahle
Matlab/Simulink
Mazda
McLaren Electronic Systems
Mennekes
Mercedes
MIT
Mitsubishi
Most Corporation
NASA
NGK Plugs
Nissan
Nvidia
Oak Ridge National Labs
Peugeot
Philips
Pico Tech/PicoScope
Pierburg
Pioneer Radio
Pixabay
Porsche

Renesas
Rolec
Rover Cars
SAE
Saab Media
Scandmec
Shutterstock
SMSC
Snap-on Tools
Society of Motor Manufacturers and Traders (SMMT)
Sofanou
Sun Electric
T&M Auto-Electrical
Tesla Motors
Texas Instruments
Thrust SSC Land Speed Team
Toyota
Tracker
Tula
Unipart Group
Valeo
Vauxhall
VDO Instruments
Volkswagen
Volvo Cars
Volvo Trucks
Wikimedia
ZF Servomatic

If I have used any information, or mentioned a company name that is not listed here, please accept my apologies and let me know so it can be rectified as soon as possible.

CHAPTER 1

How to use this book

After successful completion of this chapter you will be able to show you have achieved these outcomes:

- Understand the various icons and symbols and structure used in this book and online

- Understand how to use the learning activities and other features

DOI: 10.1201/9781003173236-1

1.1 Introduction

1.1.1 Start here **E** **1**

Introduction The associated IMI eLearning is recommended and can be used in conjunction with this book:

https://elearning.theimi.org.uk

However, it is not essential and the book is also a great standalone learning package. There are also some additional support materials here:

https://www.tomdenton.org

Qualification levels This book contains all the theory/technology content for automotive study at levels 1, 2 and 3. The materials are presented as chapters, subjects and then sections. Each section (lesson) is marked as follows:

- **E** – study only these sections if working on an Entry level qualification
- **1** – study only these sections if working on a level 1 qualification
- **2** – study these sections and those marked level 1 (unless done previously) if working on a level 2 qualification
- **3** – study these sections and those marked level 1 and level 2 (unless done previously) if working on a level 3 qualification
- **1 2 3** – if a section is marked with more than one number, then study the level you are working on in detail and skim read the higher level

Title Most of the paragraphs of text in this book with start with a title in bold as shown here! This is the title that you will see on the associated eLearning screen if you are using it. All the activities in this book can be carried out using features on the interactive website. Paper and pencils will work too in most cases so you can still work if you don't have internet access.

 When you see this icon, it means that the associated image, video or sound is available in the eLearning or other online sources. You should refer to this to, for example, add labels or complete the diagrams as necessary.

Learning activities are included for each section (lesson). The answers to these activities can be written directly in the book, on a separate notebook or, perhaps even better, as a document or in an electronic notebook. An overview of each activity is outlined below.

Activity sheets are available that match the learning activities, together with some real-world examples. These really help with the learning process and guide you through the activities. Access them and the interactive tools here:

https://www.tomdenton.org

Note: The activities suggested after each section are recommended but you can choose different ones. Remember, the more you do the more you will learn.

1.1.2 Learning activities **E** **1**

Information search Looking in other textbooks or in a library is an effective way to see the subject explained differently. Perhaps even better is to use the online search options on the interactive site

 Use a library or the web search tools to further examine the subject in this section

Media search Searching online for images, animations and videos is an excellent way to see other ways of how something works.

Use the media search tools to look for pictures and videos relating to the subject in this section

Bullets Three great tools for keeping notes electronically are Evernote, Microsoft OneNote and Google docs. My favourite currently is OneNote but I find all these tools easy to work with and they can be used online or offline, they also sync to or from a smartphone. Using any word processor is fine – as is using a pen!

Look back over the previous section and write out a list of the key bullet points from this section

Sketch Making a simple sketch to help you remember how a component or system works is a good way to learn. You can use a pencil or the online features or any drawing program – even word processors have quite good drawing tools built in.

Make a simple sketch to show how one of the main components or systems in this section operates

Word cloud A word cloud shows the most common words in a block of text in a larger font. It is a wonderful way to focus in on the important aspects of a learning screen or paragraph of text. There are a few different options available on the interactive site.

Create a word cloud for one or more of the most important screens or blocks of text in this section

Word puzzles Crossword and wordsearch puzzles are an effective way to learn new important words and the associated technologies. A good method is to work in pairs so you each create a puzzle and then swap and try to complete the answers. Hint: Use the eLearning glossary where you can copy the words and definitions (clues!). About 20 words is a good puzzle. Or construct a wordsearch grid using some key words from this section. About 10 words in a 12x12 grid is ideal. Use the online interactive tools for this activity.

Construct a crossword or wordsearch puzzle using important words from this section

Mind map A mind map can be created with pen and paper or on the whiteboard. There are also some great online tools to do this.

Create a mind map to illustrate the features of a key component or system

Information wall An information wall can be created with pen and paper or on the whiteboard. There are also some great online tools to do this. Alternatively, a flip chart or post-it notes on the wall work well.

Create an information wall to illustrate the features of a key component or system

Presentation Preparing and making a presentation to your classmates or workmates is a terrific way to learn about something new because you must study it in detail first. It can be a bit nerve-racking at first but is also good fun so don't worry. There are some great online tools for this or you can use PowerPoint (or similar) to prepare some slides you then explain in more detail.

Using images and text, create a short presentation to show how a component or system works

1.1.3 Visible thinking

Introduction Visible thinking is a method of teaching that encourages learning through observation. It is a way to encourage students to use thinking skills they have already developed outside the classroom. The routines help to promote a deeper understanding of how we think and allow for better learning. Several different routines are highlighted below and these are suggested in various places throughout the book. Teachers can help guide students but it is equally possible for a learner to use the routines directly. The routines outlined below are described as if used by a teacher.

Select a routine from section 1.13 and follow the process to study a component or system

Think-pair-share The think-pair-share routine involves posing a question to learners, asking them to take a few minutes of thinking time. After this they share their thoughts with a partner. It can be used when solving a problem, before a repair routine is started or after reading a section in this book. After discussing as a pair they then share with the whole class.

Figure 1.1 Think-pair-share

Compass points This routine works well to explore various sides and viewpoints of a proposition or idea prior to taking a stand or expressing an opinion on it. For example, a new idea about a subject can be proposed; a new type of EV battery for instance. The learners can then put their thoughts into four compass-point categories. You can use a whiteboard, flipchart, post-it notes or an electronic equivalent. Whatever works and whatever you have to hand is fine.

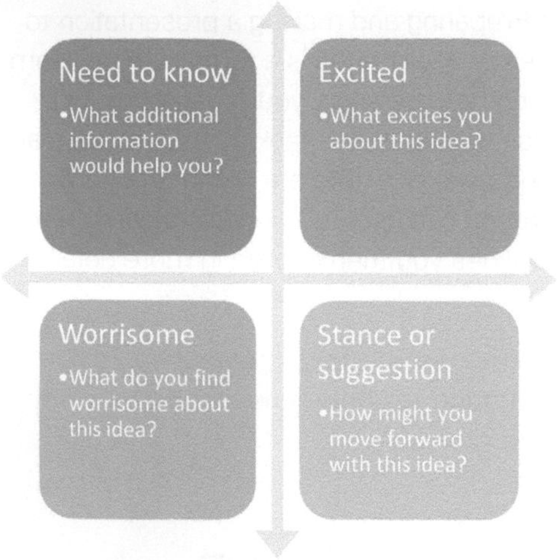

Need to know
•What additional information would help you?

Excited
•What excites you about this idea?

Worrisome
•What do you find worrisome about this idea?

Stance or suggestion
•How might you move forward with this idea?

Figure 1.2 Compass points

What makes you say that? This thinking routine asks learners to describe something such as a component or how a system works, and then support their interpretation with evidence. It can be adapted for use with almost any subject and is useful for gathering information on students' general concepts when introducing a new topic.

What's going on? → What is it that makes you say that?

Figure 1.3 What makes you say that?

I used to think but now I think This routine captures a change of opinion or perspective from what a student used to think to what they now think. It enables students to reflect on their learning, be willing to consider different ideas and to be able to acknowledge when their opinion has changed. It gives students the opportunity to reflect on why their thinking may have changed.

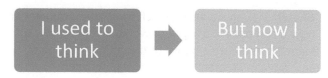

I used to think → But now I think

Figure 1.4 I used to think but now I think

I noticed this, but why? This is good for thinking further about something you have been examining in the workshop. For example, you may notice the wires connecting a starter motor are much thicker than those connected to the lights. Ask yourself why? It is also a good way to understand how a component works.

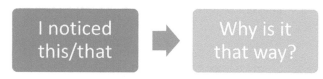

I noticed this/that → Why is it that way?

Figure 1.5 I noticed this, but why?

Question starts This is a graphic information organiser with spaces for questions which begin: why, what, how or similar. It is a visual guide to help students dig deeper into a question or topic and reach a better understanding.

Why is it like this? → What if it? → What do you know about this? → What would you change about it? → How do you know?

Figure 1.6 Question starts

3-2-1 bridge This routine is used to record initial thoughts, ideas, questions and understandings about a given topic. After reading an article or watching a video, students are then asked to rethink their initial opinions after learning more about the topic. The idea is for bridges to be built between ideas when new information is obtained.

3. Initial thoughts → 2. Initial questions → 1. Initial comparison → 3. New thoughts → 2. New questions → 1. New comparison

Figure 1.7 3-2-1 bridge

Think-puzzle-explore This routine works well at the start of a new topic taking what students already know while opening up new areas of interest. What does a student think about the topic? What puzzles them or what unresolved questions are in their minds? What and how can they explore more about the topic?

Figure 1.8 Think-puzzle-explore

Connect-extend-challenge The connect-extend-challenge routine helps students make connections between new and previous knowledge. It then acts as a guide as to how to further develop their interest. Recording students' ideas and using them at a later time can also be useful for reflecting on their understanding.

Figure 1.9 Connect-extend-challenge

1.2 Summary

In this chapter we have looked at some of the key features of this book that make learning more effective as well as more fun. There are lots of different learning activities to try out. Good luck with your studies and I hope you find this book useful.

Remember, get involved in your learning and interact, you will learn much more!

Remember, this book is designed to be used with other resources such as IMI eLearning. For this reason you will find some images do not have all the parts named or an image may be mentioned in the text that is not in the book. This is intentional and you should research the answers and add details in the textbook or in your notes – it is a good way to learn.
There is more support on the website that includes additional images and interactive features: www.tomdenton.org

CHAPTER 2

Working safely

After successful completion of this chapter you will be able to show you have achieved these objectives:

- Understand the correct personal and vehicle protective equipment to be used within the automotive environment.

- Understand effective housekeeping practices in the automotive environment.

- Understand key health and safety requirements relevant to the automotive environment.

- Understand about hazards and potential risks relevant to the automotive environment.

- Understand personal responsibilities.

DOI: 10.1201/9781003173236-2

2.1 Personal and vehicle protection

2.1.1 Personal protective equipment ❶

Personal protective equipment (PPE) such as safety clothing, is very important to protect yourself. Some people think it is clever or tough not to use protection. They are very sad and will die or be injured long before you! Some things are obvious such as when holding a hot or sharp exhaust you would likely be burnt or cut, others such as breathing in brake dust or working in a noisy area do not produce immediately noticeable effects but could affect you later in life. Fortunately the risks to workers are now quite well understood and we can protect ourselves before it is too late.

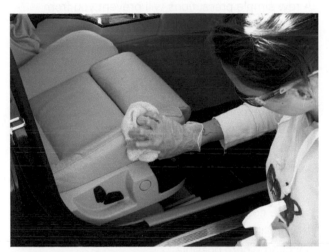
Figure 2.1 Eye protection and gloves in use

Figure 2.2 Protective clothing for spot welding

PPE In the following table, we have listed a number of items classed as PPE (Personal Protective Equipment) together with suggested uses. You will see that the use of most items is plain common sense.

Equipment	Notes	Suggested usage or examples of where used
Ear defenders	Must meet appropriate standards	When working in noisy areas or using an air chisel
Face mask	For individual personal use only	Dusty conditions. When cleaning brakes or if preparing bodywork
High visibility clothing	Fluorescent colours such as yellow or orange	Working in traffic such as when on a breakdown
Leather apron	Should be replaced if it is holed or worn thin	When welding or working with very hot items
Leather gloves	Should be replaced when they become holed or worn thin	When welding or working with very hot items and also if handling sharp metalwork
Life jacket	Must meet current standards	Use when attending vehicle breakdowns on ferries!
Overalls	Should be kept clean and be flameproof if used for welding	These should be worn at all times to protect your clothes and skin. If you get too hot just wear shorts and a T-shirt underneath
Rubber or plastic apron	Replace if holed	Use if you do a lot of work with battery acid or with strong solvents
Rubber or plastic gloves	Replace if holed	Gloves must always be used when using degreasing equipment
Safety shoes or boots	Strong toecaps are recommended	Working in any workshop with heavy equipment
Safety goggles	Keep the lenses clean and prevent scratches	Always use goggles when grinding or when any risk of eye contamination. Cheap plastic goggles are much easier to come by than new eyes
Safety helmet	Must meet current standards	Under-vehicle work in some cases
Welding goggles or welding mask	Check the goggles are suitable for the type of welding. Gas welding goggles are NOT good enough when arc welding	You should wear welding goggles or use a mask even if you are only assisting by holding something

Create an information wall to illustrate the features of a key component or system

7

2.1.2 Vehicle safety ❶

Vehicle safety and the associated regulations can be very complicated. However, for our purposes we can consider the issue across two main areas:

1. Construction of the vehicle

Before a vehicle can be constructed, a prototype has to be submitted for type approval. When awarded, this means the vehicle has passed very stringent tests and that it meets all current safety requirements. Different countries have different systems, which means some modifications to a car may be necessary if it is imported or exported. The European Union (EU) has published many 'directives' which each member country must incorporate into its own legislation. This has helped to standardise many aspects. In the UK the 'Road Vehicles (Construction & Use) Regulations 1986', is the act which ensures certain standards are met. If you become involved in modifying a vehicle, e.g. for import, you may need to refer to the details of this act. Other countries have similar legislation.

Many other laws exist relating to the motor vehicle and the environment – these are about emissions and pollution. Environmental laws change quite often and it will be important to keep up to date.

The Department of Transport states that all vehicles over three years old must undergo a safety check which ensures the vehicle continues to meet the current legislation. First set up by the Ministry of Transport, it continues to be known as the MOT test. This test now includes checks relating to environmental laws.

 Create a word cloud for one or more of the most important screens or blocks of text in this section

2. Driving and operating the vehicle

To drive a vehicle on the road you must have an appropriate driving licence, insurance, the vehicle must be taxed and it must be in safe working order.

2.1.3 Moving loads ❶

Injuries in a workshop are often due to incorrect lifting or moving of heavy loads. In motor vehicle workshops, heavy and large components, like engines and gearboxes, can cause injury when being removed and refitted.

Figure 2.3 Heavy load – correct lifting method

A few simple precautions will prevent you from injuring yourself, or others:

▶ never try to lift anything beyond your capability – get a mate to help. The amount you can safely lift will vary but for any more than you feel comfortable with, you should get help.
▶ whenever possible use an engine hoist, a transmission jack or a trolley jack.
▶ lift correctly, using the legs and keeping your back straight.
▶ when moving heavy loads on a trolley, get help and position yourself so you will not be run over if you lose control.

The ideal option in all cases is simply to avoid manual handling where possible.

Figure 2.4 Engine crane (Source: Blue-Point)

Use the media search tools to look for pictures and videos relating to the subject in this section

2.2 Housekeeping

2.2.1 Working environment

There are three main reasons for keeping your workshop and equipment clean and tidy:

1 It makes it a safer place to work.
2 It makes it a better place to work.
3 It gives a better image to your customers.

Servicing and fixing motor vehicles is in some cases a dirty job. But if you clean up after any dirty job then you will find your workshop a much more pleasant place to work.

▶ The workshop and floor should be uncluttered and clean to prevent accidents and fires as well as maintaining the general appearance.
▶ Your workspace reflects your ability as a technician. A tidy workspace equals a tidy mind equals a tidy job, which equals a tidy wage when you are qualified.
▶ Hand tools should be kept clean as you are working. You will pay a lot of money for your tools; look after them and they will look after you in the long term.
▶ Large equipment should only be cleaned by a trained person or a person under supervision. Obvious

precautions are to ensure equipment cannot be operated while you are working on it and only use appropriate cleaning methods. For example, would you use a bucket of water or a brush to clean down an electric pillar drill? I hope you answered 'the brush'!

In motor vehicle workshops, many different cleaning operations are carried out. This means a number of different materials are required. It is not possible to mention every brand name here so I have split the materials into three different types. It is important to note that the manufacturer's instructions printed on the container must be followed at all times.

Look back over the previous section and write out a list of the key bullet points from this section

2.2.2 Equipment maintenance

The cleaning and maintenance of equipment plays a big part in good housekeeping. This includes large equipment such as ramps, hoists, etc. to small hand tools. Always remember that no one should clean, maintain or use large equipment unless they have had sufficient training or are working under the supervision of an experienced and qualified person.

Figure 2.5 Hand tools

Hand tools are expensive so do look after them and in the long term they will look after you. Technicians need to learn and be aware of the following points regarding equipment:

▶ select and use equipment for basic hand tool maintenance activities
▶ storing hand tools safely and accessibly
▶ how to report faulty or damaged work tools and equipment
▶ safety when cleaning and maintaining work tools and equipment.

Material	Purpose	Notes
Detergent	Mixed with water for washing vehicles, etc. Also used in steam cleaners for engine washing, etc.	Some industrial detergents are very strong and should not be allowed in contact with your skin.
Solvents	To wash away and dissolve grease and oil, etc. The best example is the liquid in the degreaser or parts washer which all workshops will have.	NEVER use solvents such as thinners or fuel because they are highly flammable. Suitable PPE should be used, for example gloves, etc. They may attack your skin. Many are flammable. The vapour given off can be dangerous. Serious problems if splashed into eyes. Read the label.
Absorbent granules	To mop up oil and other types of spills. They soak up the spillage after a short time and can then be swept up.	Most granules are a chalk or clay type material which has been dried out.

It is important to store hand tools safely. Any hand tool left lying around can be a potential hazard to the unsuspecting person or could cause damage to a customer's vehicle. Always make sure that hand tools are stored correctly in either a tool box or in the designated place. If you think that you are probably going to need a particular tool in due course, and don't want to put it back, then be aware of where you place it. Obviously you will want it handy but at the same time you need to think of safety.

Safety also applies to the tool you are using. Don't put it down in a place where it can be damaged. Wherever you store or place a hand tool, think of the following points:

- ▶ safety of yourself and others
- ▶ protection of the customer's vehicle
- ▶ protection of other tools and workshop equipment
- ▶ protection of the tool itself.

From time to time tools and equipment will develop faults or get damaged, however careful you are with them. If you find any damage to equipment it is your duty as a technician to report it or see if it has already been reported.

Figure 2.6 Taking care

Don't leave it to someone else or assume that it must have been spotted by one of your colleagues. It is quite likely that it hasn't.

Does your workshop have a procedure for reporting faulty equipment? Does it need to be written on a report form for instance? Who is the appropriate person to report this fault to? Your supervisor or perhaps another member of staff? Whoever and however, it needs to be done quickly. Delay could possibly make the fault worse but, more importantly, if it needs to be used then work will

be held up. Would this fault be a potential safety issue? This is obviously a very important point to consider. Report it immediately so the problem can be fixed.

In the previous section I mentioned the importance of working safely when cleaning or maintaining equipment. It is important to remember that you must never clean or maintain equipment without adequate training or supervision from a qualified and competent person. Even if you are asked, politely say 'no' and explain why. It is quite likely that the person who asked you to do the task is unaware that you do not have the relevant experience.

 Create a mind map to illustrate the features of a key component or system

2.3 Health and safety

2.3.1 Introduction ❶

Health and safety law is designed to protect you. In the UK the health and safety executive (HSE) is the enforcement and legislative body set up by the government. The HSE has a very helpful website where you can get all the latest information – including a document specially developed for the motor industry. The address is: www.hse.gov.uk

Figure 2.7 UK HSE logo

Health and Safety Executive (HSE) The HSE's emphasis is on preventing death, injury and ill health in Great Britain's workplaces. However, they do have the authority to come down hard on people who put others at risk, particularly where there is deliberate flouting of the law. Since 2009, HSE has published new versions of its approved health and safety poster and leaflet. The new versions are modern, eye-catching and easy to read. They set out in simple terms, using numbered lists of basic points, what employers and workers must do and tell you what to do if there is a problem.

Employers must ensure that they are displaying the latest versions of the posters. When updates are issued then there is usually a period of time where use of older materials is allowed but it is essential that employers check this.

Employers have a legal duty under the Health and Safety Information for Employees Regulations (HSIER) to display the poster in a prominent position in each workplace or provide each worker with a copy of the equivalent leaflet outlining British health and safety laws.

Employers must meet certain criteria **but** health and safety is the responsibility of **everyone** in the workplace. The reason for the poster and the leaflets is to make everybody aware of this.

 Use a library or the web search tools to further examine the subject in this section

Figure 2.8 Health and Safety Law poster (Source: HSE)

2.3.2 **Regulations and laws** ❶

There are a number of rules and regulations you need to be aware of. Always check the details for the country in which you work. The following table lists some important areas for the UK.

Title	Rules and regulations
HSE	The Health and Safety Executive is the national independent watchdog for work-related health, safety and illness in the UK. They are an independent regulator and act in the public interest to reduce work-related death and serious injury across Great Britain's workplaces. Other countries have similar organisations.
HASAW	The Health and Safety at Work Act 1974, also referred to as HASAW, HASAWA or HSW, is the primary piece of legislation covering occupational health and safety in the United Kingdom. The Health and Safety Executive is responsible for enforcing the Act and a number of other Acts and Statutory Instruments relevant to the working environment.
COSHH	Control of Substances Hazardous to Health is the law that requires employers to control substances that are hazardous to health such as solvents.
RIDDOR	The Reporting of Injuries, Diseases and Dangerous Occurrences Regulations 1995 (RIDDOR), place a legal duty on employers, self-employed people and people in control of premises to report work-related deaths, major injuries or over-three-day injuries, work-related diseases, and dangerous occurrences (near-miss accidents).
Provision and Use of Work Equipment Regulations 1998 (PUWER)	In general terms, PUWER requires that equipment provided for use at work is: suitable for the intended use; safe for use, maintained in a safe condition and, in certain circumstances, inspected to ensure this remains the case; used only by people who have received adequate information, instruction and training; and accompanied by suitable safety measures, e.g. protective devices, markings, warnings.
Lifting Operations and Lifting Equipment regulations 1998 (LOLER)	In general, LOLER requires that any lifting equipment used at work for lifting or lowering loads is: strong and stable enough for particular use and marked to indicate safe working loads; positioned and installed to minimise any risks; used safely, i.e. the work is planned, organised and performed by competent people; and subject to ongoing thorough examination and, where appropriate, inspection by competent people.
Health and safety audit	Monitoring provides the information to let you or your employer review activities and decide how to improve performance. Audits, by company staff or outsiders, complement monitoring activities by looking to see if your company policy, organisation and systems are actually achieving the right results.

Title	Rules and regulations
Risk management and assessment	A risk assessment is simply a careful examination of what, in your work, could cause harm to people. This is done so that you and your company can decide if you have taken enough precautions or should do more to prevent harm. Workers and others have a right to be protected from harm caused by a failure to take reasonable control measures. It is a legal requirement to assess the risks in the workplace so you or your employer must put plans in place to control risks. How to assess the risks in your workplace: Identify the hazards Decide who might be harmed and how Evaluate the risks and decide on precautions Record your findings and implement them Review your assessment and update if necessary.
PPE	Personal protective equipment (PPE) is defined in the Regulations as 'all equipment (including clothing affording protection against the weather) which is intended to be worn or held by a person at work and which protects them against one or more risks to their health or safety', e.g. safety helmets, gloves, eye protection, high-visibility clothing, safety footwear and safety harnesses. Hearing protection and respiratory protective equipment provided for most work situations are not covered by these Regulations because other regulations apply to them. However, these items need to be compatible with any other PPE provided.

 Create an information wall to illustrate the features of a key component or system

2.3.3 Health and Safety law ❶

All workers have a right to work in places where risks to their health and safety are properly controlled. Health and safety is about stopping you getting hurt at work or ill through work. Your employer is responsible for health and safety, but you must help. This section is taken from the HSE leaflet for employees.

What employers must do for you:

1 Decide what could harm you in your job and the precautions to stop it. This is part of risk assessment.
2 In a way you can understand, explain how risks will be controlled and tell you who is responsible for this.
3 Consult and work with you and your health and safety representatives in protecting everyone from harm in the workplace.

Figure 2.9 Health and Safety Law leaflet (Source: HSE)

4 Free of charge, give you the health and safety training you need to do your job.
5 Free of charge, provide you with any equipment and protective clothing you need, and ensure it is properly looked after.
6 Provide toilets, washing facilities and drinking water.
7 Provide adequate first-aid facilities.
8 Report injuries, diseases and dangerous incidents at work
9 Have insurance that covers you in case you get hurt at work or ill through work. Display a hard copy or electronic copy of the current insurance certificate where you can easily read it.
10 Work with any other employers or contractors sharing the workplace or providing employees (such as agency workers) so that everyone's health and safety is protected.

What you must do:

1 Follow the training you have received when using any work items your employer has given you.
2 Take reasonable care of your own and other people's health and safety.
3 Co-operate with your employer on health and safety.
4 Tell someone (your employer, supervisor, or health and safety representative) if you think the work or inadequate precautions are putting anyone's health and safety at serious risk.

Figure 2.10 Exhaust extraction is an easy precaution to take

 Select a routine from section 1.3 and follow the process to study a component or system

2.4 Hazards and risks

2.4.1 Hazards ❶

Working in a motor vehicle workshop is a dangerous occupation – if you do not take care. The most important thing is to be aware of the hazards and then it is easy to avoid the danger. The hazards in a workshop are from two particular sources:

From you, such as caused by:

▶ carelessness – particularly whilst moving vehicles
▶ drinking or taking drugs – these badly affect your ability to react to dangerous situations
▶ tiredness or sickness – these will affect your abilities to think and work safely
▶ messing about – most accidents are caused by people fooling about
▶ not using safety equipment – you have a duty to yourself and others to use safety equipment
▶ inexperience – or lack of supervision. If in doubt – ask!

The surroundings in which you work may have:

▶ bad ventilation
▶ poor lighting
▶ noise
▶ dangerous substances stored incorrectly
▶ broken or worn tools and equipment

▶ faulty machinery
▶ slippery floors
▶ untidy benches and floors
▶ unguarded machinery
▶ unguarded pits.

The following table lists some of the hazards you will come across in a vehicle workshop. Also listed are some associated risks, together with ways we can reduce them. This is called risk management.

Hazard	Risk	Action
Power tools	Damage to the vehicle or personal injury	Understand how to use the equipment and wear suitable protective clothing, for example gloves and goggles
Working under a car on the ramp	1. The vehicle could roll or be driven off the end 2. You can bang your head on hard or sharp objects when working under the car	1. Ensure you use wheel chocks 2. Set the ramp at the best working height, wear protection if appropriate
Working under a car on a jack	The vehicle could fall on top of you	The correct axle stands should be used and positioned in a secure place
Compressed air	Damage to sensitive organs such as ears or eyes. Death, if air is forced through the skin into your blood stream	Do not fool around with compressed air. A safety nozzle prevents excessive air forces
Dirty hands and skin	Oil, fuel and other contaminants can cause serious health problems. This can range from dermatitis to skin cancer	Use gloves or a good-quality barrier cream and wash your hands regularly. Do not allow dirt to transfer to other parts of your body. Good overalls should be worn at all times
Exhaust fumes	Poisonous gases such as carbon monoxide can kill. Other gases can cause cancer or, at best, restrict breathing and cause sore throats	Only allow running engines in very well ventilated areas or use an exhaust extraction system
Engine crane	Injury or damage can be caused if the engine swings and falls off	Ensure the crane is strong enough (do not exceed its safe working load – SWL). Secure the engine with good-quality sling straps and keep the engine near to the floor when moving across the workshop

Hazard	Risk	Action
Cleaning brakes	Brake dust (especially older types made of asbestos) is dangerous to health	Only wash clean with proper brake cleaner
Fuel	Fire or explosion	Keep all fuels away from sources of ignition. Do not smoke when working on a vehicle
Degreaser solvent	Damage to skin or damage to sensitive components	Wear proper gloves and make sure the solvent will not affect the items you are washing
Spillage such as oil	Easy to slip over or fall and be injured	Clean up spills as they happen and use absorbent granules
Battery electrolyte (acid)	Dangerous on your skin and, in particular, your eyes. It will also rot your clothes	Wear protective clothing and take extreme care
Welding a vehicle	The obvious risks are burns, fire and heat damage, but electric welders, such as a MIG welder, can damage sensitive electronic systems	Have fire extinguishers handy, remove combustible materials such as carpets and ensure fuel pipes are nowhere near. The battery earth lead must also be disconnected. Wear gloves and suitable protective clothing such as a leather jacket
Electric hand tools	The same risk as power tools but also the danger of electric shocks, particularly in damp or wet conditions. This can be fatal	Do not use electric tools when damp or wet. Electrical equipment should be inspected regularly by a competent person
Driving over a pit	Driving into the pit	The pit should be covered or have another person help guide you, and drive very slowly
Broken tools	Personal injury or damage to the car. For example, a file without a handle can stab into your wrist. A faulty ratchet could slip	All tools should be kept in good order at all times. This will also make the work easier
Cleaning fluids	Skin damage or eye damage	Wear gloves and eye protection and also be aware of exactly what precautions are needed by referring to the safety data information

 Look back over the previous section and write out a list of the key bullet points from this section

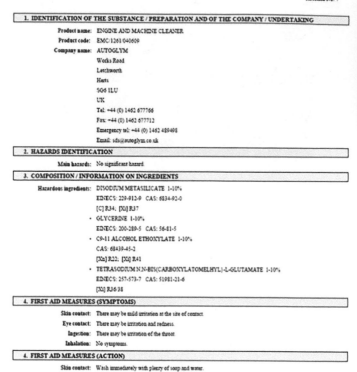

Figure 2.11 Example of a safety data sheet (Source: AutoGlym)

2.4.2 High voltage vehicles

Introduction This section covers some risks when working with electricity or electrical systems, together with suggestions for reducing them. This is known as risk assessment. The diagnostic process is no different but avoid working on high voltage vehicles unless you are trained and have specialist equipment – stay safe.

Voltages Electric vehicles (pure or hybrid) use high voltage batteries so that energy can be delivered to a drive motor or returned to a battery pack in a very short time. Voltages of 400V are now common and some are up to 700V so clearly there are electrical safety issues when working with these vehicles. High Voltage means the classification of an electric component or circuit if its working voltage is >60V and ≤1500V DC or >30V and ≤1000V AC root mean square (rms). High voltages can also be present on vehicles without high voltage batteries such as headlamps that use ballast resistors for example.

Figure 2.12 High voltage (HV) battery pack (Source: Chevrolet Media)

Components EV batteries and motors have high electrical and magnetic potential that can severely injure or kill if not handled correctly. It is essential that you take note of all the warnings and recommended safety measures outlined by manufacturers. Any person with a heart pacemaker or any other electronic medical devices should not work on an EV motor since the magnetic effects could be dangerous. In addition, other medical devices such as intravenous insulin injectors or meters can be affected.

High energy The electrical energy is conducted to or from the motor via thick orange wires connected to the battery. If these wires have to be disconnected, SWITCH OFF or DE-ENERGISE the high voltage system. This will prevent the risk of electric shock or short circuit of the high voltage system.

Chemicals Electrolytes used in EV batteries can be harmful and it is important to remember that both acid and alkali solutions are used in this technology. Understanding before you start work what type of chemical is present will enable selecting the correct procedure when dealing with a spill.

Stages of work The general advice about working on high voltage vehicles is split into four areas:

Before work: Turn OFF the ignition switch and remove the key

▶ Switch OFF the Battery Module switch or de-energise the system

▶ Wait for 5 minutes before performing any maintenance procedures on the system. This allows any storage capacitors to be discharged

▶ Isolate system and stow key in safe location only accessible to the technician

During work: Always wear correct PPE when there is risk of touching a live component

▶ Always use insulated tools when performing service procedures to the high voltage system. This precaution will prevent accidental short-circuits.

▶ Always use correct workshop equipment to define the work area, warning others of the dangers present and use a correct insulated rubber mat when appropriate.

Interruptions: When maintenance procedures must be interrupted while some high voltage components are uncovered or disassembled, make sure that:

▶ The key is only accessible with the working technician's knowledge

▶ The Battery Module switch is switched off

▶ No untrained persons have access to that area and prevent any unintended touching of the components.

After work: Before switching on or re-energising the battery module after repairs have been completed, make sure that:

▶ All terminals have been tightened to the specified torque

▶ No high voltage wires or terminals have been damaged or shorted to the body

▶ The insulation resistance between each high voltage terminal of the part you disassembled and the vehicle's body has been checked.

 Create a word cloud for one or more of the most important screens or blocks of text in this section

2.4.3 Fire 1

Accidents involving fire are very serious. As well as you or a workmate calling the fire brigade (do not assume it has been done), three simple rules will help you know what to do:

1 Get safe yourself, contact the emergency services – and shout 'FIRE!'
2 Help others to get safe if it does not put you or others at risk.
3 Fight the fire if it does not put you or others at risk.

Of course, far better than the above situation is to not let a fire start in the first place.

The fire triangle or combustion triangle is a simple model for understanding the ingredients necessary for most fires. The triangle illustrates that a fire requires three elements: heat, fuel, and an oxidizing agent (usually oxygen from the air). The fire is prevented or extinguished by removing any one of them. A fire occurs when the elements are combined in the right mixture.

Figure 2.14 Fire tetrahedron (Source: Wikimedia)

Figure 2.13 Fire triangle (Source: Wikimedia)

Without sufficient heat, a fire cannot start or continue. Heat can be removed by the application of a substance which reduces the amount of heat available to the fire reaction. This is often water, which requires heat to change from water to steam. Introducing sufficient quantities and types of powder or gas into the flame can also reduce the amount of heat available for the fire reaction. Turning off the electricity in an electrical fire removes the ignition source.

Without fuel, a fire will stop. Fuel can be removed naturally, as when the fire has consumed all the burnable fuel, or manually, by mechanically or chemically removing the fuel from the fire. The fire goes out because a lower concentration of fuel vapour in the flame leads to a decrease in energy release and a lower temperature. Removing the fuel therefore decreases the heat.

Without enough oxygen, a fire cannot start or continue. With a decreased oxygen concentration, the combustion process slows. In most cases, there is plenty of air left when a fire goes out so this is not commonly a major factor.

The fire tetrahedron is an addition to the fire triangle. It adds the requirement for the presence of the chemical reaction which is the process of fire. For example, the suppression effect of a Halon extinguisher is due to its interference in the fire by chemical inhibition. Note that Halon extinguishers are now only allowed in certain situations and are illegal for normal use.

Combustion is the chemical reaction that feeds a fire more heat and allows it to continue. When the fire involves burning metals, like magnesium (known as a class-D fire), it becomes even more important to consider

the energy release. The metals react faster with water than with oxygen and thereby more energy is released. Putting water on such a fire makes it worse. Carbon dioxide extinguishers are ineffective against certain metals such as titanium. Therefore, inert agents (e.g. dry sand) must be used to break the chain reaction of metallic combustion. In the same way, as soon as we remove one out of the three elements of the triangle, the fire stops.

If a fire does happen, your workplace should have a set procedure. So, for example, you will know:

▶ how the alarm is raised
▶ what the alarm sounds like
▶ what to do when you hear the alarm
▶ your escape route from the building
▶ where to go to assemble
▶ who is responsible for calling the fire brigade.

There are a number of different types of fire as shown in the following table:

European/ Australian/Asian	American	Fuel/Heat source
Class A	Class A	Ordinary combustibles
Class B	Class B	Flammable liquids
Class C		Flammable gases
Class D	Class D	Combustible metals
Class E	Class C	Electrical equipment
Class F	Class K	Cooking oil or fat

If it is safe to do so, you should try to put out a small fire. Extinguishers and a fire blanket should be provided. Remember, if you remove one side of the fire triangle, the fire will go out. If you put enough water on a fire, it will cool down and go out. However spraying water on an electrical circuit could kill you! Spraying water on a petroleum fire could spread it out and make the problem far worse. This means that a number of different fire extinguishers are needed. Internationally there are several accepted classification methods for handheld fire extinguishers. Each classification is useful in fighting fires with a particular group of fuel.

Fire extinguishers in the UK, and throughout Europe, are red but with a band or circle of a second colour covering between 5–10% of the surface area of the extinguisher to indicate its contents. Prior to 1997, the entire body of the fire extinguisher was colour coded.

Type	Old code	BS EN 3 colour code	Suitable for use on fire classes (brackets denote sometimes applicable)					
Water	Signal red	Signal red	A					
Foam	Cream	Red with a cream panel above the operating instructions	A	B				
Dry powder	French blue	Red with a blue panel above the operating instructions	(A)	B	C		E	
Carbon dioxide CO_2	Black	Red with a black panel above the operating instructions		B			E	
Wet chemical	Not yet in use	Red with a canary yellow panel above the operating instructions	A	(B)				F
Class D powder	French blue	Red with a blue panel above the operating instructions				D		
Halon 1211/BCF	Emerald green	No longer in general use	A	B			E	

In the UK the use of Halon gas is now prohibited, except under certain situations, such as on aircraft and by the military and police.

Figure 2.15 CO_2 and water extinguishers and information posters

 Create an information wall to illustrate the features of a key component or system

2.4.4 Signage ❶

A key safety aspect is to first identify hazards and then remove them or, if this is not possible, reduce the risk as much as possible and bring the hazard to everyone's attention. This is usually done by using signs or markings. Signs used to mark hazards are often as follows:

Function	Example	Back colour	Fore-ground colour	Sign
Hazard warning	Danger of electric shock	Yellow	Black	**Figure 2.16** Electricity
Mandatory	Use ear defenders when operating this machine	Blue	White	**Figure 2.17** Wear ear protection
Prohibition	Not drinking water	White	Red/black	**Figure 2.18** Not drinking water
First aid (escape routes are a similar design)	Location of safety equipment such as first aid	Green	White	**Figure 2.19** First aid
Fire	Location of fire extinguishers	Red	White	**Figure 2.20** Extinguisher
Recycling	Recycling point	White	Green	**Figure 2.21** The three Rs of the environment

Construct a crossword or wordsearch puzzle using important words from this section

2.5 Personal responsibilities

2.5.1 Safety procedures ❶

When you know the set procedures to be followed, it is easier to look after yourself, your workshop and your workmates. You should know:

▶ who does what during an emergency
▶ the fire procedure for your workplace
▶ about different types of fire extinguisher and their uses
▶ the procedure for reporting an accident.

Figure 2.22 Call for an ambulance if the accident is serious

If an accident does occur in your workplace the first bit of advice is: keep calm and don't panic! The HASAW states that for companies above a certain size:

▶ first-aid equipment must be available
▶ employers should display simple first-aid instructions
▶ fully trained first-aiders must be employed.

In your own workplace you should know about the above three points. The following is a guide as to how to react if you come across a serious accident:

Action	Notes
Assess the situation	Stay calm, a few seconds to think is important
Remove the danger	If the person was working with a machine, turn it off. If someone is electrocuted, switch off the power before you hurt yourself. Even if you are unable to help with the injury you can stop it getting worse
Get help	If you are not trained in first aid, get someone who is and/or phone for an ambulance
Stay with the casualty	If you can do nothing else, the casualty can be helped if you stay with them. Also say that help is on its way and be ready to assist. You may need to guide the ambulance
Report the accident	All accidents must be reported; your company should have an accident book by law. This is a record so that steps can be taken to prevent the accident happening again. Also, if the injured person claims compensation, underhanded companies could deny the accident happened
Learn first aid	If you are in a very small company why not get trained now, before the accident?

 Select a routine from section 1.3 and follow the process to study a component or system

2.5.2 Environmental protection ❶

Environmental protection is all about protecting the environment on individual, organizational or governmental levels. Due to the pressures of population and our technology, the Earth's environment is being degraded, sometimes permanently. Activism by environmental movements has created awareness of the various environmental issues. This has led to governments placing restraints on activities that cause environmental problems and producing regulations.

In a workshop, these regulations relate to many items such as solvents used for cleaning or painting, fuels, oil and many other items. Disposal methods must not breach current regulations and, in many cases, only licensed contractors can dispose of certain materials.

Failure to comply can result in heavy penalties. Make sure you are aware of your local regulations as these can change.

Finally, let's consider the three Rs:

▶ **Reduce** the amount of the Earth's resources that we use

▶ **Reuse** Don't just bin it. Could someone else make use of it?

▶ **Recycle** Can the materials be made into something new?

▶ And, as we are automotive technicians, maybe there is a fourth: **Repair?**

 Use a library or the web search tools to further examine the subject in this section

2

CHAPTER 3

Automotive industry

After successful completion of this chapter you will be able to show you have achieved these objectives:

- Understand key organisational structures, functions and roles within the automotive work environment.

- Understand the importance of obtaining, interpreting and using information in order to support your job role within the automotive work environment.

- Understand the importance of different types of communication within the automotive work environment.

- Understand communication requirements when carrying out vehicle repairs in the automotive work environment.

- Understand how to develop good working relationships with colleagues and customers in the automotive workplace.

DOI: 10.1201/9781003173236-3

3.1 The motor trade ❶

Introduction This section will outline some of the jobs that are open to you in the motor trade and help you understand more about the different types of business and how they operate. It is easy to think that the operation of a business does not matter to you. However, I would strongly suggest we should all be interested in the whole business in which we are working. This does not mean we should interfere in areas we do not understand. It means we should understand that all parts of the business are important. For example, when you complete a job, enter all the parts used so the person who writes the invoice knows what to charge!

Ask your boss to give you a 'tour' of the garage so that you can appreciate the different tasks carried out and systems that are in place – in particular, make sure you get a reasonable idea about the words and phrases in the following table. If you do not yet have a job you may be able to arrange a visit.

Figure 3.3 Modern workshop

Figure 3.1 A Volkswagen main dealer

Figure 3.4 Porsche showroom

Figure 3.2 A Ferrari and Maserati main dealer

Figure 3.5 Bodyshop

Opportunities The motor trade offers lots of opportunities for those who are willing to work hard and move forwards. There are many different types of job and you will find one to suit you with a little patience and study.

Key words and phrases Some important words and phrases are presented here:

Customer	The individuals or companies that spend their money at your place of work. This is where your wages come from.
Job card	A printed document for recording amongst other things, work required, work done, parts used and the time taken.
Invoice	A description of the parts and services supplied with a demand for payment from the customer.
Company system	A set way in which things work in one particular company. Most motor vehicle company systems will follow similar rules but will all be a little different.
Contract	An offer which is accepted and payment is agreed. If I offer to change your engine oil for £15 and you decide it is a good offer and accept it, we have a contract. This is then binding on us both.
Image	This is the impression given by the company to existing and potential customers. Not all companies will want to project the same image.
Warranty	An intention that if, within an agreed time, a problem occurs with the supplied goods or service, it will be rectified free of charge by the supplier.
Recording system	An agreed system within a company so that all details of what is requested and/or carried out are recorded. The job card is one of the main parts of this system.
Approved repairer	This can mean two things normally. The first is where a particular garage or bodyshop is used by an insurance company to carry out accident repair work. In some cases, however, general repair shops may be approved to carry out warranty work.
After sales	A term that applies to all aspects of a main dealer that are involved with looking after a customer's car after it has been sold to them by the sales team. The service/repair workshop is the best example.

Types of MV companies Motor vehicle companies can range from very small, one-person businesses to very large main dealers. The systems used by each will be different but the requirements are the same.

Figure 3.6 Accident repair reception

Figure 3.7 Motorist discount store

Figure 3.8 Tyre shop

Figure 3.9 Ford main dealer

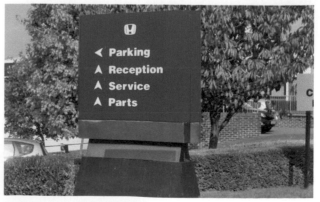

Figure 3.10 Good signs are important

Figure 3.11 A fast-fit centre

Systems A system should be in place to ensure the level of service provided by the company meets the needs of the customer. The list presented here shows how diverse our trade is:

Mobile mechanics	Servicing and repairs at the owner's home or business. Usually a one-person company.
Bodywork repairers and painters	Specialists in body repair and paintwork.
Valeter	These companies specialise in valeting – which should be thought of as much more involved than getting the car washed. Specialist equipment and products are used and proper training is essential.
Fuel stations	These may be owned by an oil company or be independent. Some also do vehicle repair work.
Specialised repairers	Auto-electrical, air conditioning, automatic transmission, and ICE systems are just some examples.
General repair workshop or independent repairer	Servicing and repairs of most types of vehicles not linked to a specific manufacturer. Often this will be a small business maybe employing two or three people. However, there are some very large independent repairers.
Parts supply	Many companies now supply a wide range of parts. Many will deliver to your workshop.
Fast-fit	Supplying and fitting of exhausts, tyres, radiators, batteries, clutches, brakes and windscreens.
Fleet operator (with workshop)	Many large operators, such as rental companies, will operate their own workshops. Also a large company that has lots of cars, used by sales reps for example, may also have their own workshop and technicians.
Non-franchised dealer	Main activity is the servicing and repairs of a wide range of vehicles, with some sales.
Main dealers or franchised dealers	Usually franchised to one manufacturer, these companies hold a stock of vehicles and parts. The main dealer will be able to provide the latest information specific

	to their franchise (Ford or Citroën for example). A 'franchise' means that the company has had to pay to become associated with a particular manufacturer but is then guaranteed a certain amount of work and that there will be no other similar dealers within a certain distance.
Multi-franchised dealer	This type of dealer is just like the one above – except they hold more than one franchise – Volvo and GM for example.
Breakdown services	The best-known breakdown services are operated by the AA and the RAC. Others include Green Flag and, of course, many independent garages also offer these roadside repair and recovery services.
Motorists shops	Often described as motorist discount centres or similar, these companies provide parts and materials to amateurs but in some cases also to the smaller independent repairers.

Company Structure A larger motor vehicle company will probably be made up of at least the following departments:

▶ Reception
▶ Workshop
▶ Bodyshop and paint shop
▶ Parts department
▶ MOT bay
▶ Valeting
▶ New and second user car sales
▶ Office support
▶ Management
▶ Cleaning and general duties

Each area will employ one or a number of people. If you work in a very small garage, you may have to be all of these people at once! In a large garage it is important that these different areas communicate with each other to ensure that a good service is provided to the customer. The main departments are explained further in the following sections.

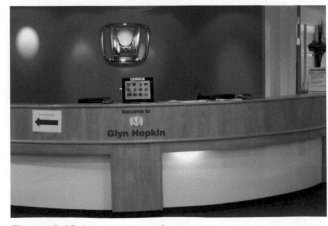

Figure 3.12 Honda reception area

23

Figure 3.13 Different departments on a main dealer site

The **role of a franchised dealer** is to supply local:

▶ new and used franchised vehicles
▶ franchised parts and accessories
▶ repair and servicing facilities for franchised vehicles.

The dealer is also a source of communication and liaison with the vehicle manufacturer.

Figure 3.14 One of the ATT company cars (we wish!)

Figure 3.15 Sales area

Figure 3.16 Displays in the Maserati dealer are used to present a nice image

Figure 3.17 Cars on sale in a showroom

Reception and booking systems The reception, whether in a large or small company, is often the point of first contact with new customers. It is very important therefore to get this bit right. The reception should be staffed by pleasant and qualified persons. The purpose of a reception and booking system within a company can be best explained by following through a typical enquiry. Your company may have a slightly different system but it will be similar.

▶ The customer enters reception area and is greeted in an appropriate way.
▶ Attention is given to the customer to find out what is required (let's assume the car is difficult to start, in this case).
▶ Further questions can be used to determine the particular problem, bearing in mind the knowledge of vehicles the customer may, or may not, have (is the problem worse when the weather is cold, for example).
▶ Details are recorded on a job card about the customer, the vehicle and the nature of the

problem. If the customer is new, a record card can be started; if this is an existing customer then a card is continued.

- An explanation of expected costs is given as appropriate. An agreement to only spend a set amount, after which the customer will be contacted, is a common and sensible approach.
- The date and time when the work will be carried out can now be agreed. This depends on workshop time availability and when is convenient for the customer. It is often better to say you cannot do the job until a certain time, rather than make a promise you can't keep.
- The customer is thanked for visiting. If the vehicle is to be left at that time, the keys should be labelled and stored securely.
- Details are now entered in the workshop diary or loading chart (usually computer based).

The **parts department** is the area where parts are kept and/or ordered. This will vary quite a lot between different companies. Large main dealers will have a very large stock of parts for their range of vehicles. They will have a parts manager and, in some cases, several other staff. In some very small garages, the parts department will be a few shelves where popular items such as filters and brake pads are kept.

Figure 3.19 Typical parts department in an independent retailer

Figure 3.20 Tyres in stock

Security is important as most parts cost a lot of money. When parts are collected from the parts department or area, they will be for use in one of three ways:

- for direct sale to a customer
- to be used as part of a job
- for use on company vehicles.

In the first case an invoice or a bill will be produced. In the second case, the parts will be entered on the customer's job card. The third case may also have a job card or, if not, some other record must be kept. In all three cases, keeping a record of parts used will allow them to be reordered if necessary. If parts are ordered and delivered by an external supplier, again they must be recorded on the customer's job card.

The motor trade – Summary To operate a modern automotive business is a complex process. However, the systems have been outlined in this section in order to show that when each part of a complex system is examined, it is much easier to understand and appreciate the bigger picture.

Figure 3.18 Typical parts department in a main dealer

Parts stock Even though the two examples given previously are rather different in scale, the basic principles are the same and can be summed up very briefly as follows:

- a set level of parts or stock is decided upon
- parts are stored so they can be easily found
- a reordering system should be used to maintain the stock.

Figure 3.21 Even the best cars need a wash

Figure 3.22 Smile, and enjoy your work!

Figure 3.23 Workshop

Figure 3.24 Independent garage

 Create an information wall to illustrate the important features of a component or system in this section.

3.2 Information and systems ❶

Estimating costs and times When a customer brings his or her car to a garage for work to be carried out, quite understandably he or she will want to know two things:

1 How much will it cost?
2 When will it be ready?

In some cases, such as for a full service, this is quite easy as the company will have a set charge and by experience will know it takes a set time. For other types of job, this is more difficult.

Figure 3.25 Customer interaction

Standard times Most major manufacturers supply information to their dealers about standard times for jobs. These assume a skilled technician with all the necessary tools. For independent garages, a publication known as the ICME manual is available. This gives agreed standard times for all the most common tasks, on all popular makes of vehicle.

To work out the cost of a job, you look up the required time and multiply it by the company's hourly rate. Don't forget that the cost of parts will also need to be included.

 Use a library or the web search tools to further examine the subject in this section

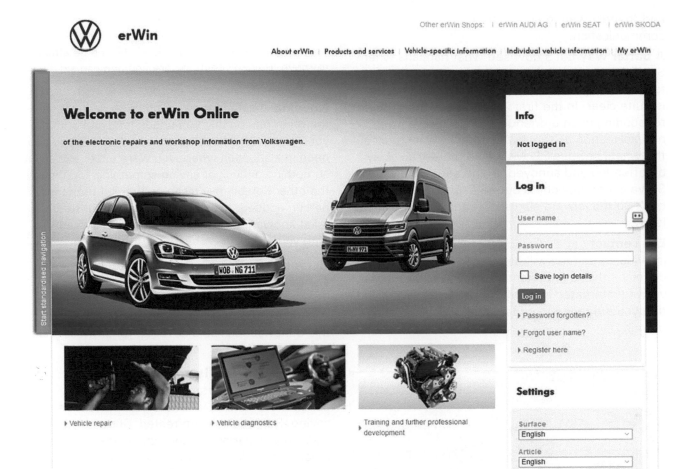

Figure 3.26 Standard times books are now replaced by online systems

3.3 Communication ❶

Introduction The way we communicate with other people is important for many reasons. If we communicate well, we are more likely to get what we want. Often disagreements and upsets are caused because communication is poor. It's easy to fall out with someone because you, the other person or both of you are not communicating well – not necessarily because you disagree.

Verbal and nonverbal Communication can be classified as one of two types:

1 Verbal communication
2 Nonverbal communication

Verbal communication always uses words, spoken and written. Nonverbal communication is any form of communication that does *not* use words. Have a look at the text below – it gives examples of both types of communication.

Positive or negative Communication can be either positive or negative. It is important for everyone in the workplace to be aware of their communication skills, i.e. try make sure their communication is positive.

The wrong way Let's now look at an example where a trainee, Steve, is asked to service and clean the wheel balancer. Steve could reply like this:

▶ 'No, it's not my job!' or even,
▶ *'No!'* and walk off.

Of course, that's perfectly true, it isn't his job as we discussed earlier but how would you feel if you were Steve's workmate and spoken to in this manner? An obvious example of negative communication. The following text shows a more professional and friendly way to approach this – positive communication.

A better way Let's now see what happens when Steve explains to Dave why he is unable to do the task. I think the difference in the two approaches is quite clear. In the first instance, apart from responding in an unfriendly manner Steve gives no explanation as to why he doesn't want to clean the wheel balancer. Dave would more than likely be offended and annoyed. In the second response Steve is not only polite but he explains why he can't do the job. He also shows that he's willing to learn.

Eye contact Let's take some of these points individually as they will be beneficial. Firstly, eye contact – this is important because it shows that you are interested in what the person is saying and that you are trying to understand. It is warm and friendly.

Figure 3.27 Eye contact

However, this can be difficult for some people and we will look further at this later. No matter how difficult, though, you should try to do it. It makes the other person feel that what they are saying is important to you. Also, if you can see their facial expressions, it is easier for you to gauge whether or not they understand you.

Body language This is important for the same or very similar reasons. Slouching, leaning, fidgeting, folding your arms in a defensive, stand-offish manner are all examples of negative body language and negative communication. On the other hand the opposite, standing or sitting upright and relaxed, is positive body language. Fidgeting can be very off-putting, not that you need to remain dead still and be afraid to move. Relaxed gestures are fine and can aid communication.

Nodding in agreement is good because it shows the other person that you understand and/or that you agree with what they are saying. Have you ever been in a situation where you were either explaining something difficult or saying something you thought the other person may disagree with? If the other person nodded, I'm sure it gave you a sense of relief knowing that they understood or agreed with you.

Nervousness Some people are naturally shy and reserved. This can be a positive thing. People generally don't like others to be too loud, constantly talking and overconfident. Quite often this behaviour can be a cover-up for shyness and insecurity.

Be interested and listen If you are nervous or self-conscious, first and foremost try to concentrate fully on what the other person is saying. Be genuinely interested. Show this by nodding and making eye contact. Being interested means you will truly be listening to what the other person is saying and this will help you to relax. It directs your attention from you to them. Really trying to listen and understand another will greatly reduce your feelings of self-consciousness.

Relaxation When you feel nervous, insecure or threatened, your breathing becomes shallow. Try to take deep, slower breaths – it will help you to relax. By feeling relaxed your confidence will grow. You will be in control of your body and voice rather than your body and voice controlling you. Remember to smile! It is so much easier to communicate with a friendly person. When you feel relaxed and confident, communication is a very natural thing. It is unforced, sounds sincere and is more effective.

Take time As with learning all things, developing new skills takes time. Some people learn certain skills more quickly than others. Practice makes perfect! Don't be too hard on yourself – trying too hard can make you feel more self-conscious and tense. Just relax.

Figure 3.28 Take time

and ask yourself, 'Will Julie/Jane/Steve/Mick be able to understand this?' This is where some very basic spelling and grammar is required.

Figure 3.29 Written report

Written communication Don't worry – you don't have to be a wonderful speller or excellent at English but it does really help if you have some basic skills. Poor written communication can take the form of illegible handwriting, bad spelling or poor grammar. This can lead to:

▶ an unprofessional image
▶ misunderstandings.

A **misunderstanding** could lead to someone's safety being at risk. Imagine that you found the cables of a hoist were frayed and your report was so poorly written that it couldn't be understood!

When **reporting problems**, always speak to the relevant person. We covered this in lines of communication. However, reporting a problem is one thing but reporting it *effectively* is another. Always report problems to the appropriate person as soon as possible (media is available online).

Verbal report 1 A technician finds that a piece of equipment is not working correctly and he needs to report this to the supervisor. Imagine that the supervisor is instructing a customer and the report is made something like this.

Verbal report 2 Pete's attention is elsewhere and it is quite likely that he may not have heard or really taken it in. The technician should wait until Pete has finished speaking with the customer and then report the problem. Also, it would be rude to the customer to interrupt like this.

Written report It may be your company's policy to have any such problems written on a report form and given to the appropriate person. It's therefore necessary to write clearly and make sure that another person would understand what you are saying. A good point of practice then is to read what you have written

Poor handwriting Look at the report on the website filled in by a technician. 'That's a bit over the top,' you may say. 'No one would write like that?' But sometimes reports are really filled in like this. Aside from the obvious poor and illegible handwriting, several boxes have been left blank. The date or time has not been entered and there is no telephone number or contact details for the client. Who is the message for?

Explain yourself clearly There is another report made from a telephone conversation with a supplier, Morris Holdings. All the necessary boxes are filled in but . . . what is it saying? If you can't read it the translation actually says 'Morris Holdings can't do cleaning delivery this Wednesday and can do it next time if we want can call Christine 01732 4691'. Well, you can probably get the gist of the message but it could be explained more clearly. On the next screen we hear this same telephone message and see a much more clearly written report. Clear in this case means handwriting, spelling, grammar and expression – saying what you mean.

Explain yourself clearly 2 'If you would pass this message on please. We're not able to make the delivery this Wednesday; unfortunately, it won't be until Wednesday of next week, the 21st. If that causes any problems you can call Christine on 01732 4671.'

Take time If you know that your handwriting is difficult to read then take greater care. Write more slowly. It may be easier if you write in capital letters.

Remember, you never became an automotive technician to be good at handwriting and English – all you need is to be understood.

Adapt what you say It's also sometimes important to be able to adapt what you say to colleagues. If you have to speak to the receptionist or one of the office staff, it's probably not a good idea to use highly technical language. 'Well, why don't they ask if they don't understand?' That's true but a big part of good communication means that you take on the responsibility of making yourself understood. Don't leave it to chance and don't rely on the other person asking you. People are much more likely to respond to you if you use their language.

Choosing your words The choice of words and how you say (or write) them is likely to be different when talking to your workmate than if you were talking to, say, a manager of the company. Is it appropriate to use first names for management in your workplace? In many cases it will be but sometimes, especially if you are part of a large organisation and don't often speak to a manager, it may be more appropriate to use Mr, Mrs, Ms or Miss.

Views and opinions Throughout life you will always meet people with different views and opinions to yourself – it will be no different in the workplace. So often, poor working relationships are caused because people think they are right and the other is wrong. Remember, no one likes a 'know it all'.

Be open and flexible One of the best ways to learn is to be open to new ideas. This doesn't mean that you always have to agree with them but it is polite and professional to listen and consider what your workmate is saying. Don't always assume that you are right. Be flexible in your opinions. Again, by listening and really considering a different point of view shows that you value another person's thoughts and feelings. This in turn leads to a good working relationship.

Figure 3.30 Good communication is important

Figure 3.31 Be open and flexible

Positive or negative approach Sometimes you may be in the right and the other person in the wrong but there are positive and negative ways of dealing with this. In this scenario Ian and Mick, two technicians, are having a discussion over a fault on a customer's car (media is available online).

Honouring commitments If you say you are going to do something, do it! One of the worst things for damaging your career is to have a reputation of making 'empty promises', i.e. saying that you will do something and then not doing it. Quite often it's not a case of being lazy or unwilling but,

- forgetting
- being too busy.

Forgetful The best way of remembering to do something is to write it down. Perhaps your workshop may have a noticeboard or whiteboard or you may want to write it on a post-it note. At the very least, decide on a time there and then when you are going to do the task. That way, if you have the time set aside and firmly placed in your mind, you are less likely to forget. Writing something down, though, is always the best option, especially if you can't do the task that same day.

Too busy Be realistic! Consider the following two points if you are asked, or before you volunteer, to do something:

- Do I have the time?
- How long will the job take?

Do I have time? Of course you will want to be seen as co-operative and willing, but sometimes you may just not have time. How do you approach this? A response like, 'No, can't, sorry don't have time' isn't likely to do you any favours. You need to explain your situation. Don't assume others will know, even if you think they should. Perhaps you may need to say that you would like to but at the moment you really are too busy with such and such a task. Through discussion you may find that you can do the job with some help, or that the new job takes priority, or that it can be done later. Be aware of your limitations.

Seriously consider If you do make the commitment to do something then do seriously consider how long it is going to take you. Think what other jobs have to be done, and give some time for the unexpected, e.g. other tasks that may arise, unforeseen problems, etc.

Summary In summary, communication can be verbal or nonverbal, spoken or written, and positive or negative. The way we communicate in the workplace can help either to build or destroy working relationships. Feeling reasonably confident and relaxed is important. And finally, Good Luck!

 Select a routine from section 1.3 and follow the process to study a component or system

3.4 Documentation ❶

Job cards and systems A managed workflow is a vital part of the workshop system in a motor vehicle company. Many companies may dispense with the 'paper' altogether and use digital systems. These can be expensive but allow fast, easy and accurate communication. Whether job cards, digital systems or even a combination are used, the principle is the same and consists of several important stages. This is often described as a four-part job card system:

1 Reception – Customers' details and requirements are collected and/or checked; this information could be stored in a paper or digital system or even both.

2 Workshop control – Jobs are allocated to the appropriate technician using a loading sheet or again via a digital solution.

3 Parts department – Parts used are added to the workshop information.

4 Accounts – Invoices are prepared from the information on the job card. Digital systems may automatically produce the invoice when the job is completed.

Invoicing As part of the contract made with a customer, an invoice for the work carried out is issued. The main parts of the invoice are as follows:

- Diagnostic charges – Applied if investigatory work is required.
- Labour charges – The cost of doing the work. Usually, the time spent multiplied by the hourly rate.
- Parts – The retail price of the parts or as agreed.
- Sundries – Some companies add a small sundry charge to cover consumable items like PPE, nuts and bolts or cable ties etc.
- Data charge – Some workshops add the cost of accessing manufacturer's data
- MOT test – If appropriate. This is separated because VAT is not charged on MOTs.

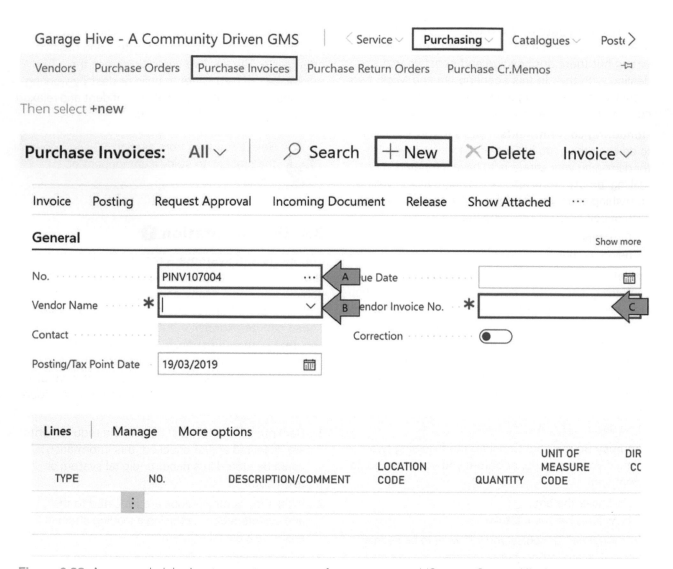

Figure 3.32 An example job sheet – most are now software generated (Source: Garage Hive)

▶ Waste – Some companies may itemize the cost of handling and disposing of hazardous waste.

▶ VAT – Value Added Tax – At the current rate, if the company is registered (all but the very small are).

Hourly rates vary quite a lot between different garages. The hourly rate charged by the company must pay a lot more than your wages – hence it will be much higher than your hourly rate! Just look round in any good workshop, as well as the rent for the premises, some of the equipment can cost tens of thousands of pounds. The money must come from somewhere.

Computerized workshop system There are several digital based workshop management systems available.

Some are specifically designed for main dealers, some for the smaller independent company. In this part, I will outline a system called Garage Hive (www.garagehive.co.uk). This system has three specific benefits:

▶ Workshop management
▶ Reporting and business intelligence
▶ Finance and accountancy

The workshop management built into Garage Hive has all the tools to run your workshop. A live dynamic work schedule, paperless jobcards, integrated Autodata repair times, technician time management, stock control and much more. You can also make informed decisions with the business intelligence & reporting application. View workshop profitability, technician efficiencies, margins and much more on

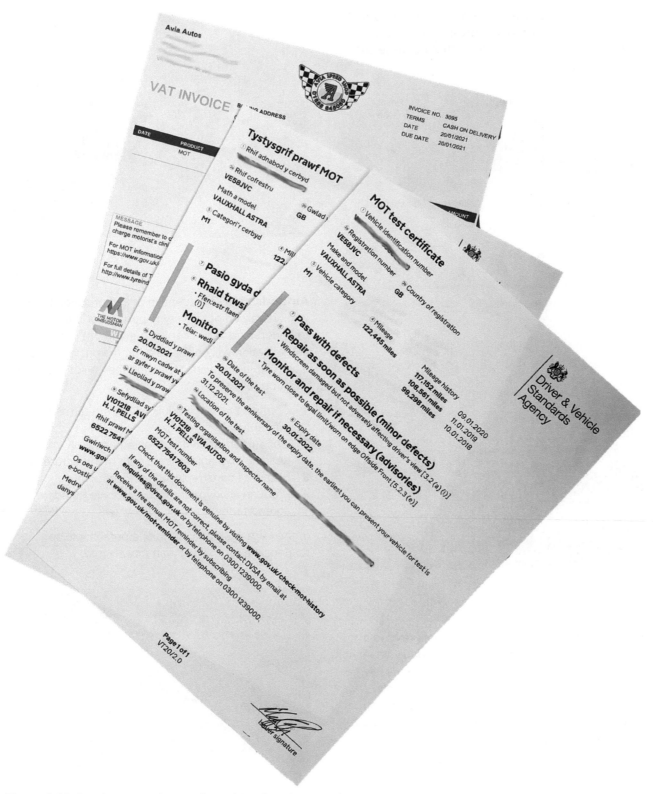

Figure 3.33 Invoice example together with other documents

all of your devices. The system offers a full integrated accountancy system. Alternatively you can take advantage of the integrations with Xero, Sage and Quickbooks.

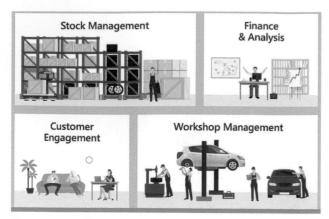

Figure 3.34 Garage Hive features

Figure 3.35 Special offers such as shown here are in addition to a warranty

Extended warranties It is also possible to have a warranty on a used vehicle or an extended warranty on a new vehicle. These often involve a separate payment to an insurance company. This type of warranty can be quite good but a number of exclusions and requirements may apply. Some examples are listed:

▶ regular servicing at an approved dealer
▶ only recommended parts must be used
▶ wear and tear is not included
▶ any work done must be authorised
▶ only recognised repairers may be used in some cases.

Figure 3.36 Mitsubishi cars

Authorisation The question of authorisation before work is carried out is very important for the garage to understand. Work carried out without proper authorisation will not be paid for. If a customer returns a car within the warranty period then a set procedure must be followed.

1 Confirm that the work is within the terms of the warranty.
2 Get authorisation if over an agreed limit (main dealers have agreements with manufacturers).
3 Retain all parts replaced for inspection.
4 Produce an invoice which relates to standard or agreed times.

In the larger garages, often one person will be responsible for making warranty claims.

Figure 3.37 If parts are replaced under warranty, they may need to be kept for inspection

Create an information wall to illustrate the features of a key component or system.

3.5 Working relationships

3.5.1 Colleagues ❶

Introduction 'He always gets the best jobs.' 'Why should I do that? It's not my responsibility.' 'I'm fed up with having to do her jobs. She's a girl, what do you expect.'

Imagine working in an environment like this or, sadly, perhaps you already do. Of course, from time to time we all have disagreements about certain things but surely the best way is to talk things through and sort out any problems in a friendly and professional manner. By studying this unit you will learn the necessary skills to maintain positive working relationships and gain the knowledge required for your qualification.

It's easy I'm sure that you will find this programme quite easy to learn, as it is mostly common sense. Hopefully it will also be quite fun. I have divided the learning programme into two sections:

1 Responsibilities, yours and your colleagues
2 Communication skills and working relationships

Your responsibilities To be a successful technician, it is essential that you fully understand your responsibilities in the organisation. That's not only the jobs that you are required to do, but also the jobs that you are not required to do and the duties that, by law, you are not allowed to do.

Job description When you start work at an organisation, you may be given a job description. At

Figure 3.38 It is important to be sure of your duties

the very least it should be made clear to you verbally if it is not written down. Either way, if you are unsure of your work duties, do ask.

Knowing your duties In the following scenario a technician is asked to do something that he is not allowed to because it is against health and safety regulations (media is available online).

Steve should politely decline this request. It is against health and safety law to clean, maintain or use any equipment without adequate training.

Your workmates' responsibilities In the previous scenario Steve's workmate may not have realised that Steve had not been trained to do this job. To avoid confusion, and breaking any laws and regulations, it is important to know and understand your work colleagues' responsibilities. It will also help avoid any bad feeling from asking a mate to do something that just isn't part of his workload, especially if he is very busy.

Giving advice and support Knowing your job role and the limits of your responsibility for giving advice and support is essential. Don't feel pressured into giving advice on something you are not qualified for. It can also be tempting to impress others with your knowledge but if you are unsure of something do make it clear that you are unsure. When the safety of someone is at stake never give advice unless it is your responsibility, and then only if you are absolutely certain of your facts.

Lines of communication How do you communicate in your workshop? Are there any specific lines of communication within your organisation? If it is a large workshop, there is almost certain to be. For instance, if you notice that you may be running low on a product, who should you inform? If someone has an accident, who is responsible for entering it into the accident book?

Job titles For the smooth running of an organisation it is necessary to follow these lines of communication. Start to get familiar with people's titles and jobs and this often will tell you the appropriate person to speak with.

 Create a mind map to illustrate the features of a key component or system.

3.5.2 Customers ❶

Introduction This section is a general overview of the type of advice and contact that you may need to have with your customers. Remember, customers pay our wages so treat them well!

35

Problems It is important to note that, generally, problems do not occur frequently during the life of a motor vehicle. Customers can be reassured that as long as their vehicle is regularly checked and serviced, they are observant of their instrumentation in the vehicle and seek advice when abnormal noises or running conditions occur, they should experience no major problems.

Service records Premature failure of the camshaft drive belt is a common problem that causes severe damage to the engine. This failure can be avoided if the camshaft drive belt is replaced at the correct interval, in addition, regular inspection and adjustment of the belt will generally highlight any problems so that the belt can be replaced before it fails. Customers will appreciate appropriate tracking of the requirement for this work so that they can be contacted and reminded when it is due.

Expensive repairs Always encourage your customers to come back to you whenever they feel that something is wrong, no matter how small. These things can often be put right before they develop into expensive repairs.

Customer records Customer records should be kept for the frequency and type of service carried out so that a check can be made on rubber hoses at appropriate time intervals. This is good practice to avoid the risk of fuel leakage and the potential for vehicle fires.

Quality replacement components It is often tempting to increase the profit on a job by using 'pattern' parts. These are copies of the original components. Some of these components, produced by well-known companies (Bosch for example) are excellent quality and are used as original equipment by many manufacturers. However, some pattern parts are cheap and you get what you pay for! One of the most irritating things for a customer is to have to return the vehicle for the same job to be repeated – and it wastes your time. Use good quality parts at all times.

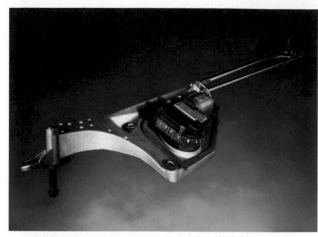

Figure 3.39 Bosch wiper components

Summary A customer, who is kept informed and treated with respect, will return and keep you in a job! Explain things to a customer when asked – it will be appreciated.

 Look back over the previous section and write out a list of the key bullet points from this section.

CHAPTER 4

Workshop skills

After successful completion of this chapter you will be able to show you have achieved these objectives:

- Understand how to select, use and care for hand tools and measuring devices in the automotive environment.

- Understand how to prepare and use common workshop equipment.

- Understand how to select materials when fabricating, modifying and repairing vehicles and fitting components.

- Understand how to apply automotive engineering, fabrication and fitting principles when modifying and repairing vehicles and components.

DOI: 10.1201/9781003173236-4

4.1 Hand tools and measuring

4.1.1 Hand tools ❶

Using hand tools is something you will learn by experience, but an important first step is to understand the purpose of the common types. This section therefore starts by listing some of the more popular tools, with examples of their use, and ends with some general advice and instructions.

Figure 4.1 Ring spanners (wrenches)

Practice until you understand the use and purpose of the following tools when working on vehicles:

Hand tool	Example uses and/or notes
Adjustable spanner (wrench)	An ideal standby tool and useful for holding one end of a nut and bolt.
Open-ended spanner	Use for nuts and bolts where access is limited or a ring spanner can't be used.
Ring spanner	The best tool for holding hexagon bolts or nuts. If fitted correctly it will not slip and damage either you or the bolt head.
Torque wrench	Essential for correct tightening of fixings. The wrench can be set in most cases to 'click' when the required torque has been reached. Many fitters think it is clever not to use a torque wrench. Good technicians realise the benefits.
Socket wrench	Often contains a ratchet to make operation far easier.
Hexagon socket spanner	Sockets are ideal for many jobs where a spanner can't be used. In many cases a socket is quicker and easier than a spanner. Extensions and swivel joints are also available to help reach that awkward bolt.
Air wrench	These are often referred to as wheel guns. Air-driven tools are great for speeding up your work but it is easy to damage components because an air wrench is very powerful. Only special, extra strong, high-quality sockets should be used.

Hand tool	Example uses and/or notes
Blade (engineer's) screwdriver	Simple common screw heads. Use the correct size!
Pozidriv, Phillips and cross-head screwdrivers	Better grip is possible, particularly with the Pozidriv, but learn not to confuse the two very similar types. The wrong type will slip and damage will occur.
Torx®	Similar to a hexagon tool like an Allen key but with further flutes cut in the side. It can transmit good torque.
Special purpose wrenches	Many different types are available. As an example, mole grips are very useful tools as they hold like pliers but can lock in position.
Pliers	These are used for gripping and pulling or bending. They are available in a wide variety of sizes. These range from snipe nose for electrical work, to engineer's pliers for larger jobs such as fitting split pins.
Levers	Used to apply a very large force to a small area. If you remember this you will realise how, if incorrectly applied, it is easy to damage a component.
Hammer	Anybody can hit something with a hammer, but exactly how hard and where is a great skill to learn!

General advice and instructions for the use of hand tools (taken from information provided by Snap-on):

▶ only use a tool for its intended purpose
▶ always use the correct size tool for the job you are doing
▶ pull a spanner or wrench rather than pushing whenever possible
▶ do not use a file, or similar, without a handle
▶ keep all tools clean and replace them in a suitable box or cabinet
▶ do not use a screwdriver as a pry bar
▶ look after your tools and they will look after you!

Note The following pages give an overview of some of the tools you will need to become familiar with – there are many more! As you study each main chapter it may be useful to refer back to the tools and equipment presented here.

Hose removal A number of tools have been developed to aid the removal of hoses. These cranked and blunt-bladed probes can be eased into the end of a hose to break the seal against the connector pipe. Blunt screwdrivers can also perform this task. Another method is to use a sharp knife to cut the hose back, but this method can only be used when the hose is to be discarded.

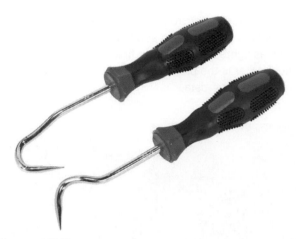

Figure 4.2 Hose remover

Hose clips There are a range of specialized hose clips that require a dedicated pair of pliers for their removal and installation. These tend to be specific to individual manufacturers, and the pliers are included in the workshop-equipment pack for the dealer. Where the special tools are not available, the clip can be replaced with a general-purpose screw-type hose clip.

Strap wrenches These are for removing and replacing fuel and oil canister type filters.

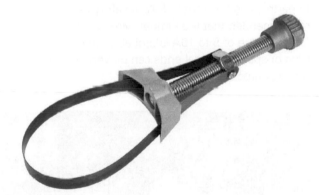

Figure 4.3 Strap wrench

Tools for exhaust removal Some useful tools for exhaust system removal are chain wrenches for twisting seized pipes, and oxyacetylene welding equipment for freeing up rusted joints. An air chisel or cutter may be necessary for cutting off components that will not be reused.

Figure 4.4 Chain cutter

Figure 4.5 Air cutter

Torque wrench A good torque wrench is an essential piece of equipment. Many types are available but all work on a similar principle. Most are set by adjusting a screwed cylinder, which forms part of the handle. An important point to remember is that, as with any measuring tool, regular calibration is essential to ensure it remains accurate.

Figure 4.6 Torque wrench

Bearing puller Removing some bearings is difficult without a proper puller. For internal bearings, the tool has small legs and feet that hook under the bearing. A threaded section is tightened to pull out the bearing. External pullers hook over the outside of the bearing and a screwed thread is tightened against the shaft.

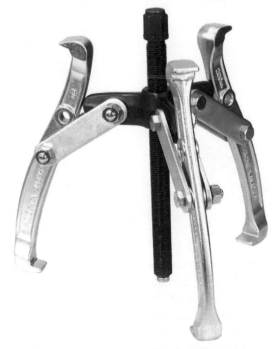

Figure 4.7 External bearing puller

4

Soft hammers These tools allow a hard blow without causing damage. They are ideal for working on driveshafts, gearboxes and final drive components. Some types are made of special hard plastics whereas some are described as copper/ hide mallets. One type has a copper insert on one side and a hide or leather insert on the other. It is still possible to cause damage, however, so you must take care!

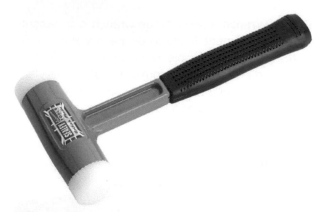

Figure 4.8 Some hammers contain metal shot to give a 'dead blow'

Brake adjusting tools On many earlier braking systems, the adjustment (gap between the shoe and drum) had to be adjusted manually during a service. Most modern systems do this automatically. However, many earlier systems are still in use so tools such as these, which are used to rotate a gear inside the drum, will be very useful. Some are made to suit particular manufacturer's systems.

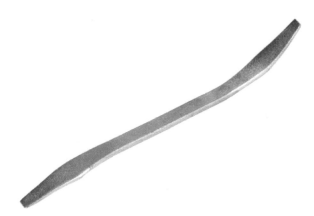

Figure 4.9 These tools are simple levers

Test lamp This is an often underrated piece of test equipment! However, it must be used with care. The advantage of a simple test lamp is that it draws some current through the circuit under test. This allows 'high resistance' faults to be located easily. However, this is also a disadvantage because drawing current

through electronic circuits can damage them. Using a test lamp for checking supplies to electrical items such as lights and motors is fine; for all other tests, a multimeter is the preferred option. If in doubt, consult manufacturers' data.

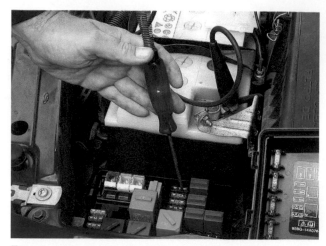

Figure 4.10 Test lamps are simple but useful

A **jumper wire** is useful for bypassing components such as switches. However, do not short supplies to earth using this method. As a safety feature, it is recommended that the jumper wire be fitted with a fuse. A value of 5 to 10A is probably ideal. Crocodile clips or spade terminal ends can be very useful for testing purposes.

Figure 4.11 Fused jump lead

Many **terminal kits** are available. They usually consist of a selection of terminals and special pliers to crimp the terminals on to the wire.

Wire strippers With practice you will be able to strip wire using side cutters. However, special tools are available to make the job easier. A number of different types are shown here.

Figure 4.12 Selection of wire strippers

Figure 4.14 Steel rule in millimetres and inches

Most **soldering irons** are electrically heated. However, there are some very good gas-powered types now available. The secret with a soldering iron is to use the right size for a specific job. One suitable for delicate integrated circuits and circuit boards will not work on large alternator diodes. More damaging would be to use a large iron on a small circuit board!

Paper clip Not found in Snap-on or other catalogues, but a very useful tool. It is not only ideal for bridging terminals as shown here, it can also be used for clipping paper together!

Figure 4.13 Paper clip in use

Use the media search tools to look for pictures and videos relating to the subject in this section.

4.1.2 Measurement 🔢

Introduction Measurement is the act of measuring or the process of being measured. It is the process of comparing a dimension, quantity, or capacity to a known value by using a suitable instrument. Simple examples of measuring instruments would be a tape measure or a steel rule (often referred to as a ruler).

Accuracy is a key aspect of measurement and generally means how close the measured value of something is in comparison with its actual value. For example, if a length of about 30cm is measured with an old tape measure, then the reading may be 1 or 2mm too high or low. This would be an accuracy of ± 2mm.

Now consider measuring a length of metal bar with a steel rule. How accurately could you measure it this time? Probably to the nearest 0.5mm if you were careful but this raises a number of issues to consider:

1 You could make an error reading the rule.
2 Do we actually need to know the length of a piece of bar to the nearest 0.5mm?
3 The rule may be damaged and not give the correct reading!

The first and second of these issues can be dispensed with by knowing how to read the test equipment correctly and also knowing the appropriate level of accuracy required. A micrometer for a tyre tread depth would be over the top and, in the same way, a tape measure for valve clearances would be nowhere near good enough. I am sure you get the idea.

Figure 4.15 Tape measure (Source: Jack Sealey)

Remember! There is more support on the website that includes additional images and interactive features: www.tomdenton.org

41

Equipment To ensure instruments are, and remain accurate, there are just two simple guidelines:

▶ look after the equipment; a micrometer thrown on the floor will not be accurate

▶ ensure instruments are calibrated regularly – this means being checked against known good equipment.

Here is a summary of the steps to ensure a measurement is accurate:

Step	Example
Decide on the level of accuracy required.	Do we need to know that the battery voltage is 12.6V or 12.635V?
Choose the correct instrument for the job.	A micrometer to measure the thickness of a shim.
Ensure the instrument has been looked after and calibrated when necessary.	Most instruments will go out of adjustment after a time. You should arrange for adjustment at regular intervals. Most tool suppliers will offer the service or in some cases you can compare older equipment to new stock.
Study the instructions for the instrument in use and take the reading with care. Ask yourself if the reading is about what you expected.	Is the piston diameter 70.75mm or 170.75mm?
Make a note if you are taking several readings.	Don't take a chance – write it down!

Straight edge and feelers A 'straight edge' is a piece of equipment with a straight edge! It is used as a reference for measuring flatness. It is placed on top of the test subject. The feeler blades are then used to assess the size of any gaps. The feeler blades are sized in either hundredths of a millimetre or thousandths of an inch.

Figure 4.16 Straight edge

Micrometer The most common uses for a micrometer in the motor trade are for measuring valve shims, brake disc thickness and crankshaft journals.

Figure 4.17 Feeler gauges

The spindle of an ordinary metric micrometer has two threads per millimetre, and therefore one complete revolution moves the spindle through a distance of 0.5mm. The longitudinal line on the frame is graduated with 1mm divisions and 0.5mm subdivisions. The thimble has 50 graduations, each being 0.01mm (one-hundredth of a millimetre). Thus, the reading is given by the number of millimetre divisions visible on the scale of the sleeve plus the particular division on the thimble which coincides with the axial line on the sleeve.

Figure 4.18 Zero to 25mm micrometer

Example Suppose that the thimble were screwed out so that graduation 5, and one additional 0.5 subdivision were visible (as shown in the image), and that graduation 28 on the thimble coincided with the axial line on the sleeve. The reading then would be 5.00 + 0.5 + 0.28 = 5.78mm.

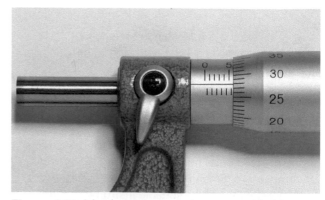

Figure 4.19 Metric system – micrometer thimble reading 5.78mm

A **dial gauge** is a device used to test the movement of something very accurately. A good example is the run out of brake discs. A plunger on the dial gauge is made to run up against the disc as it is turned. The body of the gauge is clamped in position and via accurate gears the movement of the plunger makes a hand turn on a clock dial. The face of the clock is marked off in 0.01mm increments and will usually rotate enough times to measure up to 10mm. Diesel pump timing is often set or checked with a dial gauge. The pumping plunger is made to act on the dial gauge so that, as the engine is turned, it is possible to tell the position of the injection plunger very accurately.

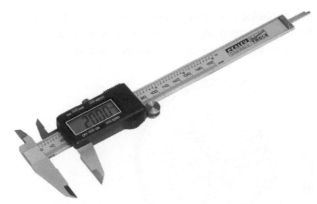

Figure 4.21 Digital caliper

Figure 4.20 Dial gauge

Reading To take a reading on the digital type is simple as it is displayed as a figure. The dial type displays the major distance on the sliding scale and then greater accuracy is achieved by adding the reading on the dial. The vernier type (Figure 4.22) is read as follows: First read the position of the pointer directly on the millimetre scale (24mm in this case). When the pointer is between two markings (as it is here), you can mentally interpolate to improve the precision of the reading (my guess would be about 24.8). However, the addition of the vernier scale allows a more accurate interpolation. All you need to do is check which mark on the vernier scale lines up exactly with one of the main scale. In this example I would say it is the one halfway between the 7 and the 8. This means the actual measurement is 24.75mm.

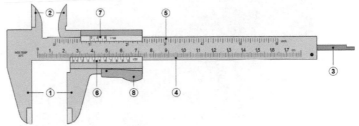

Figure 4.22 Vernier caliper – the main parts are: 1 Outside jaws: used to measure external diameter or width of an object, 2 Inside jaws: used to measure internal diameter of an object, 3 Depth probe: used to measure depths of an object or a hole, 4 Main scale: scale marked every mm, 5 Main scale: scale marked in inches and fractions, 6 Vernier scale gives interpolated measurements to 0.01mm or better, 7 Vernier scale gives interpolated measurements in fractions of an inch, 8 Retainer: used to block movable part to allow the easy transferring of a measurement

Vernier caliper There are now three types of caliper used for accurate measuring: vernier, dial, and digital. These calipers have a calibrated scale with a fixed jaw, and another jaw, with a pointer, that slides along the scale. The distance between the jaws is then read in different ways for the three types. Vernier, dial, and digital calipers can measure internal dimensions, external dimensions using the lower jaws and, in many cases, depth by the use of a probe that is attached to the movable head and slides along the centre of the body. This probe is slender and can get into deep grooves that may prove difficult for other measuring tools. The vernier scales may include metric measurements on the lower part of the scale and inch measurements on the upper, or vice versa in countries that use inches. Vernier calipers commonly used in industry provide a precision to 0.01mm, or 0.001 inch. They are available in various sizes.

Angle locator This magnetic device is used to check that the angles of a propshaft are equal. This is important because it ensures that the changing velocity effects of the universal joints (UJs) are cancelled out.

The angle locator attaches magnetically to the shaft. A dial is set to zero and then, when it is moved to a new location, the difference in angle is indicated.

Function	Range	Accuracy
Duty cycle	% on/off	0.2%/kHz
Frequency	over 100kHz	0.01%
Temperature	> 9000°C	0.3% + 30°C
High current clamp	1000A (DC)	Depends on conditions
Pressure	3 bar	10.0% of standard scale

Figure 4.23 This device checks propshaft angles

Multimeter An essential tool for working on vehicle electrical and electronic systems is a good digital multimeter (often referred to as a DMM). Digital meters are most suitable for accuracy of reading as well as other facilities.

An **oscilloscope** is a very useful measuring instrument. There were traditionally two types of oscilloscope – analog or digital. However, the digital scope is now universal. An oscilloscope draws a graph of voltage (the vertical scale or Y axis) against time (the horizontal scale or X axis). The trace is made to move across the screen from left to right and then to 'fly back' and start again. The frequency at which the trace moves across the screen is known as the time base, which can be adjusted either automatically or manually. The signal from the item under test can either be amplified or attenuated (reduced), much like changing the scale on a voltmeter.

The trigger, which is what starts the trace moving across the screen, can be caused internally or externally. When looking at signals such as ignition voltages, triggering is often external – each time an individual spark fires or each time number one spark plug fires, for example.

Figure 4.24 Multimeter and current clamp

The following list of functions broadly in order, starting from essential to desirable, should be considered:

Function	Range	Accuracy
DC voltage	500V	0.3%
DC current	10A	1.0%
Resistance	0 to 10MΩ	0.5%
AC voltage	500V	2.5%
AC current	10A	2.5%
Dwell	3,4,5,6,8 cylinders	2.0%
rpm	10 000 rpm	0.2%

Figure 4.25 Automotive Oscilloscope EV kit (Source: Pico Technology Ltd)

PicoScope The Pico automotive kit turns a laptop or desktop PC into a powerful automotive diagnostic tool for fault-finding sensors, actuators and electronic circuits. The oscilloscope connects to a USB port on a PC and can take up to 32 million samples per trace, making it possible to capture complex automotive waveforms. The scope can be used to measure and test virtually all of the electrical and electronic components and circuits in any modern vehicle including:

- ignition (primary and secondary)
- injectors and fuel pumps
- starter and charging circuits
- batteries, alternators and starter motors
- lambda, airflow, knock and MAP sensors
- glow plugs/timer relays
- CAN bus, LIN bus and FlexRay.

Waveform When you look at a waveform on a screen it is important to remember that the height of the scale represents voltage and the width represents time. Both of these axes can have their scales changed. They are called axes because the 'scope' is drawing a graph of the voltage at the test points over a period of time. The time scale can vary from a few μs to several seconds. The voltage scale can vary from a few mV to several kV. For most test measurements only two connections are needed, just like a voltmeter.

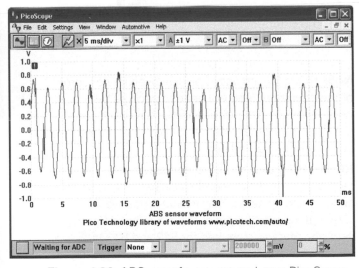

Figure 4.26 ABS waveform captured on a PicoScope

 Create an information wall to illustrate the features of a key component or system.

4.2 Equipment

4.2.1 Workshop equipment ❶

In addition to hand tools and test equipment, most workshops will also have a range of equipment for lifting and supporting, as well as electrical or air-operated tools. This table lists some examples of common workshop equipment together with typical uses:

Equipment	Common use
Ramp or hoist	Used for raising a vehicle off the floor. Figure 4.27 shows a two-post wheel-free type. Other designs include four-post and scissor types where the mechanism is built in to the workshop floor.
Jack and axle stands	A trolley jack (such as Figure 4.28) is used for raising part of a vehicle such as the front or one corner or side. It should always be positioned under suitable jacking points, axle or suspension mountings. When raised, stands must always be used in case the seals in the jack fail, causing the vehicle to drop.
Air gun	A high-pressure air supply is common in most workshops. Figure 4.30 is a typical wheel gun used for removing wheel nuts or bolts. Note that when replacing wheel fixings it is essential to use a torque wrench.
Electric drill	The electric drill is just one example of electric power tools used for automotive repair. Note that it should never be used in wet or damp conditions.
Parts washer	There are a number of companies that supply a parts washer and change the fluid it contains at regular intervals.
Steam cleaner	Steam cleaners can be used to remove protective wax from new vehicles as well as to clean grease, oil and road deposits from cars in use. They are supplied with electricity, water and a fuel to run a heater – so caution is necessary.
Electric welder	There are a number of forms of welding used in repair shops. The two most common are metal inert gas (MIG) and manual metal arc (MMA). Figure 4.29 shows the MMA process.
Gas welder	Gas welders are popular in workshops as they can also be used as a general source of heat, for example, when heating a flywheel ring gear.
Engine crane	A crane of some type is essential for removing the engine on most vehicles. It usually consists of two legs with wheels that go under the front of the car and a jib that is operated by a hydraulic ram. Chains or straps are used to connect to or wrap around the engine.
Transmission jack	On many vehicles the transmission is removed from underneath. The car is supported on a lift (perhaps similar to Figure 4.27) and then the transmission jack is rolled underneath. An example is shown as Figure 4.31.

Figure 4.27 Car lift (Source: asedeals.com)

Figure 4.30 Wheel gun (Source: Snap-on Tools)

Figure 4.28 Trolley jack and axle stands (Source: Snap-on Tools)

Figure 4.31 Transmission jack (Source: Snap-on Tools)

Note The following pages give an overview of some of the equipment you will need to become familiar with – there is a lot more! As you study each main chapter it may be useful to refer back to the tools and equipment presented here.

AC servicing unit Most modern servicing units can be used to drain, recycle, evacuate and refill air conditioning systems. Some older types would only carry out individual procedures. Note that different servicing units are required for R12 and R134a refrigerants and their oils must not be mixed with one another. If work is to be carried out on an air conditioning system, a servicing unit is essential. Refrigerant must never be released into the atmosphere.

Coil spring compressors Springs must be compressed before they are removed from suspension struts. A number of different tools are available. However, the type shown here is very

Figure 4.29 Welding process (Source: Wikimedia)

Figure 4.32 AC servicing equipment

popular. The two clamps are positioned either side of the spring using the hooked ends. The bolts are then tightened evenly until the tension of the spring is taken by the clamps.

Figure 4.33 MacPherson springs must be compressed for removal

Ball joint presses To remove a taper fitting ball joint, a splitter or press is usually needed. If the joint is to be reused, a lever- or clamp-type splitter is preferred. The tool clamps onto the arm and threaded section of the joint. A bolt is tightened, which applies a force to push the joint free.

Figure 4.34 This clamp removes tapered ball joints

Ball joint splitter Two types of ball joint splitter are in common use. One type is a simple forked wedge that is hammered in between the joint and the arm. This works well but can damage the joint. If the joint is to be reused, the lever-type splitter is preferred. This tool clamps onto the arm and threaded section of the joint. A bolt is tightened, which applies a force to push the joint free.

Figure 4.35 Lever-type splitter

A **pipe clamp** is used to block a pipe for tests or repairs to be carried out. For example, on a braking system, it can be used to prevent leakage of fluid when cylinders are replaced. Alternatively, the source of spongy brakes can be narrowed down. This is done by clamping each flexible pipe in turn and pressing the pedal. However, some manufacturers do not recommend these tools because the pipe can be damaged.

Figure 4.36 Only use recommended types

Pressure bleeder This equipment forces fluid through the reservoir under pressure. The tank is in two parts, separated by a diaphragm. The top of the tank is filled with new brake fluid and the lower part pressurised with compressed air. Using suitable adaptors, the outlet pipe, from the fluid section, is connected to the master cylinder reservoir. A valve is opened and fluid is forced out of the slave cylinders as the bleed nipples are opened. Fluid is collected in a container using a simple rubber pipe, just like when bleeding the system manually.

Figure 4.37 This equipment forces fluid through the reservoir

A **honing tool** is sometimes called a 'glaze buster'. It is used to grind the inside of a cylinder to a good, final finish. This can be done to an engine cylinder or a much smaller hydraulic brake cylinder. The tool is usually mounted in an air drill as the power source. Lubrication should be used when the equipment is operated.

Figure 4.38 Engine cylinder honing tool

Most **wheel balancers** offer facilities for measuring the wheel, and then programming this into a computer. The machines usually run from a mains electrical supply. The wheel is clamped to the machine and spun. Sensors in the machine determine the static and dynamic balance. A display states where extra weights should be added to obtain accurate balance. 'On-car' balancers have been used, but are less accurate than the later computerised types.

Tyre changer It is possible to change tyres with two levers and a hammer! However, it is much quicker and easier with an automatic changer. A lever is still needed to start the bead of the tyre lifting over the rim. An electric motor drives the wheel round as the tyre is removed or fitted. Most changers incorporate a bead breaker.

Figure 4.39 Balancer (Source: Bosch Media)

Figure 4.40 Many types of changers are available (Source: Bosch Media)

Tyre inflator This is a simple but important item of equipment. Make sure it is looked after so that the gauge remains accurate. A small difference in tyre pressure can have a significant effect on performance and wear.

Figure 4.41 The gauge must be accurate

Clutch aligner kit The clutch disc must be aligned with the cover and flywheel when it is fitted. If not, it is almost impossible, on some vehicles, to replace the gearbox. This is because the gearbox shaft has to fit through the disc and into the pilot or spigot bearing in the flywheel. The kit shown here has adaptors to suit most vehicles.

Figure 4.44 This tool is useful for removing halfshafts

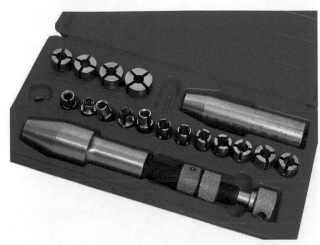

Figure 4.42 The clutch must be aligned when fitted

4.2.2 Test equipment 1 2

Pressure tester The cooling system pressure tester is both a diagnostic tool and a measuring instrument. Its use as a tool is to aid in the detection of coolant leaks by producing the operating pressure in the cooling system that replicates normal running. Any leaks will reduce the operating pressure and these can be identified.

Pilot/spigot bearing puller Removing spigot bearings is difficult without a proper puller. This tool has small legs and feet that hook under the bearing. A threaded section is tightened to pull out the bearing.

Cylinder leakage testers These can form part of an engine diagnostic system, or can be free- standing, self-contained units. Both require a compressed air supply and are operated in a similar way. The spark plugs are removed to gain access to the cylinder and the engine turned to top dead centre on the cylinder to be tested.

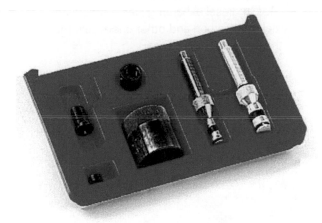

Figure 4.43 An internal bearing puller

A **slide hammer** is a form of puller. It consists of a steel rod over which a heavy mass slides. The mass is 'hammered' against a stop, thus applying a pulling action. The clamp end of the tool can screw either into or onto the component. Alternatively, puller legs with feet are used to grip under the sides of the component.

Figure 4.45 Leakage test gauge

49

Cylinder compression testers Compression testers for petrol and diesel engines are different and this should be noted. Most petrol types are hand held into the spark plug hole; others are threaded and screw into the spark plug hole. Diesel engine compression testers are always screwed into the injector pump or glow plug hole due to the higher cylinder pressures. The cylinder compression can be established by cranking or firing the engine, depending on the type of tool and the engine manufacturer's instructions. For all compression testing, follow the instructions for the type of tool being used.

Figure 4.46 Compression testers

Stethoscope or sonoscope This is a sensitive sound detector and amplifier that can be used to listen to and locate mechanical noises. It is operated by touching the probe onto the casing of mechanical units to pick up structure-borne noise. It can locate noises that travel through solid blocks, housings and casings.

Stroboscopic timing lights Stroboscopic lights are triggered by the secondary-circuit pulses to number one cylinder spark plug, and flash a strobe light onto the timing marks to show the position of the ignition spark. The stroboscopic light is used for dynamic-ignition timing checks and adjustments.

Exhaust gas analysis A special meter is needed for exhaust gas analysis. There is a range of individual gas meters as well as those that are part of an engine analyser. Some meters give carbon monoxide (CO) and hydrocarbon (HC) readings only but are still very useful. For accurate analysis of relevant exhaust gas constituents and statutory test purposes, a four-gas analyser is needed. This will provide a read out on carbon monoxide (CO), carbon dioxide (CO_2), hydrocarbon (HC) and oxygen (O_2). Some of these analysers have an additional read-out to check the operation of catalytic converters.

Figure 4.47 Four gas analyser

A **smoke meter** is used to check for excess particulates in diesel exhaust gases. Particulates are fine particles of carbon and other chemicals that form during the combustion process.

Figure 4.48 Diesel smoke meter

Engine analysers Some form of engine analyser has become an almost essential tool for fault-finding modern vehicle engine systems. The latest machines are now generally based around a personal computer. This means that more facilities can be added by simply changing the software. Whilst engine analysers are designed to work specifically with the motor vehicle, it is worth remembering that the machine consists of three parts: a multimeter, a gas analyser and an oscilloscope.

Float hydrometer This draws off a sample of coolant. The float level is marked to indicate the percentage of ethylene glycol in the coolant. Some of these hydrometers have a built-in thermometer for the coolant temperature, and include a correction chart for the actual float reading against the coolant temperature.

Figure 4.51 Balancing gauges (Source: Jack Sealey)

Pressure gauges For carrying out pressure tests, there is a range of pressure gauges with adapters to suit all vehicle and fuel system types.

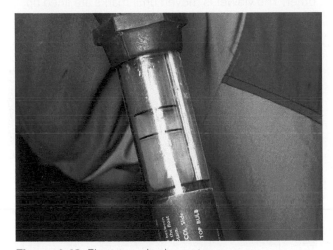

Figure 4.49 Float-type hydrometer

Drive belt tension The water pump and fan drive belt tension can be measured and adjusted with the use of a belt tension gauge. Many manufacturers recommend the use of these gauges for accurate adjustment of belt tension.

Figure 4.52 Gauges

Figure 4.50 Belt tension gauge (Source: Jack Sealey)

Airflow measurements A special vacuum gauge that fits over the air intake port of the carburettors is used to adjust the airflow through twin or multiple carburettors. This is the only reliable method of balancing the airflow through the carburettors.

A **vacuum gauge and pump** is used to test any part of a system that requires proper sealing, pressure or vacuum to operate. The pump can be used to apply a vacuum or a pressure. The gauge reads accordingly.

A **beam setter** is used to ensure that the aim of headlights is correct. This is also a legal requirement. The beam setter has a lens at the front to focus the

51

Figure 4.53 Gauge and pump kit (Source: Jack Sealey)

Figure 4.55 UV detector in use

other. The viewer is moved until marks are lined up and the tracking can then be measured.

Figure 4.56 Mirror gauge

light, which is reflected onto a translucent screen. The image shown on the screen of the beam setter is the same as would be seen on a vertical wall with the lights shining on it. Two controls are included to set the screen to the required vertical and horizontal position of the beam. Adjusters on the headlight unit are then moved to position the beam pattern accordingly. Most beam setters also include a light meter. This is used for setting the hot spot, or brightest part of the beam, to a specified direction.

Figure 4.54 Accurate alignment is important (Source: Hella-Gutmann)

Ultraviolet leak detector A popular device used when testing for AC system leaks, is an ultraviolet (UV) detector. A special dye is added to the refrigerant, which shows up under UV light.

Tracking gauges The toe-in and toe-out of a vehicle's front wheels are very important. Many types of tracking gauges are available. One of the most common uses a frame placed against each wheel with a mirror on one side and a moveable viewer on the

Alignment equipment Alignment equipment can vary in complexity from simple 'spirit level' gauges to complex laser systems. Examples of two types are shown above. On the complex systems, the principle is still very simple! Reflected light or a laser is used to set a perfect rectangular shape. How the position of the wheels deviates from this, is shown on an LCD display or a graduated scale.

Turntables allow the wheels to swivel when carrying out tests or alignment checks. Some types are simple greased plates! However, for detailed work, turntables with markings in degrees are used. Checks such as toe-out on turns are carried out on turntables.

Brake fluid tester Because brake fluid can absorb a small amount of water, it must be renewed or tested regularly. It becomes dangerous if the water turns into steam inside the cylinders or pipes, causing the brakes to become ineffective. The tester measures the moisture content of the fluid.

Figure 4.57 Simple camber gauge

Figure 4.58 Complex optical alignment system

Figure 4.59 Turntables allow the wheels to swivel

 Select a routine from section 1.3 and follow the process to study a component or system.

4.3 Materials and fabrication

4.3.1 Materials **1 2**

Introduction Different materials are used in different places on motor vehicles because of their properties. If I give a simple example it will make this clear. Cast iron is normally used for the very hot exhaust manifold but could plastic be used instead? Well, probably not, and it is obvious why. However, what if we asked if we could use aluminium instead? This time it is not obvious and more thought is required to decide on the most suitable material.

Figure 4.60 Aluminium intake pipes

Properties The following table lists several types of material together with their important properties. As a rough guide these are given as a number from 1 (best) to 5 (worst) in a kind of league table. This makes the table easier to use when comparing one material with another.

Corrosion is the eating away and eventual destruction of metals and alloys (a mixtures of metals) by chemical action. Most metals corrode eventually but the rusting of ordinary iron and steel is the most common form of corrosion. Rusting of iron or steel takes place in damp air, when the iron combines with oxygen and water to form a brown deposit known as rust. Higher temperatures make this reaction work more quickly. Salty road and air conditions make car bodies rust more quickly.

Some materials other than metals corrode or perish over a period of time; rubber-based materials, for example. Plastics have the great advantage that they appear to last for ever!

Material	Ease of shaping	Strength	Resistance to heat	Electrical resistance	Corrosion resistance	Cost	Typical MV uses
Copper	2	3	2	1	3	4	Wires and electrical parts
Aluminium	2	3	2	1	3	4	Cylinder heads
Steel	3	2	1	1	4	3	Body panels and exhausts
Cast iron	3	2	1	1	4	3	Manifolds and engine blocks
Platinum	3	1	1	1	2	5	Spark plug tips
Soft plastic	1	5	5	5	1	1	Electrical insulators
Hard plastic	1	4	4	5	1	1	Interiors and some engine components
Glass	3	5	2	5	1	2	Screens and windows
Rubber	2	4	5	5	3	2	Tyres and hoses
Ceramics	4	4	1	5	1	4	Spark plug insulators

This table is just to help you compare properties; the league table positions are only rough estimates and will vary with different examples of the same material.

Terms used to describe materials:

Property	Explanation
Hardness	Can withstand indentation (marking).
Softness	Can be easily indented.
Toughness	The ability to resist fracture.
Brittleness	Breaks or shatters under shock loads (impact).
Ductility	Plastic (deforms and stays that way) under tension or stretching.
Malleability	Plastic under compression (squeezing).
Plasticity	The ability to retain a deformation after a load is removed.
Elasticity	The ability to return to its original shape when a deforming load is removed.
Strength	The ability to withstand a load without breaking.

 Use a library or the web search tools to further examine the subject in this section.

4.3.2 Metal cutting and shaping ❶

Introduction As well as the obvious skills, such as knowledge of the systems and the ability to use normal hand tools for vehicle repairs, bench fitting and, in some cases, machining skills are also essential.

This usually involves metal-cutting operations but it can involve other materials such as wood and plastics. In this sense the word cutting is a very general term and can refer to:

▶ sawing
▶ drilling
▶ filing
▶ tapping
▶ machining.

These aspects will be examined in a little more detail in the following sections.

Fitting and machining skills may be needed to complete a particular job. In the context of the automotive engineer, we often use the term 'fitting' as a general description of the hand skills usually used on a workbench or similar to construct an item that cannot be easily purchased – for example, a support bracket for a modified exhaust or a spacer plate to allow the connection of an accessory of some type, such as additional lights.

Figure 4.61 Repairs using lathe

Machinists usually work to very small tolerances – ±0.1mm – for example, and deal with all aspects of shaping and cutting. The operations most often carried out by machinists are milling, drilling, turning and grinding. To carry out fitting or machining operations you should be familiar with:

▶ measuring tools, such as a micrometer
▶ hand tools as found in a standard toolkit
▶ machine tools, such as a bench drill
▶ work holders; for example, a vice

▶ tool holders, such as the chuck of a drill
▶ cutting tools like saws and files.

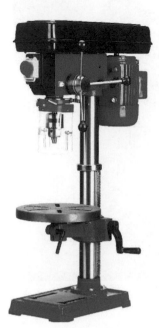

Figure 4.62 Bench drill (Source: Jack Sealey)

Filing is the process of removing material when manufacturing something; it is used mostly for finishing operations. Filing can be used on a wide range of materials as a finishing process. Emery paper may be considered as a filing tool. Files have forward-facing cutting teeth that cut best when pushed over the work piece. A process known as draw filing involves turning the file sideways and pushing or pulling it across the work. This catches the teeth of the file sideways and results in a very fine shaving action.

Figure 4.63 Hand filing (Source: South Thames College)

File types Files come in a wide variety of sizes, shapes, cuts, and tooth configurations. The most

common cross sections of a file are: flat, round, half-round, triangular and square. The cut of the file refers to how fine its teeth are. They are described, from roughest to smoothest, as: rough, middle, bastard, second cut, smooth, and dead smooth. The picture shows three common file cuts. Most files have teeth on all faces, but some flat files have teeth only on one face or edge so that the file can work against another edge without causing damage.

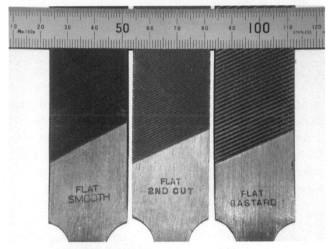

Figure 4.64 Three common types of file

Drilling is a cutting process that uses a drill bit to cut or enlarge a hole in a solid material. The drill bit cuts by applying pressure and rotation to the work piece, which forms chips at the cutting edge (see Figure 4.65). The flutes remove these chips. In use, drill bits have a tendency to 'walk' if not held very steadily. This can be minimized by keeping the drill perpendicular to the work surface. This walking or slipping across the surface can be prevented by making a centring mark before drilling. This is most often done by centre punching. If a large hole is needed, then centre drilling with a smaller bit may be necessary.

Drill bits used for metalworking will also work in wood. However, they tend to chip or break the wood, particularly at the exit of the hole. Some materials like plastics have a tendency to heat up during the drilling process – this heat can make the material expand resulting in a hole that is smaller than the drill bit used.

Cutting A hacksaw is a fine-tooth saw with a blade under tension in a frame. Handheld hacksaws consist of a metal arch with a handle, usually a pistol grip, with pins for attaching a narrow disposable blade. A screw or other mechanism is used to put the blade under tension. The blade can be mounted with the teeth facing toward or away from the handle, resulting

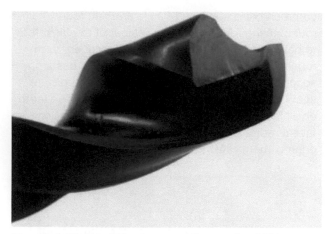

Figure 4.65 Cutting edges and flutes of a drill bit

in cutting action on either the push or pull stroke. The push stroke is most common. Blades are available in standardized lengths, usually 10 or 12 inches (15 or 30cm) for a standard hacksaw. Junior hacksaws are usually half this size. Powered hacksaws may use large blades in a range of sizes.

Figure 4.67 Sawing machine

Figure 4.66 Junior hacksaw

Tooth pitch The pitch of the teeth can vary from 18 to 32 teeth per inch (TPI) for a hand hacksaw blade. The blade chosen is based on the thickness of the material being cut, with a minimum of three teeth in the material. As hacksaw teeth are so small, they are set in a wave so that the resulting cut is wider than the blade to prevent jamming. Hacksaw blades are often brittle, so care needs to be taken to prevent fracture.

Taps and dies are cutting tools used to create screw threads. A tap is used to cut the female part of the mating pair (e.g. a nut) and a die is used to cut the male portion (e.g. a screw). Cutting threads using a tap is called tapping and using a die is called threading. Both tools can also be used to clean a thread in a process known as chasing. The use of a suitable lubricant is recommended for most threading operations.

Figure 4.68 Taps

Figure 4.69 Die

Taps A tap cuts a thread on the inside surface of a hole, creating a female surface which functions like a nut. The three taps in the picture show the three basic types:

▶ The bottoming tap has a continuous cutting edge with almost no taper, which allows it to cut threads to the bottom of a blind hole.

▶ The intermediate tap, second tap, or plug tap has tapered cutting edges, which assist in aligning and starting it into an untapped hole.

▶ The taper or starter tap is similar to a plug tap but has a longer taper, which results in a more gradual cutting action.

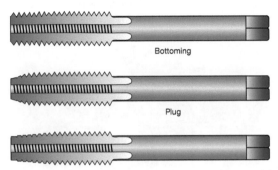

Figure 4.70 Taper (starter), plug and bottoming taps

Hole sizes The process of tapping begins with drilling and slightly countersinking a hole. The diameter of the hole is determined by using a drill and tap size chart.

Tap wrench A 'T' shaped handle is used to rotate the tap. This is often turned in steps of one turn clockwise and about a quarter turn back. This helps to break off the chips, which avoids jamming. With hard materials, it is common to start with a taper tap, because the shallower cut reduces the amount of torque required to make the threads. If threads are to be cut to the bottom of a blind hole, the taper tap is followed by an intermediate (plug) tap and a bottoming tap.

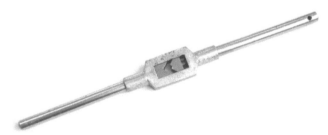

Figure 4.71 Tap Wrench

Dies The die cuts a thread on a cylindrical rod, which creates a male threaded piece that functions like a bolt. The rod is usually just less than the required diameter of the thread and is machined with a taper. This allows the die to start cutting the rod gently, before it cuts enough thread to pull itself along. Adjusting screws on some types of die allow them to

be closed or opened slightly to allow small variations in size. Split dies can be adjusted by screws in the die holder. The action used to cut the thread is similar to that used when tapping. Die nuts have no split for resizing and are made from a hexagonal bar so that a wrench or spanner can be used to turn them. Die nuts are used to clean up existing threads and should not be used to cut new threads.

Figure 4.72 Die stock

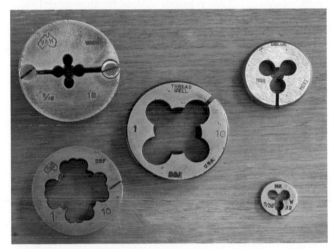

Figure 4.73 Dies (Source: Glenn McKechnie, Wikipedia)

Figure 4.74 Cylindrical rods

 Construct a crossword or wordsearch puzzle using important words from this section.

57

4.3.3 Joining ❶❷

It is very important for the correct methods of joining to be used in the construction and repair of a modern motor vehicle. Joining can cover many aspects ranging from simple nuts and bolts to very modern and sophisticated adhesives.

The choice of a joining method for a repair will depend on the original method used, as well as consideration of the cost and strength required. The table lists some typical joining methods which include the use of gaskets in some cases. An example of the use and useful notes are also given in the table.

Joining method	Example use	Notes
Pins, dowels and keys	Clutch pressure plate to the flywheel	Used for strength and alignment in conjunction with nuts or bolts in most cases.
Riveting	Some brake shoe linings	This involves metal pegs which are deformed to make the joint.
Compression fitting	Wheel bearings	Often also called an interference fit. The part to be fitted is slightly too large or small as appropriate and therefore pressure has to be used to make the part fit.
Shrinking	Flywheel ring gear	The ring gear is heated to make it expand and then fitted in position. As it cools it contracts and holds firmly in place.
Adhesives	Body panels and sound deadening	Adhesive or glue is now very popular as it is often cheap, quick, easy and waterproof. Also, when two items are bonded together the whole structure becomes stronger.
Nuts, screws, washers and bolts	Just about everything!	Metric sizes are now most common but many other sizes and thread patterns are available. This is a very convenient and strong fixing method. Figure 4.75 shows the variety of the different types.
Welding	Exhaust pipes and boxes	There are several methods of welding, oxyacetylene and MIG being the most common. The principle is simple in that the parts to be joined are melted so they mix together and then set in position.

Joining method	Example use	Notes
Brazing	Some body panels	Brazing involves using high temperatures to melt brass which forms the join between two metal components.
Soldering	Electrical connections	Solder is made from lead and tin. It is melted with an electric iron to make it flow into the joint.
Clips, clamps and ties!	Hoses cables, etc.	Hose clips, for example, are designed to secure a hose to the radiator and prevent it leaking.

Figure 4.75 A selection of joining or fastening components

Methods of joining are described as either permanent or non-permanent. The best example of the first is any form of welding. An example of the second would be nuts and bolts. In simple terms, then, the permanent methods mean that some damage would occur if the joint had to be undone.

 Look back over the previous section and write out a list of the key bullet points from this section.

4.3.4 Nuts, screws, washers and bolts ❶

Introduction A fastener is a hardware device that mechanically joins or fixes two or more objects together. There is a large number of fasteners used on automobiles, some standard and some specialist. Most of the main types are outlined in this section. Three major types of steel are used for fasteners in the automotive and other similar industries – stainless steel, carbon steel and alloy steel. Nuts and bolts, screws and variations on the theme are the most common type of fastener. They can also be described as a non-permanent fixing method (because they can be undone!).

Figure 4.76 A selection of fasteners

Nuts A nut is a type of fastener with a threaded hole. Nuts are almost always used with a corresponding bolt or stud to fasten two or more parts together. The nut and bolt are kept together by a combination of the friction of their threads, a slight stretch of the bolt and compression of the parts. In applications where vibration or rotation may work a nut loose, various locking mechanisms are used (discussed later).

Figure 4.77 A few spare nuts in case you drop one . . .

Figure 4.78 These bolts are holding the Clifton Suspension Bridge together (designed in 1830 by Isambard Kingdom Brunel)

Locking nuts Adhesives, safety pins or lock wire, nylon inserts, or slightly oval-shaped threads are used for this purpose. The most common shape for a nut is hexagonal, for similar reasons as the bolt head – six sides give a good range of angles for a tool to approach in tight spots, but more (and smaller) corners would be vulnerable to being rounded off. Other specialized shapes exist for certain needs, such as wing nuts for finger adjustment and captive nuts for inaccessible areas.

Figure 4.79 A selection of nuts – from left to right: flange nut, dome nut (for decorative purposes), wing nut (should only be hand tightened), square lock nut, Nylock nut (the nylon insert prevents loosening), extension nut, castellated nut (used with a locking pin that goes through a hole in the screw shaft)

Use of two nuts to prevent self-loosening In normal use, a nut-and-bolt joint holds together because the bolt is under a constant tensile stress called the preload. The preload pulls the nut threads against the bolt threads, and the nut face against the bearing surface, with a constant force, so that the nut cannot rotate without overcoming the friction between these surfaces. Extra preload can be created when two nuts are locked together.

Figure 4.80 Twin lock nuts

Screws/bolts A screw, or bolt, is a type of fastener characterized by a helical ridge, known as an external thread or just thread, wrapped around a cylinder. Some screw threads are designed to mate with a complementary thread, known as an internal thread, often in the form of a nut or an object that has the internal thread formed into it. Other screw threads are designed to cut a helical groove in a softer material as the screw is inserted (self-tapping). The most common uses of screws are to hold objects together and to position objects.

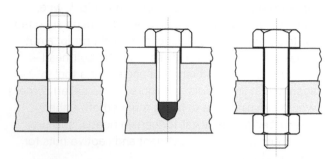

Figure 4.81 The difference between a stud, set screw (screw) and a bolt (from left to right)

Figure 4.82 Screw and a nut (with an Allen-type head)

Screw or bolt heads A screw and a bolt will always have a head, which is a specially formed section on one end of the thread that allows it to be turned, or driven. Common tools for driving screws include screwdrivers and wrenches. The head is usually larger than the body of the screw, which keeps the screw from being driven deeper than the length of the screw and to provide a bearing surface. A selection of screws is shown in Figures 4.83 to 4.85, together with some common head shapes and the many different shapes for turning tools.

Figure 4.83 A selection of screws

Figure 4.84 Screw heads from left to right: pan, dome (button), round, truss (mushroom), flat (countersunk), oval (raised head and countersunk)

Figure 4.85 Screw heads (from left to right): slot (flat), Phillips (PH), Pozidriv (PZ), square (ext), square (int), hex, hex socket (Allen), security hex socket (pin-in-hex socket), torx (T & TX), security torx (TR), Tri-Wing, Torq-set, spanner head (snake-eye), triple square (XZN), Polydrive, one-way, spline drive, double hex, Bristol, Pentalobular

A typical nut and bolt is shown here. The cylindrical portion of the screw from the underside of the head to the tip is known as the shank – it may be fully or partially threaded (screw or bolt). The distance between each thread is called the 'pitch'. Screws and bolts are usually made of steel. Where great resistance to weather or corrosion is required, materials such as stainless steel, brass, titanium, bronze, silicon bronze or other specialist materials may be used. Alternatively, a surface coating is used to protect the fastener from corrosion. Selection criteria of the screw materials include: size, required strength, resistance to corrosion, joint material, cost and temperature.

Figure 4.86 Nuts and bolts with internal star lock washers

Tightening The majority of screws are tightened by clockwise rotation, which is termed a right-hand thread. Screws with left-hand threads are used in exceptional cases. For example, when the screw will be subject to counter clockwise torque (which would work to undo a right-hand thread), a left-hand-threaded screw would be an appropriate choice. The left side wheel nuts of many heavy vehicles use a left-hand thread.

Dimensions There are many systems for specifying the dimensions of screws, but in much of the world the ISO metric screw thread preferred series has displaced the many older systems. Other relatively common systems include the British Standard

Whitworth, the BA system (British Association), and the Unified Thread Standard.

ISO metric screw thread The basic principles of the ISO metric screw thread are defined in international standards. The most commonly used pitch value for each diameter is the coarse pitch. For some diameters, one or two additional fine pitch variants are also specified. ISO metric screw threads are designated by the letter M followed by the major diameter of the thread in mm (e.g. M8). If the thread does not use the normal coarse pitch (e.g. 1.25mm in the case of M8), then the pitch in mm is also appended with a multiplication sign – for example, M8×1 where the screw thread has an outer diameter of 8mm and the pitch is 1mm. The nominal diameter of a metric screw is the outer diameter of the thread. The tapped hole (or nut) into which the screw fits has an internal diameter which is the size of the screw minus the pitch of the thread. Thus, an M6 screw, which has a pitch of 1mm, is made by threading a 6mm shank, and the nut or threaded hole is made by tapping threads into a hole of 5mm diameter (6mm–1mm).

Whitworth The first person to create a standard was the English engineer, Sir Joseph Whitworth. His screw sizes are still used, both for repairing old machinery and where a coarser thread than the metric fastener thread is required. Whitworth became British Standard Whitworth (BSW) and the British Standard Fine (BSF) thread was introduced later because BSW was too coarse for some applications. The thread angle was 55°, and the depth and pitch varied with the diameter of the thread (i.e. the bigger the bolt, the coarser the thread). Wrenches and spanners for Whitworth bolts are marked with the size of the bolt, not the distance across the flats of the screw head. So a tool marked 1/4 Whitworth has a size of over 1/2 an inch (battery lug nuts used this for many years). Whitworth sizes are not very common nowadays.

British Association A later standard established in the UK was the British Association (BA) screw threads. Screws were described as 2BA, 4BA, etc., the odd numbers being rarely used. While not related to ISO metric screws, the sizes were actually defined in metric terms, an 0BA thread having a 6mm diameter and 1mm pitch. Other threads in the BA series are related to 0BA in a geometric series. Although 0BA has the same diameter and pitch as an M6, the threads have different forms and are not compatible. BA threads are still common in certain types of fine machinery, such as moving-coil meters and clocks, and they were also used extensively in some aircraft.

The Unified Thread Standard (UTS) is most commonly used in the United States of America, but is also extensively used in Canada and, occasionally, in other countries. The size of a UTS screw is described using the major diameter D_{maj} (the nominal size of the hole through which the shaft of the screw can easily be pushed) and its pitch, P stated as the threads per inch (TPI). For sizes 1/4 inch and larger the size is given as a fraction; for sizes less than this an integer is used, ranging from 0 to 16. For most size screws there are multiple TPI available, with the most common being designated a Unified Coarse Thread (UNC or UN) and Unified Fine Thread (UNF or UF). The tool size to tighten these nuts or bolts is double the hole size (for example, a 1/2 inch socket is used on a 1/4 inch bolt).

A washer is a thin plate (usually disc-shaped) with a hole (usually in the middle) that is normally used to distribute the load of a threaded fastener, such as a screw or nut. Other uses include as a spacer, spring, wear pad, preload indicating device, locking device, and to reduce vibration (rubber washer). Washers usually have an outer diameter (OD) about twice the width of their inner diameter (ID). Automotive washers are usually metal or plastic. Washers are also important for preventing galvanic corrosion, particularly by insulating steel screws from aluminium surfaces.

Figure 4.87 Washers from left to right: spring (split), external star, wave, flat and flat

Types of washer Washers can be categorized into three main types:

▶ Plain washers, which spread a load and prevent damage to the surface being fixed or provide some sort of insulation such as electrical.
▶ Spring washers, which have axial flexibility and are used to prevent loosening due to vibrations.
▶ Locking washers which prevent loosening by preventing unscrewing rotation of the fastening device. Locking washers to prevent unwanted loosening are usually also spring washers.

Common materials include steel, stainless steel, and plastic. Hardened washers are steel washers that have been heat treated. A flange nut is a nut with an integral fixed washer. A Keps nut or K-lock nut is a nut with an integral free-spinning washer – assembly is easier because the washer is captive.

Types of spring locking washers There are several types of spring locking washers and the common ones are listed here (other images shown previously):

▶ Belleville washers, also known as a cupped spring washer or conical washer, have a slight conical shape which provides an axial force when deformed.

▶ The curved disc spring is similar to a Belleville washer except it is curved in only one direction, therefore there are only four points of contact. Unlike Belleville washers, they only exert light pressure.

▶ Wave washers have a wave in the axial direction, which provides spring pressure when compressed. Wave washers, of comparable size, do not produce as much force as Belleville washers.

▶ A split washer or a spring lock washer is a ring split at one point and bent into a helical shape. This causes the washer to exert a spring force between the fastener's head and the substrate, which maintains the washer hard against the substrate and the bolt thread hard against the nut or substrate thread, creating more friction and resistance to rotation.

▶ A star washer (external type shown here) is a variation on the split or spring lock washer where many teeth are used to lock into the nut and the item being secured.

Figure 4.88 External star spring washer

Figure 4.89 Belleville washers (plain and serrated) also known as coned-disc spring, conical spring washer, disc spring or cupped spring washer

Adhesives Another common method of securing threads is to use a locking compound such as 'Loctite'. This is in effect an adhesive which sticks the threads together. When the correct compound is applied with care, it is a very secure way of preventing important components from working loose.

Figure 4.90 Loctite® Threadlocker (Source: © 2010 Henkel AG & Co. KGaA, Düsseldorf. All rights reserved)

Self-tapping screws As the name suggests these screws are designed to create their own thread as they are used. They are only suitable therefore for light duties and with softer or thinner materials. A wide variety of sizes and head types are available – three examples are shown in Figures 4.91 to 4.93. A captive spring nut can also be used with a self-tapping screw.

Figure 4.91 Phillips head self-tapping screw

Figure 4.92 Allen head self-tapping screws

Figure 4.93 Combined slot and hex head self-tapping screw for heavier duty (may also be used with a spring-type captive nut)

Figure 4.94 Captive spring nut

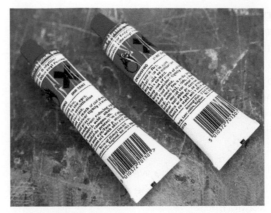

Figure 4.95 Warning signs on adhesives

Summary Nuts screws washers and bolts . . . as the old newspaper headline goes! Seriously though, the types of fasteners or joining devices we have examined here are fundamental to automotive technology as well as all aspects of engineering. Make sure when using replacements that they are of equal quality as recommended by the vehicle manufacturer.

 Use the media search tools to look for pictures and videos relating to the subject in this section.

4.3.5 Adhesives ❶❷

Introduction A very wide range of adhesives are used in today's automotive industry. The number of applications is increasing daily and tending to replace older methods such as welding. There are too many types of adhesives to cover here but most of the basic requirements are the same. It is very important to note, however, that the manufacturer's instructions must always be followed. This is because of the following:

▶ many adhesives give off toxic fumes and must be used with care
▶ most types are highly flammable
▶ adhesives are often designed for a specific application.

Adhesive terminology Adhesives also have a number of important terms associated with them:

▶ Cleanliness – Surfaces to be joined must be clean.
▶ Cure – The process of setting often described as 'going off'.
▶ Wetting – This means that the adhesive spreads evenly and fully over the surface.
▶ Thermosetting – Meaning that heat is required to cure the adhesive.
▶ Thermoplastic – Melts when heated.

▶ Contact adhesive – Makes a strong joint as soon as contact is made.
▶ 'Super glue' – Cyanoacrylate adhesive which bonds suitable materials in seconds, including skin – take care!

Figure 4.96 Loctite super glue

Advantages Adhesives have many advantages, which is why they are becoming more widely used. The following are some of the advantages:

▶ even stress distribution over the whole surface
▶ waterproof
▶ good for joining delicate materials
▶ no distortion when joining
▶ a wide variety of materials can be joined
▶ a neat, clean join can be made with little practice.

Figure 4.97 This screen will be bonded in position using adhesive

Summary As a final point in relation to adhesives, I would stress the importance of choosing the correct type for the job in hand. For example, an adhesive designed to bond plastic will not work when joining rubber to metal. And don't forget, if the surfaces to be joined are not clean you will make a very good job of bonding dirt to dirt instead of what you intended!

 Use a library or the web search tools to further examine the subject in this section.

4.3.6 Soldering, brazing and welding ❶ ❷

Soft soldering is a process used to join materials such as steel, brass, tin or copper. It involves melting a mixture of lead and tin to act as the bond. A common example of a soldered joint is the electrical connection between the stator and diode pack in an alternator. Figure 4.98 shows this process using the most common heat source, which is an electric soldering iron.

Figure 4.98 Soldering an electronic circuit

Soldering process The process of soldering is as follows:

▶ Prepare the surfaces to be joined by cleaning and using emery cloth or wire wool as appropriate.
▶ Add a flux to prevent the surfaces becoming dirty with oxide when heated, or use a solder with a flux core.
▶ Apply heat to the joint and add solder so it runs into the joint.
▶ Complete the process as quickly as possible to prevent heat damage.
▶ Use a heat sink if necessary.

Soldering, in common with many other things, is easy after some practice – take time to do this in your workshop. Note that some materials such as aluminium cannot be soldered by ordinary methods.

Brazing is a similar process to soldering except a higher temperature is needed and different filler is used. The materials to be joined are heated to red heat and the filler rod (bronze, brass or similar), after being dipped in flux, is applied to the joint. The heat from the materials is enough to melt the rod and it flows into the gap making a good, strong, but slightly flexible joint. Dissimilar metals such as brass and steel can also be joined and less heat is required than when fusion welding. Brazing is only used on a few areas of the vehicle body.

Welding is a method of joining metals by applying heat, combined with pressure in some cases. A filler rod of a similar metal is often used. The welding process joins metals by melting them, fusing the melted areas, and then solidifying the joined area to form a very strong bond. Welding technology is widely used in the automotive industry.

Figure 4.99 Welding in process

Welding processes The principal processes used today are gas and arc welding, in which the heat from a gas flame or an electric arc melts the faces to be joined. The picture shows two arc welding processes. Several welding processes are used:

▶ Gas welding uses a mixture of acetylene and oxygen which burns at a very high temperature. This is used to melt the host metal with the addition of a filler rod if required (OA or oxyacetylene).

▶ Shielded metal-arc welding uses an electric arc between an electrode and the work to be joined; the electrode has a coating that decomposes to protect the weld area from contamination and the rod melts to form filler metal (MMA or manual metal arc).

▶ Gas-shielded arc welding produces a welded joint under a protective gas (MIG or metal inert gas, TIG or tungsten inert gas).

▶ Arc welding produces a welded joint within an active gas (MAG or metal active gas).

▶ Resistance welding is a method in which the weld is formed by a combination of pressure and resistance heating from an electric current (spot welding).

Figure 4.100 TIG welding

Figure 4.101 MMA Welding

Specialised welding systems Other, specialised, types of welding include laser-beam welding, which makes use of the intensive heat produced by a light beam to melt and join the metals, and ultrasonic welding, which creates a bond through the application of high-frequency vibration while the parts to be joined are held under pressure.

Figure 4.102 Laser-beam welding

 Use the media search tools to look for pictures and videos relating to the subject in this section.

4.3.7 Shrinking ❷

Introduction When parts are to be fitted by shrinking they first have to be heated so they expand, or cooled so they contract. In both cases the component to be fitted must be made to an exact size. If parts fitted in this way are to be removed, they are usually destroyed in the process – for example, a flywheel ring gear has to be cut through with a hacksaw to remove it.

Figure 4.103 Ring gear on flywheel

Remember! There is more support on the website that includes additional images and interactive features: www.tomdenton.org

Hot shrinking For a hot shrink fitting, the part will have a smaller internal diameter than the one on which it is to be fitted. It is important not to overheat the components or damage will occur. An oven is best, but a welding torch may be used with great care. When the component has been heated, and therefore expanded, it is placed in position immediately. It will then cool and make a good, tight joint.

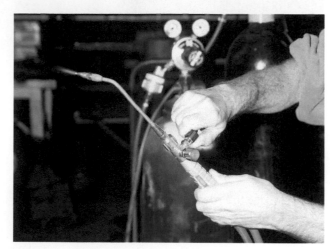

Figure 4.104 Oxy-Acetylene Welding

Cold shrinking is very similar except the component to be fitted is made very slightly larger than the hole in which it is to be fitted. A cylinder head valve insert is one example. The process is the opposite of hot shrinking. The component is cooled so it contracts, after which it is placed in position where it warms back up and expands, making a secure joint. Cold shrinking is normally a specialist job, but it is possible to buy aerosols of carbon dioxide under pressure (dry ice) which can be used to make a component very cold.

Figure 4.105 Freeze spray (Source: Arctic)

Compression fitting Many parts are fitted by compression or pressure. Bearings are the most common example. The key to compression fitting is an interference fit. This means that the component, say a bearing, is very slightly larger than the hole in which it is to be fitted. Pressure is therefore used to force the bearing onto place. Suspension bushes are often also fitted in this way. The secret is to apply the force in a way which does not let the components go together on an incorrect angle – they must be fitted true to each other.

Figure 4.106 This bearing on a gearbox shaft is held in place by compression

 Using images and text, create a short presentation to show how a component or system works.

4.3.8 Riveting ❶

Riveting is a method of joining metal plates, fabric to metal or brake linings to the shoes. A metal pin called a rivet, which has a head at one end, is inserted into matching holes in two overlapping parts. The other end is struck and formed into another head, holding the parts together. This is the basic principle of riveting but many variations are possible.

Figure 4.107 Brake lining riveted to a shoe

Pop rivets Figure 4.108 shows some pop rivets, which are one of the most common for motor vehicle repair. These are hollow rivets which are already mounted onto a steel pin. The rivet is placed through the holes in the parts to be joined and a special rivet gun grips the pin and pulls it with great force. This causes the second rivet head to be formed and when the pin reaches a set tension it breaks off, leaving the rivet securely in place. The great advantage of this method is that you can work blind. In other words, you don't need access to the other side of the hole!

Figure 4.108 Pop rivets

 Use a library or the web search tools to further examine the subject in this section.

4.3.9 Gaskets, sealants and oil seals ❶

Gaskets are used to make a fluid- or pressure-tight seal between two component faces. The best example of this is the cylinder head gasket which also has to withstand very high pressures and temperatures. Gaskets are often used to make up for less than perfect surfaces and therefore act as a seal between the two – as the temperature changes, the gasket can take up the difference in expansion between the two components. Gaskets are made from different materials depending on the task they have to perform.

Gasket material	Examples of where used
Paper or card	General purpose such as thermostat housings
Fibre	General purpose
Cork	Earlier-type rocker covers
Rubber – often synthetic	Water pump sealing ring
Plastics – various types	Fuel pump to engine block
Copper asbestos – or similar	Exhaust flange – note safety issues of asbestos
Copper and aluminium	Head gaskets
Metal and fibre compounds – with metal composites	Head gaskets

Figure 4.109 Cylinder head gaskets

Quality The general rules for obtaining a good joint, with a gasket or otherwise, are as follows:

- ▶ cleanliness of the surfaces to be joined
- ▶ removal of burrs from the materials
- ▶ use of the correct materials
- ▶ following the manufacturer's instructions (such as tighten to the correct torque in the correct sequence)
- ▶ safe working (this applies to everything you do).

Sealants Many manufacturers are now specifying the use of sealants in place of traditional gaskets. The main reason for this is a better quality of joint. Liquid sealants, often known as instant gaskets, are a type of liquid rubber which forms into a perfect gasket as the surfaces are mated together. The three major advantages of this technique are:

- ▶ easier to apply
- ▶ a perfect seal is made with a very small space being taken up
- ▶ adhesive bonding effect reduces fretting due to vibration and hence is less likely to leak.

Applying sealant The picture shows a sealant being applied. A major advantage as far as the repair trade is concerned is that with a good selection of jointing

Figure 4.110 Instant gasket and other products (Source: Loctite)

sealants, you can manufacture a gasket on the spot at any time! Note the recommendations of the manufacturers, however, as only the correct material must be used.

Figure 4.111 Loctite® sealant (Source: © 2010 Henkel AG & Co. KGaA, Düsseldorf. All rights reserved)

Oil Seals The most common type of oil seal is the neoprene (synthetic rubber) radial lip seal. The seal is fitted into a recess and the soft lip rubs against the rotating component. The lip is held in place by a spring. Figure 4.112 shows this type of seal – note how the lip faces the oil such that any pressure will cause the lip to fit tighter rather than allow oil to be forced underneath. The second picture shows a valve stem oil seal, which prevents oil from entering the combustion chamber past the inlet valves.

Figure 4.112 Oil seals (Source: © 2005 Newsad Energy Company)

 Create a mind map to illustrate the features of a key component or system.

CHAPTER 5

Maintenance

After successful completion of this chapter you will be able to show you have achieved these objectives:

- Understand the basic vehicle types, their layouts and main systems.
- Understand how to carry out routine light vehicle maintenance.
- Understand the importance of carrying out light vehicle maintenance.

DOI: 10.1201/9781003173236-5

5.1 Vehicle overview

5.1.1 Maintenance and inspections ❶

The purpose of routine maintenance is simple – it is to keep the vehicle in a good working order and in a safe condition. Manufacturers specify intervals and set tasks that should be carried out at these times. It is usually a condition of the warranty that a vehicle should be serviced according to the manufacturers' needs. The main purpose of regular inspection therefore, is to check for the following:

▶ malfunction of systems and components
▶ damage and corrosion to structural and support regions
▶ leaks
▶ water ingress
▶ component and system wear and security.

Inspections are usually:

▶ aural – listening for problems
▶ visual – looking for problems
▶ functional – checking that things work!

The main types of inspection, in addition to what is carried out when servicing, are:

▶ pre-work
▶ post-work
▶ pre-delivery inspection (PDI)
▶ used vehicle inspection
▶ special inspection (maybe after an accident, for example).

A pre-work inspection is used to find out what work needs to be carried out on a vehicle. Post-work inspections are done to make sure the repairs have been carried out correctly and that no other faults have been introduced.

A PDI is carried out on all new vehicles to check certain safety items and to, for example, remove any transport packaging such as suspension locks or similar. A used vehicle inspection is done to determine the safety and saleability of a vehicle as well as checking that everything works. After gaining experience you may be asked to carry out an inspection of a vehicle after an accident, to check the brakes' condition, for example.

In all cases, a recommended checklist should be used and careful records of your findings should be kept. Working to timescales, or reporting to a supervisor that timescales cannot be met, is essential. When a customer books a car in for work to be done they expect it to be ready at the agreed time. Clearly if this deadline can't be met the customer needs to be informed. Then, in order to make the running of a workshop efficient and

Figure 5.1 Checking data and setting up test equipment

profitable, a technician will have jobs allocated that will take a certain amount of time to complete. If for any reason this time can't be met then action will need to be taken by the workshop manager or supervisor

5.1.2 Layouts ❶

This section is a general introduction to the car as a whole. Over the years many unusual designs have been tried, some with more success than others. The most common is, of course, a rectangular vehicle with a wheel at each corner! To take this rather simple idea further, we can categorize vehicles in different ways. For example, by layout such as:

▶ front-engine driving the front wheels
▶ front-engine driving the rear wheels
▶ front-engine driving all four wheels
▶ rear-engine driving the rear wheels
▶ mid-engine driving the rear wheels
▶ mid-engine driving all four wheels.

These may be categorized further as:

▶ FWD – Front-wheel drive
▶ RWD – Rear-wheel drive
▶ AWD – All-wheel drive
▶ 4WD – Four-wheel drive.

The following paragraphs and bullet points highlight features of the vehicle layouts mentioned above.

A common layout for a standard car is the front-engine, front-wheel drive vehicle. This is because a design with the engine at the front driving the front wheels has a number of advantages:

▶ protection in case of a front-end collision
▶ easier engine cooling because of the airflow
▶ cornering can be better if the weight is at the front
▶ front-wheel drive adds further advantages if the engine is mounted sideways on (transversely)
▶ more room in the passenger compartment

▶ the power unit can be made as a complete unit
▶ the drive acts in the same direction as the steered wheels are pointing.

Figure 5.2 Front-engine FWD

Rear-wheel drive from a front engine was the method used for many years. Some manufacturers have continued its use, BMW for example. A long propeller shaft from the gearbox to the final drive, which is part of the rear axle, is the main feature. The propshaft has universal joints to allow for suspension movement. This layout has some advantages.

▶ weight transfers to the rear driving wheels when accelerating
▶ complicated constant velocity joints, such as used by front-wheel drive vehicles, are not needed.

Figure 5.3 Front-engine RWD

Four-wheel drive combines all the good points mentioned above but does make the vehicle more complicated and therefore more expensive. The main difference with four-wheel drive is that an extra gearbox, known as a transfer box, is needed to link the front- and rear-wheel drive.

The rear-engine design has not been very popular but it was used for the bestselling car of all time – the VW beetle. The advantages are that weight is placed on the rear wheels giving good grip and the power unit and drive can be all one assembly. One downside is that less room is available for luggage in the front. The biggest problem is that handling is affected because of less weight on the steered wheels. Flat-type engines are the most common choice for this type of vehicle.

Figure 5.4 Rear-engine RWD

Fitting the engine in the mid position of a car has one major disadvantage – it takes up space inside the vehicle. This makes it impractical for most 'normal' vehicles. However, the distribution of weight is very good, which makes it the choice of high performance vehicle designers. A good example is the Ferrari Testarossa. Mid-engine is the term used to describe any vehicle where the engine is between the axles, even if it is not in the middle.

Figure 5.5 Mid-engine RWD

Vehicles are also categorized by type and size as in this table:

LV	Light vehicles (light vans and cars) with a maximum allowed mass (MAM) of up to 3 500kg, no more than eight passenger seats. Vehicles weighing between 3 500kg and 7 500kg are considered as mid-sized.
LGV	A large goods vehicle known formerly, and still in common use, as heavy goods vehicle or HGV. LGV is the EU term for trucks or lorries with a MAM of over 3 500kg.
PCV	A passenger carrying vehicle or a bus (known formerly as omnibus, multibus or autobus) is a road vehicle designed to carry passengers. The most common type is the single-decker, with larger loads carried by double-decker and articulated buses and smaller loads carried by minibuses. A luxury, long-distance bus is usually called a coach.

 Look back over the previous section and write out a list of the key bullet points from this section

5.1.3 Body design ❶

Types of light vehicle can range from small, two-seat sports cars to large people carriers or SUVs. Also included in the range are light commercial vehicles such as vans and pickup trucks. It is hard to categorize a car exactly as there are several agreed systems in several different countries. Figures 5.6 to 5.14 show a number of different body types.

Figure 5.8 Hatchback (Source: Ford Media)

Figure 5.9 Coupe (Source: Ford Media)

Figure 5.6 Saloon car (Source: Ford Media)

Figure 5.10 Convertible (Source: Ford Media)

Figure 5.7 Estate car (Source: Ford Media)

Figure 5.11 Concept car (Source: Ford Media)

Figure 5.12 Light van (Source: Ford Media)

Figure 5.13 Pickup truck (Source: Ford Media)

Figure 5.14 Sports utility vehicle SUV (Source: Ford Media)

The vehicle chassis can be one of two main types – separate or integrated. Separate chassis are usually used on heavier vehicles. The integrated type, often called a monocoque, is used for almost all cars. The two main types are shown here in Figures 5.15 and 5.16.

Figure 5.15 Ladder chassis

Figure 5.16 Integrated chassis

Most vehicles are made of a number of separate panels. The online version of Figure 5.17 shows a car with the main panel and other body components named.

Figure 5.17 Body components

 Use the media search tools to look for pictures and videos relating to the subject in this section.

5.1.4 **Main systems** ❶

No matter how we categorize them, all vehicle designs have similar major components and these operate in much the same way. The four main areas of a vehicle are the engine, electrical, chassis and transmission systems.

Engine This area consists of the engine itself, together with fuel, ignition, air supply and exhaust systems. In the engine, a fuel/air mixture enters through an inlet manifold and is fired in each cylinder in turn. The resulting expanding gases push on pistons and connecting rods which are on cranks, just like a cyclist's legs driving pedals, and this makes the crankshaft rotate. The pulses of power from each piston are smoothed out by a heavy flywheel. Power leaves the engine through the flywheel, which is fitted on the end of the crankshaft, and passes to the clutch. The spent gases leave via the exhaust system.

Figure 5.18 Ford Focus engine (Source: Ford Media)

Electrical The electrical area covers many aspects such as lighting, wipers and instrumentation. A key component is the alternator which, driven by the engine, produces electricity to run the electrical systems and charge the battery. A starter motor takes energy from the battery to crank over and start the engine. Electrical components are controlled by a range of switches. Electronic systems use sensors to sense conditions and actuators to control a variety of things – in fact, on modern vehicles, almost everything.

Figure 5.19 A modern alternator (Source: Bosch Press)

Chassis This area is made up of the braking, steering and suspension systems as well as the wheels and tyres. Hydraulic pressure is used to activate the brakes to slow down or stop the vehicle. Rotating discs are gripped between pads with friction lining. The hand brake uses a mechanical linkage to operate parking brakes. Both front wheels are linked mechanically and must turn together to provide steering control. The most common method is to use a rack and pinion. The steering wheel is linked to the pinion and as this is turned it moves the rack to and fro, which in turn moves the wheels. Tyres also absorb some road shock and play a very important part in roadholding. Most of the remaining shocks and vibrations are absorbed by springs in the drivers and passengers seats. The springs can be coil type and are used in conjunction with a damper to stop them oscillating (bouncing up and down too much).

Figure 5.20 Disc brakes and part of the suspension system

Transmission In this area, the clutch allows the driver to disconnect drive from the engine and to move the vehicle off from rest. The engine flywheel and clutch cover are bolted together so the clutch always rotates with the engine and, when the clutch pedal is raised, drive is passed to the gearbox. A gearbox is needed because an engine produces power only when turning quite fast. The gearbox allows the driver to keep the engine at its best speed. When the gearbox is in neutral, power does not leave it. A final drive assembly and differential connect the drive to the wheels via axles or driveshafts. The differential allows the driveshafts, and therefore the wheels, to rotate at different speeds when the vehicle is cornering.

Figure 5.21 Differential and final drive components

The layout of a vehicle such as where the engine is fitted and which wheels are driven varies, as do body styles and shapes. However, the technologies used in the four main areas of a vehicle are similar no matter how it is described. These are the:

▶ engine system
▶ electrical system
▶ chassis system
▶ transmission system.

These areas are covered in detail and make up the four main chapters of this book.

 Create an information wall to illustrate the features of a key component or system.

5.2 Servicing and inspections

5.2.1 Introduction **1**

It is important to carry out regular servicing and inspections of vehicles for a number of reasons:

▶ ensure the vehicle stays in a safe condition
▶ keep the vehicle operating within tolerances specified by the manufacturer and regulations
▶ ensure the vehicle is reliable and reduce down time
▶ maintain efficiency
▶ extend components and the vehicle's life
▶ reduce running costs
▶ keep the vehicle looking good and limit damage from corrosion.

Figure 5.22 Servicing the brakes

In order to carry out servicing and inspections you should understand how the vehicle systems operate. It is also important to keep suitable records – this is often known as the vehicle's service history. Services and inspections of vehicles vary a little from one manufacturer to another. Servicing data and servicing requirement books are available as well as the original manufacturer's information. This type of data should always be read carefully so as to ensure that all the required tasks are completed. The following table lists some important words and phrases relating to servicing and inspection:

First service	This service is becoming less common but some manufacturers like the vehicles to be returned to the dealers after about a thousand miles or so. This is so that certain parts can be checked for safe operation and in some cases oil is changed.
Distance-based services	Ten- or twelve- or twenty-thousand-mile intervals are common distances but manufacturers vary their recommendations. Most have specific requirements at set distances.

5

Time-based services	For most light vehicles, distance based services are best. Some vehicles, though, run for long periods of time but do not cover great distances. In this case the servicing is carried out at set time intervals. This could be every six months, six weeks or after a set number of hours' run.
Inspection	The MOT test, which must be carried out each year after a light vehicle is older than three years, is a good example of an inspection. However an inspection can be carried out at any time and should form part of most services.
Records	A vital part of a service, to ensure all aspects are covered and to keep information available for future use.
Customer contracts	When you make an offer to do a service, and the customer accepts the terms and agrees to pay, you have made a contract. Remember that this is legally enforceable by both parties.

Clearly, it is important to keep a customer's vehicle in a clean condition. To do this there are a number of methods as outlined here:

▶ seat covers to keep the seats clean
▶ floor mats to protect the carpets
▶ steering wheel covers to keep greasy hand prints off the wheel
▶ wing covers to keep the paintwork clean and to prevent damage.

Figure 5.23 Bodywork protection in use during repairs

Technical information The main sources of information are:

▶ workshop manuals
▶ technical bulletins
▶ servicing schedules
▶ jobcard instructions
▶ inspection records
▶ check lists
▶ online data and repair instructions

All main manufacturers now allow online access to the full range of their data. This is accessed through special websites where you have to create an account and pay for the time.

Several companies supply information, AllData and Autodata for example.

 Use a library or the web search tools to further examine the subject in this section.

5.2.2 Rules and regulations ❶

The three main regulations that cover the repair and service of motor vehicles in the UK are:

1 The Road Traffic Act – This covers things like road signage and insurance requirements. It also covers issues relating to vehicle safety. For example, if a car suspension is modified it may become unsafe and not conform to the law.
2 Driver and Vehicle Standards Agency (DVSA) regulations – the main one of these being the annual MOT test requirements.
3 The Highway Code – All drivers must follow this and it forms part of the driving test.

Similar regulations are in place in other countries. The regulations are designed to improve safety.

Some of the main vehicle systems relating to safety are listed in the table, together with examples of the requirements. Note though that these are just examples and that specific data must be studied relating to specific vehicles.

Brakes	The foot brake must produce 50% of the vehicle weight braking force and the parking brake 16% (this assumes a modern dual line braking system). The brakes must work evenly and show no signs of leaks.
Exhaust	Should not leak, which could allow fumes into the vehicle, and it should not be noisy!
Horn	It should be noisy!
Lights	All lights should work and the headlights must be correctly adjusted.
Number plates	Only the correct style and size must be fitted. The numbers and letters should also be correctly spaced and not altered (DAN 15H is right, DANISH is wrong!).
Seat belts	All belts must be in good condition and work correctly.
Speedometer	Should be accurate and illuminate when dark.

Steering	All components must be secure and serviceable.
Tyres	Correct tread depth is just one example.
Windscreens and other glass	You should be able to see right through this one! No cracks allowed in the screen within the driver's vision.

Use a library or the web search tools to further examine the subject in this section.

5.2.3 Service sheets

Service sheets are used and records must be kept because they:

- define the work to be carried out
- record the work carried out

▶ record the time spent
▶ record materials consumed
▶ allow invoices to be prepared
▶ record stock which may need replacing
▶ form evidence in the event of accident or customer complaint.

The following table is an example of a service sheet showing tasks carried out and at what service intervals. Please note once again that this list, whilst quite comprehensive, is not suitable for all vehicles, and the manufacturer's recommendations must always be followed. Some of the tasks are only appropriate to certain types of vehicle. The table here also lists the work in a recommended order, including the use of a lift.

Use a library or the web search tools to further examine the subject in this section.

Driving vehicle into workshop	
Instrument gauges, warning/control lights and horn	Check operation
Washers, wipers	Check operation/adjust, if necessary
Inside vehicle	
Exterior and respective control lights; instrument cluster illumination	Check operation/condition
Service interval indicator	Reset after every oil change if applicable
Handbrake	Check operation/adjust, if necessary
Seat belts, buckles and stalks	Check operation/condition
Pollen filter	Renew
Warning vest	Check availability – if applicable
First-aid kit	Check availability and expiry date – if applicable
Warning triangle	Check availability – if applicable
Outside vehicle	
Hood latch/safety catch and hinges	Check operation/grease
Road (MOT) test	Check regarding next road (MOT) test due date – if applicable
Emission test	Check regarding next emission test due – if applicable
Under bonnet/hood	
Wiring, pipes, hoses, oil and fuel feed lines	Check for routing, damage, chafing and leaks
Engine, vacuum pump, heater and radiator	Check for damage and leaks
Coolant	Check antifreeze concentration: °C
Coolant expansion tank and washer reservoirs	Check/top up fluid levels as necessary – in case of abnormal fluid loss, a separate order is required to investigate and rectify
Power steering fluid	Check/top up fluid levels as necessary – in case of abnormal fluid loss, a separate order is required to investigate and rectify
Battery terminals	Clean, if necessary/grease
Battery	Visual check for leaks – in case of abnormal fluid loss, a separate order is required to investigate and rectify
Fuel filter	Drain water, if not renewed – diesel models (with drain facility)
Headlamp alignment	Check – adjust alignment, if necessary
Brake fluid	Check/top up fluid levels as necessary – in case of abnormal fluid loss, a separate order is required to investigate and rectify

Under vehicle	
Engine	Drain oil and renew oil filter
Steering, suspension linkages, ball joints, sideshaft joints, gaiters	Check for damage, wear, security and rubber deterioration
Engine, transmission	Check for damage and leaks
Pipes, hoses, wiring, oil and fuel feed lines, exhaust	Check for routing, damage, chafing and leaks
Underbody	Check condition of PVC coating
Tyres	Check wear and condition, especially at tyre wall, note tread depth: RF mm, LF mm, LR mm, RR mm, Spare mm
Brake system	With wheels off, check brake pads, discs, linings for wear and check brake cylinders for condition: check rubber components for deterioration

5.2.4 Road test ❸

Assuming you are a qualified driver, or you are able to tell a driver what you want, then a road test is an excellent way of checking the operation of a vehicle. A checklist is again a useful reminder of what should be done. A typical road test following a service or inspection would be much as follows (but remember to check specific manufacturers' requirements):

▶ fit trade plates to vehicle if necessary
▶ check operation of starter and inhibitor switch (automatic)
▶ check operation of lights, horn(s), indicators, wipers and washers
▶ check indicators self-cancel
▶ check operation of all warning indicators
▶ check foot and hand brakes
▶ check engine noise levels, performance and throttle operation
▶ check clutch for free play, slipping and judder
▶ check gear selection and noise levels in all gears
▶ check steering for noise, effort required, free play, wander and self-centring
▶ check suspension for noise, irregularity in ride and wheel imbalance
▶ check foot brake pedal effort, travel, braking efficiency, pulling and binding
▶ check speedometer for steady operation, noise and operation of mileage recorder
▶ check operation of all instruments
▶ check for abnormal body noises
▶ check operation of seat belts, including operation of inertia reels
▶ check hand brake ratchet and hold
▶ position car on lift
▶ recheck tension if drive belts have been renewed
▶ raise lift
▶ inspect engine and transmission for oil leaks
▶ check exhaust system for condition, leakage, and security
▶ lower lift: drive vehicle off lift

▶ report on Road Test findings
▶ remove car protection kit
▶ ensure cleanliness of controls, door handles, etc.
▶ remove trade plates if fitted.

 Use the media search tools to look for pictures and videos relating to the subject in this section.

5.2.5 Effects of incorrect adjustments ❶

The following table lists a selection of possible incorrect adjustments, together with their effects on the operation of the vehicle. This is intended to be an exercise to help you see why correct adjustments are so important – not so you know how to do it wrong! You must also be able to make a record and tell a customer the effects if you are unable to make the correct adjustments. This could be due to some parts being worn so that adjustment is not possible.

Remember, though, anyone can mess with a vehicle and get it wrong. As a professional you will get it right, the customer and your company will be happy and it *will* affect your pay rates in years to come. Problems can arise when a vehicle has been serviced but the work that the customer expected to be completed has not been carried out. For example, on a basic, interim or even in some cases a full service, little or no work has been carried out on the ignition system – this will not, for example, rectify a misfire. It is important that the customer is aware of what will be done as well as what was done to their vehicle. And if, during a service, you notice a fault – report it.

 Create a word cloud for one or more of the most important screens or blocks of text in this section.

Incorrect adjustment	Possible effects						
Brake	Excessive pedal and lever travel	Reduced braking efficiency	Unbalanced braking	Overheating	Skidding and a serious accident		
Drive belts	Overheating	Battery recharge rate slow	Power steering problems	AC not operating			
Fuel system	Poor starting or non-start	Lack of power or hesitation	Uneven running and stalling	Popping back or backfire	Running on or detonation	Heavy fuel usage	Fuel leaks and smells
Ignition	Poor starting or non-start	Lack of power	Hesitation	Exhaust emission	Running on		
Plug gaps	Poor starting or non-start	Lack of power	Hesitation	Uneven running	Misfire	Exhaust emissions	
Steering system	Abnormal or uneven tyre wear	Heavy steering	Pulling to one side	Poor self-centring	Wandering	Steering wheel alignment	Excessive free play
Tyre pressures	Abnormal tyre wear	Heavy steering	Uneven braking	Heavy fuel usage	Tyre lifetime is reduced		
Valve clearances	Lack of power	Uneven running	Misfire	Excessive fuel usage	Exhaust emissions	Noise from valves or camshaft	

5.2.6 Information sources

Introduction Information and data relating to vehicles are available for carrying out many forms of diagnostic work. The used to come as a book or on CD/DVD. By far the majority is now online or as part of a package. This information is essential to ensure that you find the fault – as long as you have developed the diagnostic skills to go with it of course!

Type of information available is:

▶ engine testing and tuning
▶ servicing, repair processes and times
▶ fuel and ignition systems
▶ circuit diagrams
▶ component location
▶ alignment data
▶ diagnostic routines.

Example sources There are some excellent packages that you can buy on subscription, or they are included with a diagnostic tool. Some have a pay-as-you-go option. Some example sources are:

▶ Bosch
▶ Snap-on
▶ Hella
▶ Delphi
▶ Thatcham
▶ AllData
▶ Autodata
▶ Haynes
▶ Tech4Techs
▶ REPXPERT

Manufacturers data (some of the above packages also access this). This is highly recommended because you know you have the correct and latest information. This can be bought as required from almost all manufacturers. For example, Volkswagen data is available from: https://erwin.volkswagen.de. It is necessary to register on each site but then you can access the same information that is supplied to the main dealers. At the time of writing, the Volkswagen site allowed a range of payment options, but an example was one hour of full access for 7EUR. This is more than enough time to find what you need for a specific repair or diagnostic job. Some workshops include this cost in their standard rates or add it as an extra on the customer's invoice.

Figure 5.24 V6 Mustang engine – data is essential when working on complex systems (Source: Ford Media)

Use a library or the web search tools to further examine the subject in this section.

CHAPTER 6

Engine systems

After successful completion of this chapter you will be able to show you have achieved these objectives:

- Understand how the main light vehicle engine mechanical systems operate.
- Understand how light vehicle engine lubrication systems operate.
- Understand how light vehicle engine cooling, heating and ventilation systems operate.
- Understand how light vehicle engine fuel systems operate.
- Understand how light vehicle engine ignition systems operate.
- Understand how light vehicle engine air supply and exhaust systems operate.
- Understand how to check, replace and test light vehicle engine mechanical, lubrication, cooling, fuel, ignition, air and exhaust system units and components.
- Understand how to diagnose and rectify faults in light vehicle engine systems.

DOI: 10.1201/9781003173236-6

6.1 Mechanical

6.1.1 Operating cycles **❶**

Introduction The modern motor vehicle engine is a complex machine and the power plant of the vehicle. The engine burns a fuel to obtain power. The fuel is usually petrol or diesel, although liquid petroleum gas (LPG) is sometimes used and specialist fuels have been developed for special purposes such as racing car engines.

Internal combustion engine Motor vehicle engines are known as 'internal combustion' engines because the energy from the combustion of the fuel and the resulting pressure from expansion of the heated air and fuel charge is applied directly to pistons inside closed cylinders in the engine. The term 'reciprocating piston engine' describes the movement of the pistons, which go up and down in the cylinders. The pistons are connected by a rod to a crankshaft to give a rotary output.

Figure 6.1 Internal combustion engine internal components

Air and fuel The fuel is metered into the engine together with an air charge for most petrol engines. Some use injectors that inject directly into the engine cylinder. On diesel engines, the fuel is injected into a compressed air charge in the combustion chamber.

In order for the air and fuel to enter the engine and for the burnt or exhaust gases to leave the engine, a series of ports are connected to the combustion chambers. The combustion chambers are formed in the space above the pistons when they are at the top of the cylinders. Valves in the combustion chamber at the ends of the ports control the air charge and exhaust gas movements into and out of the combustion chambers.

Poppet valves The valves are known as 'poppet' valves, having a circular plate at right angles to a central stem that runs through a guide tube. The plate has a chamfered sealing face in contact with a matching sealing face in the port. The valve is opened by a rotating cam and associated linkage. It is closed and held closed by a coil spring.

The four-stroke cycle (or Otto cycle) The opening and closing of the valves and the movement of the pistons in the cylinders follows a cycle of events called the 'four-stroke cycle' or the 'Otto cycle' after its originator.

The induction/intake stroke The first stroke of the four-stroke cycle is the induction or intake stroke. This occurs when the piston is moving down in the cylinder from top dead centre (TDC) to bottom dead centre (BDC) and the inlet valve is open. The movement of the pistons increases the volume of the cylinder so that air and fuel enter the engine.

The compression stroke The next stroke is the compression stroke when the piston moves upwards in the cylinder. Both the inlet and exhaust valves are closed and the space in the cylinder above the piston is reduced. This causes the air and fuel charge to be compressed, which is necessary for clean and efficient combustion of the fuel.

The combustion/power stroke Towards the end of the compression stroke, the fuel is ignited and burns to give a large pressure rise in the cylinder above the piston. This pressure rise forces the piston down in the cylinder on the combustion or power stroke.

The exhaust stroke Once the energy from the fuel has been used, the exhaust valve opens so that the waste gases can leave the engine through the exhaust port. To complete the exhausting of the burnt gases, the piston moves upward in the cylinder. This final stroke is called the exhaust stroke.

Four-stroke cycle The four-stroke cycle then repeats over and over again, as the engine runs. A heavy flywheel keeps the engine turning between power strokes.

The induction or intake stroke On the induction stroke of a petrol engine (most types), air and petrol enter

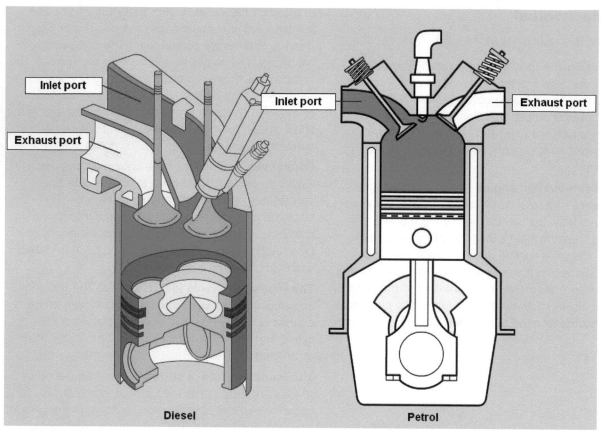

Figure 6.2 Diesel and petrol engine – pistons at BDC before start of intake strokes

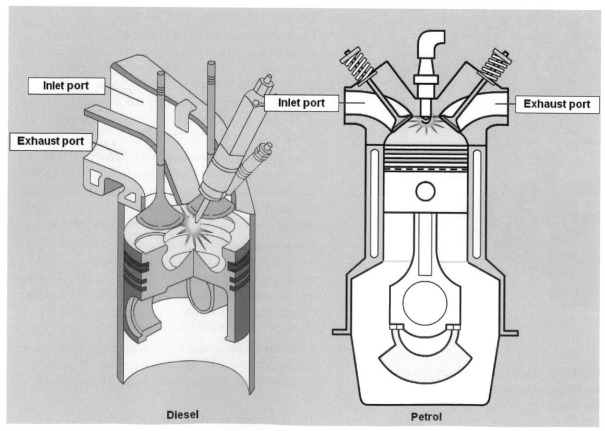

Figure 6.3 Diesel and petrol engine – pistons at TDC before start of the combustion or power strokes

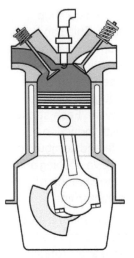

Figure 6.4 Induction – piston moving down

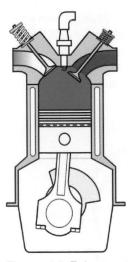

Figure 6.7 Exhaust – piston moving up

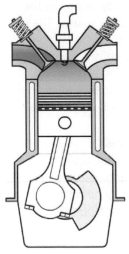

Figure 6.5 Compression – piston moving up

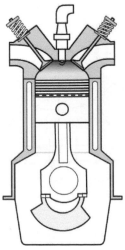

Figure 6.6 Power or combustion – piston is forced down

the cylinder so the inlet valve in the inlet port must be open. On a diesel engine, only air enters the cylinder. A rotating cam on the camshaft provides a lifting movement when it runs in contact with a follower. A mechanical linkage is used to transfer the movement to the valve stem and the valve is lifted off its seat so that the inlet port is opened to the combustion chamber.

Cylinder charge The air and fuel charge or air charge can now enter the cylinder. The inlet valve begins to open shortly before the piston reaches TDC. The exhaust valve, which is operated by its own cam in the same way as the inlet valve, is beginning to close as the piston passes TDC at the end of the exhaust stroke. Valve overlap helps clear the remaining exhaust gases from the combustion chamber. The incoming air charge fills the combustion chamber as the last quantity of exhaust gas leaves through the exhaust port. This is known as 'scavenging' and helps cool the combustion chamber by removing hot exhaust gases and gives a completely fresh air charge.

Top dead centre (TDC) and bottom dead centre (BDC) The terms top dead centre and bottom dead centre are abbreviated to 'TDC' and 'BDC' respectively. They are used to describe the position of the piston and crankshaft when the piston is at the end of a stroke and the axis of the piston and crankshaft bearing journals are in a straight line and at 0° (TDC) and 180° (BDC) of crankshaft revolution. To the abbreviations are added the letters 'A' to indicate degrees 'after' TDC or BDC and the letter 'B' to indicate 'before' TDC or BDC.

Crankshaft and camshaft The camshaft rotates once for each of the two revolutions of the crankshaft during the four-stroke cycle. The drive from the crankshaft to the camshaft has a 2:1 ratio produced by the numbers of teeth on the driven and driver gears.

Rotational data for the camshaft is usually given as degrees of crankshaft rotation and this should be considered in relation to the four-stroke cycle. The four-stroke cycle occurring over two full revolutions of the crankshaft has a 720° rotational movement.

Figure 6.8 Circular valve timing diagram

Valve timing diagram Looking at the four-stroke cycle and the relationships of the crankshaft rotation, the piston position in the cylinder and the opening and closing of the valves are best observed by looking at a valve timing diagram. This diagram is one method of providing data for valve opening and closing positions.

Valve lead, lag and overlap The terms applied to the valves when opening before and closing after the start of a stroke and when both valves are open together are called 'lead', 'lag' and 'overlap' respectively. The overlap position is often referred to as 'valves rocking' and can be used as a rough guide as to when a piston is at top dead centre (TDC).

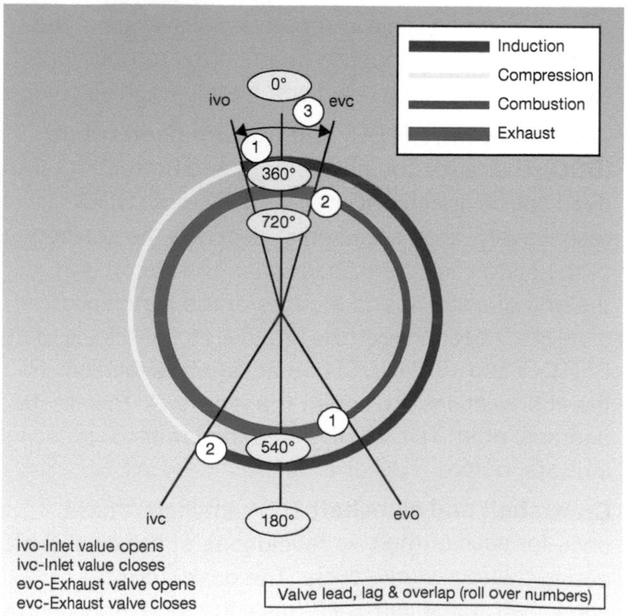

Valve timing data is given in engine workshop manuals as ° of crankshaft revolution. This can be as written data or by means of valve timing diagrams. In the most popular valve timing diagram, two circles, one inside the other, are used to represent the 720° of crankshaft rotation through which the crankshaft moves for a complete cycle. Each stroke is represented by an arc of 180° with induction and compression on the outer circle and combustion and exhaust on the inner circle. The valve opening and closing positions are marked and the duration of crankshaft rotation displayed by a thicker line.

Valve timing From the valve timing diagrams it can be seen that the valve opening and closing positions do not occur within the 180° of crankshaft rotation for each stroke of the four-stroke cycle. For instance, towards the end of the exhaust stroke, the inlet valve begins to open and this is before the exhaust valve has closed. The exhaust valve finally closes as the piston moves down on the induction stroke. The inlet valve closes as the piston is rising on the compression stroke. The exhaust valve opens before the end of the combustion stroke. The opening and closing positions of the valves are special to individual engines and are matched to other design and performance requirements.

Angular, spiral or linear diagrams Other valve timing diagrams can be straight line or spiral representations for crankshaft rotation. Valve timing data is needed for checking engines where unusual symptoms exist and if timing marks on the crankshaft and camshaft drive gears are unclear or missing.

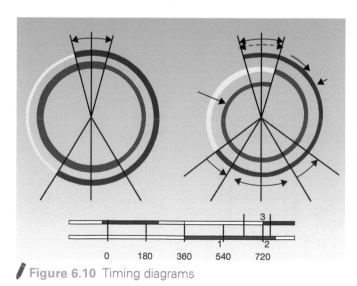

Figure 6.10 Timing diagrams

	Induction
	Compression
	Combustion
	Exhaust

ivo-Inlet value opens
ivc-Inlet value closes
evo-Exhaust valve opens
evc-Exhaust valve closes

Valve lead, lag & overlap (roll over numbers)

Figure 6.9 Valve lead, lag and overlap

Remember! There is more support on the website that includes additional images and interactive features: www.tomdenton.org

The **exhaust valve** finally closes as the piston moves down on the induction stroke.

The **inlet valve** closes as the piston is rising on the compression stroke.

Valves The exhaust valve opens before the end of the combustion stroke. The opening and closing positions of the valves are special to individual engines and are matched to other design and performance requirements.

Summary What is happening within the four stokes is more complex than their simple descriptions and therefore it is important to study them in greater depth.

 Make a simple sketch to show how one of the main components or systems in this section operates

Two-stroke petrol engine cycle All internal combustion engines have an induction, compression, expansion and exhaust process. For a four-stroke engine, each of these processes requires half an engine revolution, so the complete engine cycle takes two complete engine revolutions. That is, there is a working and a non-working (gas exchange) revolution of the engine within the cycle. A two-stroke engine combines two of the processes in each half turn of the engine; thus all processes are complete in one engine revolution and the engine has a power stroke with every revolution. In order to operate, the two-stroke petrol engine uses the crankcase (piston underside) for induction of the fuel/air mixture and transfer into the cylinder via ports in the cylinder barrel.

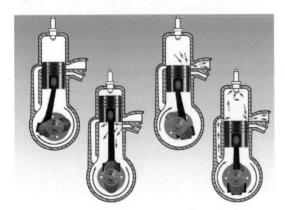

Figure 6.11 Two-stroke operating cycle

On the upstroke, the piston moves upwards towards TDC. The fuel/air charge trapped in the cylinder space above the piston is compressed and at around TDC, ignited by a spark, this is the beginning of the power stroke. As the piston rises during the upstroke, the volume in the crankcase increases and atmospheric pressure forces the fresh fuel/air charge into the crankcase (under the piston).

Downstroke – the piston moves toward the crankshaft On the downstroke, the piston moves towards BDC. As the power stroke begins, the expanding gases force the piston down the bore, producing torque at the crankshaft via the connecting rod. At the same time, the crankcase volume decreases and the fuel/air mixture is compressed under the piston. As the piston approaches BDC, the transfer port connecting the cylinder volume to the crankcase volume is uncovered by the piston. In addition, at the opposing side of the cylinder, the exhaust port is also uncovered.

This allows the fresh charge in the crankcase volume to transfer and fill the cylinder volume, at the same time forcing the exhaust gases out of the cylinder via the exhaust port. The efficiency of this scavenging process is heavily dependent on the port exposure timing and the gas dynamics. Often the piston crown has a deflector to assist this process and to prevent losing fresh charge down the exhaust. Note that two-stroke gasoline engines are normally lubricated via the provision of an oil mist in the crankcase. This is provided by oil mixed in with the fuel/air (premixed or injected), hence the oil is burnt in the combustion process which produces excessive hydrocarbon emissions.

Developments Two-stroke engines are generally more powerful for a given displacement due to the extra power stroke compared to a four-stroke engine. The problem is that the expansion stroke is short and volumetric efficiency (how easy it is to get the gases in and out of the engine) is poor; they are therefore less efficient and exhaust emissions are higher than with a four-stroke engine.

Two-stroke diesel engine cycle Some large, static diesel engines are very often two-stroke types. Note that all four operating processes are

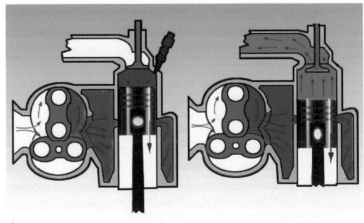

Figure 6.12 Diesel two-stroke cycle

executed in one, single engine revolution (induction, compression, expansion and exhaust). The diesel engine requires a charge of air that is compressed to raise its temperature above the self-ignition point of the fuel. This air charge is supplied by an air pump or pressure charging device (turbo or supercharger). The pressurized air from this device passes into the combustion chamber via ports in the cylinder wall. The exhaust gases leave the combustion chamber via cam-operated poppet valves. The incoming charge forces the exhaust gases out via these valves and this provides the cylinder scavenging process.

Downstroke During the downwards movement of the piston, the hot expanding gases are forcing the piston down the bore, producing torque at the crankshaft, and this is the expansion process. As the piston approaches BDC, the exhaust valve opens and the remaining pressure in the exhaust gas starts the evacuation of the gases in the cylinder via the open valves. As the piston moves further down to BDC, inlet ports are exposed around the bottom part of the cylinder bore, and these allow the pressurised, fresh air charge from the air pump (or turbocharger) to fill the cylinder, evacuating the remaining exhaust gas via the valves and completing the exhaust and induction cycles.

Upstroke At BDC, the cylinder contains a fresh air charge and the piston then begins to move up the cylinder bore. The inlet ports are closed off by the piston movement and the air charge is trapped and compressed due to the deceasing volume in the cylinder. At a few degrees before TDC, the air temperature has risen due to the compression process and fuel is injected directly into the combustion chamber, into the hot air charge, where it vaporises, burns and generates thermal and pressure energy. This energy is converted to torque at the crankshaft via the piston, connecting rod and crankshaft during the downstroke.

Rotary or Wankel engine Another variation on engine operation is the Wankel (the name of the inventor) or rotary engine. This engine has been used in a limited number of passenger car applications. The engine uses a complex geometric rotor that moves within a specially shaped housing. The rotor is connected to the engine crankshaft and turns within the housing to create working chambers. These are exposed to inlet and exhaust ports to allow a fuel/ air charge in, compress it and expand it (thus extracting work), then evacuate the waste gases and restart the cycle. The rotor has special tips to provide a gas-tight seal between the working chambers (in a similar way to piston rings).

Figure 6.13 Rotary engine (Source: Mazda Media)

Engine locations No matter what design of engine, it has to be positioned in the vehicle. There are various configurations that manufacturers have used in the configuration of their vehicle powertrains. The engine can be front, mid or rear mounted and can be installed in-line (along the vehicle axis) or transverse (across the vehicle axis).

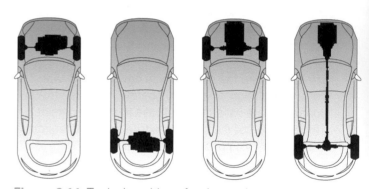

Figure 6.14 Typical positions for the engine

Engine mounting in the vehicle frame The engine mounting system is important as it supports the weight of the engine in the vehicle. In addition, it counteracts the torque reaction under load conditions. The mounting system has to isolate the vehicle from the engine structure-borne vibrations. The engine mounts consist of steel plates with a rubber sandwich between to provide the vibration isolation. The mountings have appropriate brackets and fittings to fix to the engine and vehicle frame.

Front-engine, rear-wheel-drive mountings For a front-engine, rear-drive powertrain layout, the engine mounts are often at the centre position of the engine side, approximately at the engine's centre of gravity. The engine mounts bear compression and shear forces in supporting the engine weight and torque.

Figure 6.15 Engine mountings

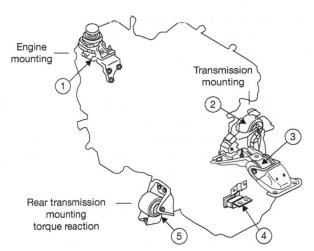

Figure 6.17 Five typical engine mountings for front-engine FWD

The rear of the engine is bolted to the transmission, which is in turn supported at the rear end via a rubber mounting system. This 3-point type mounting is very common for this powertrain configuration.

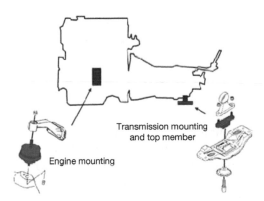

Figure 6.16 Typical engine-mountings for front engine RWD

Front transverse engine, front-drive mountings For a front-wheel drive, transverse powertrain layout, the mounting system has to cope with the weight of the engine plus the torque reaction of the wheel torque. The mounting system therefore includes mountings to support weight and counteract torque separately. These are mounted at the top or bottom of the engine respectively.

Hydraulic mountings A trend in modern vehicles is the use of hydraulic engine mountings that have superior performance with respect to noise reduction when compared to rubber types. They are often used in luxury vehicles or diesel engine installations.

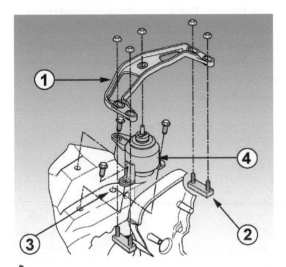

Figure 6.18 This mounting improves comfort

Summary The animation on the website shows a view from above of the four-stroke cycle operating in a four-cylinder engine. Note the firing order of 1-3-4-2 and how each cylinder runs through the four strokes: induction, compression, power and exhaust – or, suck, squeeze, bang, blow!

1 Explain what is meant by 'internal combustion'.
2 Explain the four-stroke cycle.

 Look back over the previous section and write out a list of the key bullet points from this section

6.1.2 Engine variations

Variations in engine design The following pages cover some of the many design variations that are, or have been, used on engines. The configuration of the engine depends on the number of cylinders, their relative position, the engine layout and the firing order. In addition, combustion chamber design, fuel type, valve train design, engine location and mounting position also dictate. Further parameters are engine type (reciprocating, rotary) and stroke (2 or 4).

Engine cylinder configuration The simplest engine design is the single cylinder, normally found in small engine applications. If the engine is a single-cylinder four-stroke, then the engine fires only once every other engine revolution. This gives a large variation in the torque delivery at the engine crankshaft and hence torsional vibration is significant and the engine is not very smooth. Increasing engine capacity does not increase power output directly as there are other factors that contribute to the efficiency of the engine (heat and pressure losses). Therefore, there is an optimum cylinder capacity that provides the best compromise of surface and valve area for maximum efficiency and this is approximately 0.5 litre displacement volume. For this reason, increasing the engine capacity is normally done by adding extra cylinders.

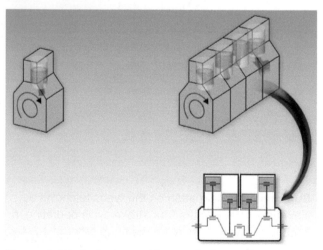

Figure 6.19 A single- and four-cylinder engine

Twin-cylinder engines can be configured as in-line, horizontally opposed or vee types. These engines have been used in car applications but are more commonly found in motor cycles. The in-line engines have been built with both pistons operating in parallel and on alternate strokes; often this depends upon whether they are two- or four-stroke engines. Horizontally opposed cylinders have pistons that move out and return in opposed directions. They are well balanced as the forces generated by the reciprocating masses (pistons, con rod, etc.) cancel out exactly. There are various vee engine configurations that have been used.

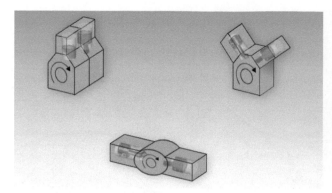

Figure 6.20 Twin cylinders

Three-cylinder Engines Three-cylinder in-line engines of around one litre displacement have been used by some manufacturers in car and motor cycle applications. The three-cylinder vee engine is a possible design being considered as it has a very compact form.

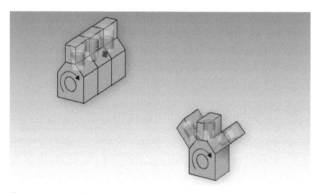

Figure 6.21 Three cylinders

Four-cylinder engines Four-cylinder in-line engines are commonly used as they provide a good compromise of performance, efficiency and smoothness with optimised individual cylinder displacements where the total engine displacement is in the range of 1 to 2.5 litres. This range is extremely common for passenger car applications. Another well-established design for this application is the opposed cylinder; also, less common, is the V4 engine.

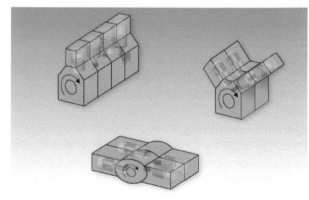

Figure 6.22 Four cylinders

Slant engine Four-cylinder passenger car engines are often designed with a slanting cylinder block. This reduces the overall height of the engine and thus a lower bonnet line is possible.

Five-cylinder engines Where five-cylinder engines have been designed and used, they are generally of in-line construction only. There is an example of a V5 engine with a very narrow vee angle, two cylinders on one bank and three on the other. The main advantage of a five-cylinder engine is that the reciprocating forces are well balanced and this provides a smooth power delivery. The vee-type five-cylinder engine has the added advantage that the overall length of the engine is reduced (compared to the in-line design) and this allows for a transverse engine powertrain layout.

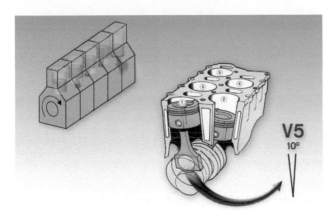

Figure 6.23 Five cylinders

Six-cylinder engines Six cylinders are very common and have been built with in-line, horizontally opposed and vee layouts. The construction and manufacturing costs of a six-cylinder engine are higher but they are well balanced and offer smoother power delivery than the equivalent four-cylinder engine. Therefore, for certain applications (for example, luxury cars) the extra cost is justified. If greater engine capacity is necessary, then a six-cylinder engine is necessary to keep the individual cylinder displacements in the optimum range.

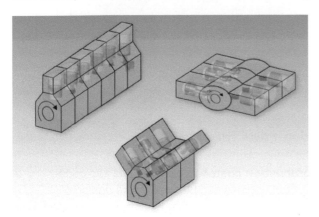

Figure 6.24 Six cylinders

Horizontally opposed and 'V' engines Horizontally opposed and vee engines have shorter crankshafts and overall length than the equivalent in-line engines. This makes them appropriate for transverse or overhung installation in the powertrain. The optimum vee angle for a six-cylinder engine is 60°.

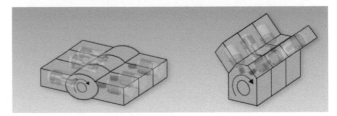

Figure 6.25 Flat and 'V' engines

Eight-, ten- and twelve-cylinder engines Eight-, ten-, twelve- and higher vee configuration engines are manufactured but less common in vehicle applications. V8 engines, though, are very common and used in most countries on larger vehicles. There are petrol and diesel engine designs that employ this layout with engine capacities greater than three litres. The optimum vee-angle for the cylinder banks of eight-cylinder vee engine is 90°.

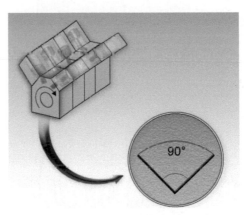

Figure 6.26 Multiple cylinders

Crankshaft construction and firing orders For multi-cylinder engines, the firing order, the crankshaft big-end journal positions, and the direction of rotation are factors that must be considered together in the engine design. Generally, in-line four-cylinder engine crankshafts have cylinders numbered one and four and two and three paired, and 180° apart. This gives two possible firing orders of 1-3-4-2 and 1-2-4-3 with alternate firing from each pair, giving a power stroke every 180° crank angle. In-line six-cylinder engine cylinders are paired: one and six, two and five, and three and four. The big-end journals are positioned at 120° intervals and this gives the most common firing order of 1-5-3-6-2-4.

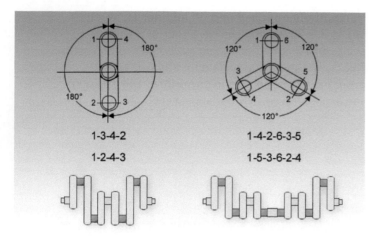

1-3-4-2

1-2-4-3

1-4-2-6-3-5

1-5-3-6-2-4

Figure 6.27 Firing order

V6 big-end journals In a V6 engine, the big-end journals carry two connecting rods each, one from each cylinder bank. The journals are positioned at 120° intervals and are either a single journal or offset journals with the two big-end crank pins offset in order to match the connecting rod angle to the journal.

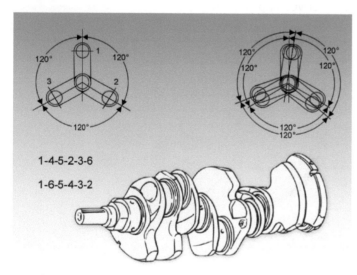

1-4-5-2-3-6

1-6-5-4-3-2

Figure 6.28 V6 crank

V6 cylinder numbering Generally V6 engine cylinder numbering is implemented in one of two ways. The first is that one bank has one, two and three and the other bank has four, five and six. The alternative numbering sequence uses alternate banks with one, three and five on one side and two, four and six on the other. The firing orders are either one bank followed by the alternate bank or one cylinder from one bank followed by a cylinder from the other bank. These variations are shown by the following diagrams.

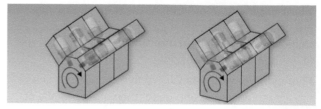

Figure 6.29 V6 engines

V8 big-end journals A V8 engine has four paired journals placed at 90° intervals to each other. Each journal carries one connecting rod from each of the opposite banks.

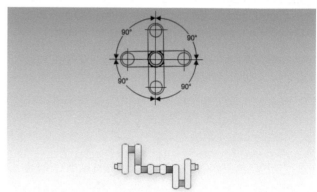

Figure 6.30 V8 crank

Cylinder numbering The cylinders can be numbered in a similar method to the six-cylinder engines with either odd numbers one side and even numbers the other, or one to four on one cylinder bank and five to eight on the other bank.

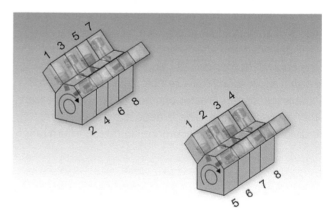

Figure 6.31 V8 engine cylinder numbers

Firing orders Typical firing orders are shown in these diagrams.

Combustion chamber designs The evolution of petrol engines can be seen in the design of combustion chambers that have been developed over the years to improve efficiency.

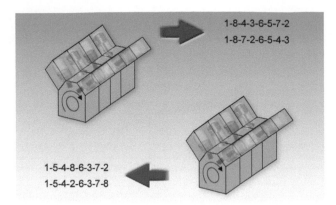

Figure 6.32 Typical firing orders

Side valve construction The earliest engines were of side valve design. They had the valves seated in the block and the combustion chamber covered the valves and part of the piston crown.

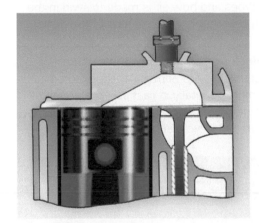

Figure 6.33 Side valve

Overhead valve engines The next major development was overhead valve engines. These employed in-line valves and bath tub combustion chambers over the piston.

Wedge-shaped chambers Improved combustion and flame propagation could be achieved with a wedge-shaped chamber. This had the valves offset from the vertical position.

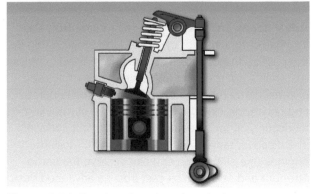

Figure 6.34 Combustion chamber design

Combustion knock (pinking) A problem associated with the wedge design is combustion knock, also known as pinking. This is caused by uncontrolled ignition of the end gases prior to ignition from the advancing flame front. It occurs due to compression of the end gases in the thin end of the wedge. This generates pockets of combustion with high pressures that damage the piston crown and land area above the compression ring.

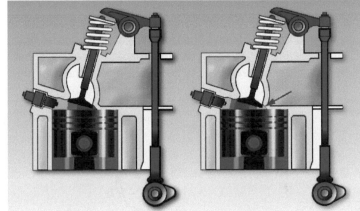

Figure 6.35 Engine knock can cause damage

Siamese ports Inlet and exhaust ports can either be on the same or opposite sides of the cylinder head, known as pre-crossflow or crossflow respectively. Also, two cylinders can share a common inlet port and this is known as a Siamese port. When a single inlet for each port is used, this is known as parallel ports. Some patterns of porting are shown in these diagrams.

Pentroof combustion chamber Most modern petrol engines use a hemispherical or pent roof combustion chamber design. This shape provides the best compromise of surface to volume ratio and this reduces heat energy loss during the expansion stroke which in turn improves the thermal efficiency of the engine. The combustion chamber design allows single or dual inlet and exhaust ports with crossflow engine breathing. This design easily accommodates two, three, four or five valves per cylinder.

Port and valve design The design of the inlet tract, including manifold, head and valves, is essential to provide the correct charge motion as the gases enter the cylinder during the induction stroke. Charge motion is important to speed up the combustion process sufficiently in order to prevent excessive exhaust emissions.

Aluminium heads Hardened valve seat inserts are required in aluminium heads. In addition, certain cast iron heads will also employ these. They are necessary

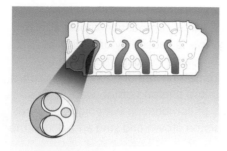

Figure 6.36 The need for a good mixture is why port and valve design is important

in order to increase the durability of the head so that it can resist the heat of the exhaust gases. Note that in older engines that ran on leaded fuel, the lead fuel additive provided an element of protection.

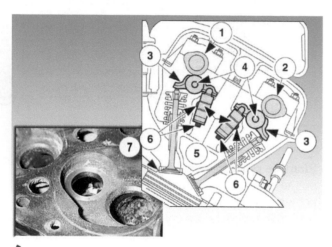

Figure 6.37 Head features

On OHV engines For OHV engines, the cam followers and push rods are encouraged to rotate and this helps to extend their life and reduce wear. In addition, in many engine designs the valves rotate for the same reason. Rotation is promoted by a slight offset or taper on the tappet or rocker face in contact with the cam lobe. Some engines may be fitted with valve rotating mechanisms which are integral with the spring retainer. These are in two parts with opposing angle faces and rollers to provide a rotational drive as the valve is operated.

Diesel combustion chambers Diesel engines have combustion chambers that fall into two main categories, either direct or indirect fuel injection. This naming convention is derived from the position where fuel is introduced into the combustion chamber. Indirect injection is more commonly found on smaller engine applications; although most modern diesel engines, for vehicle applications, employ direct injection due to the improved thermal efficiency of this design. These diagrams show the construction

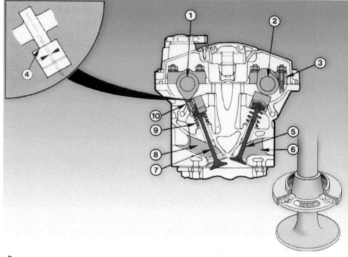

Figure 6.38 OHV features

of the two types and how air is made to swirl in the combustion chamber before injection of the diesel fuel.

Cylinder block and crankcase construction The block and crankcase are normally manufactured as a single casting, generally in cast iron or aluminium. Certain engines have a separate block and crankcase (for example, motorcycle engines).

Figure 6.39 Modern engine

Advantages and disadvantages of cast iron Cast iron has been used for cylinder block construction in the past as the cylinders can be bored directly into the material. Also, these bores can be remanufactured or repaired by reboring oversize. Cast iron is porous and hence the cylinder bore is capable of retaining lubricating oil for lubrication of the contact surfaces. The disadvantage of cast iron is weight. Modern engines use aluminium and can achieve the same strength and stiffness as cast iron via advanced design techniques.

Aluminium alloys Aluminium alloy cannot provide a suitably durable surface for piston ring contact. Therefore cylinder liners or sleeves, made from cast iron or steel, are normally fitted into an aluminium cylinder block.

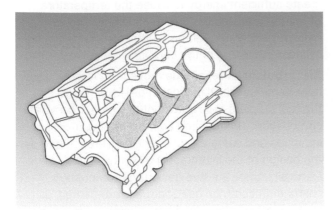

Figure 6.40 Liners in an aluminium block

Nickel phosphate and silicon carbonate Recent developments in material technology have produced a coating of nickel phosphate and silicon carbonate which provides a suitably durable surface for the cylinder bores. Note that these bores cannot be re-bored; if excessive wear occurs, the block must be replaced.

 Make a simple sketch to show how one of the main components or systems in this section operates

6.1.3 Engine operating details ❶

Atmospheric pressure The air above the Earth's surface is a fluid that exerts a pressure on all points around it. This is due to the weight of the air acting down upon the Earth's surface and this, in turn, is because of the Earth's gravitational force pulling it down. This creates a pressure known as atmospheric pressure and is 101.325 kPa, 760 mmHg, 29.92 inHg, 14.696 psi, 1013.25 millibars (let's stick with approximately 1 bar or 15 psi)!

A **naturally aspirated engine** relies on atmospheric pressure to charge the cylinder with gas (air or air/fuel mixture) ready for the combustion process. As the piston moves down the cylinder (from TDC to BDC), the volume increases and this causes the pressure in the cylinder to reduce, becoming lower than atmospheric pressure. This creates a pressure difference between the inside and outside of the cylinder and, due to this, the atmospheric pressure

(the higher pressure) forces gases into the cylinder (where there is lower pressure) until the pressure is balanced. Note that any restriction to the flow of gas will reduce the effectiveness of the cylinder charging process.

Volumetric efficiency is a measure of the efficiency of the cylinder charging process during the induction stroke. Theoretically, the cylinder should be completely filled with a mass of gas but in practice this never happens due to flow losses and inefficiencies. Therefore, the volumetric efficiency is a measure of the actual amount of gas induced compared to the theoretical amount (which is the mass required to completely fill the cylinder volume) and is expressed as a percentage. It is calculated as: (the actual mass of air/the theoretical mass of air) × 100%.

Breathing or aspiration The more efficiently the engine cylinders can fill with gas, the more air, or fuel/air, is available for the combustion process and this improves overall engine efficiency. The process of getting gases into and out of the engine is known as 'aspiration' or 'engine breathing'.

Petrol and diesel Combustion in the engine cylinder takes place because of a chemical reaction between the carbon and hydrogen in the fuel and the oxygen in the air. This reaction releases energy from the fuel in the form of heat that generates pressure in the cylinder to force movement of the piston. In order to achieve efficient combustion, the quality of the fuel/air mixture is important – that is, how evenly mixed the fuel droplets are in the induced air. Movement of the air as it enters the cylinder is important for this process and the requirements are different for petrol and diesel engines. The required air movements for each engine type are created by careful design of the components that form the inlet tract and combustion chamber.

The **inlet valve** opens and closes according to piston position and controls the incoming gas charge into the engine. It generally remains open for a small time period after the piston has reached BDC (i.e. beyond the end of the inlet stroke). This allows the energy of the moving gas column in the inlet tract to assist in the cylinder charging process which helps to increase volumetric (and engine) efficiency.

The compression stroke After the combustion chamber has been charged with gas (air or fuel/air) during the induction stroke, the cylinder inlet and exhaust valves are both closed and seal the combustion chamber. The piston begins to rise in the cylinder, thus reducing the volume of the cylinder space and hence increasing the pressure of the trapped gas charge in the cylinder prior to combustion. The opening and closing of the valves is executed in sequence via the engine valve gear, synchronized with the four-stroke cycle and piston position.

Piston rings It is important that the closed cylinder is sealed properly to maintain the appropriate pressures in the cylinder during the working cycle. Any losses in pressure would significantly reduce the efficiency of the engine. In order to seal the piston and bore, piston rings are fitted into radial grooves near the top of the piston and provide the gas-tight seal between the moving piston and the cylinder bore. When the cylinder volume is reduced during the compression stroke, the trapped gas is compressed and the amount of compression is known as the compression ratio. Compressing the charge prior to combustion allows more oxygen or fuel/oxygen in the cylinder than would otherwise have been available without compression and this improves combustion efficiency. Generally, most spark ignition engines have a compression ratio of 8:1 to 10:1. This means that the cylinder volume reduces by eight or ten times during the compression stroke.

Figure 6.41 Compression

Temperature rise during compression (petrol)

During compression of the fuel/air mixture in a petrol engine, heat energy and kinetic energy (due to gas movement) are imparted into the mixture due to the reducing volume and rising pressure. This creates a significant temperature increase and the magnitude of this increase depends upon the speed of the compression process and the amount of heat passed to the surroundings (via the cylinder combustion space, walls, head, etc.). The temperature rise elevates to a point just below the self-ignition temperature of the fuel/air charge, which will combust at or above the flashpoint when ignited via an external source (i.e. the spark plug). Note that if the temperature of the mixture

is too high, spontaneous self-ignition would occur and this would be a limiting factor for the maximum compression ratio in a petrol engine.

Temperature rise during compression (diesel)

In a diesel engine, the compression process must create sufficient energy to cause the temperature of the compressed gas (air) to rise above the self-ignition temperature of the fuel that is injected into the cylinder at the end of the compression stroke.

Air turbulence Inlet charge movement is particularly important in a diesel engine, in order to ensure that the fuel droplets have sufficient oxygen for complete combustion. The required air flows during the induction and compression processes are created by the design of the inlet tract and combustion space. Generally, there are two designs of combustion chamber in common use and these are named, as such, due to the position in the chamber where the fuel is introduced. They are known as direct and indirect injection.

Figure 6.42 Direct diesel injection

A **direct injection** combustion chamber has a 'bowl' formed in the piston crown. This is designed to promote a tumble movement of the incoming air mass; this helps to ensure good distribution of the fuel in the cylinder and reduced soot emissions.

Indirect injection The indirect type combustion chamber incorporates a pre-combustion chamber within the cylinder head. The compressed inlet charge is forced into this chamber at high velocity and pressure. This creates a swirl movement that ensures complete mixing of fuel droplets with air for maximum combustion efficiency. During the combustion process, the burning gases are ejected from this chamber with high pressure and energy. This ensures sufficient turbulence in the main combustion chamber for efficient combustion.

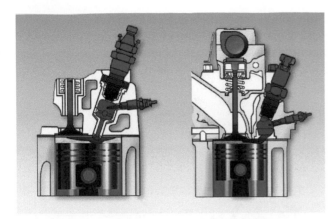

Figure 6.43 Pre-combustion chamber and swirl chamber

The **compression ratio** of a direct injection engine is typically between 16:1 and 21:1. This is sufficient to raise the induced charge temperature for self-ignition of the fuel under all engine operating conditions without creating excessive combustion noise (or diesel knock). Indirect injection engines have higher compression ratios of 22:1 to 25:1. This is necessary to generate the extra heat energy required due to losses via the increased surface area of the cylinder head. Diesel knock is less apparent in indirectly injected engines as the energy release is more controlled and less spontaneous.

The combustion or power stroke After compression of the inlet charge, combustion of the fuel creates heat and pressure energy, which is imparted to the piston to generate mechanical work. In a petrol engine, this process is initiated by the high-voltage arc at the spark plug electrodes in the cylinder.

The diesel engine is designed to produce compression pressures that generate sufficient heat in the cylinder to ignite the fuel as it is injected into the combustion chamber. This is known as compression ignition (CI). Petrol engines are generally known as spark ignition (or SI) engines. During the combustion stroke, the engine power output or work is generated, hence the name 'power' stroke in the four-stroke cycle of induction, compression, power and exhaust. Engine combustion is a fundamental process in the operation of the engine. This process must be efficiently executed and controlled via the engine subsystems (fuel, air, ignition, etc.) to ensure the best efficiency and performance, with minimum harmful exhaust emissions.

Complete combustion of fuel and oxygen
Combustion in the cylinder of an engine is a chemical reaction process between carbon and hydrogen in the fuel and the oxygen present in the induced air. The carbon and oxygen combine to form carbon dioxide (CO_2), the hydrogen combines with oxygen to form water (H_2O). Nitrogen passes through the engine as long as the combustion chamber temperatures remain below critical limits.

Incomplete combustion of fuel and oxygen If the combustion process is not efficient, incomplete combustion will result and this produces carbon monoxide (CO). If combustion chamber temperatures are high, oxides of nitrogen are produced (NOX). These are both harmful pollutants and their emissions from motor vehicles are closely regulated and controlled by environmental protection agencies and bodies around the world.

Ignition timing The combustion process should occur in a rapid but controlled manner. The flame propagation and energy release in the cylinder should have a predictable, stable behaviour depending on the engine operating conditions. The timing of the spark ignition is critical to achieve appropriate energy release for maximum efficiency in the energy conversion process that takes place in the combustion chamber. The burn duration of the fuel varies according to engine conditions; therefore, the spark must be adjusted to occur at the correct time, according to these conditions, to get the optimum torque from the engine. The optimum spark advance for a given engine condition is known as Minimum Spark Advance for Best Torque (MBT).

Figure 6.44 Ignition timing for best torque

Petrol The quality of petrol is measured by a parameter called the 'octane' rating and this gives an indication of the fuel's resistance to engine 'knock' or uncontrolled, spontaneous combustion, which causes engine damage. The higher the octane rating, the slower and more controlled the fuel burns and, hence, the greater the resistance to 'knock'. The octane rating of the fuel determines the limit of ignition advance for a given engine speed and load condition. Therefore, it is particularly important to always operate the engine on the correct fuel, to prevent damage to the engine due to 'knocking'.

Mixture strength A chemically correct air and fuel ratio mixture must exist in order to ensure that sufficient oxygen is present to completely combust all of the fuel. This is known as mixture strength and is the ratio of air mass to fuel mass. For most fuels, the correct ratio is approximately 14.7 air mass to 1 part fuel mass. If more air is present then the mixture strength is known as weak or lean. If there is less than a 14.7 air/fuel ratio, then the mixture strength is known as 'rich'. Weak and rich mixtures are less than optimum for the engine, although under certain conditions the mixture strength is adjusted by the engine control system according to demand. For example, for full power a slightly richer mixture is needed and this is provided when the engine is at full throttle. Extended running on rich or weak mixtures reduces engine efficiency and can cause damage to the engine internally and to its subsystems.

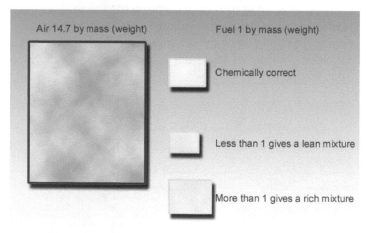

Figure 6.45 Mixture ratio

Cylinder pressure in a petrol engine The combustion process creates energy within the cylinder in the form of heat from the burning fuel/air mixture. Due to the enclosed nature of the cylinder, this heat energy creates a pressure rise in the cylinder above the piston. This pressure, applied over the piston area, in turn, creates a force pushing down on the piston and turning the crankshaft via

Remember! There is more support on the website that includes additional images and interactive features: www.tomdenton.org

the connecting rod, thus producing torque at the crankshaft. The pressure in the cylinder is shown plotted against cylinder volume in the diagram. This is known as an indicator diagram.

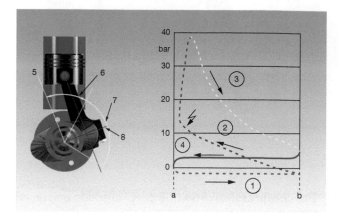

Figure 6.46 Petrol engine indicator diagram – diesel figures are approximately double

Cylinder pressure in a diesel engine The torque at the crankshaft is a function of the cylinder pressure and crankshaft angle; the maximum torque is produced when the connecting rod and crankshaft main/big-end bearings are at right angles (i.e. 90° crank rotation from TDC position). Note that at TDC, any pressure on the piston produces no work as there is no turning moment (torque), just a force pushing down on the bearings.

Ideal combustion The ignition and fuel settings of an engine are set by the manufacturer at the optimum position to achieve the best compromise of performance, economy and minimal exhaust emissions. With respect to combustion, it is important that the maximum cylinder pressure and energy release occur at the correct angle. Damage to the engine can occur if this happens too early or late in the engine cycle. An example is early ignition; this causes engine 'knock' and damages the piston if allowed to occur for any significant period of time.

Figure 6.47 Ideal, retarded and advanced ignition point

Pinking This is a characteristic noise caused by pre-ignition or early ignition of the fuel/air mixture. Early ignition causes an early pressure rise that is applied to the piston at TDC. At this crank angle, no engine torque can be produced and this means that all the combustion energy is applied directly to the engine's mechanical components (piston crown, bearings, etc.), causing them to generate the 'pinking' noise. Although the noise is quite subtle, the forces are massive and cause considerable damage to the engine.

Advanced ignition When 'pinking' occurs, the combustion energy precipitates through the engine components causing damage. In addition, heat is generated that is not dissipated normally and this causes excessive temperature of engine components (pistons, valves and valve seats, for example) and consequent heat-related damage.

Retarded ignition Over-retarded ignition causes incorrect timing of the energy release from the fuel that, in turn, means less energy to do work and therefore more energy to dissipate via the cylinder boundaries. This causes an increase in engine temperatures, damages components and reduces overall engine efficiency. This excess energy also has to be rejected via the exhaust and this causes increased exhaust gas temperatures that can damage exhaust valves and seats, as well as exhaust gas components (catalytic converter).

Diesel combustion Combustion in a diesel engine begins very rapidly as the fuel is being injected into the combustion chamber and heated. This causes a rapid energy release that generates the characteristic 'diesel' engine noise. For this reason, a simple diesel engine is noisier than the equivalent petrol engine. The combustion process is most rapid in a direct injection diesel engine and due to this, combustion losses are minimal and these are the most fuel efficient type of internal combustion engine seen in road vehicles.

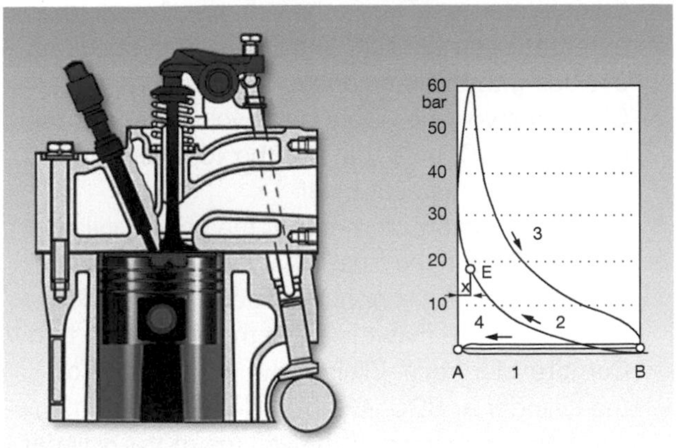

Indirect injection diesel engines Combustion is not fully completed in the pre-chamber before the combustion gases are expelled into the cylinder and continue the combustion process in the main combustion chamber. The increased surface area means that more heat from combustion is lost to the cylinder boundaries (walls, head, etc.) when compared to the direct injection type. They are, though, quieter in operation due to the longer, slower combustion process.

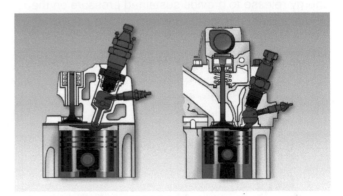

Figure 6.49 Indirect diesel injection

Exhaust gas recirculation (EGR) system Diesel engines always operate with excess air (or weak mixture) and with high exhaust gas temperatures, thus oxygen combines with nitrogen to form the pollutant NO_x (oxides of nitrogen). Exhaust gas recirculation (EGR) is often used to reduce cylinder temperatures and to reduce the amount of oxygen in the cylinder air charge, thus preventing the formation of NO_x.

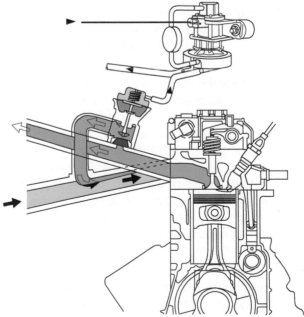

Figure 6.50 Exhaust gas recirculation (EGR)

Combustion phases For diesel engines, there are three main phases in the combustion process. The

![Figure 6.48]

Figure 6.48 Direct diesel injection

first is the delay phase as the fuel absorbs heat from the cylinder air charge and vaporises. The next phase occurs when the fuel has reached a sufficient temperature to self-ignite; this causes combustion and the flame front propagates rapidly out across the piston crown. This is where the rapid energy release occurs and causes the characteristic diesel engine noise. Once initial burning of the fuel takes place, continued injection of fuel provides a controlled burning and energy release to provide sustained pressure on the piston and good torque generation at the crankshaft.

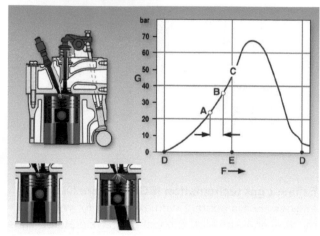

Figure 6.51 Phases of combustion

Good combustion Efficient and effective combustion promotes an engine with good power output and with minimal harmful emissions. This can only be achieved when the engine's mechanical parts are in good condition, and the engine control systems for fuel delivery, ignition and emission control are correctly optimised and set.

The exhaust stroke At the end of the power stroke, as much energy as possible has been extracted from the fuel and converted to mechanical work at the crankshaft. The next part of the working cycle is to eject the exhaust gases and their remaining heat energy from the cylinder. This process is controlled via the exhaust valves, which open around the end of the power stroke to vent the cylinder combustion gases to the atmosphere via the exhaust system. Typically, the exhaust valves open before the end of the power stroke to allow the remaining pressure energy in the end gases to assist in evacuating the cylinder.

The **exhaust valve** is opened via the engine valve gear and is synchronised with the engine operating cycle. Opening the exhaust valve before the end of the power stroke assists in the process of evacuating the cylinder volume efficiently. During the exhaust stroke, the piston rises from BDC to TDC, thus the

cylinder volume decreases and this ejects the exhaust gases out through the exhaust valve. Generally, the inlet valve opens before the end of the exhaust stroke and this creates a certain amount of time when both inlet and exhaust valves are open. This allows the kinetic energy of the exiting exhaust gases to assist in drawing the fresh gas charge into the cylinder.

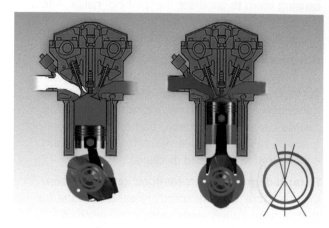

Figure 6.52 Timing diagram

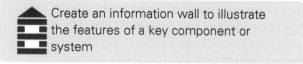

Create an information wall to illustrate the features of a key component or system

6.1.4 Terminology and systems ❶

Technical terms The following are some of the technical terms that are used to describe features of the engine:

▶ Engine capacity – This is the total, combined, displaced volume of all engine cylinders as a single value stated in units of cubic capacity. Generally this is given in cubic centimetres (cm³) or litres. In America, engine capacity is normally stated in cubic inches (in³).

▶ Swept volume – This is the volume of a cylinder bore between the TDC and BDC piston positions, excluding the volume above the piston at TDC.

▶ Clearance volume – This is the volume above the piston at TDC; note that it is the volume of the combustion chamber itself.

▶ Bore – This is the diameter of the engine cylinder.

▶ Stroke – This is the total linear distance travelled by the piston in the bore between TDC and BDC positions. Note that it is twice the crankshaft throw.

▶ Compression ratio – This is the total volume of the cylinder at BDC (swept + clearance volume), expressed as a ratio of the volume of the cylinder at TDC (clearance volume).

The information on cylinder dimension can generally be found in workshop or manufacturer manuals. In addition, these values can be measured directly or derived via calculations.

Swept volume This can be calculated via the formula:

$(l\pi d^2)/4$

where d = cylinder bore and l = stroke

Note that units of bore and stroke must be consistent. Engine volume is mostly stated in litres by manufacturers but remember that 1000cc (cubic centimetres) equals 1 litre. The total engine displacement is the sum of all cylinders' individual displacements.

Compression ratio The formula used to calculate the compression ratio (CR) is:

$CR = (V_s + V_c)/V_c$

where CR = compression ratio, V_s = swept volume and V_c = clearance volume

Note the correct order of preference when carrying out this calculation (brackets first – BODMAS).

Torque Two common terms used when expressing engine performance characteristics are 'torque' and 'power'. Torque is an expression relating to work and is a measure of the turning force provided by the engine. Torque output can vary independent of engine speed and is a measure of the load on the engine. The units of torque are Newton metres (Nm) for SI units and pounds/foot (lbs/ft) in Imperial units. Power is a derived unit and relates to the rate of work done,

Figure 6.53 Clearance and swept volumes

or the work done per unit of time. For an engine, the power is a product of torque and speed. Power output is given in kilowatts (kW) or horsepower (hp). Engine power is normally stated as measured at the flywheel, via a dynamometer or brake, hence the term 'brake horse power'.

Engine performance curves Engine manufacturers often publish performance data in a graphical form showing torque and power curves against speed. Two examples are shown above. Note that a petrol engine generally produces more power at higher speed. A diesel engine produces more torque at lower speeds.

Optimum cylinder capacity The optimum size of an individual engine cylinder is a compromise between a number of technical factors. The optimum displacement for a cylinder is generally found to

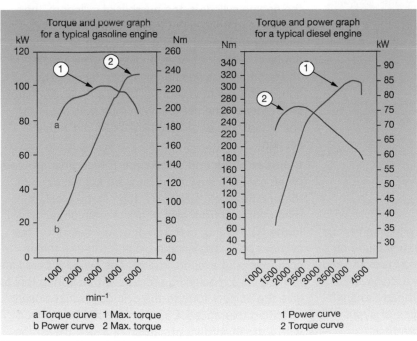

a Torque curve 1 Max. torque
b Power curve 2 Max. torque

1 Power curve
2 Torque curve

Figure 6.54 SI and CI engine curves

be between 250cm³ and 600cm³ for road vehicle applications. In this range, the combustion chamber size, surface area and individual components size (pistons, valves, etc.) produce an engine with optimum efficiency with respect to fuel consumption and emissions. Typically, engines with total displacements in the range of 1 to 2.5 litres have four cylinders. Note though that the number of cylinders also has an effect on the manufacturing costs of an engine.

Power strokes per engine revolution The number of power strokes per revolution can be found by dividing the number of engine cylinders by two (for a four-stroke engine). The greater the number of cylinders, the smoother the torque delivery due to reduced peak torque firing pulses from each cylinder and the increased number of firing strokes per revolution. Over two litres, six-cylinder engines give smooth power delivery with optimum cylinder displacement sizes. An in-line six-cylinder has a relatively long crankshaft that can be difficult to accommodate in a transverse engine installation layout, therefore by using two banks of three cylinders in a vee configuration, total length is reduced and torsional rigidity of the crankshaft is improved.

The engine's flywheel acts as an energy buffer due to its inertia. Energy stored in the flywheel maintains rotation between firing pulses and acts as a damper to smooth torque peaks as each cylinder fires.

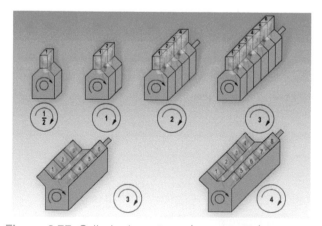

Figure 6.55 Cylinder layouts and power strokes

Component technology This section explores various different engine designs and configurations found on modern road vehicles. The most common is discussed first. This is the four-cylinder, four-stroke, transverse installation layout incorporating an overhead camshaft (OHC) petrol engine. This engine design and layout allows for a comprehensive repair programme, including a complete overhaul.

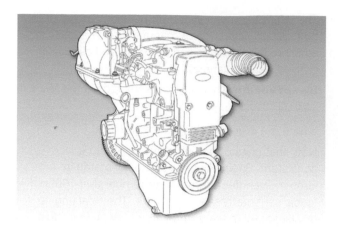

Figure 6.56 Engine designs vary

Four-cylinder, four-stroke diesel engine The next engine design to be considered is the four-cylinder, four-stroke, overhead valve (OHV) indirect injection diesel engine. This engine is of traditional design and construction and facilitates straightforward repairs and overhaul. Diesel engines are generally more strongly constructed due to the higher cylinder pressures and forces.

Special oils Diesel engine emissions produce acidic elements that can react with certain engine components. Diesel-specific engine oils are designed to neutralise this acidity via additives and should always be used. Corrosion and pitting of internal engine components can occur due to lack of regular oil changes.

Cylinder arrangement There are numerous engine configurations with respect to the arrangement of the engine cylinders, the number of cylinders, their position and firing order. In addition, combustion chamber designs and valve train layout all dictate the basic properties of an engine. Engine installation and orientation is another important factor to be considered in a road vehicle. There are also two-stroke and rotary engine designs with their own particular characteristics, all of which are explained in this section.

Cylinder block and crankcase The engine cylinders, when cast in a single housing, are known as the engine block. Generally, the engine block is manufactured from cast iron or aluminium alloy. In the latter case, cast iron or steel liners form the cylinder bore. The engine block forms the major component of a 'short' motor. The cylinder bores are formed via a machining process with a boring tool to give the correct form to the cylinder within closely specified tolerances. Cast iron is a mixture of iron with a small amount of carbon (2.5%–4.5% of the total).

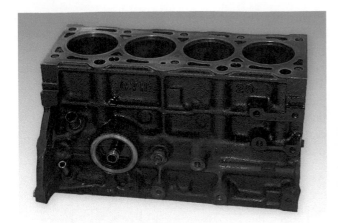

Figure 6.57 Engine block

Cast iron The carbon added to the iron gives a crystalline structure that is very strong in compression. In addition, it is slightly porous and this helps to retain a film of lubricating oil on working surfaces. This property makes cast iron particularly suitable for cylinder bores that can be machined directly into the casting.

The **crankcase** is integrated into the cylinder block and is machined in-line to form the crankshaft main bearings. This process is known as line boring. The main bearings are split in two halves; one half locates in the block, the other in the bearing cap. The bearing caps are secured before the machining process and thus each cap is matched in position with its opposite half. It is important to note this when disassembling and reassembling the bearings. The caps are located via dowels and fastened via high-tensile steel bolts. It is important to follow manufacturer guidance if the bolts are removed and refitted; replacement of the bolts and tightening procedures must be followed if specified.

Figure 6.58 Main bearing bolts

Water jacket Between the cylinder walls and the outside surface of the cylinder block, voids and

channels are formed during the casting process. This is known as the water jacket and is used for engine cooling purposes. A sand former creates this space during casting and when the cast block has cooled, the sand is evacuated via holes in the side of the block. These holes are then sealed using core plugs.

Figure 6.59 Water jacket

Figure 6.60 Core plugs

Oilways In order to supply pressurised oil to the engine moving surfaces, an oil gallery is formed along the length of the cylinder block. This has drillings to supply oil directly to the bearings in the block, crankshaft and cylinder head. Additional drillings connect the oil pump and pressure control valve to complete the oil supply system. The block is prepared, drilled and threaded in order to attach additional components like the oil sump pan and oil pump assembly.

Pistons are generally manufactured from an aluminium alloy which reduces weight and increases heat dissipation. There are numerous designs to accommodate thermal expansion according to engine type and application.

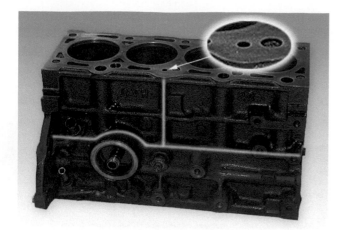

Figure 6.61 Oilways

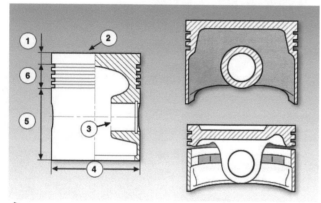

Figure 6.62 Piston features

Thermal expansion Aluminium has greater thermal expansion than the cast iron used for the block and cylinder liners. This means that the piston expands more than the block as the engine temperature increases. When the engine is cold, the working tolerances are greater to allow for expansion. The piston has design features to allow for expansion and correct tolerances at running temperatures – for example, a cold piston is slightly oval and tapered inwards towards the crown.

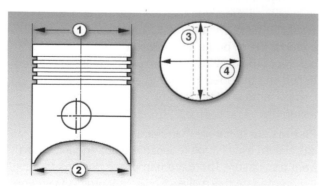

Figure 6.63 Piston dimensions

Piston pin The piston or gudgeon pin is offset by a small amount toward the thrust face of the cylinder bore which allows the thrust forces at the piston crown to maintain the piston against the cylinder wall. This has an effect when the engine is cold by reducing piston movement due to excessive clearance, which creates a noise known as 'piston slap'. Note that pistons are marked so that they can be installed correctly and this should be carefully observed.

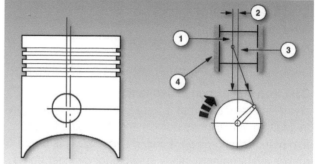

Figure 6.64 Piston operation features

Piston and piston rings Around the upper portion of the piston, grooves are cut to accommodate sealing rings, known as piston rings. Generally, there are three or four grooves and rings. The lowest is known as the oil control ring and this is used to control the amount of lubricant remaining on the cylinder bore surface to lubricate the piston. The upper rings are known as compression rings and these provide the gas-tight seal, maintaining the cylinder pressures that create force to move the piston.

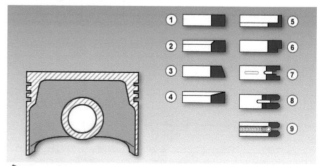

Figure 6.65 Piston rings

Piston pin bore This bore is machined into the piston to accept the piston pin, also known as gudgeon pin. The fixing mechanism of the piston pin to the piston and the connecting rod can vary. It can be an interference fit in the connecting rod, or a push fit in both the piston and connecting rod end. If the piston pin is clamped in the connecting rod, the piston pin bore is smooth. Circlip grooves are formed in the piston pin bore when a push fit piston pin is used.

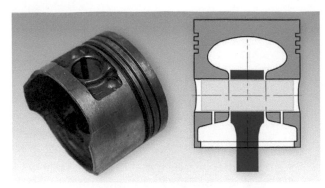

Figure 6.66 Cross section of the piston

The **piston crown** forms part of the combustion chamber and experiences the full cylinder pressure applied by the expanding gases. Many different designs are available depending on engine type; complex shapes can be formed in the piston crown to allow for valve movements and to create an effective combustion chamber space, promoting the correct charge motion for efficient combustion.

Figure 6.67 Various pistons

Piston rings These are used to seal the combustion chamber to prevent the escape of combustion gases and loss of cylinder pressure – they known as 'compression rings'. In addition, the piston rings must control the oil film on the cylinder bore surface – these are known as 'oil control rings'. Combustion pressure is allowed to act on the back of the cylinder sealing 'compression' rings to help maintain the gas-tight seal of the piston assembly.

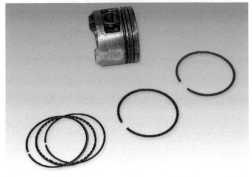

Figure 6.68 Compression and oil control rings

Compression rings are manufactured from cast iron, with a surface coating to promote fast bedding-in. This means that the rings quickly wear in to give a gas-tight seal against the cylinder pressures. It is important not to damage this coating during fitting. Note that rings have different cross sections according to their mounting position on the piston.

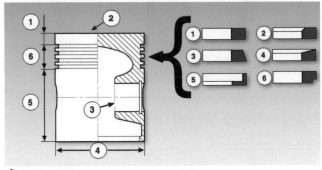

Figure 6.69 Compression rings

Oil control rings can be one of two designs. A multi-part ring consists of two thin alloy rings used in conjunction with an expander between them. A cast iron ring has a groove and slot arrangement to allow oil flow back to the sump via the ring and piston.

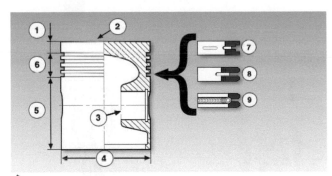

Figure 6.70 Oil control rings

Piston pin (gudgeon pin) The piston or gudgeon pin provides the mechanical link between the piston and the connecting rod. The pin locates in the piston body and the 'little end' of the connecting rod. The pin can be a clearance fit into the little end bearing or bush, and hence a corresponding interference fit, or located via circlips in the piston.

Interference fit An alternative is that the pin is an interference fit in the little end, or is clamped by the connecting rod. In this case, the piston pin bore is the bearing surface and there are appropriate drillings in the piston to allow for lubrication.

Connecting rod The main purpose of the connecting rod is to transfer the linear force from the piston and apply it to the rotating crankshaft. It is generally

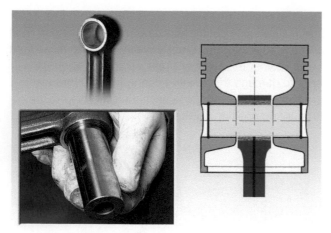

Figure 6.71 Piston pin or gudgeon pin

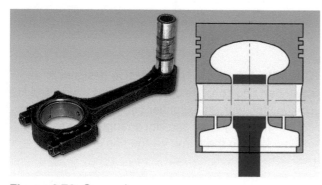

Figure 6.72 Con rod

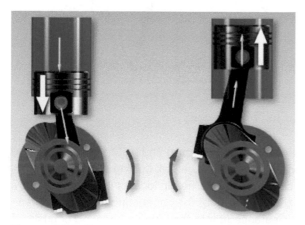

Figure 6.73 Force transfer

Figure 6.74 Big and little end bearings

Big-end bearing The crankshaft end of the connecting rod is known as the big end. This consists of a split bearing with a removable bearing cap. The bearing cap is attached to the connecting rod via bolts or nuts.

Matching connecting rod and bearing cap identification mark It is important to note that the connecting rod and bearing cap are machined as one unit and, hence, the parts are matched. Therefore, they must always be reassembled as a pair and fitted correctly oriented.

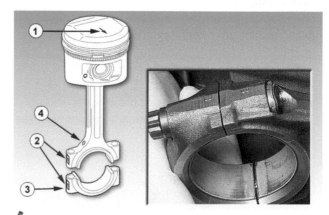

Figure 6.75 Piston and con rod and big-end bearing

manufactured from carbon steel in a process known as drop forging to form the required shape and profile.

The connecting rod is designed specifically with a high resistance to bending, compressive and tensile forces via an I-section profile. The piston end, known as the little end, has an appropriate bush, bearing or clamping arrangement for the piston pin.

Cylinder wall lubrication Drillings made through the connecting rod body provide oil spray lubrication of the cylinder walls. Often, an oil supply drilling is provided through the connecting rod to spray oil onto the underside of the piston crown to provide additional cooling.

The **crankshaft** receives the linear force of the pistons, via the connecting rods, and converts this force into a rotating torque. The crankshaft is generally manufactured from cast iron or steel alloy via a forging or casting process.

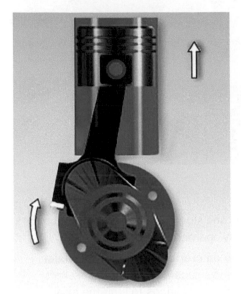

Figure 6.76 Conversion of linear to rotary motion

Crankshaft Generally the crankshaft of a four-cylinder engine has five main bearings. At the front of the crankshaft, provision is made to locate and drive the crankshaft pulley and timing gear via keyways and securing bolts. Behind this, the oil pump drive is located and then the first or front main bearing.

Figure 6.77 Crankshaft pulleys

Figure 6.78 Crankshaft main journals

Figure 6.79 Drive key

Figure 6.80 Crank-mounted oil pump

Bearing journals The big-end bearing for the first cylinder is fitted in between the crankshaft webs radiating from the main bearing journals. These webs form counterbalance weights to the big-end journal. Generally, one of the main bearings is fitted with a thrust washer to control axial movement of the crankshaft.

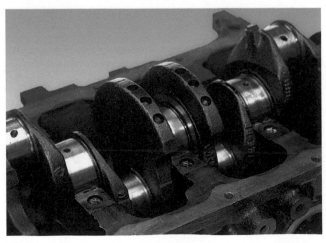

Figure 6.81 Journals

6

105

Rear main oil seal At the rear of the main bearing journal, at the back of the engine, a machined face is formed on the crankshaft as a mating surface for a sealing ring. This is the main oil seal at the back of the engine. In addition, there is a machined, threaded flange surface to accommodate the mounting of the flywheel. For a four-cylinder engine, the big-end journals are paired and set at 180°. For most four-cylinder engines the firing order is 1,3,4,2.

Figure 6.84 Signs of 'nip'

Figure 6.82 Radial lip oil seal

The **crankshaft bearings** are split-type, steel-backed shells with an alloy or coated bearing surface. Bearing types correct to engine manufacturers original specification must always be fitted.

Bearing nip Bearing shell halves, when correctly fitted and tensioned in the bearing caps, form a perfectly round profile with equidistant clearance around the bearing journal. The bearings are 'nipped' and held in position when fitted into the tightened bearing caps. The bearing shell is also fitted with a locating lug on the back that mates with a slot in the bearing locating half bores. This ensures that the bearing cannot rotate.

Oil supply holes and slots are machined in the bearing surface to supply appropriate lubrication.

Thrust bearings on crankshaft to control axial movement Axial displacement of the crankshaft is controlled by thrust bearings to limit the axial movement. These are fitted at a main bearing journal either as two semicircular rings, or as part of a main bearing shell.

Cylinder head gasket This has to form a gas-tight seal at the interface between the cylinder head and the cylinder bores. In addition, it must seal and separate the cooling water supply jacket and the oil supply and return drillings. Traditional head gaskets were constructed from copper and asbestos. Modern material technologies allow head gaskets to be made from composite materials which have superior sealing and heat transfer performance.

Figure 6.83 Bearings

Figure 6.85 Details of the bearing

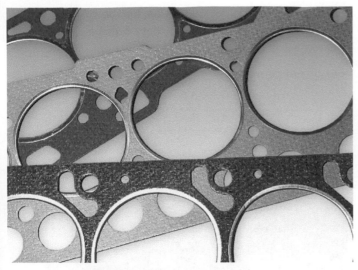

Figure 6.86 Gaskets

Head gaskets must always be replaced when the cylinder head is removed and refitted. Also, when refitting a cylinder head, it is important that the manufacturer's information is sought and applied with respect to replacement of cylinder head bolts where necessary, and correct torque and tightening sequences of cylinder head bolts.

The cylinder head is cast from aluminium alloy or cast iron. Aluminium alloy is lighter but cast iron or steel valve seats and guides must be installed in the head. Cast iron heads generally have valve seats and guides formed directly in the head material.

Figure 6.87 Aluminium head

Figure 6.88 Valve seats

The **combustion chamber** is formed in the cylinder head such that, on assembly, it is located directly over the cylinder bore in the engine block. There are numerous designs in use depending on engine type, optimisation parameters and application.

Figure 6.89 Inside the combustion chamber

Compound valve hemispherical This arrangement of combustion chamber and valves uses a hemispherical design with two valves per cylinder positioned opposite each other for crossflow movement of the intake and exhaust gases. The valves are inclined such that they sit in the curved profile of the combustion chamber space. The spark plug is mounted as close to centre as possible in the combustion chamber via an appropriate drilling. It is sealed via a compressible washer or conical sealing face.

Figure 6.90 Cross section of a cylinder head

Coolant passages The combustion chambers are surrounded by cooling water passages that are connected to the water jacket in the cylinder block. The water jacket casting holes are sealed via core plugs in a similar way to the cylinder block. On the upper surface of the cylinder head, bearing journal surfaces are formed to locate the valve operating camshafts and mechanism. Oil supply drillings ensure adequate lubrication for the camshaft bearings and valve train components.

Figure 6.91 Engine coolant passages

The **camshaft** is mounted in bearings formed in the cylinder head via an in-line boring process. The camshaft is forged from steel or cast iron and the bearings and cam surfaces are a smooth, machined finish. The camshaft has cam lobes for each valve and, to ensure the correct sequence of valve timing, the camshaft is timed and synchronised with the crankshaft position. To achieve this, the camshaft drive gear is secured to the camshaft via a keyway.

The camshaft rotates at exactly half engine speed and is marked to ensure that the correct position can be located easily when refitting.

Figure 6.92 Fitting the cam

Cam shapes The cam lobes have a specific profile that consists of a base circle and lobe to provide the correct valve opening and closing characteristics. The cam profile is not necessarily symmetric and the profile may allow progressive opening of the valve but with a sharp closing action depending on the characteristic and optimisation parameters of the engine.

Valve operating mechanisms The are various designs of mechanisms for following the cam profile and opening the valve. In this application, the engine uses pressed steel rockers to apply the force to open the valves and valve springs are used to close the valves.

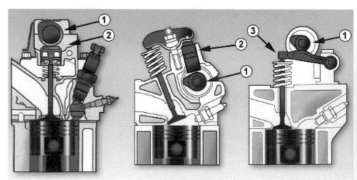

Figure 6.93 Different mechanisms

Remember! There is more support on the website that includes additional images and interactive features: www.tomdenton.org

Inlet and exhaust valves are poppet-type valves with a circular sealing face recessed in the cylinder head. The valves are located via the stem and slide inside valve guides mounted in the cylinder head. Valve heads are exposed to full, combustion chamber temperatures and pressures; the temperature of the exhaust valve can be as high as 800°C. The incoming gas charge has a cooling effect on the intake valve but, generally, heat dissipation from the valves is via the stem and guides to the cylinder head. Combustion and fuel deposits can cause problems on the valve; this can be avoided via the use of good-quality fuels and oils.

Inlet valve The total valve opening area is always greater for the inlet valves; this is to increase the volumetric efficiency of the engine due to the fact that the pressure difference across the inlet valve, when charging the cylinder, is much lower than the pressure difference across the exhaust valve when evacuating the cylinder. Hence a larger valve is needed to reduce restrictions to gas flow during the inlet stroke.

Figure 6.94 Valves

Valve seat angles Valves seats and the valve sealing face are cut at a slightly different angle. This is to ensure that a complete seal is made under working conditions as, when the valve is installed and at running temperatures, the valve head will deform slightly causing the sealing faces to meet correctly and seal efficiently. The angle of the sealing face is approximately 45°. The valves open via the force applied from the cam and valve gear and are held in the closed position via spring force. The springs are

connected to the valve via a retainer and split collets as this allows removal and refitting. In operation, the valve head rotates and this helps maintain the sealing face.

Figure 6.95 Seat angles

Camshaft drive belt (timing belt) For many engines, a toothed belt is used to drive the camshaft. The belt is manufactured from a durable, synthetic rubber with reinforcing fibres. The teeth moulded on the inside of the belt mate with the corresponding teeth on the crankshaft and camshaft pulley wheels.

Figure 6.96 Cam drive belt

Belts and pulleys The teeth formed on the belt can be trapezoidal or rounded. Note that they are not interchangeable and it is important to fit the correct type when replacing the belt.

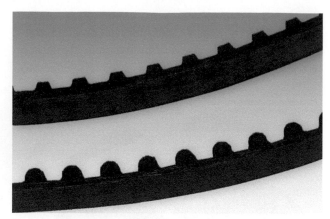

Figure 6.97 Two types of cam belt

Belt tension Correct tension of the timing belt is imperative for maximum belt life. The belt is generally tensioned by adjustable tensioner wheels. It is important to note that manufacturer and engine specific information must be sought when making adjustments in service.

Direction of rotation Often, the belt direction of rotation is marked on the belt itself, and it must be refitted in the same direction. If the belt is not marked, then the direction should be noted and marked before removal.

Covers, cases and sumps Once the main engine components have been assembled, a number of covers are fitted to enclose moving parts and retain oil. The sump is fitted on the underside and holds the oil capacity.

Rocker cover The cover that encloses the valve gear is known as the rocker or cam cover. Generally, it incorporates the oil filler cap and part of the engine breather system.

Crankshaft seals At the front of the engine, there is a casing that retains the crankshaft oil seal. In addition, a cover is fitted to enclose and protect the camshaft drive belt. At the rear of the engine, another crankshaft oil seal is fitted to a casing that is bolted to the engine and located by dowels.

Sump Note that the sump contains baffles to prevent excessive oil movement or surge. This maintains a good supply of oil around the area of the oil pick-up.

Auxiliary components In addition, attached around the engine are the components and subsystems for lubrication, cooling, ignition, fuel, air, exhaust and electrical systems.

 Construct a crossword or wordsearch puzzle using important words from this section

6.1.5 Cylinder components ❷

Cylinder liners fall into two main categories, wet and dry. Wet liners are installed such that they are in direct contact with the coolant fluid. They are fitted into the block with seals at the top and bottom and are clamped into position by the cylinder head. Spacers are fitted at the bottom to adjust the protrusion of the liner to achieve the correct clamping force.

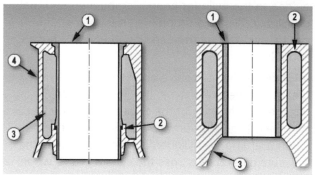

Figure 6.98 Cylinder liners

Dry liners are not in direct contact with the coolant. Generally, they are fitted into the casting mould and retained by shrinkage of the casting via an interference fit. Alternatively, they can be pressed into place in a pre-cast cylinder block. When repairing or reconditioning the engine, the former type can be re-bored whereas the latter type is replaceable.

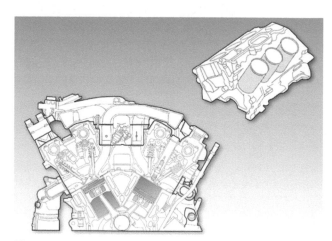

Figure 6.99 Dry liners are not in contact with the coolant

Replaceable liners Most modern engines have specific treatments applied to the cylinder bores and as such, cannot be rebored or honed. Replaceable liners mean that the liner and piston assembly can be easily replaced without the need for specialist reboring equipment. Often, commercial vehicle engines utilise replaceable liners to reduce repair times.

Figure 6.100 Liners

Separate crankcase and block For certain engine applications, separate crankcase and cylinder blocks are utilised. This is not often seen for light vehicles but is quite common on larger engines for commercial vehicles. The cylinder block can be arranged for multiple cylinders or a single block assembly per cylinder.

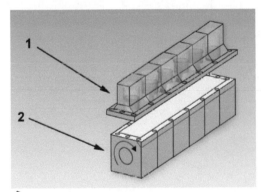

 Figure 6.101 Separate components

1 State the difference between wet and dry cylinder liners.
2 State an advantage of each type.

> Using images and text, create a short presentation to show how a component or system works

6.1.6 Valves and valve gear ❷

Overhead valve (OHV) layouts This term is used to describe the evolution of engines from side to overhead valve. It is used to describe the latter, where the valves are located in the cylinder head. The overhead valve engine valve gear is more complex as the valve motion had to be transferred to valves that are facing in the opposite direction when compared to a side valve engine.

OHV valve gear OHV valve gear transfers reciprocating motion from the cam followers and camshaft to the valves via push rods and a rocker assembly which acts on the valve stems.

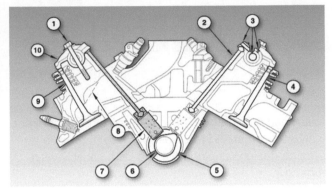

Figure 6.102 Overhead valves (OHV)

Valve clearances Adjustment of the clearance between the rocker and valve stem contact face allows for thermal expansion and correct lubrication. The adjustment is effected via screw adjusters and lock nuts. Hydraulic lifters can be used in place of cam followers to provide self-adjustment of the clearance.

Overhead cam (OHC) layout Overhead cam refers to the position of the camshaft in the engine – that is, it is positioned in the cylinder head. There are a number of designs using direct or indirect mechanisms to convert the rotating cam motion into a reciprocating motion, and then transferring this motion to the valve stems. These designs facilitate a close tolerance in operating clearances.

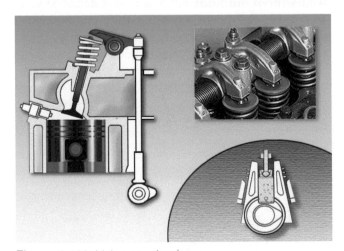

Figure 6.103 Valve mechanisms

Double overhead camshaft engine (DOHC) A further development is the use of twin or double

111

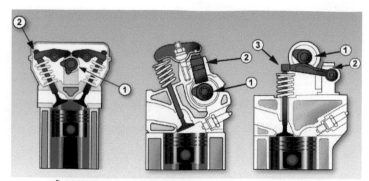

Figure 6.104 Overhead camshaft (OHC) valve operation

overhead camshafts. These can use direct or indirect valve actuation and are well suited to multivalve engine designs, including those with variable valve timing.

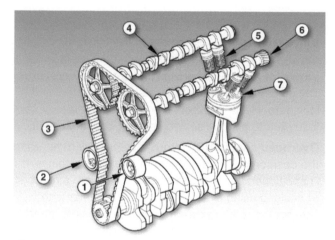

Figure 6.105 Double overhead camshaft engine (DOHC)

Adjustment methods For direct valve actuation with inverted bucket-type followers, adjustment is provided via shims or screw wedge-type adjusters. In addition, hydraulic 'self-adjusting' inverted bucket followers can be employed that require no manual adjustment task.

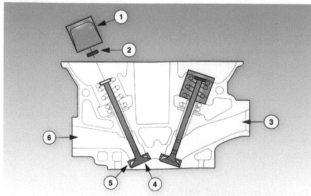

Figure 6.106 Valve components

Rocker arm systems Indirect, rocker-arm-type valve actuators incorporate close tolerance adjusters. There are two systems commonly seen – a rocker shaft and pivot stud or a rocker arm supported on a pedestal at one end and the valve stem at the other, the cam acting between these two points. A hydraulic pedestal can be employed for self-adjustment of the mechanism.

Hydraulic valve adjustment Figure 6.106 shows a typical engine oil lubrication circuit that feeds the self-adjusting followers with pressurised oil to maintain the correct valve clearances. Always refer to manufacturer's data for the service requirements of the valve train system. Often, special procedures are required when replacing and recommissioning self-adjusting valve mechanisms – these must be followed to prevent engine damage.

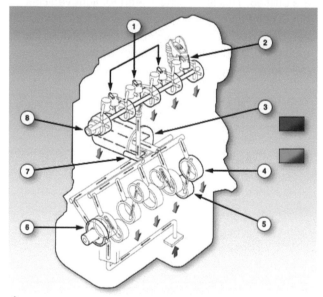

Figure 6.107 Hydraulic components

Belt-in-oil (BIO) or wet cambelts Manufacturers started using timing belts instead of chains or gears in the 1970s. Oil contamination of the belt was often a cause of failure because oil attached the rubber content. The belt would lose its teeth or snap. There are now some belts that are designed to run in oil and are known helpfully as 'belt-in-oil (BIO)'!

These BIO systems are used for oil pump drives as well as for camshaft timing drives. Belts in general have three key advantages:

▶ Absorbs and isolates crankshaft harmonics from the valvetrain
▶ Quieter in use
▶ Reduced power consumption (less friction and noise).

Figure 6.108 Belt-in-oil (Source: Robert Chalmers)

Figure 6.109 Consequences of not servicing can be very serious (Source: Donald Anthony)

Advantages BIO technology can offer up to 30% reductions in friction loss as compared to chains or dry belts. The result is lower emissions and improved fuel economy by about 1%

Improved BIO materials Wet belts use improved materials that are more temperature-resistant, less prone to stretching than conventional dry belts and have a life expectancy of up to 150,000 miles. The first automotive BIO system was introduced in 2008, in a 1.8L Ford diesel. Other manufacturers such as Volkswagen now use this technology.

Servicing There is nothing new in this advice, regular servicing with the correct materials is essential. When these belts do break or shred, other areas of the engine are seriously affected – the oil pump for example. Always refer to manufacturers recommendations.

Variable cam timing (VCT) Many engines now employ variable camshaft timing to optimise the inlet valve timing with respect to engine speed and load conditions.

Airflow oscillations in the inlet manifold As air enters the engine through the inlet manifold this forms a column of moving air that possesses kinetic energy. The pulsating nature of the engine's air consumption creates pressure waves in this air column. The energy

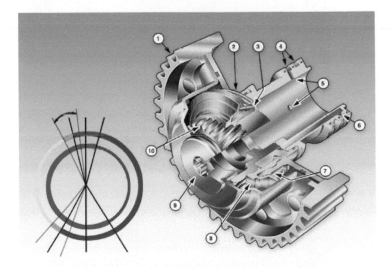

Figure 6.110 Timing is changed under electronic control

in these pressure waves can be harnessed to assist in charging the cylinder, increasing the volumetric efficiency of the engine. In order to do this, the valve opening point must be optimised according to the engine condition, and with variable valve timing this can be achieved to increase engine torque and power at various points in the operating speed range.

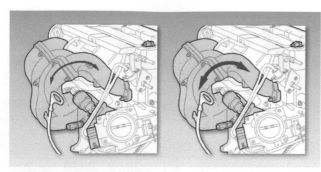

Figure 6.111 Inlet manifolds

Methods of variable cam timing (VCT) There are various technologies available to provide the required phase angle between the cam drive and the camshaft for variable valve timing. It can be generated via a hydraulic mechanism in the cam wheel that is controlled via a valve assembly from the engine ECU. Cam wheel actuators can employ 'helix' or pressure differential actuation principles. In addition, some engines have employed valve mechanisms with alternative cam profiles where the engine switches over to a different cam lift profile at certain engine speeds.

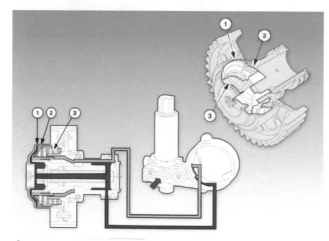

Figure 6.112 Electro hydraulic control method

 Make a simple sketch to show how one of the main components or systems in this section operates

6.1.7 Engine designs 2 3

Technical details of the four-cylinder, four-stroke OHV diesel engine This section compares the technical details of the four-stroke overhead valve diesel engine with the four-cylinder overhead camshaft engine described and studied previously.

Direct and indirect injection Diesel engines are generally categorized by the position of fuel introduction into the cylinder, either directly into the combustion chamber (direct) or into a pre-combustion chamber (indirect).

Indirect injection engine Referring to the picture, note the position of the fuel injector relative to the combustion chamber, and note the valve operating mechanism. These are the most obvious technical differences to the OHC-type engine.

Cylinder block and crankcase In the engine shown in the diagram, the cylinder block is cast iron which is structurally stronger and stiffer than an equivalent petrol engine block. The reason for this is the greater forces and pressures that occur inside a diesel engine because of the different combustion process. Note that the cylinders and crankshaft bearings are machined (bored) in a similar way to the petrol engine.

Figure 6.113 Block

Camshaft bearings In this engine design, the camshaft is housed within the cylinder block running in five ring bearings at the side of the block. Above the camshaft, bores are machined to retain the cam followers in the correct position.

Figure 6.114 Cam

The **piston** in a diesel engine is more strongly constructed than a similar displacement petrol engine piston. The construction material is silicon alloy aluminium and this provides the necessary strength and durability. The increased mass of the diesel engine piston, compared to a petrol engine piston, is a factor that limits the maximum engine speed that can be achieved. If the engine is turbocharged, further strengthening of the piston is necessary due to the increased cylinder pressures that occur.

Bowl in pistons Direct injected diesel engines have a bowl shape in the piston crown that forms the combustion chamber volume. This chamber is designed so that it creates the correct amount of swirl and turbulence in the air charge to ensure complete and efficient burning of the injected fuel.

The **piston rings** perform the same function as in petrol engines – they are similar in design and construction and both compression and oil control rings are generally used.

Figure 6.115 Followers

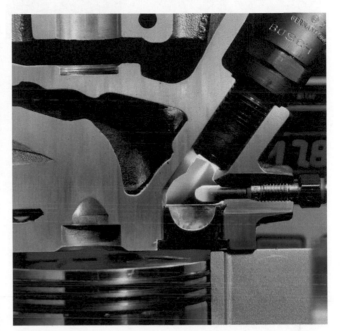

Figure 6.117 Chambers create swirl

Figure 6.116 Typical piston

Pre-combustion chamber Indirect injected diesel engines have a pre-combustion or swirl chamber in the cylinder head as part of the combustion chamber volume. This is carefully designed to give the correct air motion during the compression and expansion process. The piston crown is also designed to promote correct air movement into the pre-combustion chamber during the compression process.

Figure 6.118 Air swirl continues

115

Figure 6.119 Piston pin (gudgeon pin)

Piston pin (gudgeon pin) The piston pin is normally more heavily constructed to withstand the greater forces in the diesel engine.

Connecting rod Similar in design and function to the petrol engine, though built stronger due to the increased loading that occurs in the diesel engine. Cooling of the piston crown via oil spray on the underside is common and this is provided by an appropriate drilling in the connecting rod.

Spray nozzle An alternative to provide piston cooling is a spray nozzle, mounted in the crankcase, that provides a continuous delivery of oil to the piston underside.

Crankshaft The diesel engine crankshaft is a stronger version of that typically found in a petrol engine. For a turbocharged diesel engine, larger bearings will be used for the big end and mains due to increased forces and loading.

Figure 6.120 Crankshaft

Crankshaft bearings are split shell design, as with a petrol engine, but the bearings are manufactured from a harder, tougher alloy to withstand the increased loads. Note that bearing compounds will be specific

to the engine type – petrol, diesel or turbocharged diesel – and they are not interchangeable.

Cylinder head gasket This is manufactured appropriately, to suit the application of a diesel engine, with higher peak pressures than a similar size petrol engine. Often the number of cylinder head bolts will be greater than a petrol engine to provide the required clamping force.

Cylinder head The main constructional differences between the OHV indirect injection diesel engine and the OHC petrol engine are that pre-combustion chambers are fitted in the head, and gas expansion due to combustion takes place in the space above the piston crown against the flat face of the head. The valves are flush at the flat cylinder head surface, directly above the pistons. The valves are arranged in line with the engine axis and inlet and exhaust ports are crossflow, that is, on opposite sides of the cylinder head.

Figure 6.121 Crank bearings

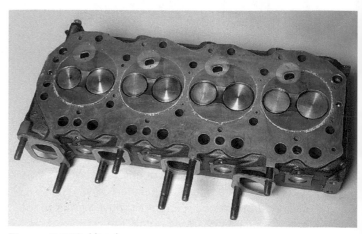

Figure 6.122 Head

Valve positions The valves are arranged in alternate positions along the cylinder head, that is, inlet – exhaust – inlet – exhaust, etc. The ports are cast and machined in the head in a parallel configuration.

Overhead valve engine (OHV) This engine uses an overhead valve type valve train mechanism. The valve gear cam lobes actuate the valves via cam followers, push rods and rockers mounted on a rocker shaft. Lubrication is supplied to the rockers via drillings in the rocker shaft pillars.

The **rocker shaft** mounting pillars are attached directly to the cylinder head. Passages formed in the cylinder head casting allow oil circulation and means of access for the push rods between the rockers and the cam followers.

The **camshaft** is driven at half engine speed by the valve train drive gears, therefore, the camshaft driven gearwheel has twice as many teeth as the crankshaft drive gearwheel.

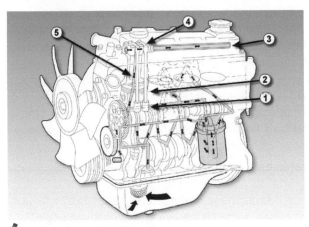

Figure 6.123 OHV engine

Figure 6.124 Cam gears

Cams The camshaft cam profiles are similar to the petrol engine and they will, of course, differ in lift and timing profiles to suit the specific engine.

Figure 6.125 Camshaft

Valve operating mechanism The camshaft is located at the side of the engine block and the valve operating mechanism has to transfer the reciprocating motion from the followers up to the rockers located in the head. This motion is transferred via the push rods. Correct adjustment of the working clearances in this mechanism is essential to allow for thermal expansion and lubrication clearances.

Figure 6.126 Rocker shaft

Oil feeds The rocker shaft and the rocker fulcrum bearings are lubricated by a direct feed from the oil gallery via drillings in the block, head and rocker shaft pillar and then through the rocker shaft which is hollow and has holes beneath each rocker. The ends of the rocker shaft are plugged.

Inlet and exhaust valves These are similar in design and form to petrol engine valves but the material composition is different.

Diesel engine covers and sump Additional components attached to the engine are the rocker cover, timing gear cover and oil pan, as well as components for the lubricating, cooling, air and exhaust systems, fuel, starting and charging systems. These units are covered in the respective learning programmes.

6

117

Technical details of a V6-cylinder four-stroke OHC petrol engine This engine is one of the latest generation of petrol engines. It has been designed to meet or exceed the latest emission regulations and it is equipped with a sophisticated engine electronic control system. Part of the legislated requirement is that this engine meets the regulations for a considerable period of its working life. In order to achieve this, replaced components and consumable parts must conform to manufacturers' specifications.

Cylinder block and crankcase The crankcase is designed using computer-aided modelling techniques in order to have the required strength but with minimum weight. Often, modern engines have a ladder or split crankcase that forms the lower half of the main bearing arrangement. Similar engines have cylinder head bolts that extend through the block and into the crankcase to provide the required load distribution within the casting. Dry liners of steel or cast iron are used within the cylinders of the aluminium alloy block. Note that manufacturing tolerances are so close that reboring is not possible.

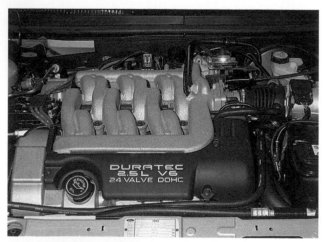

Figure 6.127 V6 engine

Pistons and piston rings Piston dimensions are closely matched to the cylinder bore, and the material used to manufacture the piston is generally aluminium with silicon and other metallic elements for durability and light weight. Transverse slits in the piston ensure the correct behaviour with respect to thermal expansion. The areas of the piston that are subjected to the highest load are reinforced with cast iron components. Note that these pistons are more compact than previous generation engine designs and this contributes to the improved performance of the engine. A specially shaped crown provides the appropriate charge motion to ensure rapid and complete combustion. Optimised piston ring dimensions and Teflon coating of the skirt reduces engine internal friction. Normally, there are two compression and one oil control rings on a piston in an engine of this type.

Connecting rod and piston pin The connecting rods are forged but new material technology is utilised in modern engines. These are sintered alloys made from heated and compressed powdered metals. This material is lightweight and strong. The big end is laser cut and fractured, known as 'cracked' type – this means that each bearing cap is unique to the connecting rod. A plain bearing bush forms the little end for a floating piston pin that is secured into the piston with circlips.

Crankshaft and bearings This crankshaft has offset big-end journals due to the fact that it has two banks of cylinders. The journal pairs are set at 120° intervals and offset to match the cylinder bank's vee angle. The firing order reflects the best balance for the cylinders, firing on each bank alternately from front to rear–1-4-2-5-3-6.

Figure 6.128 Crankcase

Figure 6.129 Crank

Crankshaft construction Crankshaft design has improved due to optimised design processes using computer aided design and simulation. In addition, improved material technologies mean that the life expectancy of a modern crankshaft is such that traditional regrinding and repair methods are not appropriate and, often, cannot be used due to the advanced material technology employed in the construction of the crankshaft.

Replacement of worn parts Often, modern engines require replacement of worn parts by a completely assembled block or short engine for crankshaft-, cylinder- and piston-related defects.

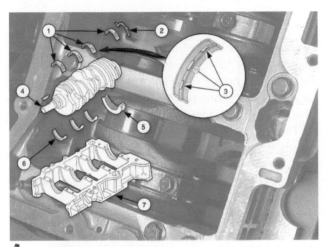

Figure 6.130 Crank details

Figure 6.131 Engine on a stand

Cylinder head gasket Improved engineering design and materials allows the production of mating engine parts with accurate joining surfaces. The need for a compliant, compressible head gasket is reduced and many modern engines use a thin, multilayered steel backed head gasket, or a single-skin metal gasket.

These have raised rings around the cylinder bores and other passages between the head and block to provide effective sealing.

Cylinder heads This engine's cylinder heads use a four-valve combustion chamber of pent roof design. The head is manufactured from a similar aluminium silicon alloy to the engine block. It is cast with passages for coolant and oil, inlet and exhaust ports, and mountings to support the two camshafts.

Vee configuration Two separate cylinder heads are generally required for a vee engine configuration. The inlet manifold is located in the 'valley' between the heads. Two exhaust manifolds are fitted, one for each bank, with one on each side of the engine.

Figure 6.132 Cam mountings

Figure 6.133 Vee layout

Hardened valve seat inserts The engine combustion chambers are fitted with hardened valve seat inserts. This is due to the fact that the aluminium alloy used in the construction of the head is not durable enough to withstand the reciprocating motion of the valve stem.

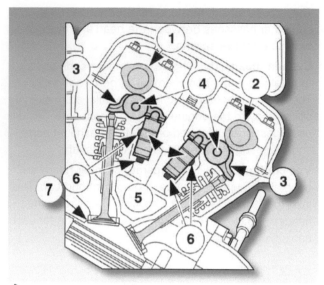

Figure 6.134 Valve details

Camshaft and valve operating mechanism This engine employs two camshafts per bank, and rocker arms transmit the required motion to the valves. There are other engine designs employing different arrangements for the valve train. These can use a single camshaft and rockers, or two camshafts with direct acting bucket tappets.

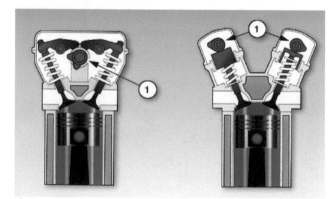

Figure 6.135 Valve mechanisms

Composite camshafts The camshafts are mounted directly into the cylinder head casting on aluminium bearings and they are retained with bearing caps. A thrust bearing, located at the sprocket end of each camshaft, controls and limits end float clearance. Composite camshafts use a hollow steel shaft with shrink-fit cam lobes and these are preferred to the conventional single-piece castings. The advantage of the composite shaft is that it has reduced weight when compared to the conventional cast type. The low weight of the camshaft reduces mass and inertia and the low frictional forces of the roller rocker arms improve the engine's operating characteristics.

Figure 6.136 Cam drive

Valve mechanisms Hydraulically operated pedestals support roller rocker arms at one end; the other end acts directly on the valve stem to open the valves against valve spring tension. The cam lobe runs above the centre of the rocker arm and acts directly upon it. Pressurised oil from the engine oil supply circuit feeds the hydraulic pedestals and these automatically provide the correct valve clearance. The valves are designed to rotate in the valve seat and guide and this extends valve and valve seat life. The rotation of the valve is achieved by a slight offset or taper on the tappet or rocker face in contact with the cam or valve. In some cases the engine is fitted with additional valve rotators. These are integral with the spring retainer and are in two parts with opposing angled faces and rollers to provide a rotational force as the valve is actuated.

Figure 6.137 Hydraulic lifters

Inlet and exhaust valves The physical design of the valves is conventional, while internally the valve stem is hollow and contains sodium, which melts when the valve is hot and assists in transferring heat from the valve face to the valve stem for dissipation via the valve guides. Cooling of the valve head can be improved significantly with a temperature reduction of up to 100°C. These valves are also known as bimetal type. Note that a sodium-filled valve of this type must be handled with care as sodium exposed to air is flammable and will self-ignite.

Camshaft drive This engine uses cam chains to drive the valve train, one chain per cylinder bank. The camshaft and sprocket form a single unit and there are two per bank, inlet and exhaust. Note that one camshaft also carries a torsional damper. Hydraulic tensioners provide automatic preloading of the chain tension.

Figure 6.140 Tensioner

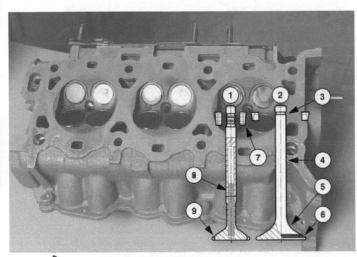

Figure 6.138 Valve details

Figure 6.139 Drive

> **Remember!** There is more support on the website that includes additional images and interactive features: www.tomdenton.org

> 🎨 Make a simple sketch to show how one of the main components or systems in this section operates

6.1.8 Cylinder deactivation ❸

Introduction The purpose of cylinder deactivation technology is to reduce fuel consumption on petrol/gasoline engines. Large-capacity engines were renowned for poor fuel consumption under low speed and torque conditions. This inefficiency is known as pumping loss and is used to describe a situation where an engine is unable to draw in a sufficient quantity of air and fuel mixture on the inlet stroke to produce high cylinder pressure during compression.

Figure 6.141 Honda engine from a hybrid car

Valve control Low pressure in an engine cylinder results in low efficiency because more fuel is required to compensate for the lack of pressure and the ability to draw fuel into the cylinder. To deactivate a cylinder of a running engine whilst retaining balance, driveability and emission targets is a challenge. The best method of deactivation therefore, is to disconnect the valve operating mechanism. If the inlet and exhaust valves of the deactivated cylinder are closed, no gases can enter or exit. This also prevents an increase in exhaust emissions. Fuel injection to the deactivated cylinder is also switched off.

Volkswagen now use a cylinder deactivation technology on smaller engines (1.4 litre, for example). Their method employs a system of cam lobes that are splined to the camshafts. In the normal working position the cam lobes are aligned with the valves. To deactivate the cylinder, a pin is engaged with a scroll that moves the cams out of alignment with the rockers and prevents the valve from opening. The pin is solenoid operated. In this state the valve rockers run on a concentric part of the shaft. After the sliding component has moved the pin is retracted. To re-activate the valves a second pin is engaged with the scroll and the cams move back to their normal position. The switchover process occurs during half of a camshaft rotation. Bentley use a similar system to Volkswagen on their somewhat larger four-litre V8 engine.

Figure 6.142 Valves operating normally (left), valve closed (right) and cylinder deactivated (Source: Volkswagen Media)

Mode of operation The VW active cylinder management (ACT) system shuts down the second and third cylinders during low and medium loads and therefore reduces fuel consumption. It is active over an engine speed range of 1 400 to 4 000 rpm, a torque output range of about 25 to 100 Nm. When the driver demands acceleration, this is detected by the pedal sensor and both cylinders are re-activated without any noticeable change. The system has no detrimental effects on the smooth running of the engine. Changes are made to ignition timing and the throttle valve position to ensure a smooth transition. If the system detects irregular driving, cylinder shutoff is deactivated. This video on the website from Volkswagen shows the operation of the system components.

Honda use a different deactivation system. This employs rockers that are connected and disconnected by hydraulically operated locking pins. One rocker always remains in contact with the profile of the cam and, when connected by the pin, operates a second rocker that in turn operates the valve. When deactivation is required, oil pressure, controlled by a solenoid valve, forces the locking pin out of engagement with the second rocker. This prevents the valve from opening. This technology has been used in hybrid models since the year 2000.

Figure 6.143 Honda cylinder deactivation mechanism

Bosch The Bosch cylinder deactivation system can facilitate the deactivation of not required cylinders within part-load operations in almost every gasoline engine. This saves fuel and thus reduces CO_2 emissions. The Bosch system shown in this video deactivates the hydraulic valve lifters to prevent the valves opening.

Summary This section has outlined three methods of cylinder deactivation:

▶ locking pins and a second rocker
▶ splined, moveable cams
▶ the deactivation of hydraulic lifters.

It is quite likely that more methods will be introduced because the system has great potential for reducing consumption and harmful emissions.

 Use the media search tools to look for pictures and videos relating to the subject in this section

6.1.9 Water injection

Introduction Even advanced petrol/gasoline engines waste roughly a fifth of their fuel. This is mostly at high engine speeds, where some of the fuel is used for cooling instead of for propulsion. Water injection means that it does not have to be that way. Particularly when accelerating quickly or driving on the motorway, the injection of additional water makes it possible to reduce fuel consumption by up to 13%.

Economy The fuel economy offered by this Bosch technology comes especially to the fore in three- and four-cylinder downsized engines: in other words, in precisely the kind of engines to be found in any average midsize car. This technology can make cars more powerful as well. Water injection can deliver an extra kick in any turbocharged engine. This is because earlier ignition angles mean that the engine is operated even more efficiently. On this basis, engineers can coax additional power out of the engine, even in powerful sports cars.

Figure 6.144 Water injection can produce extra boost for the turbocharged engine (Source: Bosch Media)

Overheating The basis of this innovative engine technology is a simple fact: an engine must not be allowed to overheat. To stop this happening, additional fuel is injected into nearly every SI engine on today's roads. This fuel evaporates, cooling parts of the engine block. With water injection, Bosch engineers have exploited the same physical principle. Before the fuel ignites, a fine mist of water is injected into the intake duct. Water's high heat of vaporization means that it provides effective cooling.

Water usage For every 100 kilometres driven, only a few hundred millilitres are necessary. As a result, the compact water tank that supplies the injection system with distilled water only has to be refilled every few thousand kilometres. If the tank should run empty, the engine will

still run smoothly, albeit without the higher torque and lower consumption provided by water injection.

 Use a library or the web search tools to further examine the subject in this section

6.1.10 Miller cycle

Introduction A traditional reciprocating internal combustion engine uses four strokes (or the Otto-cycle), of which two can be considered high-power: the compression stroke (high power flow from crankshaft to the charge) and power stroke (high power flow from the combustion gases to crankshaft). In the Miller-cycle, the intake valve is left open longer than it would be in a normal four-stroke engine. This effectively turns the compression stroke into two distinct cycles:

1 First part, when the intake valve is open
2 Second part, when the intake valve is closed.

Fifth stroke This two-stage intake stroke creates the so-called 'fifth' stroke. As the piston initially moves upwards (20-30%) in what is traditionally the compression stroke, the charge is partially expelled back out through the still-open intake valve. This loss of charge air would normally result in a loss of power. However, in the Miller cycle, this is compensated for by the use of a positive displacement supercharger (Roots or screw) as these work at low engine speed. The final 70–80% of the compression stroke (after the inlet valve has closed) is where the mixture is further compressed.

Efficiency is increased by having the same effective compression ratio but a larger expansion ratio. This allows more work to be extracted from the expanding gases as they are expanded to almost atmospheric pressure. Audi's A4 was the first model to receive a new 2.0 litre engine that starts to bring spark-ignition technology much closer to the fuel consumption and torque of compression-ignition engines. It does so via technology that is mostly on the inlet side of the cylinder. It is described as being 'comparable to the Miller cycle'.

Intake timing Focusing on the inlet side of the engine, intake timing has been greatly reduced by adopting a 140° crank angle compared to typical 190–200° timing. As mentioned above, Miller-cycle engines traditionally employ a Roots or screw-type positive-displacement supercharger, but Audi is using a turbo. The engine gets a higher boost pressure on the inlet side to provide optimal cylinder charges, despite shorter intake timing. To help achieve further performance targets, in part-load range there is an additional injection upstream from the intake valve.

6

This results in an efficient mixture formation that is already complemented by the direct injection in the combustion chamber as well as in the intake manifold. The Audi Valvelift System ensures a short intake time at part-load and a longer at higher loads.

Figure 6.145 Audi engine using features of the Miller cycle

Summary Other features of the Audi engine include controlled coolant flow control to reduce engine warm-up time. Another interesting feature is that the exhaust manifold is integrated in the cylinder head. Reduction of friction, as well as the use of low friction engine oil (OW-20), also contributes to enhanced engine efficiency. Maximum torque of the engine is claimed to be 320 Nm (236 lb ft), available from 1450 rpm to 4400 rpm.

 Using images and text, create a short presentation to show how a component or system works

6.1.11 Variable compression (VC) engine

Introduction Nissan has developed a variable-compression-ratio gasoline engine. The VC-Turbo, a 2.0-L inline four-cylinder house, features an ingenious crank train and control system that enables the effective compression ratio to be varied between 8:1 and 14:1, depending on load. Typically, VCR engines alter the compression ratio by raising or lowering the height of the piston at TDC, but the Infiniti engine achieves this in a different way. An electric motor drives a reduction gear, which moves an angled actuator arm. The arm in turn rotates a control shaft

with four aligned eccentric cams, one for each cylinder. An intermediate link with bearings at each end connects the eccentric cam at the bottom end to the multi-link at the top end. The centre of the multi-link runs in a bearing around the crankshaft journal.

Figure 6.146 Components of the Nissan (Infiniti) variable compression turbo (VC-T) engine (Source: Nissan)

Multilink A second bearing on the multilink, positioned 180° from that connecting the intermediate link effectively serves as the piston connecting rod big-end bearing. This arrangement produces a 17° offset of the con rod from the crankshaft journal centre point. The Harmonic Drive is controlled by a dedicated ECU which gathers data from engine sensors to determine the compression ratio required for the current load. Since a low compression ratio is desirable when power is required and a high compression ratio when efficiency is preferable, the piston height at TDC can be continuously varied as required by rotating the Harmonic Drive, which will determine the position of the multilink and hence the height of the piston in the cylinder bore.

Costs As with other VCR engines, complexity, mass and cost are greater than in a conventional 4-cylinder gasoline engine. On this system, there are three conrods for example. The motion described by the piston con rod big end is not circular as in a conventional engine's big-end bearing, it is more elliptical with the con rod not passing through the vertical axis between the big and small end bearings.

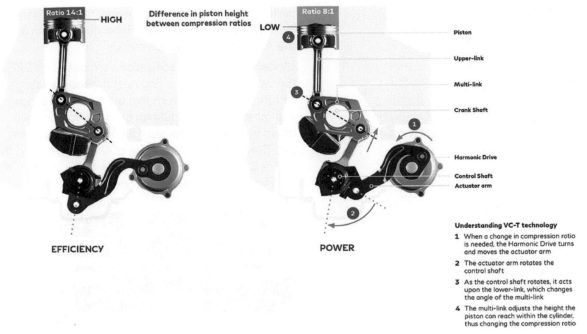

Understanding VC-T technology

1 When a change in compression ratio is needed, the Harmonic Drive turns and moves the actuator arm

2 The actuator arm rotates the control shaft

3 As the control shaft rotates, it acts upon the lower-link, which changes the angle of the multi-link

4 The multi-link adjusts the height the piston can reach within the cylinder, thus changing the compression ratio

Figure 6.147 VC-T operation at high and low compression ratios (Source: Nissan)

During the power stroke, the con rod remains almost vertical. This reduces the side force on the piston and helps to reduce the vibration.

Injection systems The development team addressed the fuelling needs generated by the range of compression ratios by fitting both multipoint injection (MPI) for low compression and direct injection (GDI) for high-compression operation. Since GDI engines inherently generate high particulate emissions, the continual phasing between GDI and MPI helps to contain particulate emissions. Both sets of injectors are brought into use under high load and engine speed conditions.

Variable valve timing (VVT) Both the inlet and exhaust camshafts are fitted with variable valve timing, electronically controlled on the inlet side and hydraulically actuated on the exhaust side. Forced induction is provided by a single scroll turbocharger, equipped with an electronic wastegate actuator.

 Select a routine from section 1.3 and follow the process to study a component or system

6.2 Lubrication

6.2.1 Friction and lubrication ❶

All types of vehicle engines incorporate metal parts that have to rub against each other, thus causing friction which creates stress, wear and heat, e.g.

cylinders in cylinder liners. These parts require lubrication to prevent the wear, keep the surfaces clean, and help to remove the heat.

Lubrication is achieved by separating the metal surfaces with a film of oil or grease. Thus the lubricating oil in the engine has traditionally been seen to have these three functions: separation, cleaning and cooling.

Oil circulates throughout the parts of an engine under pressure produced by a mechanical pump.

Another important property of a lubricant is oiliness, which can be described as the ability to adhere to the surface of materials and maintain separation of the rubbing surfaces without breaking down. This type of lubrication is called boundary lubrication and occurs in all engines during starting and before the pumped oil feed is established.

Environmental regulations Modern engines must conform to environmental regulations and modern engine oils are an important component in helping to achieve this.

Engine oil producers are responding with new blends, additives and synthetic oils to enable extended service intervals, improved wear protection, greater engine cleanliness, sludge inhibition, higher speeds and temperatures and lower oil consumption. These oils also contribute to improved performance, economy, and environmental concerns about hydrocarbon emissions into the atmosphere. They are compatible with oil-seal materials so that leakage is reduced. They also have strict limits on volatility so that vapours do not escape into the atmosphere.

Figure 6.148 Con rod big end lubrication

Use a library or the web search tools to further examine the subject in this section

6.2.2 Oils and specifications ❶

Most engine oils are refined from crude oil to which is added viscosity index enhancers, reduced-friction enhancers, anti-oxygenates, sludge, lacquer and corrosion inhibitors, and cleaning agents for carbon, acids and water.

Early specifications for engine oils defined just the physical data. New oils, which have to meet environmental and engine-performance requirements, are given specification code letters to indicate the performance level.

Synthetic and semi-synthetic oils have improved performance for environmental or special purposes.

Multigrade oils have been developed in order to modify the viscosity index and give thin oils at low temperatures that do not become excessively thin at higher temperatures.

Viscosity is a measure of an oil's resistance to flow, i.e. if thin, the oil will flow more easily than thicker oil. A viscosity index is the measure of a change in an oil's flow rate with a rise in temperature. The higher the viscosity index, the smaller the change in viscosity.

Manufacturer's recommended viscosity ratings generally reflect the lowest temperature at which the vehicle is being used and may be different for summer and winter use. The viscosity rating is not an indicator of oil quality but of oil flow under particular conditions. There are some low-grade oils that carry recommendations that limit the use of the vehicle,

particularly for high engine speeds, loads and long journeys. Good-quality oils will be labelled with at least the API and ACEA service ratings.

Modern engine oil specifications are based on SAE (Society of Automotive Engineers) viscosity ratings, API (American Petroleum Institute) service ratings and other properties defined by classifications laid down by organisations such as ACEA (*Association des Constructeurs Européans d'Automobiles*) and the earlier CCMC (*Comité des Constructeurs d'Automobile du Marché Commun*) for European vehicles.

Figure 6.149 Common specifications (SAE is now almost universal)

The API service rating classification is based on oil-performance characteristics and consists of two letters. The first letter is either 'S' for spark ignition or petrol engines, or 'C' for compression ignition or diesel engines. Originally 'S' stood for 'Service' as in Service Station and 'C' for Commercial Vehicles. The increase in diesel-engine usage in light vehicles has brought about the change of meaning. The second letter in the classification denotes the service specification, which has been updated at significant intervals and reflects the greater performance requirements of newer types of engines.

Oil-grade classifications The lowest grade of oil is SC/CC, which was suitable for engines produced during the 1960s. As the manufacturing and environmental demands have developed during recent years, improved oil performance has followed. SD and SE classifications cover the 1970s' and SF and SG cover the 1980s. This development will continue with the introduction of newer classifications. As a general rule, a later classification can be used in place of an earlier type, but not the other way round.

Diesel engine oils Diesel engine requirements are not exactly the same as those of petrol engines. Separate diesel engine oils are formulated and marketed and should be used in preference to petrol-engine oils that carry a C classification. Development of C-class oils has been slower than the S-class. Turbocharging of diesel engines is now common and these must use the appropriate grade of oil. Recent grades are CD and CE.

The ACEA classifications are divided into three groups. Group A covers petrol engines, group B covers passenger car diesel engines and Group C covers commercial vehicle diesel engines. The development of these classifications was carried out to meet the needs of European vehicles, which have different characteristics than the American engines that are used to set the API standards.

Recommended oil grades Most engine and vehicle manufacturers list the SAE, API and other classifications for engine oil for their vehicles. They frequently list oil-producer preferences, which give an indication of the co-operation that has been given by the oil producer in the design and development of the engine. Some manufacturers produce their own oils formulated specifically for their vehicles.

Engine oils are not normally biodegradable and should not be allowed to enter the environment either as vapour or liquid. Total loss lubrication systems used on small two-stroke engines, such as those on motorbikes and outboard motors, use a 'petroil' mixture of petrol and a specially formulated biodegradable oil. Other types of oil should not be used.

Figure 6.150 Two-stroke oil

 Use a library or the web search tools to further examine the subject in this section

6.2.3 Lubrication system operation ❶

Oil flow A good flow of oil from the pump provides a forced feed into the shaft bearings. The large quantity of oil in the bearing generates an oil wedge that maintains separation under the severe conditions of the combustion stroke.

The **forced-feed system** is efficient for the removal of heat and for cleaning by carrying dirt to the filter.

Drillings and oilways Oil is fed through the engine via drillings and oilways, and returns to the sump to carry

heat away. The oil pump takes the oil from the sump and feeds it through the filter where it is cleaned.

Component lubrication Other components are lubricated by splash from jets of oil, or by the flow of oil from the top of the engine back to the sump.

Full-flow forced-feed system Modern engines have all the oil flowing through a filter before entering the oil circuit of the engine. This circuit is known as the full-flow system. The full-flow, forced-feed system provides oil under pressure to critical components.

The majority of engines use a full-flow, forced-feed oil circulation and distribution system. These have a wet sump, and oil is pumped to all parts of the engine by a rotary, positive displacement pump. The oil is filtered before it enters the main gallery for distribution around the engine.

Dry sump, dry-sump system There are a few exceptions, such as some high-performance engines where a dry sump, or dry-sump system, is used.

Forced-feed pump The forced feed is provided by a pump driven by the crankshaft or camshaft. A number of different pump designs are used, but all have positive displacement. These have rotating components that sweep past inlet and outlet ports and form chambers that increase in volume, carry oil and then decrease in volume in order to pump and pressurize the oil.

Oil pressure at low speed The oil pressure is controlled so that a sufficient supply is given at low engine speeds. This means that at higher engine speeds there would be excess pressure and oil flow, but this is relieved by a pressure-relief valve that returns the excess to the sump or to the inlet side of the oil pump.

The engine lubrication system This diagram shows the components of this engine lubrication system. Look closely at the names and detail of these components on the ATT website before moving on to the next pages where they are shown individually, with information on function and construction detail.

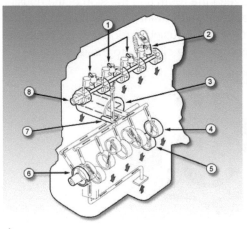

Figure 6.151 Lubrication features

Sump The sump sits below the engine and holds the main supply of engine oil. Baffles are fitted in the sump to prevent oil surges, which could cause a temporary loss of oil feed. Sumps are made from pressed steel or cast aluminium. A reinforced boss is drilled and threaded for the drain plug.

Pick-up and strainer This connects the oil pump to the oil supply in the sump. A fine mesh strainer is fitted to the supply end of the pipe to filter large particles of carbon and dirt in the oil and prevent them from entering the pump. Pick-up pipes are connected either directly to the oil pump or to the engine block in line with a drilling that feeds the pump. A good airtight seal is required at the connection. Two methods are used, a flange and gasket, or an 'O' ring on the pipe for sealing and a bracket and bolt for securing.

 Create a mind map to illustrate the features of a key component or system

6.2.4 Oil pumps and filtration ❶

Oil pumps and drive arrangements The oil pump is the heart of the system. It pumps oil from the sump into the engine. The main types of oil pump are gear, rotor, gerotor, vane and crescent.

The **gear type** uses two gears in mesh with each other. Drive is made to one gear which, in turn, drives the other. The housing has a 'figure of 8' internal shape, with one gear in each end. Ports are machined in the housing and align with the areas where the teeth move into, and out of, mesh.

Oil flow in gear-type oil pump As the teeth separate, the volume in the inlet side of the housing increases and atmospheric pressure in the sump is able to force oil into the pump. The oil is carried around inside the pump in between the teeth and the side of the housing. When the teeth move back into mesh, the volume in the outlet side of the housing is reduced, the pressure rises and this forces the oil out into the engine.

Figure 6.152 Gear pump

Rotor-type pumps use the same principle of meshing but with an inner rotor with externally formed lobes that mesh with corresponding internal profiles on the inside of an external rotor. The inner rotor is offset from the centre of the pump and the outer rotor is circular and concentric with the pump body. As the rotors rotate, the lobes mesh to give the outlet pressure of the oil supply, or move out of mesh for the intake of oil from the sump.

Gerotor pump The gerotor (gear rotor pump) is a variation on the smaller rotor pump. The gerotor pump is usually fitted around, and driven by, the crankshaft. There are inner and outer rotors, with the inner rotor externally lobed and offset from the internally lobed outer rotor. During rotation, the pumping and carrying chambers are formed by the relative positions of the lobes.

Figure 6.153 Rotor pump

The **crescent pump** is named after the solid block in the gear body. This pump is a variation on the gear pump, and it also uses gear teeth to create the pumping chambers and to carry oil from the inlet port to the outlet port of the pump.

Figure 6.154 Pump features

Operation of crescent pump The operation of this pump is based on the meshing of the gear teeth, the positioning of the ports in the housing, and alignment at each end of the crescent where the teeth move in and out of mesh. Oil is carried from the inlet port to the outlet port in the spaces between the teeth and the crescent. This pump is used for engine lubrication and for automatic transmissions.

The **vane-type pump** uses an eccentric rotor with vane plates set at right angles to the axis of the rotor and sitting in slots in the rotor. As the rotor rotates, the vanes sweep around inside the pump housing. The pump chambers increase in volume as the vanes move away from the housing walls and reduce in volume as the vanes approach the walls. Oil is carried between the vanes and the pump housing from the inlet port to the outlet port.

Figure 6.155 Vane pump

Camshaft drive arrangements Oil pumps are driven from the camshaft by gears on the camshaft and oil-pump spindle. Also, the drive gear is often used to drive the distributor for the ignition system. Another camshaft-drive arrangement is a direct drive from the end of the shaft. Some engines have used an auxiliary shaft to drive the pump and distributor, which has been driven from the crankshaft by a toothed belt or chain.

Figure 6.156 Pump drive and mounting

Direct-drive oil pump Some modern engines are now using the crankshaft to give a direct drive to the oil pump. These pumps are of the gerotor or crescent design and are fitted around the front of the crankshaft. This arrangement is used on many overhead camshaft engines because it provides a low position for the pump. Geared drives from the crankshaft are also used by some manufacturers.

Pressure-relief valve (or release valve) The oil flow and pressure at low engine speeds must be sufficient for all engine loads and, therefore, the performance of the pump is geared to low speeds. As the engine speed increases, so an excess of oil flow and pressure would occur. This would be detrimental to some engine components and, therefore, the pressure must be relieved.

Figure 6.157 Plunger and spring

Pressure-relief valve functions The pressure-relief valve is a spring-loaded conical or ball valve that opens when the pressure in the oil exceeds the spring force acting on the valve seat. When the valve opens, a return drilling is uncovered and the excess oil flows through this to return to the sump.

Oil filter Modern filters are canister types and consist of a microporous paper element in a thin steel cartridge. The paper element filters small particles of carbon and dirt that are picked up in the oil. Chemical reactions by some of the oil additives help to separate water and acids that drain into the sump. These by-products of combustion are also restricted from passing through the filter and collect on the feed side. Replacing the filter on a regular basis removes all these unwanted contaminants.

Oil flow through the filter is from the outside of the element to the inside and then into the main gallery.

Figure 6.158 Filter components

Filters screw onto a threaded sleeve in the filter housing. Sealing is made with a rubber 'O' ring.

Replaceable element-type filter Traditionally, a replaceable element in a steel bowl was used. A similar method is being introduced on many of the latest engines. This reduces the amount of material being thrown away because oil canisters are rarely recycled. The element is similar to those in canister types, and it is made from paper folded into segments to provide a round filter with a large surface area. The top and bottom of the filter are moulded to circular plates, and these provide sealing at each end so that oil is directed through the filter element from the outside to the inside and then into the main gallery.

Replaceable element This type of element fits inside a pressed steel or aluminium bowl, which is held in place by a through-bolt. A plate and spring are fitted into the base of the bowl. The plate seals the lower end and the spring holds the filter in place against the housing sealing face.

Filter housing seal The bowl is sealed by rubber rings. Where the bowl fits onto the filter housing, the rubber seal sits inside a groove cut into the housing. Small rubber and steel washers are used to seal the through-bolt where it passes through the bowl.

Oil filter bypass valve After a while, the oil filter element becomes clogged with dirt particles and the flow of oil becomes restricted. Normally, the filter is replaced before this happens, but if a blockage were to occur before replacement, the oil supply to the engine could be cut off.

Bypass valve functions In order to prevent this from happening, a bypass valve is fitted into the cartridge-filter base or the filter housing. The bypass valve works on the same principle as the pressure-relief valve. A spring-loaded plunger, ball or plate sits against a seat and is lifted by oil pressure to allow the flow of oil through the engine to be maintained. The spring tension is slightly below that of the pressure-relief valve spring tension so that the normal operating pressure is retained.

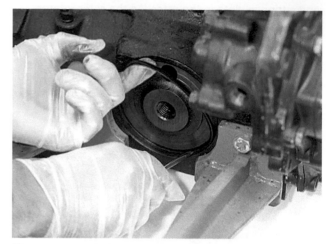

Figure 6.159 This seal is very important

Oil flow The resistance to flow through a clean filter element is less than through the bypass valve, so normal oil flow is through the element. When the filter is blocked and the bypass valve opens, the oil is no longer being filtered and dirty oil may enter the engine. Some manufacturers fit an electric warning lamp circuit with a switch fitted to the bypass valve.

 Make a simple sketch to show how one of the main components or systems in this section operates

6.2.5 Oil lubrication systems ❷

The **main gallery** consists of multiple interconnected drillings inside the engine block and cylinder head, and it is the distribution centre for oil supply to the crankshaft main (big-end) bearings and the overhead camshaft. Open ends are sealed with threaded plugs during manufacture.

The diameters of the drillings, and the use of restrictors, control the flow of oil to the different areas and components of the engine.

Oil enters the crankshaft bearings through drillings in the engine block and holes in the bearings. Radial grooves in the bearings allow a free flow of oil around the journal and bearing so that it can enter drillings in the crankshaft that connect to the big-end (main) bearings. In this way, all the crankshaft bearings receive a plentiful supply of oil.

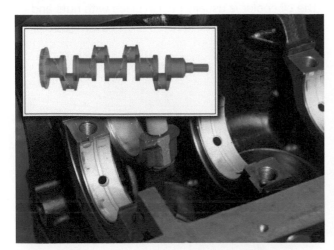

Figure 6.160 Drillings in the block and crank

Oil drillings in the connecting rods Connecting rods and big-end bearings are drilled to supply a jet of oil onto the cylinder wall. The jet is controlled so that it sprays the thrust face of the cylinder. This is achieved by using a shell bearing that does not have a radial groove, and it is the alignment of the drillings that enables the spray to be made.

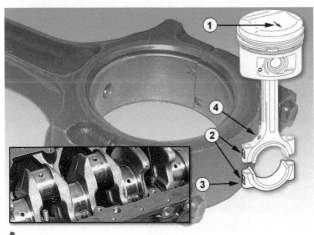

Figure 6.161 Con rod lubrication

Oil-control rings on the piston scrape the bulk of the cylinder-wall oil back through the piston and over the piston pin before returning to the sump.

Where **piston crown cooling** is required, a spray to the underside of the piston can be made either by a drilling that runs the length of the connecting rod or by a spray jet located at the base of the cylinder.

Camshaft bearing and lobe lubrication Overhead valve engines have a camshaft fitted in the side of the cylinder block. The camshaft bearings are lubricated by direct oil feed from the main gallery through a vertical drilling in the engine block and cylinder head, or from the main bearings. The oil supplies all camshaft bearings and a spray, via a hollow rocker shaft, to the valve tappets, rockers and valve stems.

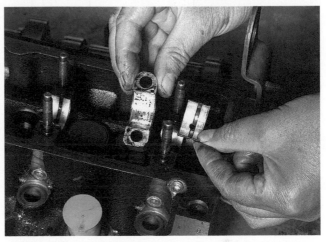

Figure 6.162 Cam bearings

Valve gear lubrication One of the rocker shaft pillars supplies oil to a hollow rocker shaft. The ends of the shaft are sealed with threaded, or pressed-in, plugs.

Each rocker is supplied on the rocker-bearing faces by oil that flows from the hollow shaft through holes drilled at appropriate positions. The oil flows over the rockers to lubricate the valve stems and the pushrod ends.

Figure 6.163 Oil drillings for the valve gear

Oil flow to the sump The oil returning to the sump (oil pan) passes down the pushrod tubes (channels) and lubricates the cam followers and cams on its way. An oil spray from the front camshaft bearing lubricates the timing chain and sprockets.

131

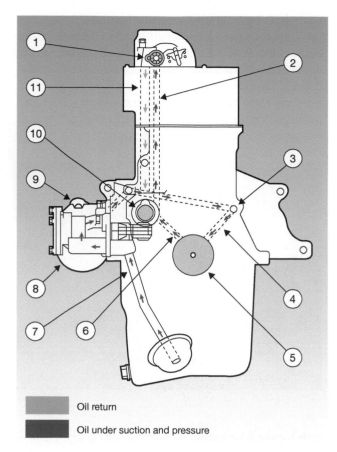

Oil return

Oil under suction and pressure

Figure 6.164 Oil flow

Oil returns to oil pan through pushrod channels

Oil returns to oil pan over the cam followers and cam lobes

Figure 6.165 Oil flow return

Make a simple sketch to show how one of the main components or systems in this section operates

6.2.6 Other lubrication components ❷

Oil cooler circuit The oil cooler commonly used is an air radiator similar to an engine cooling radiator, with tubes and fins to transfer heat from the oil to

the passing air stream. This cooler is fitted next to the cooling system radiator at the front of the vehicle. Pipes from the filter housing carry oil to and from the oil cooler radiator. These pipes have threaded union-nut fittings at each end for removal and replacement. The oil cooler is usually held in place with nuts and bolts through flexible rubber mountings.

Figure 6.166 Oil cooler

The standard fitting on all engines for checking the oil level is the dipstick marked to show the maximum and minimum acceptable levels.

Oil level indication sensor Many modern engines are now fitted with an electronic sensor that supplies information to the driver on the level of oil in the engine (low oil pressure indicating low oil level). This functions either when the ignition is first turned on and when the oil level is stable before the engine starts, or it uses a warning light covering all operating conditions when the level falls below the minimum. The warning light, or a pressure gauge in the instrument panel, indicates whether the oil level is within acceptable levels or not. The sensor is fitted into the sump or the engine block.

Oil-pressure switch and circuit A pressure-sensitive switch, fitted into the main gallery, makes an electrical contact when the pressure is below about 7 psi or 0.5 bar. The switch is fitted in the same circuit as the oil-warning lamp in the instrument panel and the live feed from the ignition switch when the ignition is on.

Function of the oil-pressure switch When the switch contacts make a connection, the lamp illuminates. This should occur before the engine is started. Once the engine is running, oil pressure builds up and the switch contacts should separate so that the circuit is broken. The warning lamp should then go out. This indicates that a minimum oil pressure is being maintained in the system.

Oil-pressure gauge Early oil-pressure gauges use capillary tubes from the oil main gallery to a bourdon, tube-type pressure gauge.

Modern gauges are electrical and use a piezo-resistive pressure sensor fitted into the main gallery and a stabilized voltage and bimetallic, or magnetic, gauge unit.

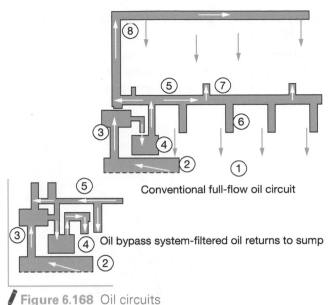

Conventional full-flow oil circuit

Oil bypass system-filtered oil returns to sump

Figure 6.168 Oil circuits

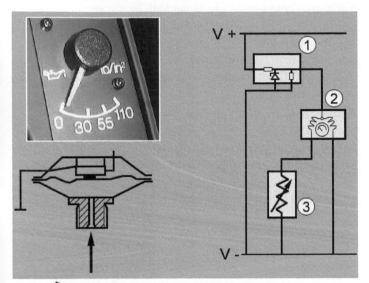

Figure 6.167 Gauge and circuit

Oil seals and gaskets These components can be classified either with the engine mechanical components or with the lubrication system. Oil seals are used to retain oil at the point where rotating shafts emerge from the engine block or cylinder head. Gaskets are used to form a seal between mating faces. Sealants are used with, or in place of, gaskets.

Bypass oil filter system Many old-type engines used a bypass oil filter circuit. In this system, the feed from the oil pump goes directly to the main gallery without being filtered. The return feed from the pressure-relief valve goes to the filter and clean oil is returned to the sump. This system is not as effective as the full-flow system in which all oil entering the main gallery is filtered.

Dry-sump system For many high-performance applications, a larger oil supply is needed so that engine heat can be removed by the engine oil as well as by the engine cooling system. A separate reservoir of oil is held in a remote tank and drawn into the main oil pump for distribution throughout the engine in the same way as a wet-sump system. The oil returns to a small sump below the engine. A scavenge pump, with a pick-up pipe in the sump, draws oil out of the sump and delivers it back to the reservoir. An oil cooler is usually fitted in this return circuit.

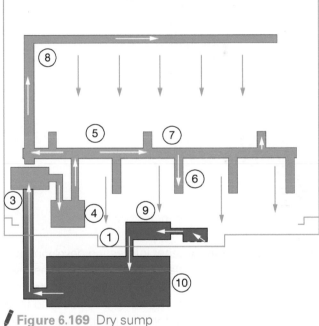

Figure 6.169 Dry sump

 Use the media search tools to look for pictures and videos relating to the subject in this section

6.2.7 Oil filter modules

Introduction The company Mahle, has developed an oil filter module that is already installed in over a million vehicles. The unit includes an integrated oil/water heat exchanger. However, this does not require plumbing into the coolant as it is designed to connect via its baseplate, which is sealed with O

133

rings, and held in place by four simple self-tapping screws.

Housing The housing is a glass fibre polyamide and has a built-in pressure relief valve. There is also a drain screw on the bottom to prevent spills as the unit is changed.

Figure 6.170 Oil filter module (Source: Mahle)

Oil change module Another interesting development in oil changing is a complete module that contains the oil and filter, and can be changed as one complete unit in just a few minutes. An electric oil pump is used to fill the unit with the old oil before it is removed, and then to return the new oil to the sump after replacement.

 Use a library or the web search tools to further examine the subject in this section

6.3 Cooling

6.3.1 Cooling introduction ❶

Introduction The main function of the cooling system is to remove the heat from the engine particularly around the cylinder walls and the combustion chambers. This should occur under all operating

conditions, including the extremes of very hot weather, hard driving and high altitude.

Emissions The engine cooling system on a modern motor vehicle has to play its part in keeping exhaust emissions to a minimum. During cold start and warm-up, an engine requires a rich petrol-to-air mixture to run smoothly.

Because a cold engine produces high levels of unwanted exhaust emissions, a rapid warm-up is needed to keep emissions to a minimum. The 'normal' running engine coolant temperature is maintained at about 90°C, which gives an engine temperature enabling clean combustion.

Emission control The control of emissions is achieved by controlling the upper-cylinder and combustion-chamber temperatures, resulting in the efficient and clean combustion of the fuel. A further reduction in harmful exhaust emissions is achieved by keeping the warm-up time to a minimum.

Figure 6.171 Engine cross section

Warm-up time Warming up to the optimum temperature as quickly as possible is important, not only because it helps to reduce exhaust emissions, but because it also helps to prevent the formation of water particles in the combustion chamber and exhaust when the engine is cold. Any water that does not evaporate can enter the engine and contaminate the engine oil, or remain in the exhaust system and cause premature corrosive damage.

The **water jacket** is cast into the cylinder block and cylinder head. Casting sand is used to shape the inside, or core, of the casting for the water passages. The sand is removed after casting through a series of holes in the sides, ends and mating faces of the cylinder block and head.

Water passages The holes in the sides and ends of the block and head are machined to provide an accurate location for the core plugs that complete the outside watertightness of the water jacket. The holes in

Figure 6.172 Engine block with core plugs

the mating faces are aligned to allow coolant flow from the cylinder block to the cylinder head. These components are also machined for the fitting of the water pump and a water outlet to the radiator.

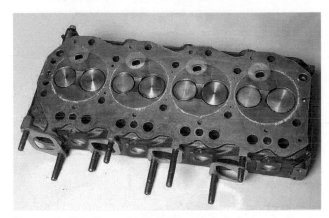

Figure 6.173 Core plugs and coolant holes

Coolant flow The internal designs of the head and block vary to give different coolant-flow patterns. An even flow to all areas of the engine is very important. The main areas where cooling is needed are around the combustion chambers and the upper-cylinder walls.

Cylinder head The need for inlet ports, exhaust ports and valves makes cooling of these regions difficult. These areas are prone to cracking and other deterioration from overheating, freezing, and the use of incorrect, or old, antifreeze solutions.

Air cooling Some older engine designs used an air-cooling system. Modern engines use water-cooling because this is capable of giving the precise engine temperature control needed for exhaust emission regulations.

Figure 6.174 Air-cooled system

Figure 6.175 Water-cooled system

Bypass system Recent developments in coolant circulation give improved control of engine temperature. Mixing cold and hot water as it enters the engine achieves this, as opposed to the cold fill of earlier systems. Both the old and new systems are covered in this learning programme.

Service life Cooling system components must have a service life that is comparable with the engine mechanical components. However, some are subject to wear and natural deterioration and need to be replaced at scheduled service intervals.

Heating and cooling systems The cooling system also provides heat to the vehicle interior for the comfort and safety of the occupants. In some cases, heat and/or cooling are provided for other engine systems such as the inlet manifold. An oil cooler for automatic transmission fluid may also form part of the cooling system.

Remember! There is more support on the website that includes additional images and interactive features: www.tomdenton.org

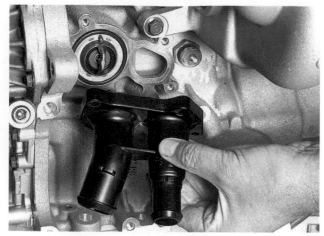

Figure 6.176 Ford engine coolant ports

Manifold heating Many engines use a heated inlet manifold that has a coolant flow from the engine water jacket running continuously through it. As soon as an engine is started, some heat is produced and this rises into the inlet manifold very quickly. The heat vaporizes the fuel in the air stream into the engine. This improves atomization and fuel distribution in the new air and fuel charge.

Figure 6.179 Inlet manifold

Figure 6.177 Ford engine coolant circuit

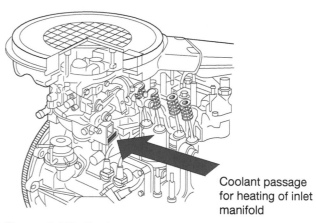

Coolant passage for heating of inlet manifold

Figure 6.180 Coolant passage

Heat exchangers for cooling the engine oil and heating diesel fuel are fitted into an adapter between the oil, or diesel, filter and the filter housing. A coolant supply may also be provided to exhaust gas turbochargers to cool the spindle and bearings.

The **coolant** must be able to resist freezing and boiling. Contamination and corrosion of engine and cooling system components must be kept to a minimum.

Heat energy Heat is a form of energy that can be sensed as a change in temperature. The engine uses chemical energy in the fuel. A combustion process converts the energy into heat and then into movement.

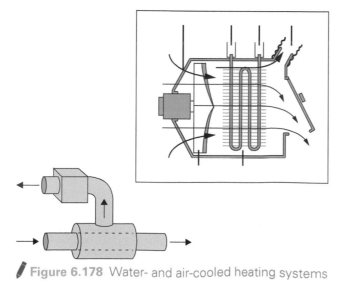

Figure 6.178 Water- and air-cooled heating systems

The **energy conversion** process is not very efficient and only about 30% is converted into movement energy. Of the remaining heat, up to 50% goes out of the exhaust and the rest heats the engine. Excessive heating of the engine must be controlled to prevent damage.

Expansion Components expand with heat and, at high temperatures, this expansion can cause seizure and burning of pistons and valve seats. High temperatures would also produce rapid deterioration of the engine oil.

Overheating A result of overheating is a change in the nature of the combustion process. The combustion time reduces which, in turn, leads to a rapid rise in the pressure and force acting on the piston crown, connecting rod and crankshaft. A 'pinking' sound may be heard and premature failure of these components is likely. There is also an increase in temperature to a point at which high levels of nitrogen oxides are produced and these are harmful to the environment.

Cooling system design Cooling systems are designed to maintain engines at an optimum temperature. This allows the design of components that expand on heating to form very tight fits and running tolerances. The adjustment of ignition and fuel settings are equated to the optimum temperature required for the clean and efficient combustion of fuel.

Air cooling Air-cooled systems have the air stream passing directly over the cylinder heads and cylinders to remove heat from the source. Fins are cast into the cylinder heads and cylinders to increase the surface area of the components, thus ensuring that sufficient heat is lost.

Figure 6.181 Cooling fan and radiator

Liquid cooling systems use a coolant to carry heat out of the engine and dissipate the heat into the passing air stream. The liquid coolant is contained in a closed system and is made to circulate almost continuously by the impeller in the water pump. Heat is collected in the engine and dissipated from the radiator into the passing air stream.

 Create a word cloud for one or more of the most important screens or blocks of text in this section

6.3.2 **Components and operation** ❶

The **coolant** is a mixture of water and antifreeze. The antifreeze is needed to prevent water expanding as it freezes. The force from that expansion would be sufficient to cause engine cylinder blocks and radiators to burst apart.

Antifreeze Sufficient antifreeze is needed for the climate in which the vehicle is operated. Modern antifreeze formulae are also designed to give year-round protection by increasing the boiling point of the coolant for hot-weather use.

Heat transfer All three forms of heat transfer are used in the cooling system:

Convection occurs in the water jacket, creating flows of internal coolant from the cylinder block to the cylinder head.

Conduction occurs through the cylinder and combustion chamber surfaces as heat passes to the coolant.

Radiation of heat occurs from the radiator and cooling fins when heat is dissipated to the atmosphere.

Rate of heat transfer The amount of heat transfer is dependent on four main factors:

1 The temperature difference between the engine and coolant.
2 The temperature difference between the coolant and the air stream passing through the radiator.
3 The surface area of the radiator tubes and fins.
4 The rate of flow of air and coolant through the radiator.

Thermostat Liquid cooling systems traditionally used a thermostat in the outlet to the top hose to control engine temperature.

A thermostat is a temperature sensing valve that opens when the coolant is hot and closes as the coolant cools down. This allows hot coolant to flow from the engine to the radiator where it cools down and returns to the engine. The cooled coolant in the engine acts on the thermostat and it closes.

6

Figure 6.182 Thermostat

Coolant flow The coolant reheats in the engine and the thermostat opens and the cycle of hot coolant flow to the radiator and cool coolant returning to the engine repeats itself. Although this system provides a reasonably effective method of engine temperature control, it does produce a fluctuating temperature. However, a steady temperature is required for very clean and efficient combustion.

Bypass mixing cooling system Modern engine design is moving towards a system with the thermostat in the radiator bypass channel. When the thermostat opens, it allows cold water from the radiator to mix with the hot-water flow in the bypass as it enters the water pump. This system provides

a steady engine temperature and prevents the fluctuating temperature cycle of the earlier system. The modern system is shown in Figure 6.179 with arrows indicating the coolant flow.

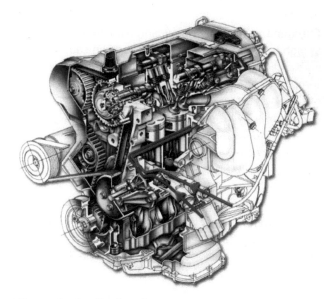

Figure 6.184 Coolant flow

Heat distribution The heat distribution within the engine needs to be controlled. The temperature around all the cylinders and combustion chambers should be identical. The heat removed by the cooling system has, therefore, to be consistent for all areas of the engine. All modern engines have a fairly rapid coolant circulation within the engine so that an even temperature distribution is achieved.

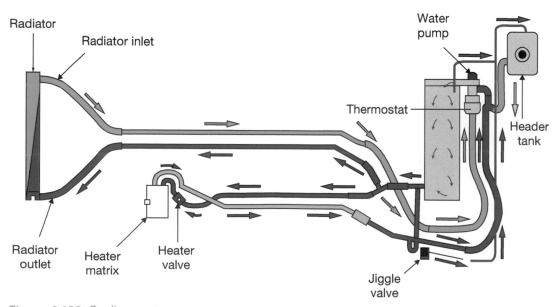

Figure 6.183 Cooling system

The **water (or coolant) pump** draws the coolant through a radiator bypass channel when the engine is cold and from the radiator when the engine is hot. The impeller in the water pump drives the coolant into the engine coolant passages or water jacket. Water jacket passages are carefully designed to direct the coolant around the cylinders and upwards, over and around the combustion chambers.

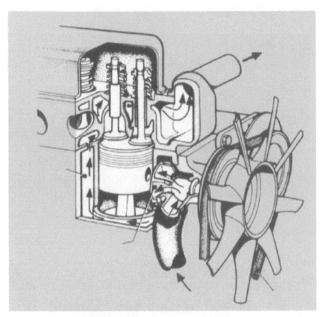

Figure 6.185 Water pump action

Coolant density The density of coolant falls as it heats up and as the temperature approaches boiling point, bubbles begin to form. These bubbles can create areas in the water jacket where the coolant is at a lower density and the actual mass of coolant in those areas is reduced. The reduced mass of coolant therefore cannot effectively absorb heat efficiently in order to cool the engine.

Another problem of poor heat transfer and lowered coolant density occurs when the rapid flow of coolant into and out of restrictions in the water jacket induces a phenomenon known as 'cavitation'. This results in localized drops in pressure and density in the coolant.

Heat distribution The two causes of localized coolant density change – bubble formation and cavitation – can seriously affect the performance of the cooling system. This is because an even heat distribution around the cylinders and combustion chambers is not maintained.

To overcome these problems, all liquid cooling systems are pressurized. When hot, most modern systems have an operating pressure equivalent to about one atmosphere (1 bar, or 100 kPa).

Expansion The pressure is obtained by restricting the loss of air above the coolant in a radiator header tank or an expansion tank. As coolant heats up it expands. If the air above the coolant has less space to occupy, and it cannot immediately escape, it increases in pressure.

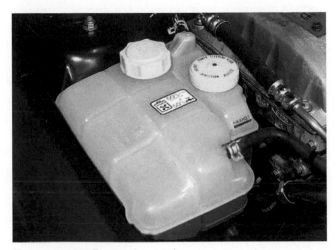

Figure 6.186 Expansion tank

Radiator pressure cap A pressure-sensing valve in the radiator cap allows this higher pressure to escape but retains the operating pressure.

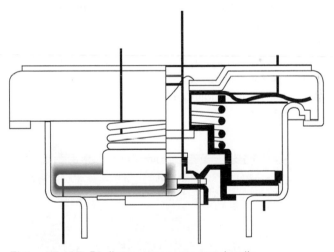

Figure 6.187 Radiator pressure cap details

Pressure cap vacuum valve As the engine cools down, the coolant contracts and the pressure drops. A vacuum valve in the pressure cap allows air to return to the system. This prevents depressurization below atmospheric pressure and also the risk of the inward collapse of components. An early sign of the failure of this valve to open is a top hose that has collapsed.

The pressure in the system acts on the coolant to increase the density, which would otherwise have fallen without the increase in pressure. This helps to reduce the risk of cavitation and to increase the boiling point of the coolant under pressure. The advantages are a more efficient cooling system, with a higher safe operating temperature. It can also be used at high altitudes without the need for modification.

Summary A cooling system is needed to prevent engine damage caused by overheating. It also helps to reduce emissions by shortening the engine warm-up time. Heat is used from the cooling system to operate the heater.

 Use the media search tools to look for pictures and videos relating to the subject in this section

6.3.3 Cooling and heating ❶

In a liquid cooling system, the coolant carries heat from the engine to the radiator. Airflow through the radiator dissipates the heat into the atmosphere. Air is forced through the radiator by the forward movement of the vehicle or is assisted by a fan fitted behind the radiator.

Cooling fan The fan can be driven by an electric motor or by a belt from the crankshaft. Traditional engines had the fan mounted on the front of the water pump with a 'V' belt driving the fan and pump.

Fan design A number of energy-saving fan designs have been used such as variable-pitch and viscous-hub types. Vehicles that are regularly used for carrying loads, or for towing, can be fitted with secondary fans to improve cooling efficiency and prevent engine overheating.

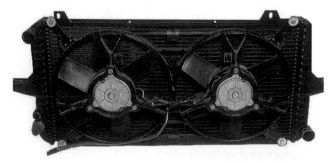

Figure 6.188 Twin electric fans

In-car heating Some of the surplus heat from the cooling system is used for in-car heating. Pipes and hoses from the water jacket carry hot coolant to a heater radiator or matrix fitted into the heater housing.

Heater controls Two methods of heat control are used. One uses a water valve to control coolant flow through the heater. The other, which has a continuous coolant flow, uses control flaps to mix hot and cold air in the heater housing.

Fresh air ducts Ducts into and out of the heater direct air to the screen, side screens and passenger compartment. This is for demisting, defrosting and warming the passenger compartment. Control flaps in the heater direct the airflow to the ducts. Fresh air vents in the fascia can direct either hot or cold air into the vehicle interior.

Figure 6.189 Heater air control flaps

Figure 6.190 Fresh air fascia vents

Component design Some of the cooling system and heater components have different designs. Many of these have been developed to improve the efficiency of the system, or because of changes in vehicle design.

 Create an information wall to illustrate the features of a key component or system

6.3.4 Antifreeze ❶

Coolant The coolant is a mixture of water, antifreeze and inhibitors. The antifreeze is usually ethylene glycol, which needs inhibitors to prevent corrosion and foaming. These inhibitors have a lifespan of about two years, which means that the coolant should be changed at biennial intervals. Selection of the correct coolant mixture must be made to meet the manufacturer's specifications. Aluminium alloy engines are more prone to corrosion than cast iron engines.

Antifreeze is mixed to a specified ratio with water. Many manufacturers specify a 50/50 mixture of water and antifreeze, which allows higher engine temperatures before the coolant boils and prevents freezing.

Ethylene glycol An ethylene glycol antifreeze solution has an added advantage. It forms a semisolid wax solution prior to solidification and this enables any expanding ice crystals to move within the water passages.

Frost protection A 50/50 coolant mixture will increase the boiling point to 106°C (223°F) and provide protection down to −34°C (−30°F). For colder temperatures down to −65°C (−90°F), a maximum mixture of 65% ethylene glycol can be used. Higher concentrations begin to freeze at higher temperatures and therefore no more than 65% ethylene glycol should be used.

Hard water areas Many areas have 'hard' water that contains calcium or chalk. This separates from the water when it is heated. Deposits can form inside the water jacket or radiator where they can block small water passages. Frequent topping up with mains water in hard water areas should be avoided. In these areas, it may be recommended to use distilled water, or water from outside the area.

 Construct a crossword or wordsearch puzzle using important words from this section

6.3.5 Components ❷

Thermostat The modern thermostat uses a wax pellet in an enclosed cup. Inside the wax is a rubber sleeve enclosing a pin. The pin is connected to a plate that acts as the valve. All these components are held in the thermostat body, together with a spring to hold the valve closed when the coolant is not hot. The thermostat body includes a flange that fits into a housing in the coolant outlet from the cylinder head, or a radiator bypass channel.

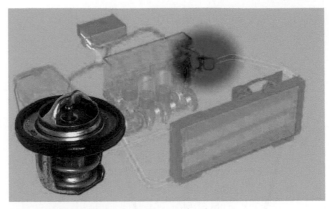

Figure 6.191 Thermostat positioning

'Wax stat' When the temperature of the coolant acting on the wax pellet reaches the operating temperature, there is sufficient heat to cause the wax to expand, press on the pin, and force it out of the cup to open the valve. Coolant is then free to flow through the valve.

Thermostat fitting The wax pellet must always be fitted so that it sits on the hot side of the coolant flow through the thermostat.

Air bleed valve Some thermostat flanges are fitted with a small sub-valve to allow air to flow through the thermostat as the system is filled with coolant. This small valve must be fitted towards the top if the thermostat is fitted on its side.

Thermostat position Some manufacturers have fitted the thermostat in the radiator top hose. The thermostat may also be fitted directly into its own housing and, if so, has to be replaced as a complete assembly.

Radiator construction There are a number of different designs and manufacturing materials used for radiators.

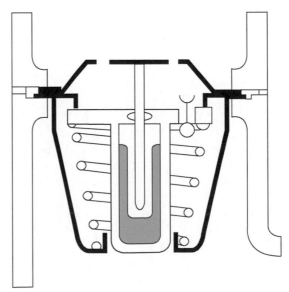

Figure 6.192 Thermostat closed

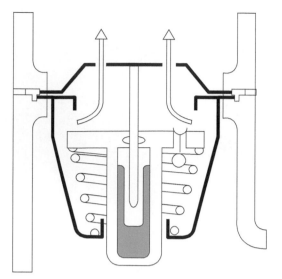

Figure 6.193 Thermostat open

Figure 6.194 Thermostat with its wax pellet in the hot-coolant area

They all consist of a series of small tubes through which the coolant flows. Very thin sheets of metal are used to form a large surface area surrounding the small tubes. This large surface area makes radiators efficient heat exchangers for engine cooling purposes.

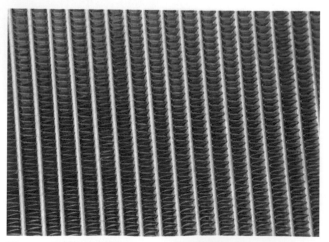

Figure 6.195 Radiator core construction

Figure 6.196 Radiator tubes

Radiator types The radiator tubes are fitted to tanks at each end, and these tanks are fitted with connections for the top and bottom hoses. The traditional radiator had the core tubes set vertically and the coolant flowing downward from the header tank to the bottom tank. The air space required for expansion of the coolant could be either in the header tank or in a separate expansion tank.

Automatic transmission radiator If a vehicle is fitted with automatic transmission, an extra set of pipes, running in the bottom radiator tank, may be used to cool the transmission fluid.

Crossflow radiator With the lower frontal area of modern vehicles, a different radiator layout is needed. The crossflow radiator has tubes and thin sheet fins forming the core. The core tubes run across the vehicle, and the coolant flows from one side to the other. The tanks at each end of the radiator are joined to the core and have connections for the top and bottom hoses. Crossflow radiators usually have a remote expansion tank to which the pressure cap is fitted. Some crossflow radiators have an integral expansion tank.

Figure 6.197 Crossflow radiator

Figure 6.198 Remote expansion tank

Radiator developments Traditionally, radiators were made from copper and brass and soldered together. Modern radiators are constructed from an aluminium core with nylon, or plastic, end-tanks that are cinched together. This is a method of folding the edges of the radiator-core ends over a sealing ring and a lip on the end-tank. Aluminium radiators are lighter and cheaper to produce than copper/brass radiators.

The **pressure cap** was traditionally called the radiator cap because it was fitted to the radiator. On modern vehicles, the cap is fitted to the expansion, or overflow, tank. There are a number of different designs and operating pressures. Many new vehicles are fitted with a plastic, or nylon, cap that is specific to one manufacturer.

Figure 6.199 Pressure caps – bayonet and screw types

Cap operation The main parts of all pressure caps are the sealing ring, pressure valve, vacuum valve, and a bayonet, or screw, fitting. The pressure valve consists of a spring-loaded seal that rests on a seat, either in the filler neck or in the cap. The vacuum valve allows air to return to the system as it cools. It is fitted in the centre of the pressure valve. Both the pressure valve

and the vacuum valve are one-way valves and operate in opposite directions. The pressure valve allows air out, and the vacuum valve allows air in.

Bayonet and ring-cap fittings Bayonet-fitting pressure caps are tightened on a ring cam under the lip of the filler neck. A safety stop is provided to prevent the cap from coming off. For removal, the cap has to be pushed down and turned to pass the safety stops. The cap should be turned fully clockwise when fitting to ensure the correct tension on the pressure-release spring of the valve.

The **water pump** is usually fitted into the water jacket of the cylinder block. However, there have been some engines where it has been fitted into the cylinder head. An external water pump is used on some engines and connected to the water jacket by pipes or hoses. The water pump is driven from the engine crankshaft by a belt.

Figure 6.200 Water pump in water jacket

Figure 6.201 External water pump

Pump construction Running through the centre of the water pump is a spindle mounted on a bearing.

The bearing is pre-packed with grease and fitted with seals for retaining the grease and keeping the coolant in the engine. The drive pulley is fitted to the spindle on the outside of the pump.

Figure 6.204 Multi-V belt and pulley

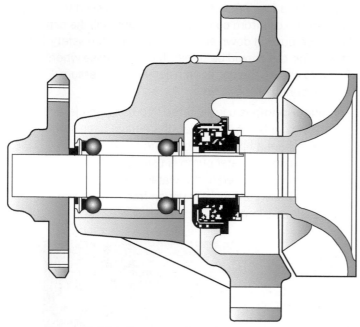

Figure 6.202 Cross section of a water pump

Impeller The movement of the impeller creates a coolant flow through the water jacket. Water pumps are supplied as a replacement part fully assembled in a housing holding the bearing, spindle, impeller and drive flange for the pulley.

Pump drive belts The drive components for the water pump usually consist of a 'V' belt that also drives the alternator, and 'V' pulleys on the crankshaft and water pump. Multi-V belts are also in common use. The toothed camshaft drive belt drives some pumps. An adjuster for the belt is provided on the alternator mounting, or as a separate tensioner.

Cooling fan The fan is used to ensure an adequate airflow through the radiator when this is not provided by the forward speed of the vehicle. The fan was traditionally fitted to the front of the water pump and attached with the same bolts as the drive-belt pulley. Many longitudinal engines still use this system, but the fan, formerly a pressed-steel component, now incorporates a thermostatic viscous hub and nylon fan blades.

Viscous coupling The viscous hub is a fluid clutch using silicon oil. The operation of the clutch is temperature controlled with a bimetallic valve. When the airflow temperature over the viscous hub is cool, the valve remains closed, and the clutch is inoperative. When the airflow temperature over the viscous hub increases, the valve in the hub opens and the viscous fluid is driven outwards by centrifugal force. The increased force in the fluid locks the plates in the hub together to engage the clutch drive to the fan.

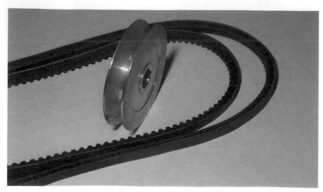

Figure 6.203 'V' belt and pulley

Figure 6.205 Viscous fan hub

Electric fan An alternative temperature-sensing arrangement is for the fan to be driven by an electric motor mounted on a cowl frame attached to the radiator. A plastic fan is fitted to the motor spindle, and this operates when a temperature-sensitive switch closes.

Fan operation The supply for the electric fan is direct from the battery on some makes of vehicle, and it can run at any time whether the ignition is on or off. Other makes are connected into the ignition circuit. The electrical supply to the motor may be connected directly to the switch, or be connected through a relay.

Two-speed fans Some vehicles, particularly those fitted with air-conditioning, may have two-speed fan circuits. These have a control circuit to switch the motor (or motors) to half speed at 95°C and full speed at 100°C. This arrangement can be operated by the engine management system.

Figure 6.206 Two-speed fan

Hoses and clips Hoses are manufactured from fabric-reinforced rubber, and are moulded to suit the vehicle application. Connectors are cast, or formed, with a raised lip on the pipes leading into, and out of, other components. The hoses are held with round clips that can be drawn tight to give a watertight seal (Jubilee clips).

 Using images and text, create a short presentation to show how a component or system works

6.3.6 Heater ❷

Heater The vehicle interior heater is made from an air box with a heat exchanger inside. The heat exchanger, called a heater matrix, is very similar to the cooling radiator in that it consists of a similar series of tubes and fins.

Heater matrix Hot coolant from the engine flows through the matrix, thus heating the tubes and fins.

Air flows across the outside and collects some of the heat for distribution inside the vehicle.

Figure 6.207 Heater matrix

Air supply Air is drawn into the heater through ducts on the vehicle exterior. The design of the ducts provides a dust and water trap, and usually the ducts have an outlet hose for water drainage. Many new vehicles have a pollen filter fitted in the air intake ducts. The filter is a microporous paper element that traps pollen and dust particles.

Air distribution The distribution of air inside the vehicle is provided by a series of ducts and outlets. These are positioned on the underside of the fascia, at fascia level, and adjacent to the front and side screens. The outlets can be selected by operating the control levers to the required positions. The control levers are connected to flaps in the heater air box by a cable or vacuum system. The flap position directs air to the appropriate outlet.

Heat control Temperature selection is achieved by regulating the coolant flow through the heater by means of a valve, or by a flap in the heater air box that directs how much air flows through the heater matrix. The

Figure 6.208 Heater intake ducts with pollen filter

145

water valve, or flap, is connected to a control lever by a cable, or vacuum-control system. Thermostatic devices are used to control air temperature on some vehicles.

Figure 6.209 Heater control with flaps

A **fresh air supply** through fascia vents is fitted to most vehicles. The air supply can be independent of the heater system, so the heater can supply hot air to the footwell and cool air from the fascia outlets. On other vehicles, the fascia vents are integral with the heater system and cannot supply different temperatures of air.

Figure 6.210 Fascia blower box

Heater blower motor A blower motor in the air intake duct boosts airflow through the heater. The motor is usually fitted with a series of resistors in order to provide a range of speeds. The motor switch routes the electric current through the appropriate resistor for the speed selected on the switch.

Figure 6.211 Wide fan

Summary A cooling system is needed to prevent engine damage caused by overheating. It also helps to reduce emissions by shortening the engine warm-up time. Heat is used from the cooling system to operate the heater.

 Use the media search tools to look for pictures and videos relating to the subject in this section

6.3.7 Thermal management system (TMS)

Introduction Thermal management is about controlling air and coolant flows intelligently. However, it goes beyond cooling system technology because it considers all the heat flow systems on the vehicle. With this approach it is possible to improve efficiency, reduce emissions and improve passenger comfort.

Intelligent systems Giving intelligence to the cooling system of a vehicle is the first step. This can be locally or in microprocessor-controlled systems. This intelligence will, for example, operate shutters (air control), coolant thermostats, bypass or mixing valves, and electrically driven coolant pumps (water/glycol control). Warm up times are faster and aerodynamic drag can be reduced by using radiator shutters. There is a reduction in energy consumed, passenger compartment warm up time is reduced, and temperature control of the engine is far more precise. Taking further steps, it is possible to warm up and control the temperature of transmission fluids using engine coolant. Service life and efficiency of the transmission are therefore improved.

Figure 6.212 Simple representation of a thermal management control system

Powertrain Thermal management systems need to be integrated with powertrain control systems, and a holistic view of the whole system should be taken. The range of potential applications appropriate for thermal management is considerable; here are some example systems where temperature can be controlled:

▶ Coolant pump(s)
▶ Thermostat
▶ Radiator air flow using shutters
▶ Cooling fan viscous clutch

▶ Coolant-cooled intercooler
▶ Exhaust gas cooling (EGR system)
▶ Transmission oil temperature
▶ HVAC
▶ Turbocharger
▶ Throttle body.

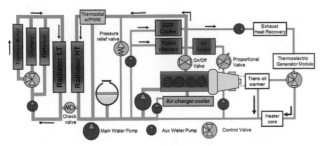

Figure 6.213 Hybrid engine thermal management system (Source: Solvay)

Summary In passenger vehicles the potential fuel saving could be as much as 7.5%, along with the consequential reduction in CO_2 emissions. Just as the use of electronic components has increased in the engine, the same is true of the TMS. Modern engines demand more extensive cooling, but in a controlled manner. This is delivered in part by adding electronic functionality to the TMS, and in part by redesigning TMS components.

Figure 6.214 Audi coolant pump and control valves (Source: Solvay)

6.4 Fuel

6.4.1 Introduction ❶

Introduction The fuel supply systems of petrol engine vehicles have been extensively developed over the last 50 years. There have been many developments in the traditional petrol supply and air mixing methods used in carburettors. However, the introduction of mechanical and electronic petrol injection systems has made carburettors almost obsolete. Fuel injection is now fitted to all petrol engine vehicles in order to meet the latest requirements for reducing harmful exhaust gas emissions.

Developments Many of the developments have been introduced to provide improved engine performance with higher power outputs and lower fuel consumption. Meeting environmental protection regulations has been equally important. Diesel fuel pumps and injectors have seen similar developments and the introduction of electronic control of fuel metering and full electronic control systems. Because of these developments, a wide range of fuel systems exists. In this section, the main systems are explained in some detail. The older systems such as carburettors are therefore only examined briefly.

Chemical action All fuel delivery systems have to supply a quantity of fuel that matches the amount of oxygen that is in the air entering the engine. For petrol engined vehicles, the quantities of hydrogen and carbon in the fuel and the oxygen content of the air should be chemically correct to give a complete chemical change during combustion. The chemical change for clean combustion is $CH + O \rightarrow CO_2 + H_2O$ or carbon and hydrogen plus oxygen equals carbon dioxide and water.

Combustion For complete and clean combustion, the ratio of air to fuel should be as close as possible to the stoichiometric value. This is where λ (lambda) equals one. This is a ratio of 14.7:1 by mass of air to fuel. In petrol engines, the optimum for clean combustion is for these quantities to be delivered accurately to each cylinder in the engine. Refer to the 'Engine Mechanical Repair' and the 'Air Supply, Exhaust and Exhaust Emission Control' learning programmes, for additional information on exhaust gas constituents.

Spark and compression ignition Petrol is ignited in the combustion chamber by a spark arcing across the electrode gap of a spark plug. Diesel fuel ignites following injection into the high temperature air charge. The high temperature is obtained by compression of the air charge. The air charge on petrol engines is matched to the amount of fuel delivered. On diesel engines, a full air or gas charge is required in order to raise the temperature by compression.

Compression ignition On diesel engines, the air charge is not balanced to the fuel delivered and this gives, under most conditions, a surplus of oxygen. This surplus of oxygen can combine with chemicals in the cylinder to form harmful gases such as nitrogen oxides. However, the use of exhaust gas recirculation (EGR) can keep the surplus oxygen to a minimum. The mix of air and exhaust gas helps to reduce the production of nitrogen oxides.

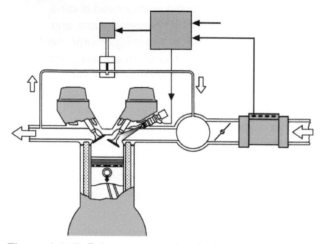

Figure 6.215 Exhaust gas recirculation

Exhaust gas The carbon dioxide and water in the exhaust are not directly harmful. Carbon dioxide is a greenhouse gas and is believed to contribute to global warming. Exhaust gases are not normally clean, as internal engine combustion is not a simple process. This makes the use of catalytic converters essential.

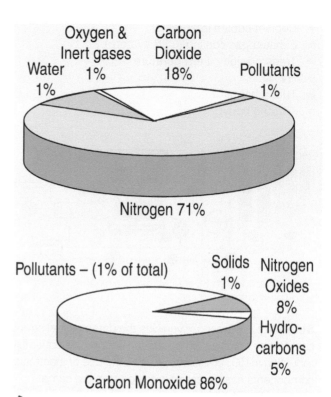

Figure 6.216 Exhaust gas constituents

Catalytic converters help to change any residual fuel, carbon monoxide or nitrogen oxides into harmless substances. The fuel may have impurities or additives

Remember! There is more support on the website that includes additional images and interactive features: www.tomdenton.org

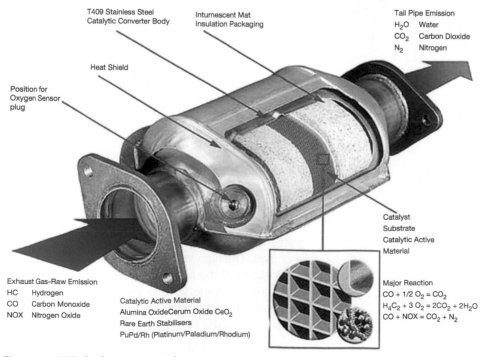

Figure 6.217 Action of catalytic converters (Source: www.cateran.com.au)

to aid combustion, which is the case with leaded fuel. There are other gases in the air which also undergo chemical reaction during the burning of the fuel in the engine. Poor distribution of the fuel in the air charge will give local variations in the combustion process and harmful exhaust gas constituents.

Fuel additives The impurities and additives have been considerably reduced over recent years. The most significant impurity was low quantities of sulphur, which combines with atmospheric gases to form sulphuric acid. Leaded petrol has tetraethyl lead added as an octane enhancer. This additive was used to prevent knocking by slowing the combustion process to a specific value. However, lead is a substance that is damaging to people's health and is being phased out of petrol in favour of other chemicals, which can perform the same task.

Octane rating The rate of combustion of petrol is a measurable value. Octane is the fuel that is used as a comparative for all petrol fuels. The combustion characteristics of octane are given a value of 100. This value is then used to measure other petrol fuels. The octane rating of a fuel is used to set, among other things, the engine compression ratio and ignition timing at the design stage. It is then used in service as a guide for the fuel that should be used. Typical petrol octane ratings are 93 and 97. The equivalent substance for diesel fuels is cetane and typical cetane ratings are 48 to 50.

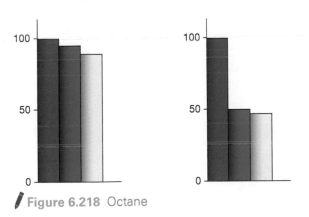

Figure 6.218 Octane

Fuel additives for petrol and diesel include substances that will prevent or reduce some of the harmful effects of raw fuel. Examples are gum and tar formation and the corrosion of components. Antifreeze is added to petrol for very cold climates and to diesel for all conditions where temperatures fall below freezing. Diesel fuel contains paraffin, which forms a wax at cold temperatures. This wax prevents the fuel from flowing. All winter grade diesel fuel contains additives to provide appropriate protection for the area in which it is sold.

Fuel supply The fuel on the vehicle is held in a tank fitted in a safe position. Recent construction legislation requires that the tank is unlikely to be ruptured in a vehicle collision. The positioning and protection of the tank is considered at the design stage of the vehicle and tested during development. The tank is fitted with a filler neck and pipework from the filler cap to the tank. Also fitted are the outlets to the atmospheric vent or evaporative canister and the fuel feed and return pipes to the engine. The fuel gauge is located in the fuel tank. Fuel supply and return lines are made from steel pipes, plastic pipes and flexible rubber joining hoses depending on the application and the type of fuel used.

Fuel pumps A pump to supply fuel to the engine is fitted into or near to the tank on petrol injection vehicles. On carburettor vehicles, a mechanical lift pump is fitted to the engine and is operated by a cam on the camshaft or crankshaft, or an electric pump is fitted in the engine compartment. Diesel engined vehicles using a rotary fuel injection pump, may use the injection pump to lift fuel from the tank. Alternatively, they may have a separate lift pump similar to the ones used on carburettor engines. A separate priming pump fitted in the fuel line may also be used.

Figure 6.219 Mechanical fuel pump

Figure 6.220 Fuel injection supply pump

Mechanical lift pump The online diagram shows the operation of a typical mechanical fuel lift pump. Refer to vehicle manuals for details of specific fuel pumps. Modern pumps are sealed units and have to be replaced as a complete unit if they become defective.

149

Select a routine from section 1.3 and follow the process to study a component or system

6.4.2 Electronic fuel injection ❶ ❷

Injection methods Petrol/gasoline fuel injection systems can be classified into two main categories:

▶ Single point
▶ Multipoint injection.

The multipoint systems (used by almost all cars now) can then be further divided into:

▶ Manifold or port injection
▶ Direct injection (into the combustion chamber)

The three methods are shown here. Depending on the sophistication of the system, idle speed and idle mixture adjustment can be either mechanically or electronically controlled.

Electronic Fuel Injection (EFI) Systems Electronic petrol injection systems have been in use for many years, first on expensive and sports vehicles and now as standard equipment on all vehicles. The tougher standards of exhaust emission regulations have made the use of microelectronic control systems for fuel delivery a virtual necessity. There are many different manufacturers of electronic fuel systems and this programme covers the main points of the systems.

Electronic control unit At the heart of electronic fuel injection (EFI) systems is the fuel control or electronic control unit (ECU) with a stored map of operating conditions. Electronic sensors provide data to the microprocessor in the ECU, which calculates and sends the output signals to the system actuators – the fuel pump, fuel injectors and idle air control units. The ECU will also switch some of the exhaust emission and auxiliary system components.

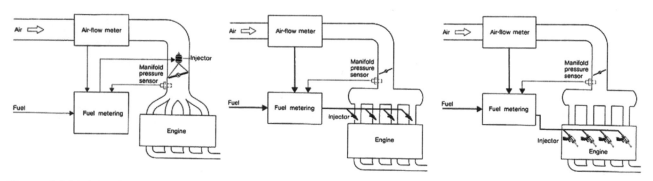

Figure 6.221 Injection methods (single point, multipoint port, multipoint direct)

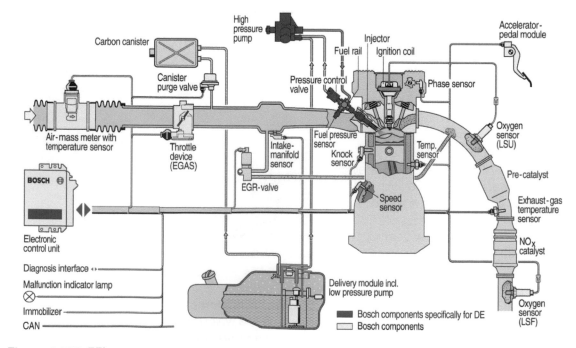

Figure 6.222 EFI system

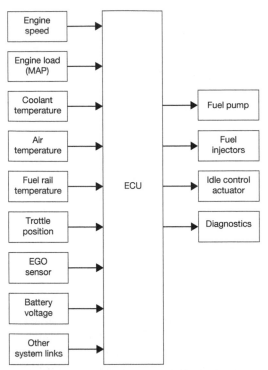

Figure 6.223 ECU with example inputs and outputs

Fuel injection methods Electronic fuel injection (EFI) systems are named by the position and operation of the fuel injectors. There is a range of throttle body injection (TBI) systems. They are also known as single point (SPI) or central point (CPI) systems. However they are named, the injector is positioned in a housing fitted on the inlet manifold. This is in the same position as the carburettor was traditionally fitted.

Simultaneous and sequential injection The port fuel injection (PFI) or multi-point (MFI) systems have individual injectors for each cylinder. The injectors are fitted so that fuel is sprayed into the inlet ports. Port fuel injection systems are either simultaneous, where all injectors operate at the same time, or sequential, where each injector operates in turn.

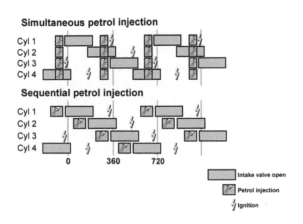

Figure 6.224 Types of fuel injection systems

Gasoline Direct Injection (GDi) A recent development has been the Introduction of a direct injection petrol engine where the fuel is injected into the combustion chamber.

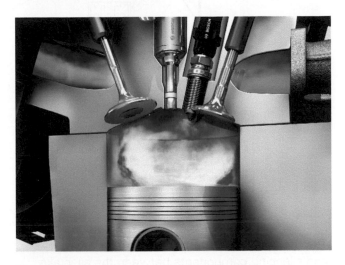

Figure 6.225 GDi (Source: Bosch Media)

Engine Control Module Modern petrol injection systems are linked to the ignition systems and are controlled by an engine control module (ECM). The latest developments have all electronic systems linked to form a power train control module (PCM). This is also described as a vehicle control module (VCM). All modern fuel injection systems have closed loop electronic control using an exhaust gas oxygen sensor. For clarity, each electronic control unit will be referred to as an ECU.

Inputs and outputs The components for any electronic fuel injection system can be divided into four groups:

1 The air supply components.
2 The fuel supply components.
3 The electronic control unit (ECU), together with the power supply and system harness.
4 The sensors which provide data to the ECU.

Figure 6.226 EFI components

6

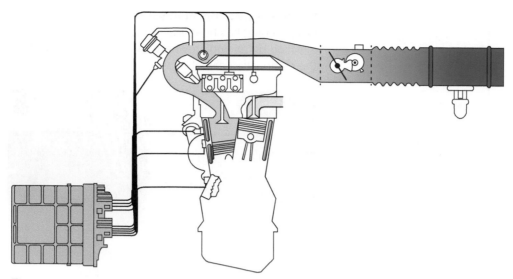

Figure 6.227 Air supply

The **air supply components** consist of ducting and silencing components between the air intake and the inlet manifolds. This will also include an air filter, a throttle body, throttle plate assembly and idle control components. The air supply components must provide sufficient clean air for all operating conditions. The airflow into the engine would be noisy and unbalanced between cylinders without the use of resonators and plenum chambers. A plenum chamber is a large-volume air chamber that can be fitted either in front of or behind the throttle plate housing.

Air filters on most modern petrol engine vehicles consist of a plastic casing with a paper filter element. Airflow into the filter is upwards so that dust and dirt particles drop into the dust chamber, or is rotary so that dust and dirt is thrown out before the air enters the engine. Crankcase ventilation and the air supply or pulse air exhaust emission systems are also connected to the filter assembly.

Figure 6.228 Air filter on a modern vehicle

Throttle body The throttle is a conventional circular plate in an air tube. For fast idle and warm-up, an auxiliary air valve is fitted to bypass the throttle plate, or an electromechanical link is made to the throttle plate spindle. An auxiliary air valve, idle air control valve (IAC) or idle speed control valve (ISC) is operated from signals from the ECU.

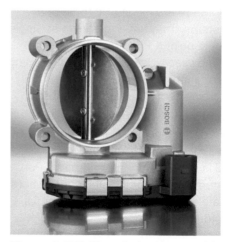

Figure 6.229 Throttle body assembly

Auxiliary air valve Early designs of the auxiliary air valve use a disc with a calibrated aperture for closing or opening the bypass air channel. The disc is held closed by a pull off spring and opened by a bimetal spring. When the engine is cold, the bimetal spring bends to open the valve. With the engine running, an electrical heating current acts on the bimetal strip. This causes it to bend and allow the pull off spring to close the air channel.

Rotary air valve A later development of the auxiliary air valve is the rotary air valve. This has a special electric motor to move and hold the valve in position. The position is based on the electrical signals supplied by the ECU. Two electric windings in the motor work in opposition to each other so that the motor is variable over a 90° arc.

Solenoid valve Other designs of auxiliary or 'extra air' valves have graduated opening values based on the strength of current supplied from the ECU. These valves operate to hold the idle speed to the stored data specification for engine temperature and load conditions. The valve consists of a solenoid valve with a spring-loaded armature connected to the valve in the air channel.

Figure 6.230 Solenoid air valve

Air bypass The amount of air allowed to flow through the bypass channel of the auxiliary air valve is regulated by the position of the valve. At idle, the valve is continuously adjusted to stabilise the speed. When the throttle is closed during deceleration, the valve plate is adjusted to control exhaust emissions. During engine starting, the valve is open and when the engine is switched off the valve is in the rest position.

Idle speed control can also be provided by direct action onto the throttle spindle. Electric solenoids or stepper motors are used for this method of control. The solenoids can be single-position or multi-position types and can be used for not only cold start and warm-up control but also to open the throttle when high-load systems, such as the air conditioner, are switched on. Stepper motors give graduated positions depending on the supply current to a number of electric windings. Sensors in the idle control mechanisms provide feedback signals to the ECU to provide data on operation and position.

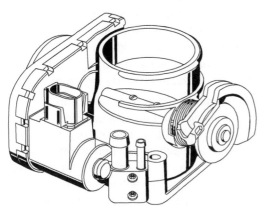

Figure 6.231 Throttle idle speed control

The **fuel supply** from the fuel tank to the injector valves for all electronic systems follows the same basic layout. The delivery of fuel at the injector valves is also based on a similar function for all systems. A basic layout of fuel supply components is shown in Figure 6.232. A fuel pump is fitted either in, or close to, the fuel tank. A fuel filter is fitted in the delivery fuel lines from the tank to the fuel rail. A fuel pressure regulator is fitted on either the housing for throttle body injector systems or the fuel rail for port fuel injection systems. The return fuel lines run from the pressure regulator to the fuel tank.

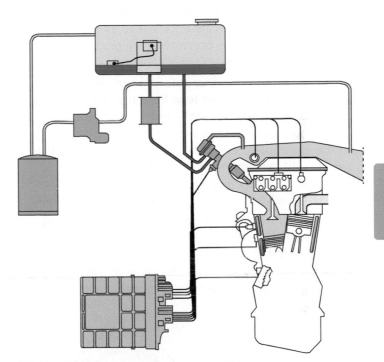

Figure 6.232 Fuel supply components

The **fuel pump** is a roller cell pump driven by a permanent magnet electric motor. Fuel flows through the pump and motor but there is no risk of fire as there is never an ignitable mixture in the motor. The delivery pressure is set by a pressure-relief valve, which allows fuel to return to the inlet side of the pump when the operating pressure is reached. There is a non-return valve in the pump outlet. Typical delivery pressures are between 300 and 400 kPa (3 to 4 bar).

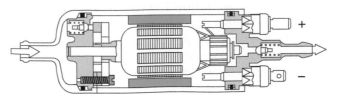

Figure 6.233 Roller cell pump

Roller cell pump The rollers in the roller cell pump are thrown out by centrifugal force when the motor armature and pump rotor spindle rotate. The rotor is fitted eccentrically to the pump body, and as the rollers seal against the outer circumference, they create chambers that increase in volume to draw fuel in. They then carry the fuel around and finally discharge it as the chamber volume decreases.

Pump electrical supply The fuel pump electrical supply is live only when the engine is being cranked for starting or is running. The fuel pump electric feed is from a relay that is switched on with the ignition. Safety features are built into the electric control feed to the relay so that it operates only to initially prime the system or when the engine is running. The control functions of the fuel pump relay are usually provided by the fuel control module.

Inertia switch A further safety feature is the use of an inertia switch in the feed from the relay to the fuel pump. This operates, in the event of an accident, to cut the electric feed to the fuel pump and to stop the fuel supply. It is an impact-operated switch with a weight that is thrown aside to break the switch contacts. Once the switch has been operated it has to be manually reset.

The **fuel filter** is an in-line paper element type that is replaced at scheduled service intervals. The filter uses microporous paper that is directional for filtration. Filters are marked for fuel flow with an arrow on the casing and correct fitting is essential.

Figure 6.234 An in-line paper element type

Fuel pressure regulation The fuel pressure regulator is fitted to maintain a precise pressure at the fuel injector valve nozzles. On port fuel injection systems, a fuel rail is used to hold the pressure regulator and the fuel feed to the injector valves. The injector valves usually fit directly onto or into the fuel rail. The fuel rail holds sufficient fuel to dampen fuel pressure

fluctuations and keep the pressure applied at all injector nozzles at a similar level.

Figure 6.235 Fuel pressure regulator

Figure 6.236 Regulator on throttle body

Operation of the fuel pressure regulator Fuel regulators are sealed units with a spring-loaded diaphragm and valve on the return outlet to the fuel tank. Fuel is pumped into the regulator and, when the pressure is high enough, it acts against the diaphragm and compression spring to open the valve. Surplus pressure and fuel is allowed to return to the fuel tank. Once the pressure in the fuel regulator is reduced, the valve closes and the pressure builds up again. Throttle body injection systems operate in the region of 1 bar, and port fuel injection systems in the region of 2.5 bar.

Inlet manifold vacuum On port fuel injection systems, inlet manifold vacuum acts against the compression spring in the fuel pressure regulator. This is required in order to maintain a constant pressure differential between the fuel rail and the inlet manifold. With a constant pressure differential, the amount of fuel delivered during a set time will be the same irrespective of inlet manifold pressure.

Turbocharger For vehicles fitted with a turbocharger or supercharger, inlet manifold pressure is applied to the diaphragm and regulator valve. When the inlet manifold pressure rises above a certain value, the regulator valve is closed so that the full pump delivery pressure is applied to the injector valve nozzles. This raises the amount of fuel delivered to match the boosted air charge.

The **injector valves** spray finely atomised fuel into the throttle body or inlet ports depending on the system. The electromagnetic injection valves are actuated by signals from the ECU. The signals are of a precise duration depending on operating conditions but within the range of about 1.5 to 10 milliseconds. This open phase of the injector valve is known as the 'injector pulse width'.

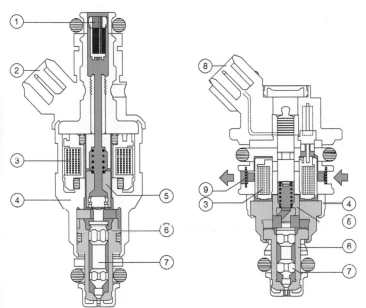

Figure 6.237 Injector features: 1-Fuel supply and filter, 2-Electrical connection, 3-Solenoid (winding), 4-Injector body, 5-Armature (moving part), 6-Valve body, 7-Needle, 8-Electrical connection, 9-Filter

Solenoid injectors There is a range of individual injector valve designs but all have the same common features. These are an electromagnetic solenoid, with a spring-loaded plunger, connected to a jet needle in the injector valve nozzle. The electrical supply to the solenoid is made from the system relay or ECU. Earthing or grounding the other connection energises the solenoid. This lifts the plunger and jet needle so that fuel is injected for the duration that the electric current remains live. As soon as the electrical supply is switched off in the ECU, a compression spring in the injector valve acts on the solenoid plunger to close the nozzle.

Throttle body injector A fuel injector valve for a throttle body system is shown here. This cross-sectional view shows the housing, the magnetic coil for the electric solenoid and the jet needle and nozzle.

Multi-point injector A top feed fuel injector valve for port fuel injection systems is shown here. This type of valve is generally used on earlier systems. One problem experienced with this fuel feed arrangement is fuel vaporisation and bubbles forming in the fuel rail. The bubbles can cause starting and running problems. To overcome this problem lateral or side or bottom feed injectors are used. An example of this type is shown here. When fitted in the fuel rail it can be seen that any bubbles that may form will be at the top of the rail. They will, therefore, be flushed out through the regulator as soon as the fuel pump is actuated.

Cold start injector valve with a thermo-time switch Early electronic fuel injection systems used a cold start injector valve with a thermo-time switch control circuit. The cold start injector valve is an electromagnetic solenoid valve. It is energised during starter motor operation but subject to the condition of the thermo-time switch. The cold start injector valve operates only when the engine is cold and for a maximum time of about ten seconds (when the ambient temperature is below −20°C). The time progressively shortens above this temperature. Later electronic fuel injection systems have the cold start enrichment calculated in the ECU. The pulse width of the fuel injector valves is increased to provide the extra fuel needed during start up.

6

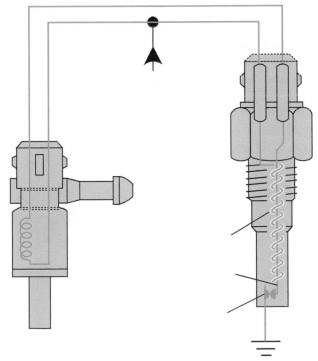

Figure 6.238 Cold start injector and circuit (early system)

Electronic control unit (ECU) The electronic control unit is an electronic microcomputer with a central processing unit (CPU) or microprocessor. Inside the CPU are software programs that compare all sensor input data with a fixed map of operating conditions. It then calculates the required output signal values for the injection valves and other actuators. A computer program that demonstrates the operation of a fuel ECU and system is available from the automotive technology website.

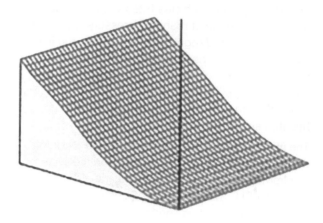

Figure 6.241 Transition map

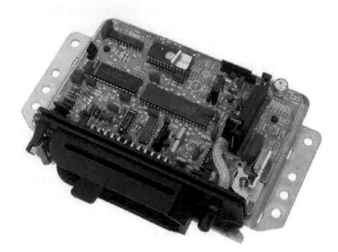

Figure 6.239 Inside an ECU

ROM and RAM The fixed map of operating conditions specific for each engine is held in a fixed value memory or read only memory (ROM). The operating data store of input values from the sensors is held in a random access memory (RAM). A 'keep alive' memory (KAM) of specific data such as adjustments, faults and deviations in component performance may also be used. The RAM data is erased when the ignition is switched off. The KAM data is erased when the battery is disconnected. New data is replaced in the RAM and KAM during engine start up and operation.

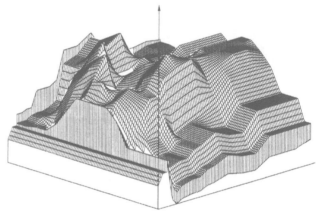

Figure 6.242 Timing map

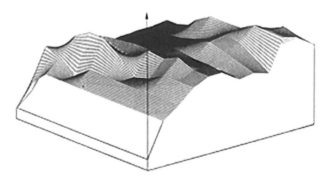

Figure 6.243 Injection map

Cold start and warm-up phases During the cold start and warm-up phases of engine operation the computer operates in an open loop mode based on the sensor data. Once the engine reaches a certain temperature and the signals from the exhaust gas oxygen (or lambda) sensor are logical, the computer operates in a closed loop mode based on the data from this sensor.

CPU operation Other programs in the CPU monitor the system and sensor data. They provide fault diagnosis and limp home or a limited

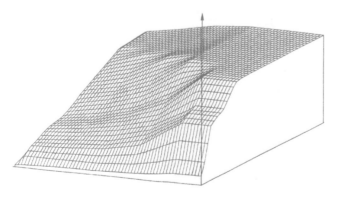

Figure 6.240 Dwell map

operation strategy in the event of any defects being detected. Other components in the ECU provide signal amplification and pulse shaping. This includes analog to digital (A/D) converters for DC voltages and pulse formers for AC voltages. The CPU requires digital signals for all processing functions. On the output side, power transistors are used for switching the actuator supply voltages either to the components or to an earth or ground point.

Emission control The ECU also operates the emission control components at appropriate times depending on the engine operating conditions. Typical emission control actuators are the canister purge solenoid valve, the exhaust gas recirculation (EGR) valve and the secondary air solenoid valve. Secondary air is provided by either the air injection reactive or pulse air systems.

The **electrical harness** for the engine management system is a complex set of cables and sockets. Cables have colour and/or numerical coding and the sockets are keyed so that they can be connected in one way only. Special low-resistance connectors are used for low-current sensor wiring. Follow manufacturer's data sheets for further technical detail.

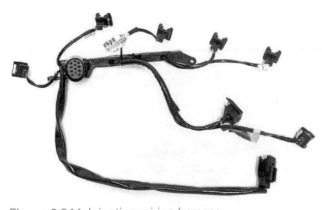

Figure 6.244 Injection wiring harness

Sensors provide data to the ECU. The engine speed and load conditions are used to calculate the base time value (in milliseconds) for the injector pulse width. A range of correction factors are added to or subtracted from the base time value to suit the engine operating conditions occurring at all instances of time.

Engine speed and position On early electronic fuel injection systems, the engine speed was provided from signals obtained from the ignition low-tension primary circuit. On engine management systems, the engine speed and position is required for the ignition and fuel systems.

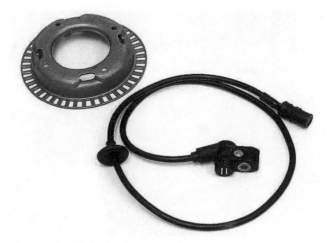

Figure 6.245 Inductive speed sensor

Continuous injection systems On continuous injection systems, the fuel is injected on the induction strokes when one of the inlet valves is open. On the other stroke, fuel is injected into the inlet port where it remains until the valve opens. On sequential systems, a single fuel charge is injected during the inlet valve open stage. Accurate engine speed and position sensing is required for this to occur correctly.

Simultaneous petrol injection

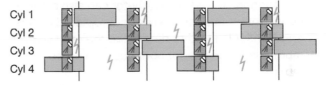

Sequential petrol injection

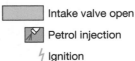

Figure 6.246 Injection process

Methods of engine speed and position sensing
There are two methods of engine speed and position sensing. The older system is a conventionally geared distributor with an inductive or Hall effect generator. This provides an alternating signal current that is used by the ignition system. It is also used for engine speed sensing in the fuel electronic control module.

157

Figure 6.247 Early distributors were used for engine speed and position sensing

Inductive pulse generators Most of the latest systems have inductive pulse generators mounted close to, and responding to, a toothed wheel attached to the crankshaft pulley or flywheel. There is an air gap between the toothed wheel and the inductive generator, and as the teeth pass the inductive generator, an alternating electric current is produced. The waves of the alternating current are used to measure engine speed. For position sensing, a missing or different size of tooth or mask opening on the sensor ring is used. A distributor can also provide a reference for number one cylinder at top dead centre. When a sensor is fitted to determine the crankshaft position, this is suitable for continuous injection systems.

For **sequential injection**, a camshaft position sensor or phase sensor in the distributor is used to recognise

the position of number one cylinder. The ECU is then able to follow the engine firing order.

AC voltage pulse Inductive sensors produce an output pulse each time a lobe or tooth passes the inductive coil. The frequency and pattern of the pulses is used by the ECU to determine the engine speed and position.

Airflow meter The fuel requirement is calculated in the ECU from the engine speed and load conditions. An airflow meter is one method of measuring the engine load conditions. A variable voltage, corresponding to the measured value at the airflow meter, is used by the ECU to calculate the amount of fuel needed to give a correct air/fuel ratio.

Engine load can also be determined from the inlet manifold absolute pressure (MAP) and this is used on some systems to provide the data. In these systems, an airflow meter is not used.

Airflow metering There are two main types of airflow meter. These are the vane type (VAF) and the resistive type (MAF). The vane-type airflow meter consists of an air passage and damping chamber into which is fitted a fixed pair of flaps (or vanes), which rotate on a spring-loaded spindle. The spindle connects to and operates a potentiometer and switches.

Figure 6.249 Vane-type airflow meter

Flap-type airflow meter Airflow through the meter acts on the intake air flap to move it in opposition to the spring force. The integral damper flap moves into the sealed damper chamber to smooth out the intake pulses. The degree of flap movement and spindle rotation is measurable at the potentiometer as a variable voltage dependent on position. The voltage

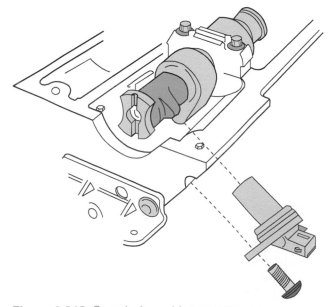

Figure 6.248 Camshaft position sensor

signal, together with other signals, is used in the ECU to calculate the fuel requirement.

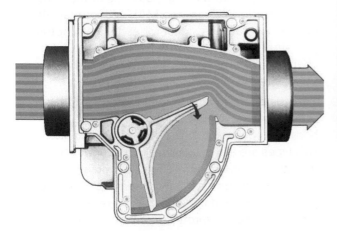

Figure 6.250 Action of the vane airflow meter

A **bypass air duct** is built into the housing. This provides a means of starting without opening the throttle, a smooth airflow during engine idle and a means to adjust the idle mixture.

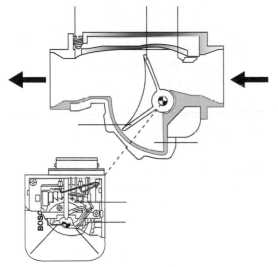

Figure 6.251 Air bypass

Mass airflow meters Mass airflow meters are fitted with two similar resistors inside an air tube. A measurement resistor is heated and often referred to as a hot wire. The other resistor is not heated. It provides a reference value for use in the calculation of the air mass. The control circuit maintains the temperature differential between the two resistors. The signal sent to the ECU is proportional to the current required to heat the measurement resistor and maintain the temperature differential. The output signal from some mass airflow meters is similar to the air vane types. However, some produce a digital output signal.

Figure 6.252 Hot wire/film airflow meter

Manifold absolute pressure sensor On some EFI systems manifold absolute pressure (MAP) sensor signals are used by the ECU to calculate the fuel requirements. These systems do not have an airflow meter. The signals from manifold absolute pressure, engine speed, air charge temperature and throttle position are compared in the ECU to calculate the injector pulse width.

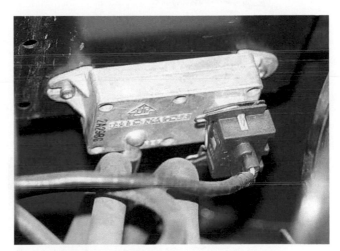

Figure 6.253 MAP sensor

The **MAP sensor** is a pressure-sensitive component consisting of a diaphragm and piezoelectric circuit. It can be a component fitted in the engine compartment or be integral with the ECU. It is connected by a rubber hose to the inlet manifold. The ECU supplies a stabilised reference voltage, usually 5V, to the sensor. This voltage is adjusted by the MAP sensor electronics to provide an output signal proportional to the sensed absolute atmospheric pressure.

Sensor output The actual pressure in the manifold is read as a proportional voltage typically from

159

Figure 6.254 Throttle potentiometer

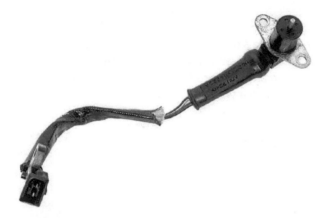

Figure 6.255 Crank sensor

Figure 6.256 Air temperature sensor

4.5V at high pressure, to 0.1V at low pressure. The electronic circuitry in some MAP sensors converts the reference voltage to a frequency signal that is fed back to the ECU. This is as a

proportional frequency (80 to 165Hz), depending on the vacuum or pressure in the inlet manifold. When the pressure is high, such as at the wide open throttle position, the MAP sensor may also provide a reference signal for actual barometric pressure. This is used as a correction value for changes in altitude, which could lead to poor performance.

Throttle position sensor Two types of throttle position sensor are used. Both are fitted to the throttle body and operated by the throttle plate spindle. The two types are a 'throttle switch assembly' and a 'throttle potentiometer'. A throttle switch assembly has two switches, one to indicate the closed throttle or idle position and the other for the wide open throttle position. A throttle potentiometer is a variable resistor with a rotary sliding contact. The sliding contact is moved along the rotary resistance track to provide changes in voltage proportional to the position of the throttle.

The **throttle potentiometer** signals are used in the ECU for a number of functions. At the closed throttle position, idle speed and deceleration fuel cut-off are controlled. In the part open throttle position (about 5% to 70% open) there is normal operation with close control of fuel delivery and exhaust emissions. In the wide open throttle position (70% to 100%), full load enrichment is provided. During rapid movement of the throttle plate there is acceleration enrichment, depending on the rate of change of the throttle plate and signal voltages from the sensor.

Air intake temperature sensor In order for the ECU to correctly calculate the required fuel for a correct mixture ratio, an accurate figure for air mass is necessary. However, air volume and density are affected by changes in temperature. As the temperature rises, the air density falls. The air flow, or manifold absolute pressure measurement, therefore, must be corrected for temperature. The sensor is a temperature dependent resistor with a negative temperature-coefficient (NTC).

The **engine coolant temperature sensor** is a negative temperature coefficient (NTC) thermistor. It is of a similar type to the air temperature sensor. It is fitted into the water jacket close to the thermostat or bypass coolant circuit passages. The sensor measures the engine coolant temperature and provides a signal voltage to the ECU. This is used for cold start and warm-up enrichment as well as fast idle speed control through the idle speed control valve.

Figure 6.257 Coolant thermistor

Exhaust gas oxygen sensor The Greek letter (λ) lambda is used as the symbol for a chemically correct air to fuel ratio. This is the stoichiometric ratio of 14.7 parts of air to 1 part of fuel by mass. This letter is used to name the sensor that controls the amount of fuel delivered, maintaining a very close tolerance to the stoichiometric ratio.

The **lambda sensor** is often known as an exhaust gas oxygen sensor. Some of these sensors are electrically heated. Preheating allows the sensor to be fitted lower down in the exhaust stream and prolongs the life of the active element. The sensor measures the presence of oxygen in the exhaust gas and sends a voltage signal to the engine electronic control unit.

Figure 6.258 An exhaust gas oxygen sensor

Oxygen content More fuel is delivered when oxygen content is detected and less fuel when it is not. In this way, an accurate fuel mixture, close to the stoichiometric ratio, is maintained. This produces the correct exhaust gas constituents for chemical reactions in the catalytic converter. Exhaust gases pass over the active element and when the oxygen

concentration on each side is different, an electric voltage is produced. Voltages of about 0.8V for little or no exhaust oxygen and 0.2V for higher content are typical outputs.

Systems in operation sensors The sensors for power steering and air conditioning are pressure or mechanically operated switches. They provide a voltage signal when the system is in operation. The ECU uses these signals to increase the engine idle speed to accept the increased engine load.

Automatic transmission Switches are used in the automatic transmission. They include the neutral drive switch, which is used for idle speed control, the kick-down switch for acceleration control, and the brake on/off switch, which is used to ensure that the torque converter lock-up clutch is released. This is to prevent the engine stalling as the vehicle comes to rest.

Exhaust gas recirculation A transducer measures exhaust gas pressure. It uses a ceramic resistance transducer, which responds to the exhaust gas pressure applied through a pipe connection to the exhaust system. The signal voltage from the electronic pressure transducer is used to regulate the EGR valve. The valve may be operated directly from the ECU if electromechanical, or by vacuum through a solenoid vacuum switch.

Other correction factors are fuel temperature, octane rating, remote CO adjustment and the service plug or OBD connections. These sensors, variable resistors, switches and multi-plugs provide additional data to the ECU. The fuel temperature sensor is fitted in the fuel rail. At a preset value, a bimetal strip bends to close the signal circuit to the ECU. This signal, together with other signals, is used by the ECU for optimum fuel delivery during hot engine starting.

Malfunction indicator light Service and on-board diagnostic (OBD) plugs are used for diagnostic and corrective actions with scan tools, dedicated test equipment and other test equipment. If faults are detected, the system malfunction indicator lamp on the vehicle fascia will come on. Alternatively, it will fail to go out after the preset time duration after switching on the engine. All faults should be investigated as soon as possible. Many electronic systems have a limp home or limited operation strategy programme, which allows the vehicle to be driven to a workshop for repair.

Remember! There is more support on the website that includes additional images and interactive features: www.tomdenton.org

 Select a routine from section 1.3 and follow the process to study a component or system

6.4.3 Bosch DI-Motronic ❸

Gasoline direct injection Bosch's high-pressure injection system for petrol engines is based on a pressure reservoir and a fuel rail, which a high-pressure pump charges to a regulated pressure of up to 120 bar. The fuel can therefore be injected directly into the combustion chamber via electromagnetic injectors.

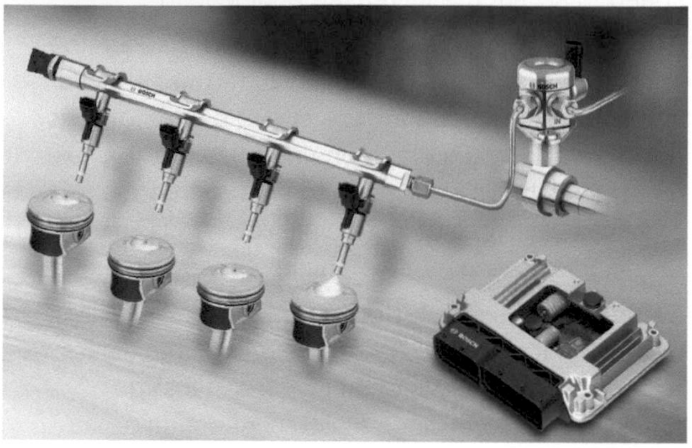

Figure 6.259 Components of gasoline direct injection

The **air mass** drawn in can be freely adjusted through the electronically controlled throttle valve and is measured with the help of an air mass meter. For mixture control, a wide-band oxygen sensor is used in the exhaust, before the catalytic converters. This sensor can measure a range between lambda = 0.8 and infinity. The electronic engine control unit regulates the operating modes of the engine with gasoline direct injection in three ways:

1. Stratified charge operation – with lambda values greater than 1
2. Homogenous operation – at lambda = 1
3. Rich homogenous operation – with lambda = 0.8.

Injection time The expertise of Bosch is reflected in the high-pressure injectors of the DI-Motronic. Compared to the traditional manifold injection system, the entire fuel amount must be injected in full-load operation in a quarter of the time. The available time is significantly shorter during stratified charge operation in part-load. Especially at idle, injection times of less than 0.5 milliseconds are required due to the lower fuel consumption. This is only one-fifth of the available time for manifold injection.

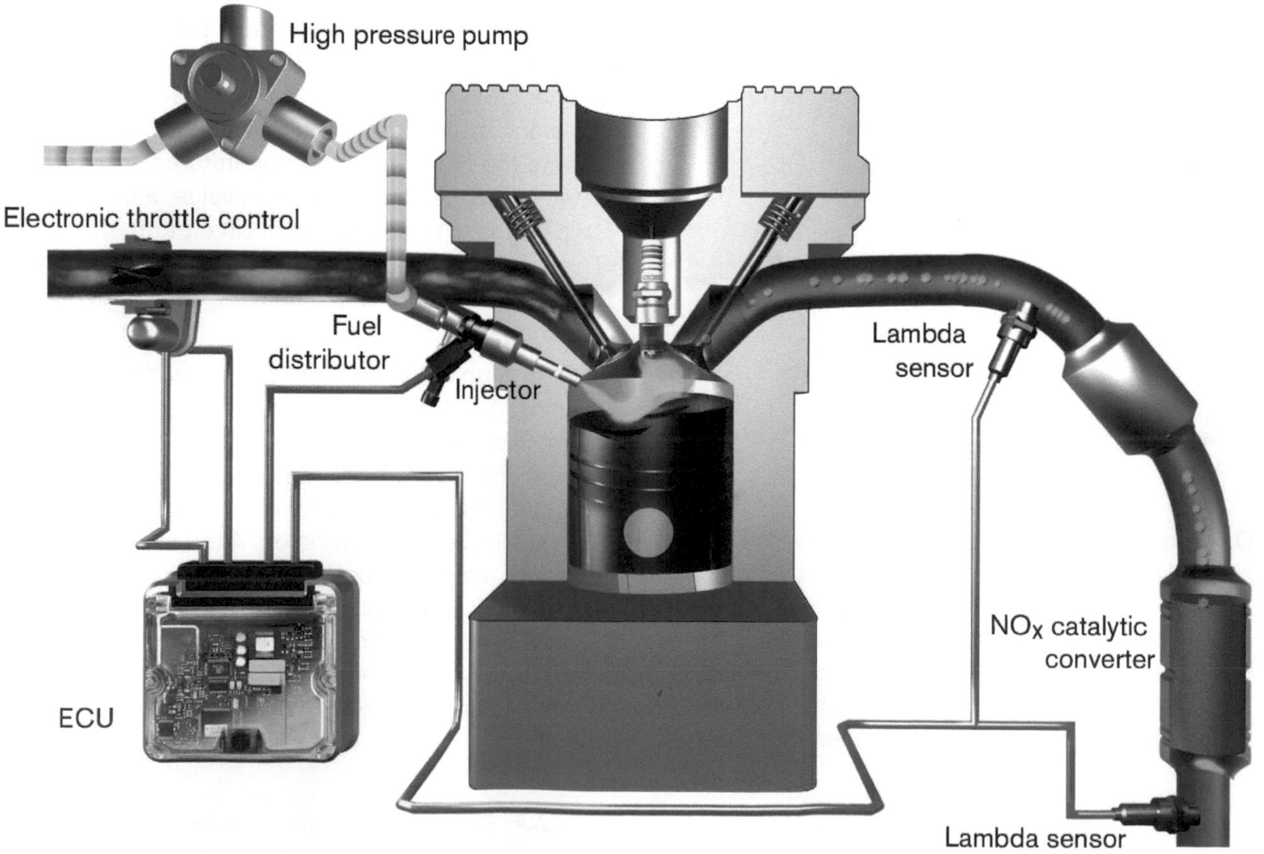

Figure 6.260 Bosch DI-Motronic

Figure 6.261 Injector for direct injection under test 4

Atomization The fuel must be atomized very finely in order to create an optimal mixture in the brief moment between injection and ignition. The fuel droplets for direct injection are on average smaller than 20μm. This is only one-fifth of the droplet size reached with the traditional manifold injection and one-third of the diameter of a single human hair. This improves efficiency considerably. However, even more important than fine atomization is even fuel distribution in the injection beam. This is done to achieve fast and uniform combustion.

Figure 6.262 Fuel droplet size is important

The function of Bosch DI-Motronic Conventional spark ignition engines have a homogenous air/fuel mixture at a 14.7 to 1 ratio, corresponding to a value of lambda = 1. Direct injection engines, however, operate according to the stratified charge concept in the part-load range and function with high excess air. In return, very low fuel consumption is achieved.

Injection timing With retarded fuel injection, a combustion chamber split into two parts is an ideal condition, with fuel injection just before the ignition point and injection directly into the combustion chamber. The result is a combustible air/fuel mixture cloud on the spark plug, cushioned in a thermally insulated layer, composed of air and residual gas. This raises the thermodynamic efficiency level because heat loss is avoided on the combustion chamber walls. The engine operates with an almost completely opened throttle valve, which avoids additional alternating charge losses.

Stratified mode   **Homogeneous mode**

Figure 6.263 Operating modes4

Fuel savings With stratified charge operation, the lambda value in the combustion chamber is between about 1.5 and 3. In the part-load range, gasoline direct injection achieves the greatest fuel savings with up to 40% at idle compared to conventional petrol injection processes.

Homogenous operation at higher load With increasing engine load, and therefore increasing injection quantities, the stratified charge cloud becomes even richer and emission characteristics become worse. Like diesel engine combustion, soot may form. In order to prevent this, the DI-Motronic engine control converts to a homogenous cylinder charge at a predefined engine load. The system injects very early during the intake process in order to achieve a good mixture of fuel and air at a ratio of lambda = 1.

163

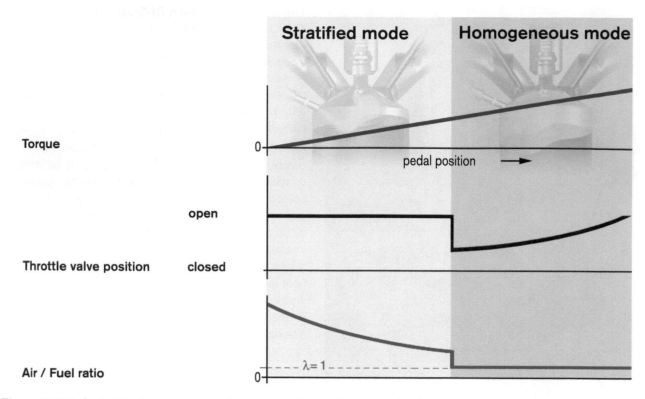

Figure 6.264 Switching between operating modes depending on engine load

Fuel calculation As is the case for conventional manifold injection systems, the amount of air drawn in for all operating modes is adjusted through the throttle valve according to the desired torque specified by the driver. The DI-Motronic ECU calculates the amount of fuel to be injected from the drawn-in air mass and performs an additional correction via lambda control. In this mode of operation, a torque increase of up to 5% is possible. Both the thermodynamic cooling effect of the fuel vaporizing directly in the combustion chamber, and the higher compression of the engine with gasoline direct injection, play a role in this.

Injection point For these different operating modes, two central demands are raised for engine control:

▶ The injection point must be adjustable between 'late' (during the compression phase) and 'early' (during the intake phase) depending on the operating point.

▶ The adjustment for the drawn-in air mass must be detached from the throttle pedal position in order to permit un-throttled engine operation in the lower load range. However, throttle control in the upper load range must also be permitted.

With optimal use of the advantages, the average fuel saving is up to 15%.

Nitrogen oxides In stratified charge operation the nitrogen oxide (NOX) segments in the very lean exhaust cannot be reduced by a conventional, three-way catalytic converter. The NOX can be reduced by approximately 70% through exhaust returns before the catalytic converter. However, this is not enough to fulfil the ambitious emission limits of the future. Therefore, emissions containing NOX must undergo special treatment. Engine designers are using an additional NOX accumulator catalytic converter in the exhaust system. The NOX is deposited in the form of nitrates (HNO_3) on the converter surface, with the oxygen still contained in the lean exhaust.

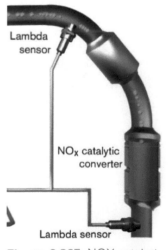

Figure 6.265 NOX catalytic converter

Emissions The capacity of the NOX accumulator catalytic converter is limited. As soon as it's exhausted, the catalytic converter must be regenerated. In order to remove the deposited nitrates, the DI-Motronic briefly changes over to its third operating mode (rich homogenous operation with lambda values of about 0.8). The nitrate together with the carbon monoxide (CO) is reduced in the exhaust to non-harmful nitrogen and oxygen. When the engine operates in this range, the engine torque is adjusted according to the gas pedal position via the throttle valve opening. Engine management has the difficult task of changing between the two different operating modes, in a fraction of a second, in a way not noticeable to the driver.

Figure 6.266 Vehicle under test conditions

Summary The continuing challenge, set by legislation, is to reduce vehicle emissions to very low levels. Bosch is a key player in the development of engine management systems. The DI-Motronic system, which is now or will soon be used by many manufacturers, continues to reflect the good name of the company.

Create a mind map to illustrate the features of a key component or system

6.4.4 Gasoline DI electronic control

Introduction Achieving highly precise engine control demanded by today's high-end engines is only possible with gasoline direct injection. By and large the biggest challenge is processing more and more data in shorter time intervals, with an increasing proportion of model-based control. Infineon microcontrollers, for example, feature outstanding computing performance.

Benefits The benefits of this system include:

▶ Flexible and scalable product portfolio tailored to the performance and real-time needs of the premium and value segment
▶ Conforms with the latest emission legislation while delivering the highest possible fuel efficiency

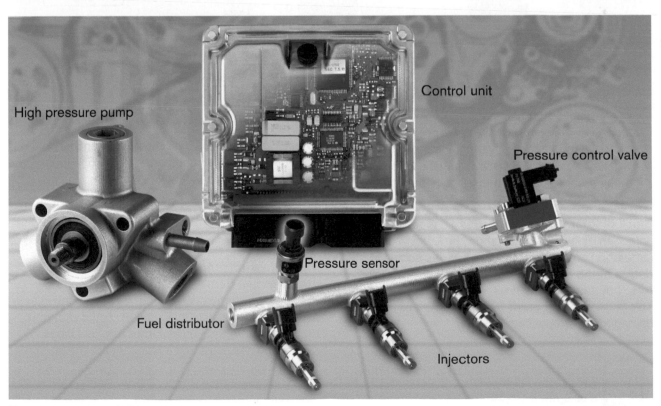

Figure 6.267 Bosch DI-Motronic components

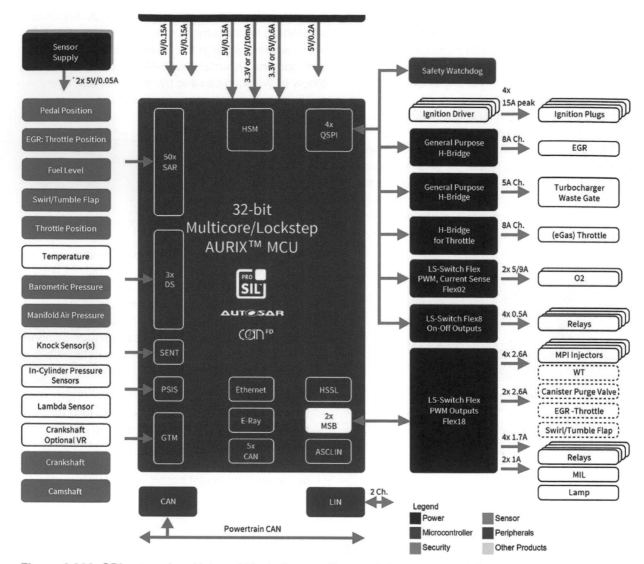

Figure 6.268 GDi external and internal block diagram (Source: Infineon, www.infineon.com)

▶ Benchmark-setting real-time performance facilitates down-sizing, direct injection, turbocharging and highly efficient after-treatment

▶ New sensor families provide enhanced measurement precision (e.g. ignition control, misfire detection)

Block diagram The block diagram uses the normal approach of inputs (sensors) on the left and outputs (actuators) on the right. It also includes additional details of the driver circuits and the microcontroller. Note how the outputs have stated maximum currents. This is because different output stages can be damaged if this is exceeded.

 Look back over the previous section and write out a list of the key bullet points from this section

6.4.5 Diesel introduction ❶

Diesel fuel injection systems Diesel engines have the fuel injected into the combustion chamber where it is ignited by heat in the air charge. This is known as compression ignition (CI) because no spark is required. The high temperature needed to ignite the fuel is obtained by a high compression of the air charge.

High-pressure pump Diesel fuel is injected, under high pressure from an injector nozzle, into the combustion chambers. The fuel is pressurised in a diesel injection pump. It is supplied and distributed to the injectors through high-pressure fuel pipes. Some engines use a unit injector where the pump and injector are combined in a single unit. The high pressure generation is from a direct acting cam or a separate pump.

The **airflow** into a diesel engine is usually unobstructed by a throttle plate so a large air charge is always provided. Throttle plates may be used to provide control for emission devices. Engine speed is controlled by the amount of fuel injected. The engine is stopped by cutting off the fuel delivery. For all engine operating conditions a surplus amount of air is needed for complete combustion of the fuel.

Direct and indirect injection Small high-speed diesel engine compression ratios are from about 19:1 for direct injection (DI) to 24:1 for indirect injection (IDI). These compression ratios are capable of raising the air charge to temperatures of between 500°C and 800°C. Very rapid combustion of the fuel occurs when it is injected into the hot air charge.

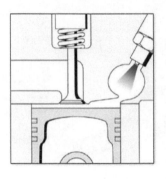

Figure 6.269 Indirect and direct injection

Diesel combustion process The combustion process follows three phases. These are the ignition delay, flame spread and controlled combustion phases. In addition, an injection lag occurs in the high pressure pipes as the pressure builds up just before injection.

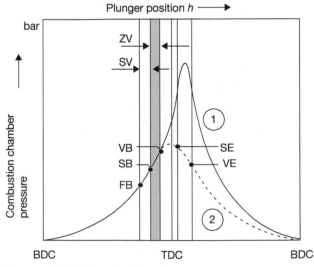

Figure 6.270 Combustion phases

Phases of diesel fuel combustion The most important phase of controlled combustion is when fuel is being injected into a burning mixture. This must be at a rate that maintains an even combustion pressure onto the piston throughout the critical crankshaft rotational angles. This gives maximum torque and efficient fuel usage because temperatures remain controlled and the heat lost to the exhaust is minimised. The low temperatures also help to keep nitrogen oxide emissions (NOX) to a minimum.

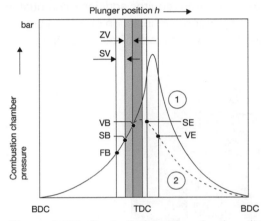

Figure 6.271 Controlled combustion in cylinder

Flame spread The speed of flame spread in a diesel engine is affected by the air charge temperature and the atomisation of the fuel. These characteristics are shared with the delay period. A sufficiently high air charge temperature, of at least 450°C, is a minimum requirement for optimum ignition and combustion.

The **delay phase** or ignition lag for diesel fuel combustion lasts a few milliseconds. It occurs immediately on injection as the fuel is heated up to the self-ignition temperature. The length of the delay is dependent on the compressed air charge temperature and the grade of fuel. The air charge temperature is also affected by the intake air temperature and the engine temperature.

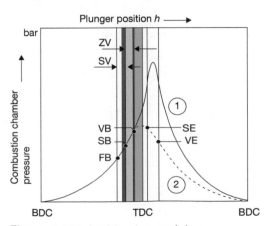

Figure 6.272 Ignition lag or delay

167

Diesel knock A long delay period allows a high volume of fuel to be injected before ignition and flame spread occurs. In this situation diesel knock is at its most severe. When a diesel engine is cold, there may be insufficient heat in the air charge to bring the fuel up to the self-ignition temperature. When ignition is slow, heavy knocking occurs.

Cold start devices To aid starting and to reduce diesel knock, cold start devices are used. For indirect injection engines, starting at lower than normal operating temperatures requires additional combustion chamber heating. For direct injection engines, cold start devices are only required in frosty weather.

Figure 6.273 Glow plugs

Initial delay An initial delay, known as injection lag, occurs in the high-pressure fuel lines. This occurs between the start of the pressure rise and the point when pressure is sufficient to overcome the compression spring force in the injectors.

Diesel fuel injection timing Ignition of the fuel occurs in the combustion chamber at the time of injection of fuel into the heated air charge. The injection point and the ignition timing are therefore, for all purposes, the same thing.

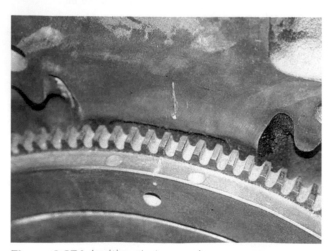

Figure 6.274 Ignition timing mark

Injection timing Diesel engine injection timing is equivalent to the ignition timing for petrol engines. Injection timing must fall within a narrow angle of crankshaft rotation. It is advanced and retarded for engine speed and load conditions. Injection timing is set by accurate positioning of the fuel injection pump. Incorrect timing leads to power loss. An increase in the production of nitrogen oxides (NOX) when too far advanced, or an increase in the hydrocarbon (HC) emissions when too far retarded, also occurs.

Figure 6.275 Diesel pump being timed on an old engine

Particulates Another exhaust gas constituent is particulate emissions. These result from incomplete combustion of the fuel. Particulates are seen as black carbon smoke in the exhaust under heavy load or when fuel delivery and/or timing is incorrect. White smoke may also be visible at other times, such as when the injection pump timing is incorrect. It also occurs when compression pressures are low or when coolant has leaked into the combustion chambers.

Direct and indirect injection Direct injection (DI) is made into a combustion chamber formed in the piston crown. Indirect injection (IDI) is made into a pre-combustion chamber in the cylinder head. Direct injection engines are generally more efficient but the indirect types are quieter in operation. The internal stresses in the engine are very high. Direct injection produces a higher detonation stress than indirect injection and therefore the smaller engines tended, until recently, to be the indirect type.

Electronic control Recent developments in electronic diesel fuel injection control have made it possible to produce small direct injection engines. It is probable that all new designs of diesel engine will be of this type. Diesel engines are built to withstand the internal stresses, which are greater than other engines. Diesel engines are particularly suitable for turbocharging. This improves power and torque outputs.

Figure 6.276 Common rail injection

Exhaust gas recirculation Exhaust gas recirculation (EGR) has two advantages for diesel engine operation. EGR is usually used to reduce nitrogen oxide (NOX) emissions and this is true for diesel engines. Additionally, a small quantity of hot exhaust gas in the air charge of a cold engine helps to reduce the delay period and the incidence of cold engine diesel knock.

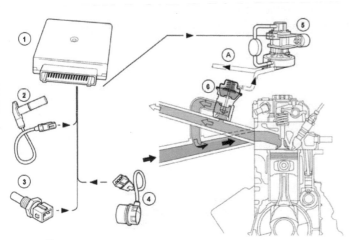

Figure 6.277 EGR

Catalytic converters Many modern diesel engine vehicles are fitted with oxidation catalytic converters that work in conjunction with other emission components to reduce hydrocarbon and particulate emissions. Turbocharging, EGR and catalytic converters are described in the 'Air Supply, Exhaust and Emission Control' learning programme.

Figure 6.278 Inside a catalytic converter

Injection pressures The fuel systems for direct and indirect injection are similar and vary only in injection pressures and injector types. Until recently, all light, high-speed diesel engines used rotary diesel fuel injection pumps. These pumps produce injection pressures of over 100 bar for indirect engines. However, these can rise up to 1000 bar at the pump outlet for turbocharged direct injection engines.

Pressure differential Injectors operate with a pulsing action at high pressure to break the fuel down into finely atomised parts. Atomisation is critical to good fuel distribution in the compressed air charge. The air charge pressure may be in excess of 60 bar. The pressure differential, between the fuel injection pressure and air charge pressure, must be sufficient to overcome the resistance during injection. This also gives good fuel atomisation and a shorter injection time.

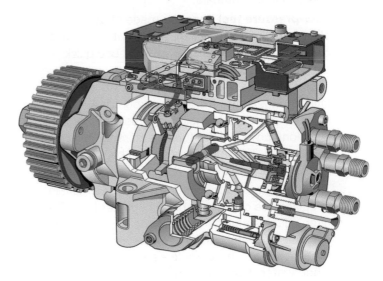

Figure 6.279 Bosch injection pump

Swirl An aid to good fuel distribution in the air charge is the swirl in the airflow induced in the inlet manifold. This is created by the combustion chamber design.

Airflow into and out of the pre-combustion chamber produces a swirl in the chamber. These chambers are often referred to as swirl chambers. The 'bowl in piston' combustion chambers, of direct injection engines, are shaped to maintain the induction air swirl during compression and combustion.

Diesel fuel injection components The main components of a diesel fuel system provide for either the low pressure or the high pressure functions. The low pressure components are the fuel tank, the fuel feed and return pipes and hoses, a renewable fuel filter with a water trap and drain tap, and a priming or lift pump. Fuel heaters may be fitted in the filter housing to reduce the risk of paraffin separation and waxing at freezing temperatures.

High pressure components The high pressure components are the fuel injector pump, the high pressure pipes and the injectors. Other components provide for cold engine starting. Electronically controlled systems include sensors, an electronic diesel control (EDC) module and actuators in the injection pump.

Low pressure components The fuel tank is a pressed steel, sealed unit, treated both inside and out with anti-corrosion paint. The inside is treated in order to resist corrosion from water that accumulates at the bottom of the tank. Some modern tanks are manufactured from a plastic compound that is burst-proof in an accident. They are also unaffected by diesel fuel, which can attack some plastic materials.

Low-pressure fuel lines are made of steel or hard plastic and connections are made with short hoses clamped at each end. New vehicles use quick coupling connections for ease of service and assembly operations. The feed lines run from the tank to the filter and then onto the injection pump. A low-pressure return line is used to maintain a fuel flow through the injection pump and the fuel injectors for lubrication and cooling. The return carries fuel back to the filter housing or the fuel tank.

The **fuel filter** is a microporous paper element in a replaceable canister or detached filter bowl. The filter includes a water and sediment trap and tap for draining the water. Many vehicles have a sensor in the water trap. This completes a warning lamp circuit when water is detected above a certain level. All diesel fuel entering the injection pump and injectors must be fully filtered. The internal components of the pump and injectors are manufactured to very fine tolerances. Even very small particles of dirt could be damaging to these components.

Figure 6.280 Filter and housing

Fuel heating may be provided from the engine coolant or by an electric heater element in the filter housing. The fuel is lifted from the fuel tank to the injection pump by the transfer pump in the injection pump on some vehicles. This is possible where the distance and height of lift are of small dimensions.

Fuel lift pump For improved delivery and for priming the injection pump, another pump may be necessary. A conventional fuel lift pump driven from the engine camshaft is a common method. These pumps, in some instances, have an external operating lever. Hand-operated priming pumps are fitted for use when the vehicle runs out of fuel. They are also used for service operations such as when the filter is changed. Many modern injector pumps are self-priming.

Figure 6.281 Heated fuel filter housing

Figure 6.282 Fuel priming pump

Injection pump The injector pump shown is a rotary distributor type pump. These pumps are filled with diesel fuel, which provides not only fuel for the engine but also for full lubrication and cooling of the pump. These pumps are made from specially manufactured materials with surface treatments. Parts are lapped together to give very fine tolerances. Only clean and filtered fuel should be used to avoid damage to these parts.

Types of pump There are two types of pump with different internal operation. Two major original designers make or license the manufacture of these pumps. The Lucas DP and Bosch VR pumps are radial-piston designs. They use opposing pistons or plungers inside a cam ring to produce the high pressure. Bosch V series pumps are axial-piston designs, having a roller ring and cam plate attached to an axial piston or plunger in the distributor head to generate the high pressure.

Pump operation The operation of the two types of injector pump are quite different and explained separately further on in this section. Later versions of these pumps have electrical and electronic control. The latest versions have full electronic control.

Figure 6.283 Bosch pumps

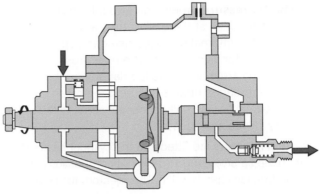

Figure 6.284 Bosch VE pump cam plate and plunger

High pressure outlets All types of pumps have delivery valves or pressure valves fitted to the high pressure outlets, which feed to each cylinder in turn. The delivery valves control the generation of pressure waves in the high pressure pipes. They do this by initially giving a quick pressure drop and then by retaining a lower residual pressure. The delivery valve consists of a conical valve held closed by a compression spring and opened by hydraulic pressure when injection pressures are produced in the injection pump.

Delivery valve Closure of the delivery valve, when the injector pump pressure drops, allows a quick pressure drop in the pipe and injector. This means the injectors will close fully. Without the valve, pressure waves would oscillate in the pipe and force the injector to reopen. This would cause unwanted fuel to be injected. The retained low pressure helps to prevent fuel dribble from the injector nozzle during the non-injection period. It also aids lubrication through the leak-off pipes.

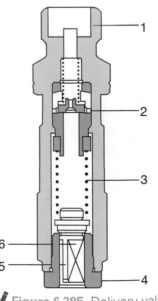

Figure 6.285 Delivery valves

The **high pressure pipes** are of double thickness steel construction and are all of the same length. This is so that the internal pressure rise characteristics are identical for all cylinders. The high pressure connections are made by rolled flanges on the pipe ends and threaded unions securing the rolled flanges to convex, or occasionally concave, seats in the delivery valves and injectors.

The **fuel injectors** are fitted into the cylinder head with the nozzle tip projecting into the pre-combustion (IDI) or combustion chamber (DI). The injectors for indirect combustion are of a pintle or 'pintaux' design and produce a conical spray pattern on injection. The injectors for direct injection (DI) are of a pencil-type multi-hole design that produces a broad distribution of fuel on injection.

Figure 6.286 DI injector

Injector operation Fuel injectors are held closed by a compression spring. They are opened by hydraulic pressure when it is sufficient to overcome the spring force on the injector needle. The hydraulic pressure is applied to a face on the needle where it sits in a pressure chamber. The fuel pressure needed is in excess of 100 bar (1500 psi). This pressure lifts the needle and opens the nozzle so that fuel is injected in a fine spray pattern into the combustion chamber.

Figure 6.287 IDI injector

Injector spray The pressure drops when fuel is injected and the spring force on the needle closes the injector. This is immediately followed up by a build-up of pressure that again opens the nozzle. This results in a cycle of oscillations of the needle to give a finely atomised and almost continuous spray. The spray continues until the pump pressure is reduced at the end of the delivery stroke.

New types of injector The newer types of injector have two springs of different value in order to provide a small initial charge for ignition and then the main charge for controlled burning. These injectors reduce diesel knock on direct injection engines and give a smoother engine performance. Fuel injectors are carefully matched to the type of engine and pump.

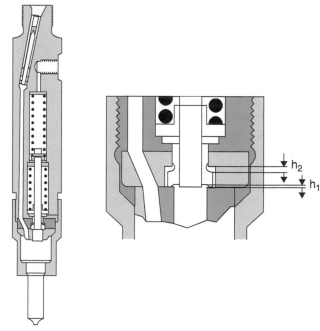

Figure 6.288 Two-spring direct injection injector

Glow plugs There are two types of cold starting devices used on diesel engines. These are glow plugs and flame start devices. Glow plugs are used mainly on indirect injection engines although they are used on some small direct injection engines. Flame start devices are used on many, but not all, direct injection engines. Glow plugs are fitted in the combustion chamber. Their purpose is to help to ignite the fuel during injection.

Flame start devices are fitted in the inlet manifold. They preheat the intake air so that it achieves a high temperature on the compression stroke. The online diagram shows the components of a flame start device and the electrical control circuit.

 Create a word cloud for one or more of the most important screens or blocks of text in this section

6.4.6 Bosch VR system ❷

VR Pumps with electronic control The Bosch VR pumps are used on high-speed direct injection diesel engines for cars and light commercial vehicles. They are radial-piston distributor injection pumps having opposing plungers that are forced inwards by cam lobes on the inside of a cam ring in order to produce high pressure, which can be up to 1400 bar in some applications. The cam is located in the pump body and the plungers are in the rotor driven by the pump spindle. Four-cylinder engines have two plungers and four cam lobes. Six-cylinder engines have three plungers and six cam lobes. The pump is driven from the engine at half crankshaft speed.

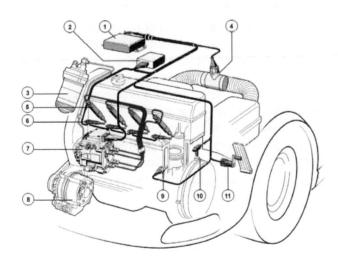

Figure 6.289 Bosch VR pump system

Low pressure feed A low pressure feed to the injection pump is provided by a submerged electrical pump in the fuel tank. This provides for priming and positive pressure in the injection pump. In common with all diesel fuel systems, a fuel filter and water trap are used to ensure that only very clean fuel is delivered to the pump. Return pipes are used for excess fuel leakage, for purging the pump and for lubrication of the injectors.

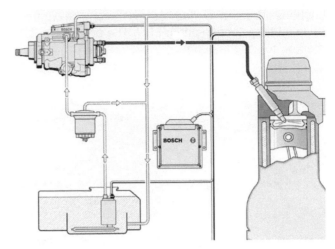

Figure 6.290 Fuel system

Vane pump Inside the distributor pump is a vane-type pump, which is used to produce the pump body pressure. Pump body pressure is used for charging the high-pressure chamber between the plungers and for injection advance. A pressure control valve is used to prevent excessive pressure. It is a spring-loaded plunger that is lifted by hydraulic pressure to expose ports in the valve bore. This will then allow fuel to flow back to the inlet side of the vane-type pump.

Low pressure stage An overflow throttle valve, in the pump housing, is used to allow a defined quantity of fuel to flow back to the fuel tank at all times. This provides some cooling in the pump and venting of air during pump priming. A second larger overflow bore in the valve opens at a given pressure to allow a flow of fuel from the distributor head.

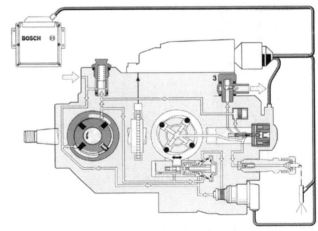

Figure 6.291 Distributor pump – low pressure stage

Electronic control The Bosch VR pump has full electronic control for fuel metering and for injection advance. The electronic diesel control unit consists of two electronic control modules to perform the control functions. These two modules are the engine control ECU and the injection-pump ECU. The pump ECU is fitted on top of the pump.

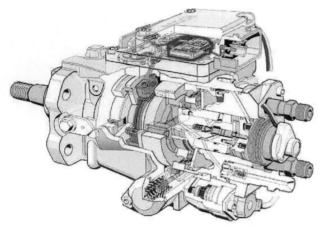

Figure 6.292 Pump ECU

Fuel metering is controlled by the high-pressure solenoid valve. This is an electrically actuated valve set centrally inside the distributor rotor. There are connecting bores in the distributor rotor for filling of the high-pressure circuit, through the inlet port at pump body pressure, and for delivery at high pressure to the fuel injectors. These are either connected or separated by the position of the valve.

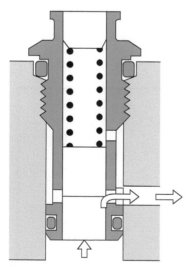

Figure 6.293 Pressure control valve

Solenoid valve The high-pressure solenoid valve is closed by an electrical signal from the pump electronic control unit. When the valve is closed, fuel under high pressure passes from the high-pressure pump chamber, through the bores in the rotor and distributor head, the return-flow throw throttle valve (delivery valve) and out to the injectors. It is then injected

into the engine combustion chambers. The few microseconds of time that the valve remains closed are referred to as the delivery or injection period.

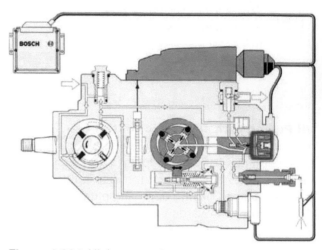

Figure 6.294 High pressure stage

Injection period The delivery or injection period starts when the solenoid valve is closed. This occurs at the beginning of the injection period when the high-pressure plungers are at bottom dead centre on the cam lobes. At this point, they are just beginning to be forced inwards for high-pressure generation. Fuel injection continues whilst the valve is closed.

End of the delivery The end of the delivery or injection period occurs when the solenoid valve is opened. A return compression spring in the rotor acts on the valve to open it when the signal current from the pump electronic control unit is switched off. The pressure in the high-pressure chamber, rotor bores and delivery pipes to the injectors is dissipated to the inlet side of the high-pressure pump. Pressure surges are controlled by an accumulator in the low-pressure inlet.

Engine ECU and pump ECU The quantity of fuel that is metered for injection at any time is computed by the engine ECU, which sends signals to the injection pump ECU for control of the high-pressure solenoid valve. The electrical current for operating this valve is high and the two electronic control units are separated, in order to avoid high current interference, in the more electronically vulnerable engine ECU.

Electronic diesel control units The electronic diesel control units are provided with data signals from sensors and switches attached to the engine, the pump and other vehicle systems. The sensors are used for comparisons to programmed operating parameters and for calculations for metering the amount of fuel delivered and for controlling the injection advance.

Injection advance mechanism Injection advance is obtained by rotation of the cam ring by pump body pressure in the injection advance mechanism. The injection advance mechanism consists of a transverse timing device piston and control components and an electrical solenoid valve. Maximum advance is 40° of crankshaft rotation.

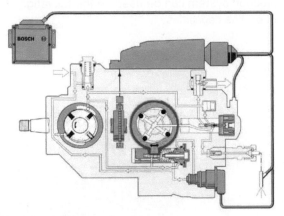

Figure 6.295 Cam ring

The **timing device piston** has a cut out section that locates a ball pivot on the cam ring. Hydraulic pressure from the vane-type supply pump is used to move and control the movement of the timing device piston. Transverse movement of the piston causes the cam ring to move radially. The plungers, in the high-pressure pump, now come into contact with the cam lobes at an earlier or later time. This advances or retards the point of injection.

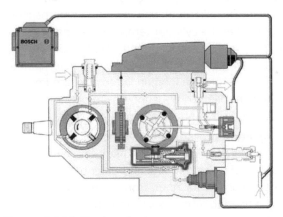

Figure 6.296 Timing device

Injection points Movement of the timing device piston occurs because of the increase in pump body pressure, which occurs with an increase in engine speed. The pressure is applied, as control pressure through a restriction, to the ring chamber of the hydraulic stop. When the solenoid valve is closed, hydraulic pressure is applied to the control plunger, which shifts to push the control collar to a position that allows fuel pressure

to be applied to the timing device piston. Movement of the timing device piston is opposed by a return spring. Once this resistance is overcome, the piston moves to advance the point of injection.

Injection advance Control of the injection advance is made by regulating the hydraulic pressure in the ring chamber of the hydraulic stop. Variation of the pressure affects the relative position of the control plunger to the spill ports in the piston control bore. This regulates the amount of pressure available behind the piston to move it in an advance direction.

Ring chamber Pressure control of the ring chamber pressure is made by releasing the pressure through the needle valve of the timing device solenoid valve. This valve opens and closes from actuating electrical pulses from the pump ECU. The signal pulses are calculated in the pump ECU and are based on data received from the system sensors.

The **needle motion sensor** sends a signal to the engine ECU at the instant of opening of the injector. This point, relative to the crankshaft rotational angle before top dead centre, is used for load and speed injection timing calculations and for control of the exhaust gas recirculation valve.

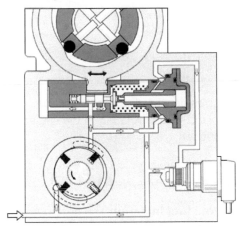

Figure 6.297 Timing device solenoid valve

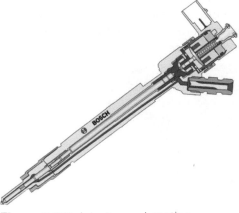

Figure 6.298 Injector and motion sensor

175

Angular position of the cam ring The angle of rotation sensor is fitted inside the pump. It consists of a finely toothed trigger wheel on the pump spindle and an inductive sensor on the cam ring. The sensor signals are used to define the actual angular position of the cam ring. The large tooth gaps on the trigger wheel are positioned to provide cam lobe to plunger position for start of injection for each cylinder. The inductive pick-up of the angle of rotation sensor is fitted to the ring cam so that the data provided is always specific to the point of injection. Its main function is to control the actuation of the high-pressure solenoid at the start of injection.

Summary The Bosch, VR electronic diesel control system, uses a number of sensors and control actuators. This allows it to achieve optimum performance. A range of other sensors and actuators are used.

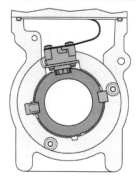

Figure 6.299 Angle of rotation sensor

 Use the media search tools to look for pictures and videos relating to the subject in this section

6.4.7 Bosch CR system ❸

Bosch common rail fuel injection The development of diesel fuel systems is continuing, with many new electronic changes to the control and injection processes. One of the latest developments is the Bosch CR 'common rail' system, operating at very high injection pressures. It also has piloted and phased injection to reduce noise and vibration.

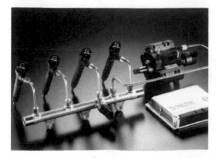

Figure 6.300 Common rail system

The **common rail system** has made it possible, on small high-speed diesel engines, to have direct injection when previously they would have been of indirect injection design. These developments are showing improvements in fuel consumption and performance of up to 20% over the earlier indirect injection engines of a similar capacity. The common rail injection system can be used on the full range of diesel engine capacities.

Figure 6.301 Four-cylinder system

Pilot injection The combustion process, with common rail injection, is improved by a pilot injection of a very small quantity of fuel, at between 40° and 90° BTDC. This pilot fuel ignites in the compressing air charge so that the cylinder temperature and pressure are higher than in a conventional diesel injection engine at the start of injection. The higher temperature and pressure reduces ignition lag to a minimum so that the controlled combustion phase during the main injection period is softer and more efficient.

Fuel injection pressures are varied, throughout the engine speed and load range, to suit the instantaneous conditions of driver demand and engine speed and load conditions. Data input, from other vehicle system ECUs, is used to further adapt the engine output to suit changing conditions elsewhere on the vehicle. Examples are traction control, cruise control and automatic transmission gearshifts.

The **electronic diesel control** (EDC) module carries out calculations to determine the quantity of fuel delivered. It also determines the injection timing based on engine speed and load conditions. The actuation of the injectors, at a specific crankshaft angle (injection advance), and for a specific duration (fuel quantity), is made from signal currents from the EDC module. A

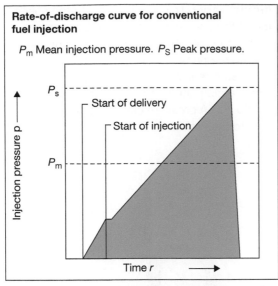

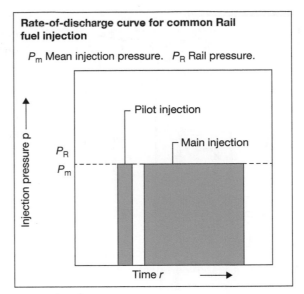

Figure 6.302 Conventional system compared to Common rail system

further function of the EDC module is to control the accumulator (rail) pressure.

Common rail component groups The Bosch CR common rail diesel fuel injection system, for light vehicles, consists of four main component areas. These are the low pressure delivery, high pressure delivery with a high-pressure pump and accumulator (rail), the electronically controlled injectors and electronic control unit and associated sensors and switches.

Figure 6.303 Signal processing in the ECU

Low pressure delivery The low pressure delivery components are the fuel tank, a pre-filter, pre-supply (low pressure) pump, a fuel filter and the low pressure delivery pipes to the high-pressure pump and for excess fuel return. The low pressure pump, depending on application, can be of the roller cell type and be fitted in either the fuel tank or in-line where it is mounted to the vehicle body close to the fuel tank. Where the pump is fitted in the fuel tank, it includes a pre-filter and has the fuel gauge sender unit attached to the same attachment flange on the side or top of the fuel tank.

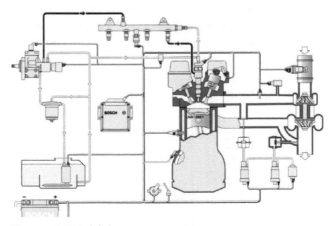

Figure 6.304 Main components

Fuel pump The electrical supply to the fuel pump is made either directly or through a relay from the electronic diesel control module. An inertia switch is generally used to cut the electrical current to the pump motor in an accident. On some vehicles, a gear-type

177

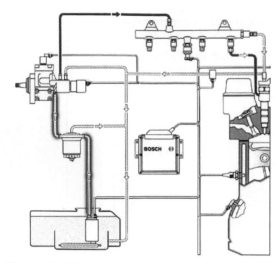

Figure 6.305 Low pressure components

pump may be incorporated into the high-pressure pump and be driven from a common drive shaft. It can be a separate pump attached to the engine with a geared drive from the camshaft or crankshaft. The low pressure delivery pipes connect to a fuel filter and water trap. A continuous flow of fuel runs through the filter and primes the high-pressure pump or returns to the fuel tank.

The **high-pressure pump** is driven from the engine crankshaft through a geared drive at half engine speed and is fitted where a conventional distributor pump would be. It can also be fitted on the end of the camshaft housing and be driven by the camshaft. It is lubricated by the diesel fuel that flows through it.

High-pressure fuel injection The pump has to produce all of the high pressure for fuel injection. It is a triple piston radial pump, with a central cam for operation of the pressure direction of the pistons and return springs to maintain the piston rubbing shoes in contact with the cam. The pump has a positive displacement with inlet and outlet valves controlling the direction of flow through the pump.

The **pump delivery rate** is proportional to the speed of rotation of the engine so that it meets most engine speed requirements. To meet the engine load requirements, the pump has a high volume. To meet the high pressure requirements for fine atomisation of the fuel on injection, the pump can produce pressures of up to 1350 bar. A pressure control valve returns excess fuel to the fuel tank.

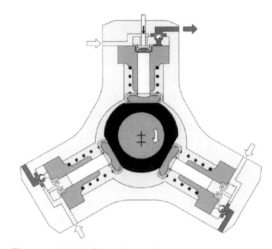

Figure 6.307 Pumping elements

Controlling pressure The pressure control valve is a mechanical and electrical unit. It is fitted on the pump or the high pressure accumulator (rail). The mechanical part of the valve consists of a compression spring that acts on a plunger and ball valve. The electrical component is a solenoid that puts additional and variable force to the ball valve. The solenoid is actuated on signal currents from the EDC module. When the solenoid is not actuated, the ball valve opens at 100 bar against the resistance of the compression spring. This spring valve smooths out some of the high-frequency pressure fluctuations produced by the pump.

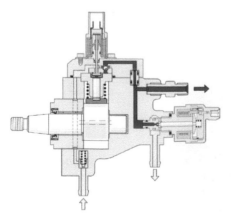

Figure 6.306 High-pressure pump

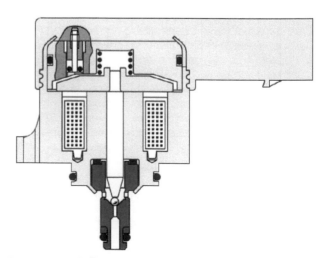

Figure 6.308 Pressure control valve

Setting a variable mean pressure The solenoid in the pressure control valve is used for setting a variable mean pressure in the high pressure accumulator (rail). The pressure in the rail is measured by a sensor and compared with a stored map in the EDC module for the current engine operating conditions. In order to increase the fuel rail pressure an electrical alternating current is applied to the solenoid. The energising current is varied by the EDC module, so that the additional force on the ball valve produces the required fuel rail pressure.

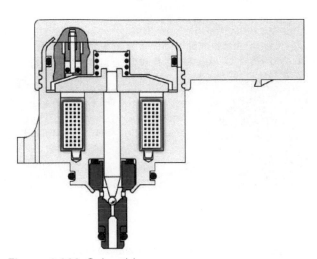

Figure 6.309 Solenoid

High pressure accumulator rail The high pressure accumulator (rail) is common to all cylinders and derives its name 'common rail' from this. This term is used in preference to fuel rail, which is used for petrol engines. The rail is an accumulator because it holds a large volume of fuel under pressure. The volume of fuel is sufficient to dampen the pressure pulses from the high-pressure pump. Fitted to the high pressure accumulator are an inlet from the high-pressure pump and flow limiter valves in each of the outlets for the cylinders. A pressure sensor and a pressure limiter valve that returns fuel to the fuel tank, if excess fuel pressure occurs in the rail, are also fitted.

Preventing continuous injection There is a flow limiter valve on the high pressure outlets to the injectors for each of the engine cylinders. The flow limiter valve is needed to ensure that continuous injection cannot occur if an injector nozzle remains open. The flow limiter valve consists of a spring-loaded hollow plunger, with narrow throttle holes between the inlet and outlet ports.

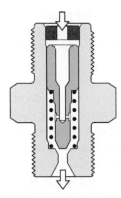

Figure 6.310 Flow limiter valve

Normal operation of the injector During normal operation, the injector supply pipe, flow limiter valve and fuel rail are under the same pressure. When the injector opens, the pressure in the injector supply pipe drops, and the rail pressure pushes the flow limiter plunger towards the outlet and the seat of the valve. At the end of injection, fuel flows through the narrow throttle holes and the pressure equalises, so that the return spring pushes the plunger back to its stop. This cycle continues for normal injection.

Drop in pressure If a large leak in the high-pressure fuel pipe or the injector occurs, the flow limiter valve will operate. The drop in pressure on the outlet side will cause a pressure differential, which is sufficient to push the plunger into the closed position of the valve. It will hold it in this position until the fuel rail pressure is released.

Rail pressure The rail pressure sensor is a very accurate electronic unit, with an error value of less than 2%. It consists of an integrated electronic element with a diaphragm and evaluation circuit. The diaphragm, which is open to fuel rail pressure, is a layered semiconductor. It distorts when fluid pressure is applied to it. The evaluation circuit measures the resistance changes in the diaphragm. When it distorts, it converts the electrical value to a range between 0.5 and 4.5V. The signal is supplied to the EDC module.

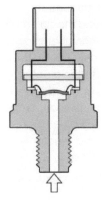

Figure 6.311 Rail pressure sensor

Monitoring of rail pressure The fuel rail pressure is closely monitored in the EDC module. This is because the changes in fuel pressure affect the quantity of fuel delivered within a set time range. The EDC module sends actuating signal currents to the pressure control valve solenoid to hold the instantaneous pressure at the appropriate map value. The pressure limiter valve is fitted in the end of the fuel rail. It is a spring-loaded plunger-type valve and opens at a pre-determined value set by the compression of the return spring. Fuel pressure opens the valve so that excess fuel is able to return to the fuel tank through the fuel return pipes.

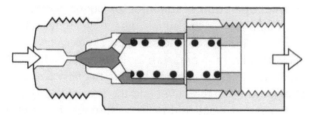

Figure 6.312 Pressure limiter valve

Electronic controlled injectors The injectors on the common rail system have nozzles that are similar to all other diesel injectors for direct injection engines. The nozzle needle seats in the nozzle to obstruct the holes in the tip where the fuel is injected into the combustion chamber. The nozzle needle is held closed by a compression spring and opened by hydraulic pressure. Opening and closing of the injector is controlled, not by a high-pressure fuel pulse from an injector pump, as in a conventional rotary distributor pump, but by actuation of an electrical solenoid in the injector body. This is controlled by the electronic diesel control module. A permanent high pressure is maintained in the injector at the same pressure as the rail.

Plunger and nozzle needle operation The plunger and nozzle needle are normally held closed by a compression spring. An electrical signal current from the EDC module actuates the solenoid so that the control valve opens. Fuel, under pressure, leaves the valve control chamber and passes through the bleed orifice into a large chamber around the solenoid armature. The pressure in the control chamber drops and the pressure in the pressure chamber below the control plunger is now greater. This lifts the plunger and needle. The injector nozzle opens and fuel is injected into the combustion chamber. As soon as the electrical signal current from the EDC module ceases, the control valve closes. Operation of the injector is controllable for very small intervals of time.

Summary The electronic control of the common rail diesel injection system has three main component groups. These are the sensors and switches and other system ECUs that provide data, the EDC module that analyses and calculates the system requirements, and the actuators for diesel fuel delivery and emission control functions. The main sensors for calculation of the fuel quantity and injection advance requirements are the accelerator pedal sensor, crankshaft speed and position sensor, air-mass meter and the engine coolant temperature sensor.

Figure 6.313 CR direct injection

 Create an information wall to illustrate the features of a key component or system

6.4.8 Diesel injection pressure

Higher injection pressure saves fuel and increases performance and torque. It is expected that most diesel engines will soon work with injection pressures of around 2,000 bar. Although 3,000 bar is not unrealistic, it will be limited to racing cars and high-performance diesel engines.

Piezo injector The common-rail system featured is Bosch's first piezo injector for passenger vehicles that works with an injection pressure of 2,500 bar. With their higher injection pressure, the new piezo models from Bosch are at the technological vanguard. The optimized fuel injection system atomizes the fuel more finely, improving combustion. Lower consumption is just one advantage of this technology.

The pressure generated by a common-rail system is roughly equivalent to the pressure a 2,000kg rhinoceros would exert standing on a fingernail. The compressed fuel is then finely dispersed at the speed of a supersonic jet.

Diesel facts

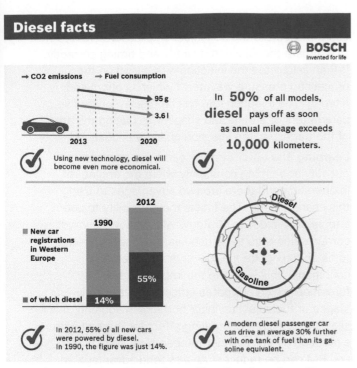

BOSCH
Invented for life

→ CO2 emissions   → Fuel consumption

95 g
3.6 l

2013   2020

✓ Using new technology, diesel will become even more economical.

In **50%** of all models, **diesel** pays off as soon as annual mileage exceeds **10,000** kilometers.

✓

■ New car registrations in Western Europe

1990   2012

55%

■ of which diesel   14%

✓ In 2012, 55% of all new cars were powered by diesel. In 1990, the figure was just 14%.

Diesel

Gasoline

✓ A modern diesel passenger car can drive an average 30% further with one tank of fuel than its gasoline equivalent.

Figure 6.314 Diesel facts. The diesel engine offers the ideal combination of fuel economy and driving pleasure, particularly appreciated by business travellers and commuters.

Power and torque A higher injection pressure generates greater specific power and increases torque. Therefore, increasing an engine's injection pressure makes it more powerful: the time available for combustion is extremely limited as soon as an engine is running at full load and high engine speed. This means the fuel must be injected into the engine very quickly at high pressure to achieve optimum power yield.

Figure 6.315 Bosch CRS3-25 with 2,500 bar pressure.

Under pressure The more air there is in the combustion chamber, the higher the injection pressure must be. A large amount of fuel must be introduced within a short space of time to achieve a combustible air-fuel mixture. Multiple turbocharged engines, particularly bi-turbo and tri-turbo models, benefit from injection pressures in excess of 2,000 bar. A higher injection pressure is a key factor in reducing an engine's untreated emissions. Indeed, in compact-class vehicles it can often even help to avoid the need for exhaust gas treatment. The greater the injection pressure, the more finely both the injector and injection nozzle can be constructed. This improves atomization and results in a better air-fuel mixture, meaning that optimum combustion is achieved and no soot can form.

Summary A higher injection pressure requires more than just a re-engineered injector. With its comprehensive diesel systems competence, Bosch is able to assemble a finely tuned system comprising not only the control unit, but also the fuel pump, the common-rail system and the injector.

> Use the media search tools to look for pictures and videos relating to the subject in this section

6.4.9 Common rail intelligent injectors

Introduction There have been some interesting developments in fuel system technology for common rail diesels. Denso has recently addressed pressure wave phenomena. This has taken the intelligence in diesel engine fuel systems to the next level with the introduction of the Intelligent Accuracy Refinement Technology (i-ART). This technology features a fuel-pressure sensor with an integrated microcomputer which monitors injection pressure based on various input data.

Integration The whole assembly is integrated into the top of each fuel injector. The closed-loop system precisely manages injections of fuel to match specific drive cycle conditions. It replaces the single pressure sensor typically positioned in the fuel rail. Denso engineers have stated that i-ART can improve fuel efficiency by 2%, compared with open-loop systems. It was developed to enable diesel engines to meet Euro 6 regulations with a reduced after-treatment burden.

6

Figure 6.316 The new range of Volvo power units includes Denso i-ART technology (Source: Volvo)

Injection quantity detection A conventional injection system could only detect an injection quantity based on indirect methods such as combustion or an engine rotation fluctuation. The i-ART system enables direct detection of the injection quantity; each injector is equipped with a built-in fuel pressure sensor to measure injection pressure inside the injector itself. Based on the information from the built-in pressure sensor, the engine ECU reads fuel pressure values for each injection rapidly and calculates an actual injection quantity and timing for each cycle based on this information, using a rapid waveform processing technique. The learning value for the injection quantity and timing calculated with the i-ART system is applied to subsequent injections and adapted throughout its lifetime.

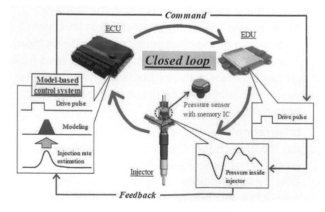

Figure 6.317 System overview - i-ART intelligent injectors and feedback data flow (source: Denso)

Pressure wave form The actual pressure wave form generated by the i-ART pressure sensor is shown

here. The system performs a pre-processing by compensation to the non-injection pressure waveform to estimate the injection quantity and timing correctly. It then calculates the injection rate based on the processed pressure waveform which is optimised by filtering. The injection rate can be expressed by five parameters of a trapezoid shape. Calculating the area of the trapezoid, the injection quantity is obtained.

Learning The i-ART system learns the injection quantity and timing constantly while the engine is in operation and there are two advantages to using this characteristic. The first is the possibility to use a triple pilot injection strategy, which allows a lower compression ratio to be used as less heat is needed to be able to ignite the fuel under all operating conditions. This is due to the improved mixture formation which promotes efficiency in the early stages of fuel injection/initial burning. In addition, this allows a sufficient preheating effect for the fuel with a reduced overall cylinder temperature, such that NOx and PM can be reduced. As a second advantage, in conjunction with cetane number detection, a stable combustion with minimised combustion noise can be achieved irrespective of the variation of cetane number with fuels in certain markets.

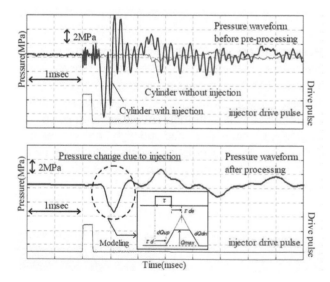

Figure 6.318 Fuel pressure waveforms at the i-ART injector (source: Denso)

Summary This technology is a big leap for common rail diesels, but also a significant step forward for measurement technology that can now be employed in production. There are significant advantages to being able to establish the fuel pressure directly at each injector, at the point of injection, as this helps considerably in being able to model the injection rate and fuel mass per stroke. The goal is to develop an injector where the rate and quantity of injection can be varied without a step and within a cycle. This would then facilitate the ability to truly control the combustion and energy release in a diesel engine, with high precision, on a cycle-by-cycle basis.

 Using images and text, create a short presentation to show how a component or system works

6.4.10 Coding components

Introduction Fitting a new component to a vehicle used to be a simple matter of obtaining the correct item and them following the appropriate procedures to 'bolt' and 'wire' it in place. In many cases it still is but on some vehicles certain parts have to be coded. In other words, the vehicle has to be 'told' what has changed. This can apply to all sorts of parts such as headlight modules but fuel injectors are most important. Another example would be fitting a tow bar – you must inform the car that it has one!

Figure 6.319 Bosch diesel injector

Diesel injectors Diesel injectors were generally 'plug and play'. However, since the more widespread use of common rail diesel systems, it can be confusing as to whether an injector needs coding or not. Common rail diesel systems have a number of advantages compared to earlier diesel systems. For example, improved performance, lowered fuel consumption, quieter engines and reduced emissions. Coding therefore is essential if the system requires it.

Injector code An injector code is usually known as an IMA code (Bosch and Siemens) or a calibration code (Delphi). Its purpose is to ensure accurate injection control. The IMA coding is now an industry standard, and stands for Injector Menge Abgleichung (Injector Quantity Offset). When an injector is tested during

manufacture, it generates an IMA code. This code identifies where the needle and nozzle assembly are in a tolerance range. When the injector is coded to the ECU it will be able to control the fuel accordingly and ensure optimal engine performance.

Figure 6.320 Injector code - a somewhat unusual one! (Source: Simon McCormac)

Failure to code If injectors are not coded correctly, this can result in several issues:

▶ Reduced performance, increased fuel consumption or even black smoke
▶ On some systems if you do not code in the injectors the vehicle will not start
▶ Engine management light is likely to come on

How to determine if the injector(s) need coding
Refer to manufacturer's information but the following is a general guide:

▶ Delphi common rail injectors always require coding. Usually the code is 16 alphanumeric characters (C2i Injectors) or 20 characters for newer applications (C3i injectors). The code is usually found on injector head.
▶ VDO/Siemens common rail injectors when used on VW applications require coding. Other vehicles will too so always check. Usual length is 6 digits and found on the injector head.

▶ DENSO common rail injectors all require coding except for some very early ones. The usual code length is 16-24 alphanumeric characters dependent upon vehicle manufacturer. The code, as with other types, is found on the injector head. A QR code on the injector may also be used.

▶ BOSCH common rail injectors all require coding except for some very early ones. The code is usually 6-10 digits and found on the injector head.

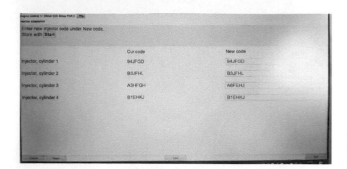

Figure 6.321 Changing injector codes (Source: Charlie Thompson)

Summary As a general rule, if an IMA code is present then the injector will need coding. A remanufactured injector should also have a new calibration code, which is usually on a sticker on the injector. Fuel pumps might need to be coded in a similar way to injectors and for the same reasons.

Figure 6.322 New and old injectors (Source: Charlie Thompson)

 Using images and text, create a short presentation to show how a component or system works

6.5 Ignition

6.5.1 Ignition overview ❶

Purpose The purpose of the ignition system is to supply a spark inside the cylinder, near the end of the compression stroke, to ignite the compressed charge of air/fuel vapour. For a spark to jump across an air gap of 1.0mm under normal atmospheric conditions, (1 bar) a voltage of 4 to 5kV is required. For a spark to jump across a similar gap in an engine cylinder, having a compression ratio of 8:1, approximately 10kV is required. For higher compression ratios and weaker mixtures, a voltage up to 20kV may be necessary. The ignition system has to transform the normal battery voltage of 12V to approximately 8 to 20kV and, in addition, has to deliver this high voltage to the right cylinder, at the right time. Some ignition systems will supply up to 40kV to the spark plugs.

Figure 6.323 Combustion taking place (Source: Ford Media)

Conventional ignition is the forerunner of the more advanced systems controlled by electronics. It is worth mentioning at this stage, however, that the fundamental operation of most ignition systems is very similar. One winding of a coil is switched on and off causing a high voltage to be induced in a second winding. The basic types of ignition system can be classified as shown in the table.

Type	Conventional	Electronic	Programmed	Distributorless
Trigger	Mechanical	Electronic	Electronic	Electronic
Advance	Mechanical	Mechanical	Electronic	Electronic
Voltage source	Inductive	Inductive	Inductive	Inductive
Distribution	Mechanical	Mechanical	Mechanical	Electronic

Engine management Modern ignition systems now are part of the engine management, which controls fuel delivery, ignition, and other vehicle functions. These systems are under continuous development and reference to the manufacturer's workshop manual is essential when working on any vehicle. The main ignition components are the engine speed and load sensors, knock sensor, temperature sensor and the ignition coil. The ECU reads from the sensors, interprets and compares the data, and sends output signals to the actuators. The output component for ignition is the coil.

Figure 6.324 Electronic ignition and injection

Developments Ignition systems continue to develop and will continue to improve. However, keep in mind that the simple purpose of an ignition system is to ignite the fuel/air mixture every time at the right time. And, no matter how complex the electronics may seem, the high voltage is produced by switching a coil on and off.

Generation of high voltage If two coils (known as the primary and secondary) are wound on to the same iron core, then any change in magnetism of one coil will induce a voltage in to the other (see Chapter 7 for more details). This happens when a current is switched on and off to the primary coil. If the number of turns of wire on the secondary coil is more than the primary, a higher voltage can be produced. This is called transformer action and is the principle of the ignition coil.

The value of this 'mutually induced' voltage depends upon:

▶ primary current
▶ turns ratio between primary and secondary coils
▶ speed at which the magnetism changes.

The two windings are wound on a laminated iron core to concentrate the magnetism. This is how all types of ignition coil are constructed.

Ignition timing For optimum efficiency, the ignition timing (or advance angle) should be such as to cause the maximum combustion pressure to occur about 10° after TDC. The ideal ignition timing is dependent on two main factors, engine speed and engine load. An increase in engine speed requires the ignition timing to be advanced. The cylinder charge, of air/fuel mixture, requires a certain time to burn (something like 2ms). At higher engine speeds the time taken for the piston to travel the same distance reduces. Advancing the time of the spark ensures full burning is achieved.

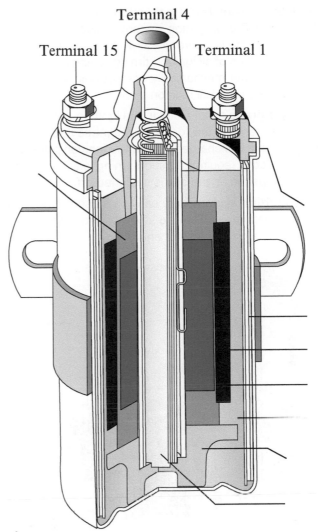

Terminal 4

Terminal 15   Terminal 1

Figure 6.325 Traditional ignition coil sectioned

Engine load A change in timing due to engine load is also required as the weaker mixture used on low-load conditions burns at a slower rate. In this situation further ignition advance is necessary. Greater load on the engine requires a richer mixture, which burns more rapidly. In this case some retardation of timing is necessary. Overall, under any condition of engine speed and load an ideal advance angle is required to ensure maximum pressure is achieved in the cylinder just after top dead centre. The ideal advance angle may also be determined by engine temperature and any risk of detonation.

Spark advance is achieved in a number of ways, the simplest of these being the mechanical system comprising of a centrifugal advance mechanism and a vacuum (load sensitive) control unit. Manifold depression is almost inversely proportional to the engine load. However, I prefer to consider manifold pressure instead of vacuum or depression even though it is lower than atmospheric pressure. The manifold absolute pressure (MAP) is therefore proportional to engine load. Digital ignition systems adjust the timing in relation to the temperature as well as speed and load. The values of all ignition timing functions are combined either mechanically or electronically in order to determine the ideal ignition point.

Energy storage takes place in the ignition coil. The energy is stored in the form of a magnetic field. To ensure the coil is charged before the ignition point, a dwell period is required. Ignition timing is at the end of the dwell period as the coil is switched off.

Early ignition system Very early cars used something called a magneto, which is a story for another time, but here is a nice picture of one anyway!

Figure 6.326 First Bosch high-voltage magneto ignition system with spark plug in 1902 (Source: Bosch Media)

Mechanical switching For many years ignition systems were mechanically switched and distributed. The following table gives an overview of the components of this earlier system.

Figure 6.327 Contact breaker system

Spark plug	Seals electrodes for the spark to jump across in the cylinder. Must withstand very high voltages, pressures and temperatures
Ignition coil	Stores energy in the form of magnetism and delivers it to the distributor via the HT lead. Consists of primary and secondary windings
Ignition switch	Provides driver control of the ignition system and is usually also used to cause the starter to crank
Contact breakers (breaker points)	Switches the primary ignition circuit on and off to charge and discharge the coil. The contacts are operated by a rotating cam in the distributor
Capacitor (condenser)	Suppresses most of the arcing as the contact breakers open. This allows for a more rapid break of primary current and hence a more rapid collapse of coil magnetism which produces a higher voltage output
Distributor	Directs the spark from the coil to each cylinder in a preset sequence
Plug leads	Thickly insulated wires to connect the spark from the distributor to the plugs
Centrifugal advance	Changes the ignition timing with engine speed. As speed increases the timing is advanced
Vacuum advance	Changes timing depending on engine load. On conventional systems the vacuum advance is most important during cruise conditions

Figure 6.328 Traditional ignition coil

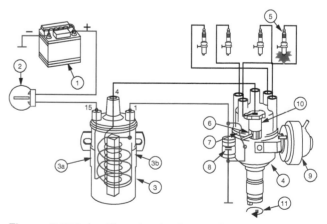

Figure 6.329 Ignition circuit of an early system: 1 Battery, 2 Ignition key switch, 3 Coil, a: primary, b: secondary winding, 4 Distributor body containing centrifugal (speed) advance/retard mechanism, 5 Spark plugs, 6 Cam (with a lobe for each cylinder), 7 Contact breakers (points), 8 Condenser (capacitor), 9 Vacuum (load) advance/retard mechanism

Figure 6.330 Traditional system using a distributor

Modern systems All current vehicle ignition systems are electronically switched and most are now digitally controlled as part of the engine management system. However, there are many vehicles out there still using conventional electronic ignition so the next main section will give an overview of these systems.

 Construct a crossword or wordsearch puzzle using important words from this section

6.5.2 Electronic ignition ❶ ❷

Early ignition systems had some major disadvantages:

▶ Mechanical problems with the contact breakers not least of which is the limited lifetime.
▶ Current flow in the primary circuit is limited to about 4A or damage will occur to the contacts – or at least the lifetime will be seriously reduced.
▶ Legislation requires stringent emission limits which means the ignition timing must stay in tune for a long period of time.
▶ Weaker mixtures require more energy from the spark to ensure successful ignition, even at very high engine speed.

These problems were overcome by using a power transistor to carry out the switching function and a pulse generator to provide the timing signal.

Figure 6.331 Power transistor

Dwell The term 'dwell' when applied to ignition is a measure of the time during which the ignition coil is charging, in other words when primary coil current is flowing. The dwell in traditional systems was simply the time during which the contact breakers were closed, and in these early electronic systems it is the time that the transistor is switched on. Whilst this was a very good system in its time, constant dwell still meant that, at very high engine speeds, the actual time available to charge the coil would only produce a lower power spark. Note that as engine speed increases dwell angle or dwell percentage remains the same but the actual time is reduced. All systems nowadays are known as constant energy, ensuring high performance ignition even at high engine speed.

Constant energy In order for a constant energy electronic ignition system to operate, the dwell must increase with engine speed. This will only be of benefit, however, if the ignition coil can be charged up to its full capacity in a very short time (the time available for maximum dwell at the highest expected engine speed). To this end, constant energy coils are very low resistance so a high current will flow quickly. Constant energy means that, within limits, the energy available to the spark plug remains constant under all operating conditions.

Pulse generator This was achieved by using a pulse generator in the distributor to 'tell' an ignition module the engine position and speed so that the module could determine the switch on (start of dwell) and switch off point (end of dwell and ignition timing spark). Two types of pulse generator (sensors) were most common:

1 Hall Effect
2 Inductive

Figure 6.332 Distributor with ECU fitted (Source: Bosch Media)

6

Hall Effect As the central shaft of the Hall Effect distributor rotates, the chopper plate attached under the rotor arm alternately covers and uncovers the Hall chip. The number of vanes corresponds with the number of cylinders. In constant dwell systems, the dwell is determined by the width of the vanes. The vanes cause the Hall chip to be alternately in and out of a magnetic field. The result of this is that the device will produce almost a square wave output, which can then easily be used to switch further electronic circuits. The three terminals on the distributor are marked '+', '0' and '–'; the terminals + and – are for a voltage supply and terminal 0 is the output signal.

Figure 6.333 Hall effect distributor

Hall sensor output Typically the output from a Hall Effect sensor will switch between 0V and a few volts (systems vary). The supply voltage is taken from the ignition ECU and on some systems is stabilised at about 10V to prevent changes to the output of the sensor when the engine is being cranked. Hall Effect distributors are very common due to the accurate signal produced and their long term reliability. They produced a kind of square wave output signal.

Inductive pulse generators use the basic principle of induction to produce a signal. Many forms exist but all are based around a coil of wire and a permanent magnet. The distributor shown in Figure 6.335 has the coil of wire wound on the pick-up and, as the reluctor rotates, the magnetic flux varies due to the peaks on the reluctor. The number of peaks or teeth on the reluctor corresponds to the number of engine cylinders. The gap between the reluctor and pick-up can be important and manufacturers have recommended settings. These systems produce a form of sinewave output.

High energy Due to the high-energy nature of constant energy ignition coils, the coil cannot be allowed to remain switched on for more than a certain time. This is not a problem when the engine

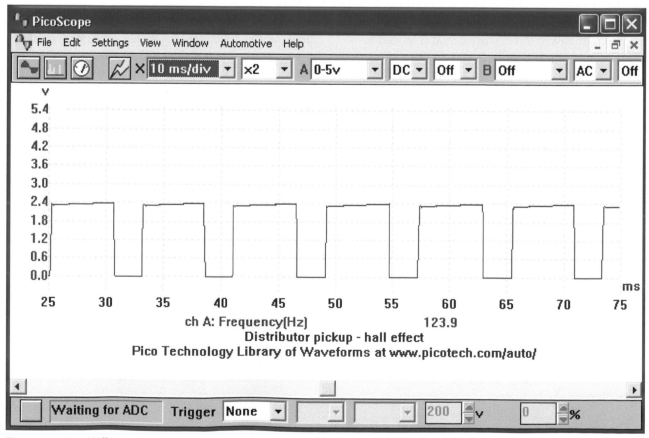

Figure 6.334 Hall sensor output

Figure 6.335 Inductive distributor

Figure 6.337 Electronic ignition module (Source: Bosch Media)

is running as the variable dwell or current limiting circuit prevents the coil overheating. Some form of protection must be provided for, however, when the ignition is switched on but the engine is not running. This is known as stationary engine primary current cut off.

Create a word cloud for one or more of the most important screens or blocks of text in this section

6

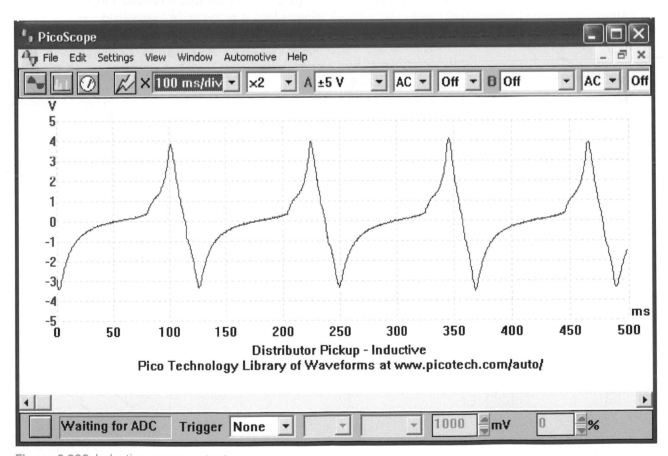

Figure 6.336 Inductive sensor output

6.5.3 Electronic spark advance ❷

Overview Constant energy electronic ignition was a major step forwards and is still used on most vehicles. However, limitations lie in having to rely upon mechanical components for speed and load advance characteristics. In many cases this does not ideally match the requirements of the engine.

Electronic spark advance (ESA) ignition systems have a major difference compared with earlier systems in that they operate digitally. Information about the operating requirements of a particular engine is programmed into the memory inside the electronic control unit. The data for storage in ROM are obtained from rigorous testing on an engine dynamometer and from further development work on the vehicle under various operating conditions. ESA ignition has several advantages.

▶ The ignition timing can be accurately matched to the individual application under a range of operating conditions.
▶ Other control inputs can be utilized such as coolant temperature and ambient air temperature.
▶ Starting is improved and fuel consumption is reduced, as are emissions, and idle control is better.
▶ Other inputs can be taken into account such as engine knock.

▶ The number of wearing components in the ignition system is considerably reduced.

ESA (also referred to as programmed ignition), can be a separate system but is now most likely to be included as part of the full engine management system.

Sensors and input information A typical early ESA system is shown in Figure 6.338. In order for the ECU to calculate suitable timing and dwell outputs, certain input information is required.

Engine speed and position – crankshaft sensor This sensor is a reluctance sensor positioned as shown in Figure 6.338. The device consists of a permanent magnet, a winding and a soft iron core. It is mounted in proximity to a reluctor disc. The disc has 34 teeth, spaced at 10° intervals around the periphery of the disc. It has two teeth missing, 180° apart, at a known position before TDC (BTDC). Many manufacturers use this technique with minor differences. As a tooth from the reluctor disc passes the core of the sensor, the reluctance of the magnetic circuit is changed. This induces a voltage in the winding, the frequency of the waveform being proportional to the engine speed. The missing tooth causes a 'missed' output wave and hence the engine position can be determined.

Engine load – manifold absolute pressure sensor Engine load is proportional to manifold pressure in that high-load conditions produce high pressure and lower load conditions – such as cruise – produce

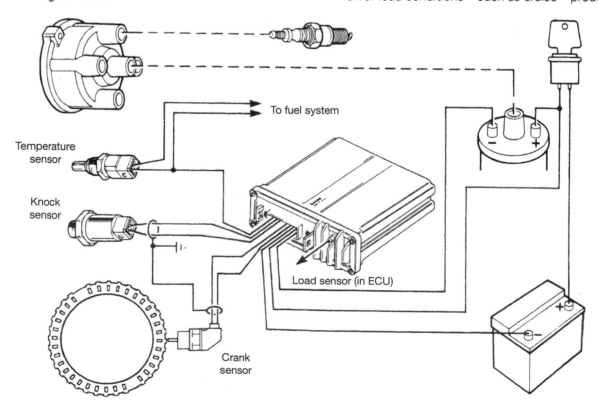

Figure 6.338 Programmed ignition or Electronic Spark Advance (ESA)

lower pressure. Load sensors are therefore pressure transducers. They are either mounted in the ECU or as a separate unit and are connected to the inlet manifold with a pipe. The pipe often incorporates a restriction to damp out fluctuations and a vapour trap to prevent petrol fumes reaching the sensor.

Engine temperature – coolant sensor Coolant temperature measurement is carried out by a simple thermistor, and in many cases the same sensor is used for the operation of the temperature gauge and to provide information to the fuel control system. A separate memory map is used to correct the basic timing settings. Timing may be retarded when the engine is cold to assist in more rapid warm-up.

Figure 6.339 Temperature sensor

Detonation Combustion knock can cause serious damage to an engine if sustained for long periods. This knock, or detonation, is caused by over-advanced ignition timing. At variance with this is that an engine will, in general, run at its most efficient when the timing is advanced as far as possible. To achieve this, the data stored in the basic timing map will be as close to the knock limit of the engine as possible. The knock limit is also known as the detonation border line (DBL). The knock sensor provides a margin for error.

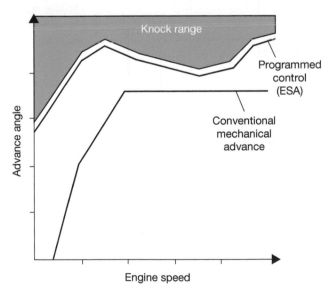

Figure 6.340 Ideal timing angle for an engine is close to the knock limit

Knock sensor The sensor itself is an accelerometer often of the piezoelectric type. It is fitted in the engine block between cylinders two and three on in-line four-cylinder engines. Vee engines require two sensors, one on each side. The ECU responds to signals from the knock sensor in the engine's knock window for each cylinder – this is often just a few degrees each side of TDC. This prevents clatter from the valve mechanism being interpreted as knock. The signal from the sensor is also filtered in the ECU to remove unwanted noise. If detonation is detected, the ignition timing is retarded on the fourth ignition pulse after detection (four-cylinder engine) in steps until knock is no longer detected. The steps vary between manufacturers, but about 2° is typical. The timing is then advanced slowly in steps of, say, 1° over a number of engine revolutions until the advance required by memory is restored. This fine control allows the engine to be run very close to the knock limit without risk of engine damage.

Figure 6.341 Knock sensor

Battery voltage Correction to dwell settings is required if the battery voltage falls, as a lower voltage supply to the coil will require a slightly larger dwell figure. This information is often stored in the form of a dwell correction map.

Electronic control unit As the sophistication of systems has increased, the information held in the memory chips of the ECU has also increased. The earlier versions of a programmed ignition system achieved accuracy in ignition timing of 1.8° whereas a mechanical distributor is 8°. The information, which is derived from dynamometer tests as well as running tests in the vehicle, is stored in ROM. The basic timing map consists of the correct ignition advance for a range of engine speeds and load conditions. This is shown using a cartographic representation. A separate three-dimensional map is used that has speed and temperature-related settings. This is used to add corrections for engine coolant temperature to the basic timing settings. This improves drivability and can be used to decrease the warm-up time of the engine.

Remember! There is more support on the website that includes additional images and interactive features: www.tomdenton.org

Ignition output The output of a system, such as this programmed ignition, is very simple. The output stage, in common with most electronic ignitions, consists of a heavy-duty transistor that forms part of, or is driven by, a Darlington pair. This is simply to allow the high ignition primary current to be controlled. The switch

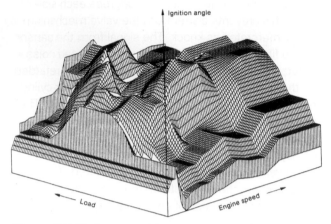

Figure 6.342 Cartographic map representing how ignition timing is stored in the ECU

off point of the coil will control ignition timing and the switch on point will control the dwell period.

HT distribution The high tension distribution is similar to a more conventional system. The rotor arm however is mounted on the end of the camshaft with the distributor cap positioned over the top.

Select a routine from section 1.3 and follow the process to study a component or system

6.5.4 Distributorless ignition ❷

Distributorless ignition systems (DIS) use a special type of ignition coil, which outputs to the spark plugs without the need for an HT distributor.

Figure 6.343 Distributorless ignition coil in position

Lost spark The basic principle is that of the 'lost spark'. The distribution of the spark is achieved by using two double-ended coils, which are fired alternately by the ECU. The timing is determined from a crankshaft speed and position sensor as well as a load (MAP) sensor and other corrections such as engine temperature. When one of the coils is fired, a spark is delivered to two engine cylinders, either 1 and 4, or 2 and 3. The spark delivered to the cylinder on the compression stroke will ignite the mixture as normal. The spark produced in the other cylinder will have no effect as this cylinder will be just completing its exhaust stroke.

Operation Because of the low compression, and the exhaust gas in the 'lost spark' cylinder, the voltage used for the spark to jump the gap is only about 3kV. The spark produced in the compression cylinder is therefore not affected. An interesting point here is that the spark on one of the cylinders will jump from the earth electrode to the spark plug centre. Many years ago this would not have been acceptable as the spark quality when jumping this way would not have been as good as when it jumps from the hotter centre electrode. However, the energy available from modern constant energy systems will result in a spark of high quality regardless of its polarity.

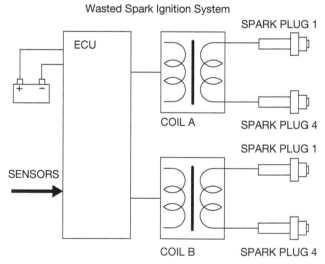

Figure 6.344 DIS simplified circuit (wasted spark ignition system)

DIS components The DIS system consists of three main components – the electronic control unit (ECU), a crankshaft position sensor and the DIS coil. A manifold absolute pressure sensor is integrated in the module or mounted separately. The module uses an electronic spark advance system. Data on ideal dwell and timing is held in memory maps for a wide range of speed, load and voltage conditions. This can be described as an electronic spark advance (ESA) system

as appropriate if a six-cylinder engine). On most cars now the ignition system is combined with the fuel system so that even more accurate control of outputs is possible and input data from sensors can be shared.

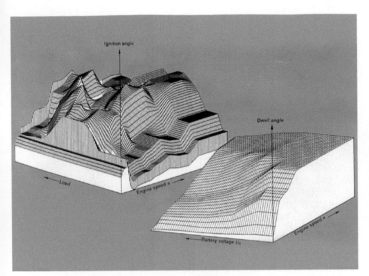

Figure 6.345 Timing and dwell maps

Figure 6.347 DIS coil and plug leads

Use the media search tools to look for pictures and videos relating to the subject in this section

6.5.5 Coil on plug ignition ❷

Crank position sensor (CPS) The crankshaft position sensor is similar in operation to the one described in the fuel section. It is an inductive sensor and is positioned against the front of the flywheel or against a reluctor wheel just behind the front crankshaft pulley. The tooth pattern usually consists of 35 teeth. These are spaced at 10° intervals with a gap where the 36th tooth would be. The missing tooth is positioned at 90° BTDC for numbers one and four cylinders. This reference position is placed a fixed number of degrees before top dead centre, in order to allow the timing or ignition point to be calculated as a fixed angle after the reference mark.

Coil on plug (COP) or direct ignition is a further improvement on distributorless ignition. This system utilises an inductive coil for each engine cylinder. These coils are mounted directly on the spark plugs. The use of an individual coil for each plug ensures that the charge time is very fast (full coil charge in a very small dwell angle). This ensures that a very high voltage, high-energy spark is produced. This voltage, which can be in excess of 40kV, provides efficient initiation of the combustion process under cold starting conditions and with weak mixtures.

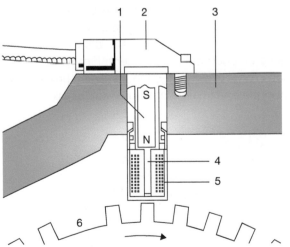

Figure 6.346 Inductive sensor: 1 Magnet, 2 Cover, 3 Engine, 4 Core, 5 Winding

Ignition timing and dwell are controlled in a manner similar to the previously described electronic spark advance (ESA) system. The one important addition to this on most systems is a camshaft sensor to provide

Coil The primary winding is supplied with battery voltage to a centre terminal. The appropriate half of the winding is then switched to earth in the module. The high tension windings are separate and are specific to cylinders one and four, or two and three (or

Figure 6.348 Six direct ignition coils in position

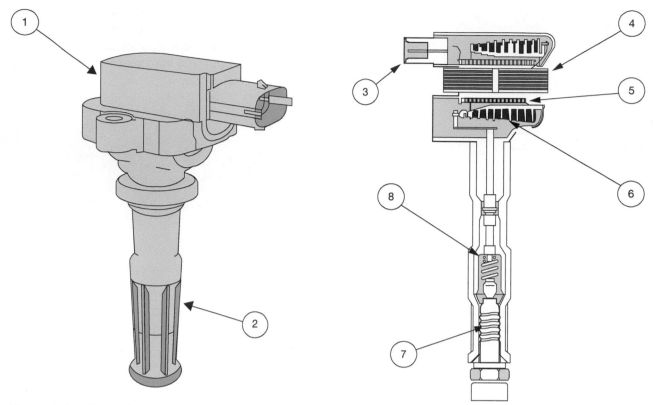

Figure 6.349 Direct ignition coil features: 1 Direct ignition coil, 2 Spark plug connector, 3 Low voltage connection, outer, 4 Laminated iron core, 5 Primary winding, 6 Secondary winding, 7 Spark plug, 8 High voltage connection, inner, via spring contact

information as to which cylinder is on the compression stroke. A system which does not require a sensor to determine which cylinder is on compression (engine position is known from a crank sensor) determines the information by initially firing all of the coils. The voltage across the plugs allows measurement of the current for each spark and will indicate which cylinder is on its combustion stroke. This works because a burning mixture has a lower resistance. The cylinder with the highest current at this point will be the cylinder on the combustion stroke.

Flooding A further feature of some systems is the case when the engine is cranked over for an excessive time making flooding likely. The plugs can all fire with multi-sparks for a period of time after the ignition is left on to burn away any excess fuel. During difficult starting conditions, multi-sparking is also used by some systems during 70° of crank rotation before TDC. This assists with starting and then, once the engine is running, the timing will return to its normal calculated position.

 Create a word cloud for one or more of the most important screens or blocks of text in this section

6.5.6 Spark plugs and leads **1** **2**

Overview The simple requirement of a spark plug is that it must allow a spark to form within the combustion chamber to initiate combustion. In order to do this, the plug has to withstand a number of severe conditions. It must withstand severe vibration and a harsh chemical environment. Finally, but perhaps most importantly, the insulation properties must withstand voltage pressures up to 40kV.

Figure 6.350 Modern high-performance spark plug

Standard spark plug The centre electrode is connected to the top terminal by a stud. The electrode is constructed of a nickel-based alloy. Silver and platinum are also used for some applications. If a copper core is used in the electrode this improves the thermal conduction properties. The insulating material is ceramic based and of a very high grade. Flash over or tracking down the outside of the plug insulation is prevented by ribs which effectively increase the surface distance from the terminal to the metal fixing bolt, which is of course earthed to the engine.

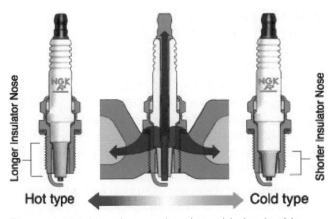

Figure 6.352 Heat-loss paths: the cold plug is able to transfer heat more easily so is suitable for a hot engine

The **heat range** of a spark plug is a measure of its ability to transfer heat away from the centre electrode. A hot running engine will require plugs with a higher thermal ability than a colder running engine. Note that hot and cold running of an engine in this sense refers to the combustion temperature, not to the cooling system.

Spark plug **electrode gaps**, in general, have increased as the power of the ignition systems driving the spark has increased. The simple relationship between plug gap and voltage required is that as the gap increases so must the voltage (leaving aside engine operating conditions). Further, the energy available to form a spark at a fixed engine speed is constant, which means that a larger gap using higher voltage will result in a shorter duration spark. A smaller gap will allow a longer duration spark. For cold starting an engine and for igniting weak mixtures, the duration of the spark is critical. Likewise the plug gap must be as large as possible to allow easy access for the mixture to prevent quenching of the flame. The final choice is therefore a compromise reached through testing and development of a particular application. Plug gaps in the region of 0.6 to 1.2mm seem to be the norm at present.

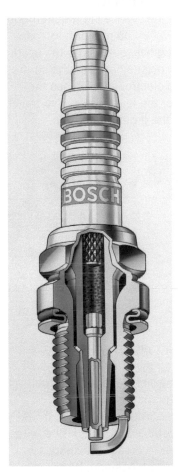

Figure 6.351 Cutaway section of a spark plug

Temperature Due to many and varied constructional features involved in the design of an engine, the range of temperatures a spark plug is exposed to can vary significantly. The operating temperature of the centre electrode of a spark plug is critical. If the temperature becomes too high then pre-ignition may occur where the fuel/air mixture may be ignited due to the incandescence of the plug electrode. If the electrode temperature is too low, then carbon and oil fouling can occur as deposits are not burnt off. The ideal operating temperature of the plug electrode is between 400 and 900°C.

Figure 6.353 A range of spark plugs (Source: Bosch Media)

High tension (HT) is just an old fashioned way of saying high voltage. HT components, such as plug leads, must meet or exceed stringent ignition product requirements, such as:

▶ insulation to withstand 50 000V
▶ temperatures from 40°C to 260°C (40°F to 500°F)
▶ radio frequency interference suppression
▶ 160 000 km (100 000 mile) product life
▶ resistance to ozone, corona, and fluids
▶ ten-year durability.

HT cables must meet the increased energy needs of lean-burn engines without emitting electromagnetic interference (EMI). The cables shown in Figure 6.358 offer metallic and non-metallic cores, including composite, high-temperature resistive and wire-wound inductive cores. Conductor construction includes copper, stainless steel, Delcore, CHT, and wire-wound. Jacketing materials include silicone.

Resistive ignition cable

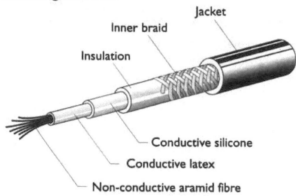

Wire-wound ignition cable

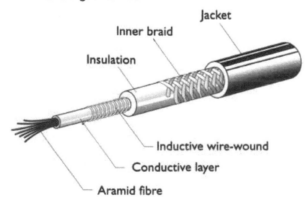

Figure 6.354 Ignition plug leads

 Look back over the previous section and write out a list of the key bullet points from this section

6.5.7 Spark plug electrode designs ❸

Introduction Every time the plug sparks, minute particles of material are worn away from the electrodes. This phenomenon is called spark erosion. Over time, this process increases the spark plug gap between the centre and the earth/ground electrode. If the gap becomes too large, misfiring will occur.

To extend the service interval of vehicles, the service life of the spark plug must be increased. Some manufacturers are fitting multi electrode spark plugs as original equipment to achieve this. Multi electrode spark plugs can have two, three or four ground electrodes depending on the service life requirement of the manufacturer. However, no matter how many ground electrodes the plug has, every time the spark plug fires, only one spark occurs between the centre electrode and the ground electrode which has the lowest required voltage or the least distance to travel between the centre and the ground electrode.

Figure 6.355 Spark plug electrodes, left to right: multi-electrode, iridium wire, hybrid (Source: NGK Plugs)

Ignition quality The spark plug plays a vital role in the quest to improve ignition quality, engine performance, reduce emissions and reduce fuel consumption. Spark plugs that employ small diameter centre and sometimes ground electrodes can offer benefits in several areas. These fine wire plugs require less voltage to create the spark, have a more consistent spark position, better gas flow around the firing position and experience less quench effect than other designs. As the electrodes erode during use it is necessary to compensate for the use of smaller electrodes by some means otherwise the plugs would have an unacceptable (short) service life.

By using small chips of special precious metals such as platinum or even iridium which are welded to the tips of the electrodes it is possible to increase the service life significantly whilst maintaining the highest ignition performance. These metals are extremely hard and have very high melting points thus making them ideal for use in this hostile environment.

Injection Many modern vehicles use a direct fuel injection system and these vehicles demand high ignition quality and extreme anti-fouling performance. NGK has developed a plug that has several special features designed to offer the required performance. Essentially a very projected fine wire spark plug with platinum electrodes is combined with a semi-surface discharge design. The resulting plug has three ground electrodes, two of which are mostly redundant unless in extreme circumstances the plug becomes very carbon fouled. At this point the spark will discharge across the insulator nose to one of the side electrodes preventing a misfire and unburned fuel reaching the catalyst. This type of plug must only be use in the specific applications as listed in the NGK catalogues.

 Make a simple sketch to show how one of the main components or systems in this section operates

6.5.8 Dynamic skip fire ❸

Introduction Dynamic Skip Fire (DSF) is an interesting extension of cylinder deactivation technology in which any of the cylinders for automobile engines are fired or skipped (deactivated) on a continuously variable basis. The engine control system commands the appropriate number and sequence of fired cylinders to deliver the instantaneous torque demanded from the engine. This enables fully optimized combustion and reduced engine pumping losses, thereby increasing fuel efficiency.

Figure 6.356 DSF Operation tracking engine torque demand (Source: Tula, www.tulatech.com)

Operation Operating the engine in a dynamic skip fire manner alters the torque excitations on the vehicle powertrain, which could lead to unacceptable NVH characteristics. Tula's novel firing decision and control algorithms manage noise and vibration algorithmically to maintain a high-quality driving experience.

This interesting technology from Tula integrates advanced signal processing with sophisticated powertrain controls to create the ultimate variable displacement engine. The result is optimal fuel efficiency at the lowest cost.

 Use the media search tools to look for pictures and videos relating to the subject in this section

6.6 Air supply, exhaust and emissions

6.6.1 Air pollution ❶

Atmospheric pollution has become a serious problem for the environment and the health of people. Many urban areas are now heavily polluted, with people suffering medically from the effects of vehicle exhaust pollution.

Fossil fuels There have been many changes in climatic conditions in the world. Many of these have occurred over a long period and animals and plants have adapted to the changes naturally. However, the rapid burning of fossil fuels during this century has increased carbon dioxide levels in the atmosphere.

Vehicle designs are concentrating on weight reduction, aerodynamics, reducing rolling resistance, and on fuel-efficient engines. Alternative fuel sources to reduce fossil-fuel usage and to conserve the world's stock of these fuels have also been developed.

Carbon dioxide allows the sun's heat in but reduces the ability of the heat to radiate outward, causing the Earth to warm up. Many studies of the warming process indicate that the rate of Earth warming is increasing too quickly and preventing animals and plants from adapting. During the history of the Earth, rapid changes like this have caused the extinction of some species of animals and plants.

Weather patterns As a result of warming, weather patterns change. Arid areas become wet and wet areas become dry. Drought conditions become common in heavily populated areas and other areas suffer severe flooding. Because the distribution of populations and agricultural production are linked, they end up in the wrong climatic conditions. The consequences are severe shortages of water and poor agricultural production.

Ozone layer A layer of ozone in the stratosphere filters harmful radiation. Ozone, or trioxygen

(O_3), is a form of oxygen with three oxygen atoms. Vehicle emissions and other industrial chemicals, such as the CFCs used in refrigeration, air-conditioning and aerosols, rise up into the stratosphere and chemically combine with the ozone. This causes it to break down into less beneficial substances. The deterioration of the ozone layer allows an increase in the harmful radiation that reaches the Earth's surface, which can cause skin and other cancers.

Environmental regulations are now in place to find safer alternatives, or to reduce the production and use of the most harmful pollutants. Other regulations and agreements are seeking to reduce the production of carbon dioxide by improving the efficiency of fossil-fuel burners. For retaining the energy produced, improvements will also be introduced, such as the use of insulation and other methods.

Lead has, until recently, been used as an additive in petrol in order to slow down the combustion process. This was to eliminate knocking or pinking in the engine. It made engines more efficient but the lead did not burn and was, instead, passed into the atmosphere from the exhaust and produced airborne concentrations that were capable of causing many physical disabilities, including brain damage.

For this reason, lead additives are no longer used and modern engines are now designed to run on lead-free fuel. There may be a small portion of naturally occurring lead in some fuels but, because this is very low, the description 'lead-free' is more precisely a statement that lead additives have not been used.

Sulphur Another naturally occurring substance in fossil fuels, particularly diesel, is sulphur. This does not burn but, during combustion, chemically reacts with oxygen in the air to form sulphur dioxide (SO_2). This passes from the engine exhaust into the atmosphere where it combines with water to form sulphuric acid (H_2SO_4) and falls back to Earth as acid rain, which destroys trees, plants, other vegetation and aquatic life in streams, rivers and lakes. Fuel suppliers remove, or reduce, the amount of sulphur during the refining process.

Nitrogen oxides Air consists of approximately 80% nitrogen which, under normal circumstances, is an inert gas. An inert substance is one that has very little chemical reaction and does not burn, or mix easily, with other chemicals. Nitrogen, however, will mix with oxygen at high temperatures to form nitrogen oxides

(NOX). These combine in exceptional geographical and meteorological conditions to form smog, acids and increases in low-level ozone. This serves to make a very unpleasant atmosphere in which to live. Many respiratory and asthmatic fatalities occur under these conditions.

 Create an information wall to illustrate the features of a key component or system

6.6.2 Worldwide harmonized light vehicles test procedure ❸

Introduction The worldwide harmonized light vehicles test procedure (WLTP) is a global standard for determining levels of pollutants and CO2 emissions, fuel or energy consumption, and electric range. It is used for passenger cars and light commercial vans. Experts from the EU, Japan, and India, under guidelines of UNECE World Forum for Harmonization of Vehicle Regulations, developed the standard. It was released in 2015.

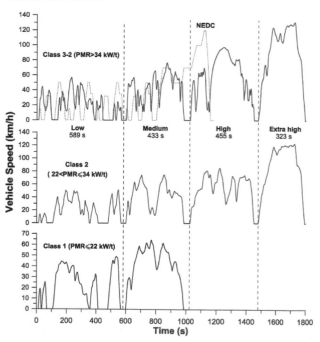

Figure 6.357 WLTP test cycles

Test cycles Like all previous test cycles, it has its drawbacks, but it is a good attempt to make the test more realistic. The key thing is that a standardised test allows vehicles to be compared accurately, even if the figures differ from real-world driving. It includes strict guidance regarding conditions of dynamometer tests and road load (motion resistance), gear changing, total car weight (by including optional equipment, cargo and passengers), fuel quality, ambient temperature, tyres and their pressure. Three different cycles are used depending on vehicle class defined by power/weight ratio PWr in kW/Tonne (rated engine power / kerb weight):

▶ Class 1 - low power vehicles with PWr <= 22
▶ Class 2 - vehicles with 22 < PWr <= 34
▶ Class 3 - high-power vehicles with PWr > 34

Light vehicle class Most modern cars, light vans and busses have a power-weight ratio of 40-100 kW/t, so belong to class 3, but some can be in class 2. In each class, there are several driving tests designed to represent real world vehicle operation on urban roads, extra-urban roads, motorways, and freeways. The duration of each part of the tests is fixed between classes, the difference is that the acceleration and speed curves are shaped differently. The sequence of tests is also restricted by maximum vehicle speed (Vmax).

Gear shifting Because there is a wide range of manual gearboxes with 4, 5, 6 and 7, or even 8 gears, it is impossible to specify fixed gear shift points. To overcome this, the WLTP uses an algorithm for calculating optimal shift points, which considers total vehicle weight and full load power curves, within normal engine speeds. This covers the wide range of rpm and engine power available on current vehicles. To reflect normal use, and a fuel efficient driving style, gear changes are filtered out if they occur in less than 5 seconds.

WLTC (worldwide harmonized light vehicles test procedure) is an improvement over the New European Driving Cycle (NEDC), but transitions are still very slow. For example, the most rapid 0 to 50 km/h (0 to 30 mph) time is 15 seconds. Most drivers in Western Europe accelerate from rest to 50 km/h in 5 to 10 seconds. There is no hill climbing in the cycle. Perhaps there should be as even modest gradients increase pollutant emissions because the engine load increases 2 to 3 times. Nonetheless, it is a step in the right direction.

Summary The WLTC driving cycle for a Class 3 vehicle is divided in four parts:

1 Low
2 Medium
3 High
4 Extra High speed

If V_{max} < 135 km/h, the Extra High speed part is replaced with Low speed part.

 Use a library or the web search tools to further examine the subject in this section

6.6.3 **Euro 6 overview** ❸

Introduction It was expected that Euro 6 would make diesel cars as clean as petrol cars, but this is a contentious issue in some cities. The main thrust of the regulation is to set lower limits for vehicle emissions of particulates and nitrogen oxides. Since September 1, 2014, diesel vehicles must emit no more than 80 mg of nitrogen oxides per kilometre (petrol vehicles: 60 mg per kilometre). This replaces the previous limit of 180 mg per kilometre. From January 1, 2015, all new vehicles sold had to meet the Euro 6 limits.

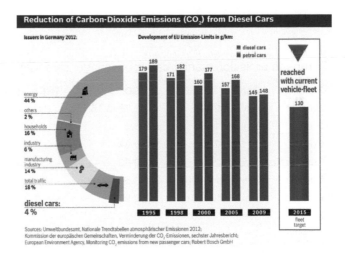

Figure 6.358 Reduction of Carbon dioxide in diesel cars

Technical advances Since the Euro 1 regulation was introduced in 1993, emissions from road traffic have been drastically reduced. Advanced automotive technology reduces emissions of substances such as CO_2, nitrogen oxides and particulates. Technical advances in powertrains are also having an effect: since 1990, particulate emissions from diesel engines have been reduced by around 99%, while modern diesels emit some

98% less nitrogen oxide than comparable vehicles from the early 1990s. As electrification of the powertrain continues to progress, emissions will fall even further.

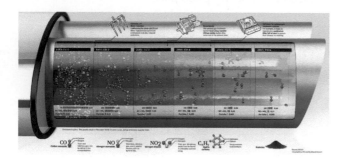

Figure 6.359 Development stages of technology to meet Euro 6 legislation

Exhaust-treatment Diesel vehicles need a perfectly tuned exhaust-treatment system to meet the lower limits set out in Euro 6. For vehicles weighing up to around 1,70kg, a low-cost NOx storage catalytic converter is sufficient. In heavy vehicles, an SCR catalytic converter with AdBlue will be needed. This system injects AdBlue, an odourless urea solution, which converts the nitrogen oxides into harmless

water vapour and nitrogen. AdBlue is refilled at regular service intervals.

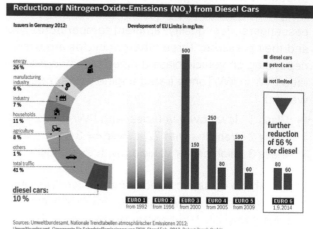

Figure 6.360 Reduction of Nitrogen oxides in diesel cars

Euro legislation limits emissions of nitrogen oxides (NOx), total hydrocarbon (THC), non-methane hydrocarbons (NMHC), carbon monoxide (CO) and particulate matter (PM) for most vehicle types. The overview is presented in the tables below:

European emission standards for passenger cars (Category M*), g/km - Diesel

Tier	Date	CO	THC	NMHC	NOx	HC+NOx	PM	PN [#/km]
Euro 1†	July 1992	2.72 (3.16)	-	-	-	0.97 (1.13)	0.14 (0.18)	-
Euro 2	January 1996	1.0	-	-	-	0.7	0.08	-
Euro 3	January 2000	0.64	-	-	0.50	0.56	0.05	-
Euro 4	January 2005	0.50	-	-	0.25	0.30	0.025	-
Euro 5a	September 2009	0.50	-	-	0.180	0.230	0.005	-
Euro 5b	September 2011	0.50	-	-	0.180	0.230	0.005	6×101^1
Euro 6	September 2014	0.50	-	-	0.080	0.170	0.005	6×101^1

European emission standards for passenger cars (Category M*), g/km – Petrol/Gasoline

Tier	Date	CO	THC	NMHC	NOx	HC+NOx	PM	PN [#/km]
Euro 1†	July 1992	2.72 (3.16)	-	-	-	0.97 (1.13)	-	-
Euro 2	January 1996	2.2	-	-	-	0.5	-	-
Euro 3	January 2000	2.3	0.20	-	0.15	-	-	-
Euro 4	January 2005	1.0	0.10	-	0.08	-	-	-
Euro 5	September 2009	1.0	0.10	0.068	0.060	-	0.005**	-
Euro 6	September 2014	1.0	0.10	0.068	0.060	-	0.005**	6×101^1***

* Before Euro 5, passenger vehicles > 2500 kg were type approved as light commercial vehicles N1-I
** Applies only to vehicles with direct injection engines
*** 6×101^2/km within first three years from Euro 6 effective dates
† Values in parentheses are conformity of production (COP) limits

Summary At the time of writing Euro 6 is the current regulation. However, Euro 7 is coming and many manufacturers have already started work on new technologies. It is expected around 2025 and may include a review of the emissions of the previous year rather than an 'on the day' test.

 Use a library or the web search tools to further examine the subject in this section

6.6.4 Engine combustion

Combustion The combustion of fuel inside the engine is a chemical process that combines the carbon and hydrogen in the fuel with oxygen in order to release energy. Slightly less than 20% of air is made up from oxygen. Complete combustion produces carbon dioxide (CO_2) and water (H_2O). Neither of these is directly harmful. Both are naturally occurring substances in large concentrations in the atmosphere. However, carbon dioxide concentrations are increasing and contributing to the greenhouse effect.

Incomplete combustion Incomplete combustion leaves some of the carbon and oxygen not fully combined. The product of this is carbon monoxide (CO), which is toxic. Small quantities of carbon monoxide molecules are dangerous because they attach themselves to red blood cells. This reduces the oxygen that the cells normally carry around the body. The result is oxygen deprivation, brain damage and fatality.

Unburnt fuel Another product of incomplete combustion is particles of fuel that have not been burnt. These are carried, with the exhaust gases, into the atmosphere and are called un-burnt hydrocarbons (HC). Very small amounts of hydrocarbons in the atmosphere can cause respiratory problems.

Engine oil drawn into the combustion chamber, either from the inlet valve stem or by bypassing the pistons, can also be a source of hydrocarbon pollution. Oil vapours form in the engine crankcase and can escape into the atmosphere. A positive crankcase ventilation system is now used to draw the vapours into the engine so that they are burnt to form water and carbon dioxide.

Evaporative emissions Previously, vapour in the tank was directly vented to the atmosphere. This is no longer the case, but the fuel tank must still be vented to the atmosphere to allow air to flow into the tank as fuel is used. A charcoal filter is now used to prevent the loss of fuel vapour and for the expansion of the fuel when the weather is hot. The fuel vapour in the charcoal canister is drawn into the engine and burnt.

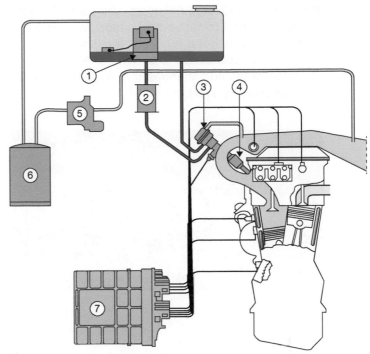

Figure 6.361 Emission-control system

Air/fuel ratio Good fuel economy is obtained with a lean air-to-fuel mixture. However, this mixture produces higher combustion temperatures and greater risks of nitrogen oxides being formed. In order to prevent, or reduce to a minimum, the formation of nitrogen oxides, the combustion temperature has to be kept as cool as possible and the amount of oxygen limited to match the quantity of fuel delivered.

Exhaust gas In order to reduce the amount of oxygen in the air charge, a gas that is low in oxygen can be introduced. This maintains the total air-charge mass to give good compression pressures and efficient operation of the engine. The available gas is the exhaust gas that has already used up its oxygen content during combustion. The addition of a regulated charge of exhaust gas reduces the oxygen content of the new charge to suit the amount of fuel delivered. This in turn reduces the combustion temperature and limits the formation of nitrogen oxides. The catalytic conversion of any remaining harmful gases can give a clean exhaust gas.

Use the media search tools to look for pictures and videos relating to the subject in this section

6.6.5 Environmental protection

Reducing pollution Vehicle engine and component manufacturers have put a great deal of effort into reducing pollution. Lead is no longer needed in petrol because other, less damaging substitutes have been found. The changes in the fuel have necessitated the use of hardened valves and valve seats and changes to the ignition timing and fuel delivery systems.

Figure 6.362 Valves

Figure 6.363 Air filter

Air intake systems Air intake systems have been developed from a simple ducting to a complex airflow design adapting to the changing speed and load conditions of the engine. Filtration is also an important aspect.

Electronic control Electronic control of the combustion process has achieved reductions in CO, NOX and HC emissions. Exhaust gases are monitored in an electronic engine control module from signals sent from a lambda, or oxygen, sensor in the exhaust. This then allows fuel and air supplies to be accurately merged for near-perfect combustion.

Figure 6.364 Lambda sensor

Pollutant control The remaining pollutants in the exhaust gases, which cannot be controlled by the electronic systems, can be converted into less harmful substances. This is achieved by using air injection into the exhaust and/or a catalytic converter.

Figure 6.365 Air injection

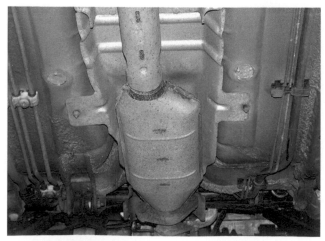

Figure 6.366 Catalytic converter

Atomization Developments to improve the atomization and the mixing of the fuel in the incoming air stream include heating the inlet manifold, or heating the air as it enters the inlet manifold. This can be achieved with a heater element below the carburettor, or by preheating the air by ducting the air supply over the exhaust manifold.

Oil and fuel vapours Oil and fuel vapours are trapped and routed through the engine to be burnt. Positive crankcase ventilation and a charcoal filter in an evaporative canister are used for this purpose. Nitrogen oxide formation is reduced with exhaust gas recirculation (EGR).

Figure 6.367 Air temperature control valve

Figure 6.368 EVAP canister

Supercharging and turbocharging Engine performance has been increased, without an increase in weight, by the use of supercharging and turbocharging. Other emission control devices that correct the ignition timing and fuel delivery are covered in the appropriate learning programmes. These devices improve the performance of those systems as well as reducing harmful exhaust emissions.

Figure 6.369 Turbocharger

 Select a routine from section 1.3 and follow the process to study a component or system

6.6.6 Air supply ❶

Introduction The air supply system has to provide clean air in sufficient quantity to the engine. Also, it must supply equal quantities of air to each cylinder. This will assist fuel vaporization and an even mixture distribution. Creating a swirl in the airflow as it enters the cylinders is also desirable. A system of warm air for cold starts, followed by temperature-controlled air for normal running, is essential. Finally, the system must silence the airflow and provide a flame trap in the event of fire in the inlet manifold.

System components The air supply systems for most vehicles are similar. They consist of an air intake duct, an air temperature control mechanism, an air cleaner housing and filter, an inlet manifold and inlet ports. A position for an exhaust gas recirculation system may also be included. For multi-point, or port, fuel injection engines, the system will also include a throttle body housing and an airflow meter.

Figure 6.370 Throttle body

Clean air is required in the engine to prevent particles of dust and grit from damaging, or blocking, engine and fuel supply components. Air is filtered through an element in the air cleaner. Most air cleaner elements are made from microporous paper, which allows a good flow of air but traps airborne dust. Other elements have included oiled wire gauze and foam rubber. The air cleaner housing and the filter elements are cleaned, or replaced, at scheduled service intervals.

Figure 6.371 Air filter

Figure 6.372 Airflow meter

Figure 6.373 Ducting

Paper elements are folded to provide a large surface area and long service life. The element can be wrapped to form a circular element if required. The outside edges are sealed with an integral, or separate, rubber sealing ring.

Figure 6.374 Air filter

Air cleaner housings have internal ducting to distribute the air over the full surface of the filter. The airflow in some filter housings is made to swirl so that airborne dirt is thrown out and falls into a dust trap in the base of the filter. The airflow into flat filters is from the underside so that dirt falls out from below, rather than into the top of, the filter.

Figure 6.375 Filter in its housing

Inlet manifolds on modern engines are usually of the same length and diameter to enable all cylinders to be supplied with the same volume and airflow characteristics. Early engines, with manifolds using pipes of differing lengths, often produced slightly different combustion patterns in each of the cylinders.

At the entrance to the inlet manifold is the throttle plate controlling the flow and quantity of air entering the engine. Diesel engines do not use a throttle plate unless a vacuum is required for the control, or operation, of other systems.

Intake air heating for cold engines Mixture composition occurs in the inlet manifolds of carburettor and monopoint fuel injection engines. These manifolds are heated to aid the atomization and distribution of the fuel in the air charge. This is particularly important when the engine and the air supply are cold. Inlet manifolds were made from aluminium, which readily conducts heat and warms evenly and quickly. However, thermoplastic is now being used more often.

Manifold heating On older types of engine with the inlet manifold positioned over the exhaust manifold, an exchange of heat was provided by connecting the two manifolds together. This design is not suitable for crossflow and 'V' engines. One method of inlet manifold heating on these engines uses the engine cooling system. Water passages in the manifold are connected to the water jacket so that coolant flows as soon as the engine is started.

Figure 6.376 Plastic inlet manifold

Electrical heating Another method, which does not use the cooling system, has an electric heater element under the centre of the manifold operating when the engine is started from cold. A temperature-sensing switch in the engine coolant cuts off the electrical supply when the engine temperature rises.

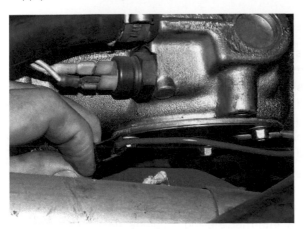

Figure 6.377 Electric manifold heater

Heated air supply On some engines, the incoming air supply is heated. Two designs have been used for this. One heats the air below the carburettor, and the other before it enters the air cleaner. The older fuel-evaporative system, used on some American-vehicle engine designs, had an electric heater element below the carburettor to heat the air flowing into the manifold. The heater element was supplied with an electric current through a relay and controlled by an engine temperature-sensing switch.

Air temperature control Heating the air entering the inlet duct assists in atomization and fuel distribution in the air charge. To warm the air, it is passed over the exhaust manifold before being drawn into the air duct. This is only necessary when the air is cold. When the engine temperature increases, the air density, and therefore mass, would be reduced if heating of the air were continued. At an engine temperature of about 50°C, the full air supply is drawn from a cold position in the engine compartment, or from the front of the vehicle. Between a cold engine and 50°C, progressive mixing occurs.

Figure 6.378 Pick-up for hot air on an exhaust manifold

Flap control The ducting of warm, or cool, air is controlled by a flap in the air cleaner intake. This provides either a normal airflow or one from over the exhaust manifold. Two designs of thermostatically controlled air cleaner operation are used. One type uses a vacuum motor and bimetallic vacuum valve and the other uses a wax-pellet actuator.

Vacuum system The layout of the vacuum system is shown here. The bimetallic valve responds to the temperature of the incoming air stream and opens or closes the vacuum supply from the inlet manifold to the vacuum motor. The motor reacts to the vacuum supply to move the flap and mix warm and cool air.

Wax pellet actuator The wax pellet actuator is set in the warm air supply duct and, when cold, holds the flap across the cool air duct. As the wax pellet is heated up by the warmed air, it expands and forces out the insert pin, or piston. The pin is connected to a lever which pulls the flap open to allow in a cool airflow. The lever and flap are held by a calibrated return spring and actuated by the force of the expanding wax pellet.

Air intake for multi-point fuel injection The air supply components for multi-point fuel injection engines have additional items for the control and measurement of the air supply. Sensors for the fuel injection electronic control unit are included in the air supply system. An actuator for controlling the engine fast idle speed is also included. The fuel injectors are fitted into the inlet ports or in a special housing between the inlet manifold and the inlet ports.

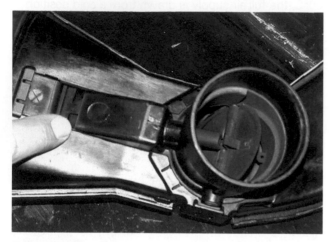

Figure 6.379 Wax pellet warm-air control

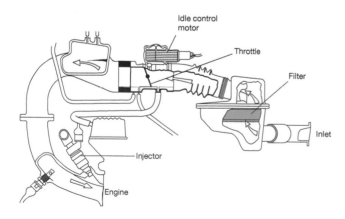

Figure 6.380 Air supply system

Inlet manifolds The air supply to each cylinder passes through the inlet manifolds which are of equal length and diameter. Feeding the manifold tubes is a plenum chamber, which holds a large volume of air so that each intake tube receives an equal air supply. The airflow is made to swirl in the intake tubes, and careful design of the shape and direction of the tubes is required to make this happen. Another factor affecting the swirl is the volume and speed of the airflow.

Variable length inlet manifolds Many engines have a dual-intake system that responds to low and high engine speeds. These systems have valves that open at higher engine speeds to balance the pressure in the two intake manifolds, or open to enable a secondary air supply to provide an adequate airflow for the higher engine speed. These systems have been developed to meet the changing airflow and swirl characteristics occurring with increases in air mass and speed.

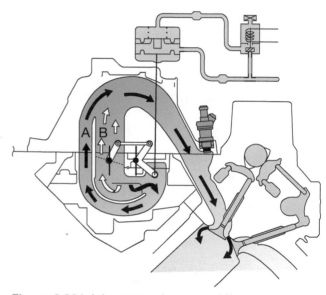

Figure 6.381 Inlet tracts change at different speeds

Make a simple sketch to show how one of the main components or systems in this section operates.

6.6.7 Exhaust systems ❶

System requirements The exhaust system has to carry the exhaust gases out of the engine to a safe position on the vehicle, silence the exhaust sound and cool the exhaust gases. It also has to match the engine gas flow, resist internal corrosion from the exhaust gas and resist external corrosion from water and road salt.

System components The exhaust system consists of the exhaust manifold, silencers, mufflers, expansion boxes and resonators. It also has down or front pipes, intermediate and tail pipes, heat shields and mountings. Also included are one or more catalytic converters, one or two lambda sensors and an outlet for the exhaust gas recirculation system.

High temperatures The exhaust gases are at a very high temperature when they leave the combustion chambers and pass through the exhaust ports. The exhaust manifold is made from cast iron in order to cope with the high temperatures. The remainder of the exhaust system is made from steel, which is alloyed and treated to resist corrosion.

Figure 6.382 Exhaust system

The **down pipe, or front pipe**, is attached to the manifold with a flat, or ball, flange. This joint is subject to bending stresses with the movement of the engine in the vehicle. To accommodate the movement and reduce stress fractures, many flange connections have a flexible coupling made from a ball flange joint and compression springs on the mounting studs.

Exhaust movement Another system to accommodate movement is a flexible pipe constructed from interlocking stainless steel coils, or rings. Where a flexible joint is not required, the front pipe may be supported by a bracket welded to the pipe which is bolted to a convenient position on the engine or gearbox. Where a catalytic converter is used, it is fitted to the front pipe so that the exhaust heat is used to aid the chemical reactions taking place within the catalytic converter. The front pipe connects to an expansion box or silencer. The exhaust gases are allowed to expand into this box and begin to cool. They contract on cooling and slow down in speed.

Silencers or mufflers are constructed as single- or twin-skin boxes, and there are two main types – the absorption type, which uses glass fibre or steel wool to absorb the sound; and the baffle type, which uses a

> **Remember!** There is more support on the website that includes additional images and interactive features: www.tomdenton.org

series of baffles to create chambers. In the baffle type, the exhaust gases are transferred from a perforated inlet pipe to a similarly perforated outlet pipe. These silencers have a large external surface area so that heat is radiated to the atmosphere. Additional pipes and silencers carry the exhaust gas to the rear.

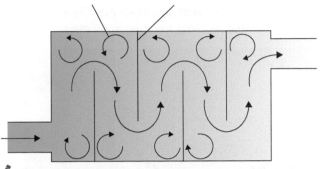

Figure 6.383 Baffle silencer

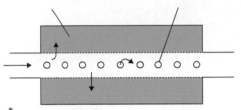

Figure 6.384 Absorption silencer

Joints Pipes are joined together by a flange, or clamp, fitting. Flange connections have a heat-resistant gasket and through-bolts to hold the flange together. Clamp fittings are used where pipes fit into each other. The larger pipe is toward the front and the smaller pipe fits inside. A ring clamp, or 'U' bolt and saddle, are tightened around the pipes to give a gas-tight seal. An exhaust paste is usually applied to improve the seal of the joint. The exhaust system must be sealed to prevent toxic exhaust gases from entering the passenger compartment.

Figure 6.385 Flange connections

Exhaust mountings A small water-drain hole may be used on the underside of some silencers. This is to reduce internal corrosion from standing water in the silencer body forming on short journeys. The exhaust is held underneath the vehicle body on flexible mountings. These are usually made from a rubber compound and many are formed as a large ring that fits on hooks on the vehicle and the exhaust pipe brackets. Other mountings are bonded-rubber blocks on two steel plates.

Figure 6.386 Pipe connections

Figure 6.387 Flexible . . .

Figure 6.388 . . . mountings

Heat shields are fitted to the exhaust, or to the vehicle floor, to prevent the ignition of sound-deadening and anti-corrosion materials. Catalytic converters become very hot during operation. It is important, therefore, that all heat shields are correctly fitted and positioned to insulate the vehicle from the high temperature of the catalytic converter.

 Look back over the previous section and write out a list of the key bullet points from this section

6.6.8 Catalyst systems ❷

Introduction There is a range of devices and systems that are used to control the constituents of the exhaust gases and vapour emissions from engine oil and fuels. One or more of these systems will be used. The actual systems chosen by a manufacturer are suited to particular engines and the environmental protection regulations currently in force although, in practice, manufacturers develop their vehicles to be over and above the standards imposed by the regulations. The main systems are described in this section.

Figure 6.389 Catalytic converter in position

A **catalyst** is a substance that will precipitate chemical changes in other substances without effect to itself. The purpose of the catalyst is to convert potentially harmful chemicals in the exhaust gas into harmless water vapour, carbon dioxide, nitrogen and oxygen.

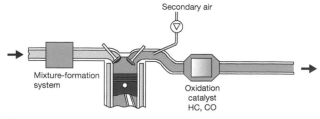

Figure 6.390 Additional air

There are several types of catalytic converters used on motor vehicles. The main types are oxidation, three-way and three-way/oxidizing. The main catalytic materials used consist of a mixture of platinum, palladium and rhodium, but less expensive materials are being investigated and developed. Titanium oxide is one substance that is now being used in oxidation catalysts.

Catalytic material is applied as a thin coat to ceramic, or stainless steel, 'honeycomb' or pellets. The exhaust gases flow freely through the honeycomb, or pellets, where the catalytic chemical reactions take place. The operating temperature of the catalyst is high, and the catalyst must be heated before it becomes effective. Exhaust heat is used for this. Catalysts allow chemical reactions to take place at significantly lower temperatures than might otherwise be required.

Figure 6.391 Honeycomb substrate

Figure 6.392 Steel substrate

Oxidation catalysts require surplus oxygen in the exhaust gases for use in the conversion of hydrocarbons (HC) and carbon monoxide (CO) to water (H_2O) and carbon dioxide (CO_2). Oxidation catalysts are suitable for engines that run with a surplus of oxygen, such as diesel engines, and where additional air and, therefore, oxygen can be supplied.

Three-way catalysts Three-way and three-way/oxidizing catalysts convert the HC and CO to H_2O and CO_2 and, additionally, reduce the nitrogen oxides. In these catalytic converters, the nitrogen oxides react with carbon monoxide to give nitrogen (N_2) and carbon dioxide (CO_2). The nitrogen oxides also react with hydrogen to give nitrogen and water vapour. The performance of catalytic converters relies on the correct exhaust gas constituents being produced. Modern engines use an electronic closed loop control with a lambda sensor in the exhaust manifold. This measures the amount of oxygen in the exhaust gas.

The **lambda sensor** is named after the Greek letter lambda, which is used as the symbol for a chemically correct air-to-fuel ratio, or stoichiometric ratio of 14.7 parts of air to 1 part of fuel. This sensor is also known as a heated exhaust gas oxygen sensor (HEGO) when it is pre-heated. The sensor measures the presence of oxygen in the exhaust gas and sends a voltage signal to the engine electronic control module (ECM).

Lambda control More fuel is delivered when an oxygen content is detected and less fuel when it is not. In this way, an accurate fuel mixture close to the stoichiometric, or lambda, ratio is maintained. This produces the correct exhaust gas constituents for chemical reactions in the catalytic converter.

Correct exhaust gas constituents Maintenance of the efficiency of the catalytic converter relies on correct exhaust gas constituents. Leaded fuel must not be used. The engine tuning for ignition and fuel delivery must be accurately maintained. The engine should not be turned on the starter, or by towing, for longer than is normal for starting because un-burnt petrol can enter and damage the catalyst when it ignites as the engine finally starts. Running the engine with a misfire will also produce an excess of un-burnt fuel. The fuel will burn in the converter and produce considerably higher temperatures than normal, and this can damage the catalyst.

Cold start and warm-up During cold start and warm-up, during acceleration, and at high engine speeds, an engine requires a richer mixture than normal. The exhaust will contain increased levels of un-burnt hydrocarbons and carbon monoxide. These harmful gases will escape to the atmosphere on vehicles without a catalytic converter. On vehicles with a catalytic converter, but without closed loop control, this will also be a problem during the warm-up phase. The catalyst has to reach operating temperature before it functions correctly. This is known as the 'light-off' temperature.

Complete combustion Injection of air into the exhaust ports creates a chemical reaction. This converts the hydrocarbons to water vapour and carbon dioxide, and the carbon monoxide to carbon

6

dioxide. This occurs when the additional oxygen in the air continues and completes the combustion of the excess fuel in the exhaust gases.

Air pump The air injection reactive system (AIR) uses an air pump driven by the engine and air injectors in the exhaust ports. The air pump draws air from the air cleaner. The air is pumped to the air-diverter valve for distribution to the exhaust ports through a non-return check valve, or returned to the air cleaner. The air-diverter valve is an electric solenoid and plunger operated from the engine electronic control unit (ECU). When the engine is cold, or on full throttle, the air is routed to the exhaust manifold and, at other times, to the air cleaner.

The **pulse air system** uses gas flow pulses in the exhaust to draw air into the exhaust ports through non-return valves. The pulse air system is controlled by the engine ECU and actuated by a vacuum motor and electric solenoid. A simple form of pulse air system has a set of pipes that fit into the cylinder head exhaust ports. It has pipe-union connectors at one end and a feed to an inlet valve and air cleaner assembly at the other end. This system works automatically with the pulses created by the exhaust gas. The valves in the inlet box allow airflow into an exhaust port when the pressure wave in the port area is negative.

Figure 6.393 Heated exhaust-gas oxygen sensor

Figure 6.394 Air pipes in the exhaust manifold

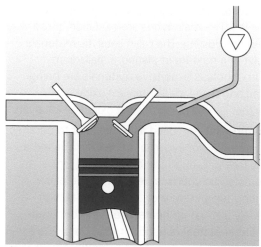

Figure 6.395 Air is fed into the manifold

 Create an information wall to illustrate the features of a key component or system

6.6.9 Diesel particulate filters (DPF) and selective catalytic reduction (SCR) ❸

Introduction The main approach to lowering of diesel engine emissions involves internal engine improvements. This is because improved fuel combustion prevents, as far as possible, the formation of pollutants and reduces fuel consumption. In this respect, automobile manufacturers and their component suppliers have already achieved a great deal. A diesel particulate filter (DPF) is a device designed to remove diesel particulate matter or soot from the exhaust gas of a diesel engine. Wall-flow diesel particulate filters usually remove 85% or more of the soot and under certain conditions can attain soot removal efficiencies of close to 100%.

Types of filter The most common filter is made of cordierite (a ceramic material that is also used as catalytic converter cores). Cordierite filters provide excellent filtration efficiency, are (relatively) inexpensive and have thermal properties that make packaging them for installation in the vehicle simple. The major drawback is that cordierite has a relatively low melting point (about 1200°C) and cordierite substrates have been known to melt down during filter regeneration. This is mostly an issue if the filter has become loaded more heavily than usual, and is more of an issue with passive systems than with active systems unless there is a system break down.

Cordierite filter cores look like catalytic converter cores that have had alternate channels plugged – the plugs force the exhaust gas flow through the wall and the particulate collects on the inlet face.

Figure 6.396 Cordierite filter cores

Silicon carbide, or SiC, has a higher (2700°C) melting point than cordierite, however it is not as stable thermally, making packaging an issue. Small SiC cores are made of single pieces, while larger cores are made in segments, which are separated by special cement so that heat expansion of the core will be taken up by the cement, and not the package. These cores are often more expensive than cordierite ones, however they are manufactured in similar sizes and one can often be used to replace the other.

Ceramic fibre filters are made from several different types of ceramic fibres that are mixed together to form a porous media. This media can be formed into almost any shape and can be customized to suit various applications.

Lifespan The particulate filter shown here is made of sintered metal and lasts considerably longer than current ceramic models since its special structure offers a high storage capacity for oil and additive combustion residues. The filter is designed in such a way that the filtered particulates are very evenly deposited, allowing the condition of the filter to be identified more reliably and its regeneration controlled far better than with other solutions. This diesel particulate filter is designed to last as long as the vehicle itself.

The two main DPF systems are those with an additive and those without. To enable a vehicle to operate without an additive the particulate filter must be fitted close to the engine. Because the exhaust gases will not have travelled far from the engine, they will still be hot enough to burn off the carbon soot particles. In these systems, an oxidising catalytic converter will be integrated into the particulate filter. In other systems, the particulate filter is fitted some distance from the engine and as the exhaust gases travel along the exhaust they cool. The temperatures required for ignition of the exhaust gas can only be achieved using an additive.

Figure 6.398 Different fitting methods for DPF

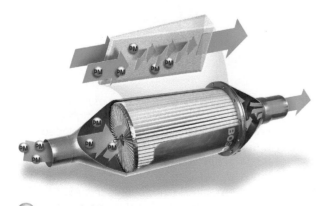

PM = Particulate matter

Figure 6.397 Diesel particulate filter (Source: Bosch Media)

Use of an additive lowers the ignition temperature of the soot particles and, if the engine management ECU raises the temperature of the exhaust gas, the filter can be regenerated. Regeneration is usually necessary after between 300 and 450 miles, depending on how the vehicle is driven. The process takes about 5–10 minutes and the driver shouldn't notice it is occurring, although sometimes there may be a puff of white smoke from the exhaust during regeneration. The additive is stored in a separate tank and is used at a rate of about 1 litre of additive to 3000 litres of fuel. It works by allowing the carbon particles trapped in the particulate filter to burn at a significantly lower temperature than would usually be required (250–450°C rather than 600–650°C).

On-board active filter management can use a variety of strategies:

▶ Engine management to increase exhaust temperature through late fuel injection or injection during the exhaust stroke (the most common method)

▶ Use of a fuel borne catalyst (the additive) to reduce soot burn-out temperature

▶ A fuel burner after the turbo to increase the exhaust temperature

▶ A catalytic oxidizer to increase the exhaust temperature, with after injection

▶ Resistive heating coils to increase the exhaust temperature

▶ Microwave energy to increase the particulate temperature

Passive regeneration Not running the regeneration cycle soon enough increases the risk of engine damage and/or uncontrolled regeneration (thermal runaway) and possible DPF failure. There are two types of regeneration – passive and active. Passive regeneration takes place, automatically, on motorway-type runs in which the exhaust temperature is high. If the exhaust is hot enough to ignite the soot particles, the regeneration process can carry on continuously and steadily across the platinum coated catalytic converter.

Once the storage capacity of the particulate filter has been exhausted, the filter must be regenerated by passing hot exhaust gases through it which burn up the deposited particulates. To produce the necessary high exhaust gas temperatures, the EDC alters the amount of air fed to the engine as well as the amount of fuel injected and the timing of injection. In addition, some unburnt fuel can be fed to the oxidizing catalytic converter by arranging for extra fuel to be injected during the expansion stroke. The fuel combusts in the oxidizing catalytic converter and raises the exhaust temperature even further.

Figure 6.399 Sectioned view of a new filter

Active regeneration A significant number of people don't use motorways so passive regeneration will be possible only occasionally. In the case of a filter without additives when the soot loading reaches about 45% the ECU switches off the EGR and increases the fuel injection period so there is a small injection after the main injection. These measures help to raise the engine exhaust temperature to over 600°C which is high enough to burn off the soot particles.

DPF warning light A warning light is triggered at a 55% soot loading. In such circumstances the car needs to be driven hard in a lower gear so the temperature in the particulate filter will be sufficient to burn off the soot. If the driver ignores the warning and continues to use the car as normal, the soot will continue to build until it reaches 7%. Additional warnings will then be given using the malfunction indication lamp (MIL). Now it will not be possible to clear the DPF by driving and it may need to be replaced. If loading reaches 95% then the DPF will need to be replaced.

Exhaust Gas Recirculation (EGR) valve If the EGR valve malfunctions, this may overload the DPF causing issues with regeneration. A malfunctioning EGR may present a light on the dash without immediate noticeable change to driving, a situation that can lead to a domino effect of component failure if left unaddressed.

Selective catalytic reduction (SCR) is a method of converting nitrogen oxides (NOx) with the aid of a catalyst into nitrogen (N_2), and water (H_2O). A gaseous reductant, typically anhydrous ammonia, aqueous ammonia or urea, is added to a stream of exhaust gas and then into a catalyst. Carbon dioxide (CO_2) is a reaction product when urea is used as the reductant. A typical SCR fluid is known as AdBlue.

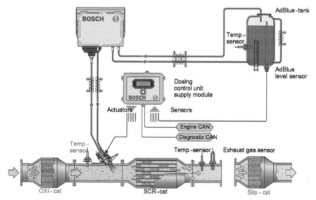

Figure 6.400 NOx reduction Bosch Denoxtronic 2.2 (Source: Bosch Media)

Summary Emissions become an area of increasing concern as the acceptable limits become tighter to meet legislative requirements. This has developed areas of specific services for the maintenance and repair of these systems due to the expense of the components and possibility of MOT failure. Strategies to help protect emissions components include premium fuel, tank additives and specialist cleaning machines developed for aggressively cleaning carbon that could cause component failure. The use of SCR is now essential to meet the Euro 6 diesel emissions standards for heavy trucks, and also for cars and light commercial vehicles. In many cases, emissions of NOx and PM (Particulate Matter) have been reduced by upwards of 90% as compared with vehicles of the early 1990s.

 Using images and text, create a short presentation to show how a component or system works

6.6.10 Emission control systems 3

Crankcase ventilation Oil vapour builds up in the engine crankcase because of heat, spray and the churning action of engine components as the engine is running. A fine mist of oil vapour is always present in a running engine. The engine crankcase pressure is never constant. Slight leakages into and from the combustion chambers, and the movement of the pistons, are responsible for most of the pressure variations.

Open ventilation A vent to atmosphere system was once used for ventilating pressure variations in the engine. This simple vent allowed a large quantity of oil vapour to escape. By fitting an oil separator, the quantity of oil was reduced but unacceptable quantities of oil vapour were still emitted. Developments since that time have seen the introduction of a positive crankcase ventilation system. This takes any escaping oil vapour into the engine for combustion.

Positive ventilation The main features of all systems include an oil separator, pipes and hoses from the crankcase, a valve or restrictor and hoses into the intake manifold and/or the air cleaner. The direction of airflow through the system depends on the variations in crankcase pressures at different phases of the engine operation.

Fuel vapour control Fuel vapour, and particularly petrol vapour, is harmful. It is given off from petrol at quite low ambient temperatures. Fuel is stored in underground tanks in order to reduce vapour formation. However, during fill up and when fuel is in the tank, vapour can escape into the atmosphere. Preventing vapour loss during fill up is difficult to control. Modern vehicles are fitted with fuel systems that prevent vapour loss from the vehicle.

Figure 6.401 Fuel tank cap

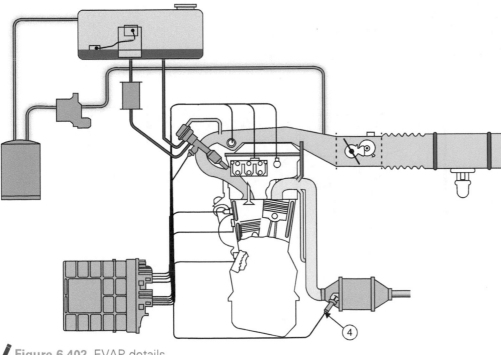

Figure 6.402 EVAP details

Evaporative emission control system This evaporative emission control system (EVAP) has a sealed tank and fuel lines and is vented for expansion and use of the fuel through a charcoal or carbon filter canister. The filter forms a barrier between fuel vapour in the system and the atmosphere. Air can pass through but the fuel vapour is trapped. In order to prevent the filter becoming saturated it is cleaned, or in technical terms, purged by drawing air through the filter in the opposite direction.

EVAP operation The air collects the deposited fuel vapour and carries it through pipes into the inlet manifold and engine where it is burnt. To prevent vapour loss through this route when the engine is not running, a canister purge control valve is fitted into the fuel vapour line. The valve is closed when the engine is stationary and during warm-up. When the engine is at normal running temperature the valve opens and the inlet manifold vacuum is able to cause an airflow through the canister. This draws vapour out of the filter and into the inlet manifold.

The **vacuum purge** is operated by a thermostatic vacuum switch, fitted in the engine coolant. Alternatively, it is actuated by a solenoid valve and the engine ECU, based on engine temperature. The purge vacuum is applied once the engine

temperature has risen to about 50°C. A control vacuum is applied to a diaphragm valve for the evaporative canister. This opens the route to the inlet manifold.

The **evaporative canister** can be fitted in almost any location in the vehicle. It may be near to the fuel tank, in the engine compartment or under a body panel. Also fitted in the vent pipes are expansion traps to catch excess fuel volume from hot weather expansion. Fuel traps, to prevent fuel loss if the vehicle turns over in an accident, are also fitted.

Exhaust gas recirculation (EGR) has become a common feature on petrol and diesel engines. The addition of exhaust gas to a fresh air and fuel charge lowers the combustion temperature and reduces the formation of nitrogen oxides (NOX). EGR operates during normal engine temperature and high vacuum conditions.

EGR valve Exhaust gases are piped from the exhaust manifold, to the inlet manifold, through a vacuum or electronically operated valve. Vacuum is applied to the EGR valve, on simple systems, through a thermostatic vacuum switch fitted into the engine coolant. Inlet manifold vacuum is applied to a diaphragm in the EGR valve. Attached to the diaphragm is the stem of the valve, which opens internal ports to allow exhaust gases to flow.

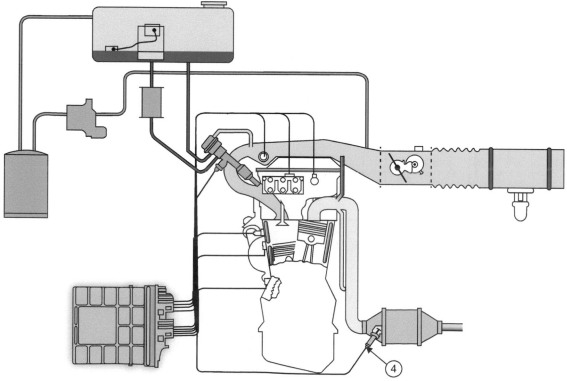

/ **Figure 6.403** ECU operates the valve

Figure 6.404 Charcoal canister

EGR valve designs Negative back pressure, ported and linear valves are used, according to the year of manufacture and type of engine management. The amount of exhaust gas introduced into the air supply is usually less than 15% of the total charge. However, where closed loop control is used, up to 50% can be used under some conditions on diesel engine systems. Some systems use a one-piece electrical solenoid valve in place of the separate electronic vacuum regulator and valve. Some valves have a switch fitted above the valve so that the ECU is able to monitor the opening performance.

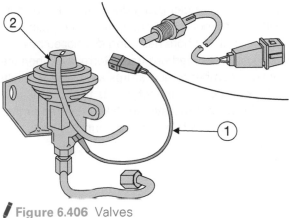

Figure 6.406 Valves

6

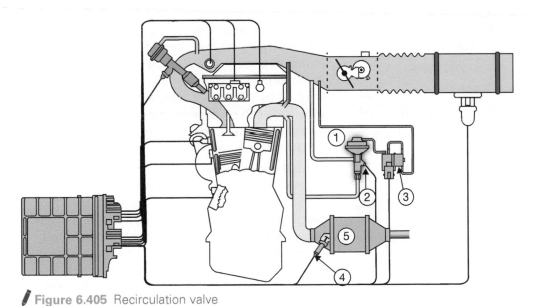

Figure 6.405 Recirculation valve

Create a mind map to illustrate the features of a key component or system

215

6.6.11 Cooled EGR ❸

Introduction In petrol and diesel engines, exhaust gas recirculation (EGR) is a technique used to reduce nitrogen oxide (NOx) emissions. It works by recirculating a small portion of an engine's exhaust gas back to the engine cylinders. This reduces the amount of oxygen in the incoming air stream and provides gases inert to combustion to act as absorbents of combustion heat. As NOx is produced in a narrow band of high cylinder temperatures and pressures, it can be reduced by this method.

Cooling the gas during EGR can result in a further reduction of NOx. Cooler gas reduces its volume so it is possible to 'fit more in' to the combustion chamber. The system outlined here has been developed by a company known as Pierburg from Germany. When the combustion chamber temperatures are lower, it results in less NOx production. In fact, this is the primary purpose of EGR, but this system takes it a step further by using a module with integrated control, bypass valves and a cooler.

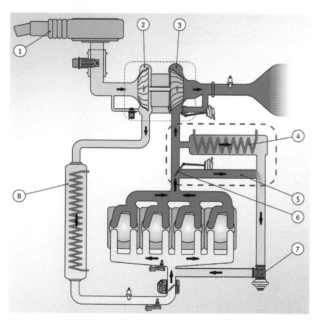

Figure 6.407 Cooled gas EGR block diagram. 1. Air filter, 2. Turbocharger compressor, 3. Turbocharger turbine, 4. EGR cooler, 5. Bypass duct, 6. Bypass flap (vacuum controlled in this example), 7. EGR valve, 8. Charge air cooler (Source: Pierburg)

Operation The flap (5) allows exhaust gases to bypass the cooler during warm-up so that the engine and catalytic converter can reach their ideal operating temperature as quickly as possible. This also results in reduced HC emissions and less diesel knock. An additional advantage is that the cooler can be bypassed during the regeneration phase of a diesel

particulate filter, when high temperatures are used for a short time.

Figure 6.408 EGR cooler module with integrated EGR valve and bypass flap (Source: Pierburg)

> Look back over the previous section and write out a list of the key bullet points from this section

6.6.12 Pressure charging ❸

Turbochargers and superchargers Supercharging is a method of increasing the performance of petrol and diesel engines by boosting the air charge with an air pump. The most popular method is turbocharging, as this uses some of the lost energy in the exhaust gas flow. Supercharging of an engine is not strictly an emission control device but a method by which an increase in power and fuel efficiency can be obtained from a smaller engine. At the same time, there are improvements in exhaust emissions.

Forced air induction has advantages over natural aspiration because cylinder charging is more consistent over the full engine speed range. This helps to give high torque and power over a wider speed range, improved overall performance and improved fuel consumption. Greater power from an engine, with only a small increase in weight, improves the engine and vehicle power to weight ratios. Superchargers are driven from the engine crankshaft. Engine power is therefore used to drive the charger.

Exhaust **turbochargers** use waste energy in the exhaust gas flow for power. This method of air boost charging is suitable for all types of engine. However, applications on small petrol engines are usually found only on high-performance vehicles.

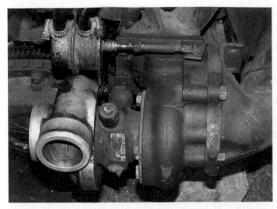

Figure 6.409 Turbocharger

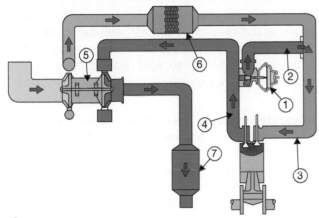

Figure 6.410 Air and exhaust flow

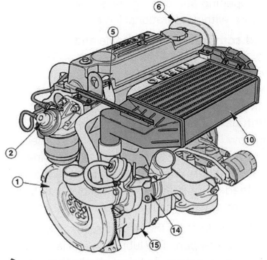

Figure 6.411 Ford turbo diesel with intercooler

Figure 6.412 Turbo operation

Figure 6.413 Waste gate actuator

Diesel engines Turbocharging of small high-speed diesel engines, used in cars and light vans, is now becoming popular. Diesel engine cylinder charging can be increased from about 60% for naturally aspirated engines to about 90% with exhaust turbocharging. The increased volume of air means that a corresponding increase in fuel can be delivered and more torque and power can be obtained per litre of engine capacity.

Exhaust gas energy Turbochargers use the energy in the exhaust gas to drive a turbine. The turbine is connected by means of a shaft to a compressor wheel in the engine air intake tract. The greater the flow of exhaust gas, the greater the speed of the turbine and compressor wheel and therefore the amount of additional charging.

Boosted air pressure is from 0.2 to 0.9 bar, depending on compressor speed. The maximum boost pressure is regulated by splitting the exhaust gas stream so that the excess gas flow and energy bypasses the turbine through a waste gate. The waste gate is a pressure-operated poppet or plate valve, which normally remains closed.

Waste gate valve When the boost pressure in the inlet air stream rises, it is applied to the waste gate valve. The pressure acts on a diaphragm connected to the waste gate valve and when it reaches the maximum operational pressure, the valve opens. This allows exhaust gases to bypass the turbine. With the reduced gas flow, the turbine and compressor slow down, the pressure reduces and the waste gate closes. This opening and closing cycle maintains the boost pressure within operational limits.

Turbine and compressor The turbine and compressor fan wheels are radial flow types. The exhaust flows towards the centre and then out. The inlet air flows in at the centre and outwards to the engine air intake duct. The air is forced out by the rotary centrifugal action of the compressor wheel. Air and gas flow is directed by spiral ducting in the turbocharger body. The spindle carrying the turbine and compressor wheel is mounted on special bearings with forced feed oil lubrication, which allows rotation with a minimum of metal-to-metal contact.

The **oil feed** is made from the engine main oil gallery and returns to the oil pan. The lubricating oil is used for cooling in the turbocharger. Lubricating oils that meet turbocharging specifications must be used. These oils have the ability to withstand the high temperatures of turbocharged engines without breaking down. If this occurred, it would deposit lacquer in the turbocharger bearings and elsewhere in the engine.

Switching off and intercooling Turbochargers must be allowed to slow down and to cool down before switching off the engine. Usually about 30 seconds to a minute is required for this. The charged air increases in temperature through the turbocharger and becomes

less dense and of lower mass. In order to overcome this loss, an intercooler is often fitted between the turbocharger and the inlet manifold. The intercooler is similar in construction to a coolant radiator but is an air-to-air heat exchanger.

Superchargers are mainly of Roots blower or radial flow types. The radial flow types are similar to the compressor on the exhaust turbocharger. However, they are driven by belts and gears from the engine crankshaft. Vane-type radial superchargers have been used but are less popular.

The **Roots blower** uses two or three lobe intermeshed rotors to pump air. The rotors have helical rotor vanes to reduce noise and improve efficiency. The rotor vanes are driven and matched together with a pair of gears, so that they rotate in mesh with each other. They run on ball or needle roller bearings at each end of the rotor spindles. They must be lubricated with high performance grease or synthetic oil in the bearing cases.

Integrated systems The air intake, exhaust and emissions systems are integrated with other engine systems. All these systems work together to minimise environmental pollution and to reduce noise and vibration in the vehicle. This is good for the environment and for comfort.

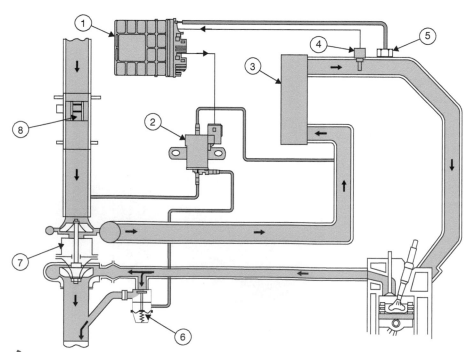

Figure 6.414 Boost control system

Figure 6.415 Turbine blades

Figure 6.416 Damaged turbo!

 Use the media search tools to look for pictures and videos relating to the subject in this section

6.6.13 Electrical pressure chargers ❸

Introduction A UK company Aeristech, have developed an eSupercharger. It claims that its proprietary electric motor technology delivers an engine's full load air requirements in under one second, without the cooling issues associated with conventional motors. This could solve a significant challenge facing the use of downsized engines in pursuit of low CO_2 emissions.

Figure 6.418 eSupercharger (Source Aeristech)

Complexity For serious engine downsizing multi-stage or mechanical superchargers are not the answer. This is because an engine using them would lack low speed power. An electric supercharger with enough power to supply all the low speed boosting needs of the engine could be the answer. Multi-stage turbocharging, which combines large and small turbos, makes downsized engines easier to drive across a wider speed range. However, other issues such as cost, complexity, thermal management and catalyst performance become problematic. The catalyst issue is likely to be more of a problem as tailpipe emissions are further restricted. The thermal mass of a two-stage turbocharger system located between the engine and the catalyst makes light-off more difficult during light loads or low engine speeds.

Voltage An electric supercharger with sufficient power to meet all low speed boosting needs could be achieved with the move to 48V systems. Aeristech uses permanent magnet motor technology, providing a faster response without the cooling challenges arising from the alternative switched reluctance (SR) motors. With a motor control strategy that separates commutation and power control, it means the electrical switching frequency need be no higher than running speed. The strategy would make permanent

Figure 6.417 Twin Roots blowers

magnet technology cost-effective, meaning that an eSupercharger could run continuously at boost levels of 2.5 bar or more.

Summary The company claims the eSupercharger motor will accelerate to 150,000 rpm with a transient response of idle-to-target speed in under 0.4 s. The company successfully subjected its technology to independent evaluation by Ricardo and Mahle Powertrain UK.

Audi's 2018 V6 diesel included a new electrically-enhanced turbocharger system, as well as a new integrated NOx and PM after treatment suite. Many diesels could be electrically turbocharged due to the uptake of diesel engines in the USA.

Figure 6.419 Audi V6 Diesel with electric turbo (Source: Audi Media)

Performance It is claimed that using the e-turbo will easily gain you a lead of at least two car lengths in the first two seconds at the traffic lights! This illustrates the clear benefit of the system in that turbo lag is all but eliminated as the turbo compressor can be

brought up to speed almost instantly. Other new features are to be incorporated:

- ▶ Piston rings have been optimized for minimal friction
- ▶ Crankcase and new cylinder heads have separate coolant loops
- ▶ A new thermal management system improves efficiency
- ▶ Turbocharger and the fully variable-load oil pump have been updated.

Emissions The engine will include a NOx storage catalytic converter which has been combined with a diesel particulate filter and SCR injection in a single unit. The engine package satisfies the most stringent of emissions legislation, including Euro 6, and reduces CO_2 emissions by an average of 15 g/km. Emission and economy issues are what manufacturers are working towards with all their new developments.

Summary Electrically drive turbo/superchargers are likely to become more common, particularly on vehicles with a 48V system. This is because they offer an increase in performance, a reduction of emissions and reduced costs.

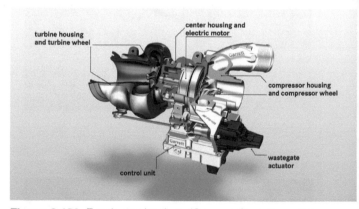

Figure 6.420 E-turbo technology (Source: Garrett media)

 Create a mind map to illustrate the features of a key component or system

CHAPTER 7

Electrical systems

After successful completion of this chapter you will be able to show you have achieved these objectives:

• Understand light vehicle electrical and electronic principles.

• Understand how light vehicle batteries, starting and charging systems operate.

• Understand how light vehicle auxiliary electrical systems operate.

• Understand how to check, replace and test light vehicle electrical systems and components.

• Understand how to diagnose and rectify faults in auxiliary electrical systems.

DOI: 10.1201/9781003173236-7

7.1 Electrical and electronic principles

7.1.1 Electricity and the atom ❶

What is electricity? To understand electricity properly we must start by finding out what it really is. This means we must think very small! The molecule is the smallest part of matter that can be recognised as that particular matter. Subdivision of the molecule results in atoms. The atom is the smallest part of matter and consists of a central nucleus made up of protons and neutrons. Around this nucleus, electrons orbit, like planets around the sun. The neutron is a very small part of the nucleus. It has an equal positive and negative charge. It is therefore neutral and has no polarity. The proton is another small part of the nucleus; it is positively charged. As the neutron is neutral and the proton is positively charged, this means the nucleus of the atom is positively charged.

The **electron** is an even smaller part of the atom and is negatively charged. It is held in orbit around the nucleus by the attraction of a positively charged proton. When atoms are in a balanced state, the number of electrons orbiting the nucleus equals the number of protons. The atoms of some materials have electrons which are easily detached from the parent atom and join an adjacent atom. In so doing they move an electron (like polarities repel) from this atom to a third atom and so on through the material. These are called free electrons.

Conductors and insulators Materials are called conductors if the electrons can move easily. However, in some materials it is difficult to move the electrons. These materials are called insulators.

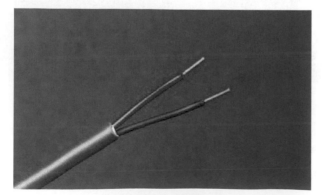

Figure 7.1 Insulated conductors

If an **electrical pressure** (voltage) is applied to a conductor, a directional movement of electrons will take place. There are two conditions for electrons to flow – a pressure source (e.g. from a battery or generator) and a complete conducting path for the electrons to move (e.g. wires).

An electron flow is termed an electric current. Shown in the online animation is a simple electric circuit. The battery positive terminal is connected, through a switch and lamp, to the battery negative terminal. With the switch open, the chemical energy of the battery will remove electrons from the positive terminal to the negative terminal via the battery. This leaves the positive terminal with less electrons and the negative terminal with a surplus of electrons. An electrical pressure exists between the battery terminals. With the switch closed, the surplus electrons on the negative terminal will flow through the lamp back to the electron-deficient positive terminal. The lamp will therefore light until the battery runs down.

The movement from negative to positive is called the electron flow. However, it was once thought that current flowed from positive to negative. This convention is still followed for practical purposes. Therefore, even though it is not correct, the most important point is that we all follow the same convention. We say that current flows from positive to negative.

Effects of electricity When a current flows in a circuit, it can produce only three effects: heat, magnetic and chemical. The heating effect is the basis of electrical components such as lights and heater plugs. The magnetic effect is the basis of relays and motors and generators. The chemical effect is the basis for electroplating and battery charging. The three effects are reversible. For example, electricity can make magnetism, and magnetism can be used to make electricity. The number of electrons through a lamp every second is the rate of flow. The cause of electron

Figure 7.2 Heating effect

Figure 7.3 Chemical effect

Figure 7.4 Magnetic effect

Figure 7.5 Reversibility – motor and generator

flow is the electrical pressure. The lamp produces an opposition to the rate of flow set up by the electrical pressure. Power is the rate of doing work or changing energy from one form to another. All these quantities are given names.

If the voltage applied to the circuit was increased but the lamp resistance stayed the same, then current would increase. If the voltage was maintained but the lamp was changed for one with a higher resistance, the current would decrease. This relationship is put into a law called Ohm's Law. This law states that in a closed circuit the current is proportional to the voltage and inversely proportional to the resistance. Any one value can be calculated if the other two are known.

When voltage causes current to flow, energy is converted. This is described as power. The unit of power is the watt. As with Ohm's law, any one value can be calculated if the other two are known. See the online resources for more details about Ohm's law and how to calculate values.

 Make a simple sketch to show how one of the main components or systems in this section operates

7.1.2 Basic circuits and magnetism ❶

Conductors, insulators and semiconductors
All metals are conductors. Silver, copper and aluminium are among the best and are frequently used. Liquids which will conduct an electric current are called electrolytes. Insulators are generally non-metallic and include rubber, porcelain, glass, plastic, cotton, silk, wax, paper and some liquids. Some materials can act as either insulators or conductors depending on conditions. These are called semiconductors. They are used to make transistors and diodes.

Figure 7.6 Insulated cable

Figure 7.7 Semiconductor

Figure 7.8 Conductor

Figure 7.9 Electrolyte

Factors affecting resistance of a conductor The amount of resistance offered by a conductor is determined by a number of factors:

▶ length – the greater the length the greater the resistance
▶ cross-sectional area – the larger the area the smaller the resistance
▶ the material – the resistance offered by a conductor will vary according to the material from which it is made
▶ temperature – most metals increase in resistance as temperature increases.

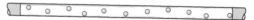

Figure 7.10 Conductor resistance

Series circuits When resistors are connected so that there is only one path for the same current to flow through each resistor they are connected in series. In a series circuit:

▶ current is the same in all parts of the circuit
▶ applied voltage equals the sum of the volt drops around the circuit
▶ total resistance of the circuit equals the sum of the individual resistance values.

Parallel circuits When resistors are connected such that they provide more than one path for the current to flow in, and have the same voltage across each component, they are connected in parallel. In a parallel circuit:

▶ voltage across all components of a parallel circuit is the same
▶ total current from the source is the sum of the current flowing in each branch. The current splits up depending on each component resistance
▶ total resistance of the circuit is the sum of the reciprocal (one divided by the resistance) values.

Magnetism and electromagnetism can be created by a permanent magnet or by an electromagnet. The space around a magnet in which the magnetic effect can be detected is called the magnetic field. Flux lines or lines of force represent the shape of magnetic fields in diagrams. Electromagnets are used in motors, relays and fuel injectors, to name just a few instances. Force on a current-carrying conductor in a magnetic field is caused because of two magnetic fields interacting. This is the basic principle of how a motor works.

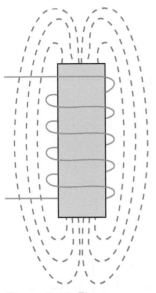

Figure 7.11 Electromagnet

Electromagnetic induction When a conductor cuts or is cut by magnetism, a voltage is induced in the conductor. The direction of this voltage depends upon the direction of the magnetic field and the direction in which the field moves relative to the conductor. The size is proportional to the rate at which the conductor cuts or is cut by the magnetism. This effect of induction, meaning that voltage is made in the wire, is the basic principle of how generators such as the alternator on a car work. A generator is a machine that converts mechanical energy into electrical energy.

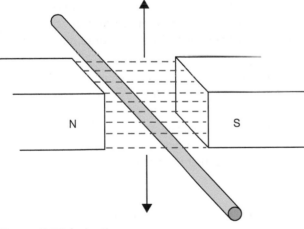

Figure 7.12 Induction

Mutual induction If two coils, primary and secondary, are wound on to the same iron core, any change in magnetism of one coil will induce a voltage in the other. This happens when the primary current is switched on and off. If the number of turns of wire on the secondary coil is more than the primary, a higher voltage can be produced. This is called transformer action and is the principle of the ignition coil.

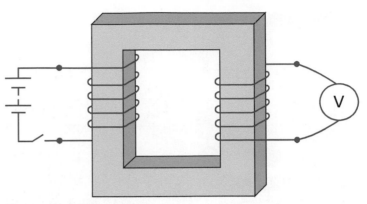

Figure 7.13 Transformer

Figure 7.14 Ignition coil

 Create a mind map to illustrate the features of a key component or system

7.1.3 **Electrical components** ❶

A switch is a simple device used to break a circuit; that is, it prevents the flow of current. A wide range of switches is used. Some switches are simple on/off devices such as an interior light switch on the door pillar. Other types of switch are more complex. They can contain several sets of contacts to control, for example, the indicators, headlights and horn. These are described as multifunction switches.

Resistors Good conductors are used to carry the current with minimum voltage loss due to conductor resistance. Resistors are used to control the current flow in a circuit or to set voltage levels. They are made of materials that have a high resistance. Resistors to carry low currents are often made of carbon. Resistors for high currents are usually wire wound.

Figure 7.15 Resistor

A **relay** is a very simple device. It can be thought of as a remote-controlled switch. A very small electric current is used to magnetise a small winding. The magnetism then causes some contacts to close, which in turn can control a much heavier current. This allows small delicate switches to be used to control large current users, such as the headlights or the heated rear window.

Figure 7.16 Simple 'cube' relay

A **capacitor** is a device for storing an electric charge. In its simplest form, it consists of two plates separated by an insulating material. One plate can have excess electrons compared to the other. On vehicles its main uses are for reducing arcing across contacts and for radio interference suppression circuits as well as in electronic control units.

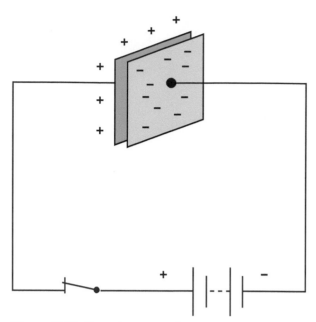

Figure 7.17 Capacitor operation

Some form of circuit protection is required to protect the electrical wiring of a vehicle and to protect the electrical and electronic components. It is now common practice to protect almost all electric circuits with a fuse. A fuse is the weak link in a circuit. If an overload of current occurs then the fuse will melt and disconnect the circuit before any serious damage is caused. Automobile fuses are available in three types: glass cartridge, ceramic and blade type. The blade type is now the most popular choice due to its simple construction and reliability. Fuses are available in a number of rated values. Only the fuse recommended by the manufacturer should be used.

A **fuse** protects the device as well as the wiring. A good example of this is a fuse in a wiper motor circuit. If a value were used, which is much too high then it would still protect against a severe short circuit. However, if the wiper blades froze to the screen, a large value fuse might not protect the motor from overheating.

Current rating	Colour code
3	Violet
4	Pink
5	Clear/Beige
7.5	Brown
10	Red
15	Blue
20	Yellow
25	Neutral/White
30	Green

Fusible links in the main output feeds from the battery protect against major short circuits in the event of an accident or error in wiring connections. These links are simply heavy-duty fuses and are rated in values such as 50, 100 or 150A.

losing ground to the smaller blade terminals. Circular multi-pin connectors are used in many cases, the pins varying in size from 1mm to 5mm. With any type of multi-pin connector, an offset slot or similar is used to prevent incorrect connection.

Figure 7.18 These links connect to the battery

Figure 7.20 Terminals and connectors in use

Circuit breakers Occasionally circuit breakers are used in place of fuses, this being more common on heavy vehicles. A circuit breaker has the same rating and function as a fuse but with the advantage that it can be reset.

Protection against corrosion of the connector is provided in a number of ways. Earlier methods included applying suitable grease to the pins to repel water. It is now more usual to use rubber seals to protect the terminals although a small amount of contact lubricant can still be used. Many multi-connectors use some kind of latch not only to prevent individual pins working loose but also to ensure that the complete plug and socket is held securely.

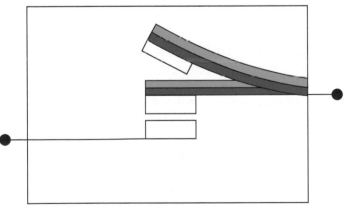

Figure 7.19 A bimetal strip is the main component

Terminals and connectors Many types of terminals are available. These have developed from early bullet-type connectors into the high-quality waterproof systems now in use. A popular choice for many years was the spade terminal. This is still a standard choice for connection to relays for example, but is now

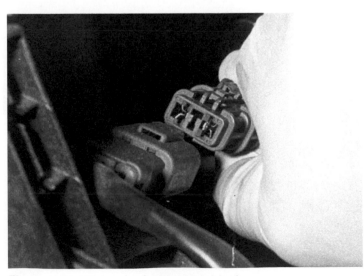

Figure 7.21 Waterproof connector block

Wires Cables or wires used for motor vehicle applications are usually copper strands insulated with PVC. Copper, beside its very low resistance, has ideal properties such as ductility and malleability. This makes it the natural choice for most electrical conductors. For the insulation, PVC is ideal. It not only has very high resistance, but is also very resistant to fuel, oil, water and other contaminants.

Figure 7.22 Cables in a wiring harness

Cable size The choice of cable size depends on the current it will have to carry. The larger the cable used, the better it will be able to carry the current and supply all of the available voltage. However, it must not be too large or the wiring becomes cumbersome and heavy! In general, the voltage supply to a component must not be less than 90% of the system supply. Cable is available in stock sizes, but a good 'rule of thumb' guide is that one strand of 0.3mm diameter wire will carry 0.5 amps safely.

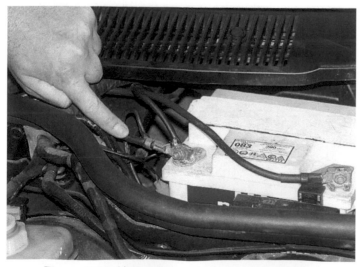

Figure 7.23 Heavy-duty cable

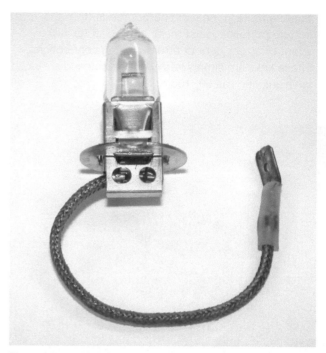

Figure 7.24 Light-duty cable

UK Colour Code The UK system uses twelve colours to determine the main purpose of the cable. Tracer colours further define its use.

A 'European' system used by Ford, VAG, BMW and other manufacturers is based broadly on the online table. Please note that there is no connection between the 'Euro' system and the British standard colour codes. In particular, note the use of the colour brown in each system!

A popular system is the terminal designation. This helps to ensure correct connections are made on the vehicle, particularly in after-sales repairs. It is important however to note that the designations are not to identify individual wires but are to define the terminals of a device.

Symbols and circuit diagrams The selection of symbols shown in Figure 7.25 is intended as a guide to some of those in use. Many manufacturers use their own variations. The idea of a symbol is to represent a component in a very simple but easily recognisable form.

The conventional type of diagram shows the electrical connections of a circuit but does not attempt to show the various parts in any particular order or position.

A **layout circuit diagram** attempts to show the main electrical components in a position similar to those on the actual vehicle. Due to the complex circuits and

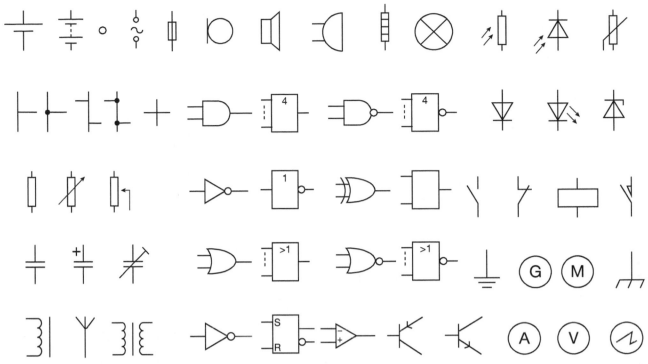

Figure 7.25 Symbols

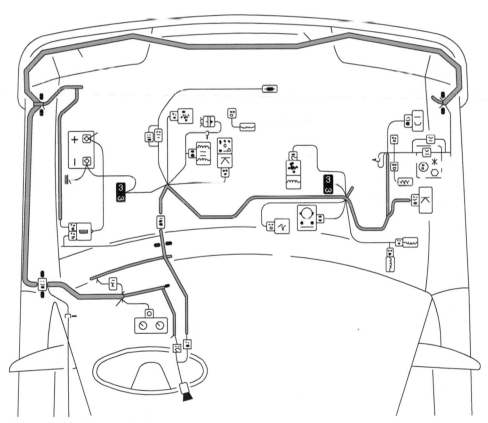

Figure 7.26 Layout circuit

the number of individual wires, some manufacturers now use two diagrams – one to show electrical connections and the other to show the actual layout of the wiring harness and components.

Terminal circuit diagram A terminal diagram shows only the connections of the devices and not any of the wiring. The terminal of each device, which can be represented pictorially, is marked

229

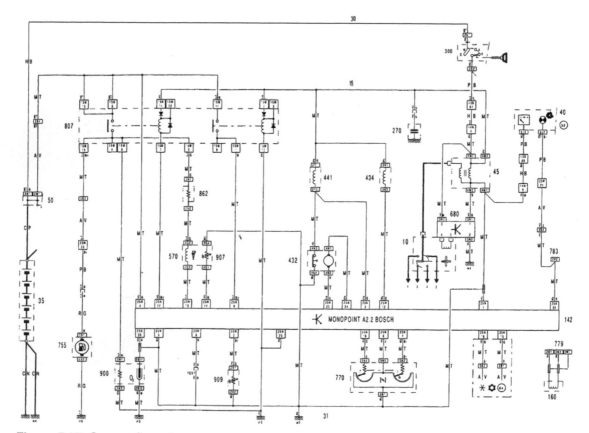

Figure 7.27 Current flow diagram

with a code. This code indicates the device terminal designation, the destination device code and its terminal designation and in some cases the wire colour code.

Figure 7.27 is laid out such as to show current flow from the top of the page to the bottom. These diagrams often have two supply lines at the top of the page marked 30 (main battery positive supply) and 15 (ignition-controlled supply). At the bottom of the diagram is a line marked 31 (earth or chassis connection).

Shown online is a basic lighting circuit. Click on each switch in turn to make the circuit operate. Notice the effect of some switches being connected in series.

Describing electric circuit faults Three descriptive terms are useful when discussing electric circuits:

▶ open circuit – the circuit is broken and no current can flow
▶ short circuit – a fault has caused a wire to touch another conductor and the current uses this as an easier way to complete the circuit
▶ high resistance – a part of the circuit has developed a high resistance (such as a dirty connection), which will reduce the amount of current that can flow.

Figure 7.28 Short circuit

Figure 7.29 Open circuit

Limits of the wiring systems The complexity of modern wiring systems has been increasing steadily. However, in recent years it has increased dramatically. The size and weight of the wiring harness is a major problem. The number of separate wires required on a top of the range vehicle can be in the region of twelve hundred. The wiring loom required to control all the functions in or from the driver's door can require up to 50 wires. This is clearly becoming a problem as, apart from the obvious issue of size and weight, the number of connections and number of wires increases the possibility of faults developing.

Create an information wall to illustrate the features of a key component or system

7.1.4 Electronic components ❷

Introduction This section describes the principles of various electronic components and circuits. It is not intended to explain their detailed operation. The intention is to describe briefly how circuits work and where they may be utilised in vehicle applications. An understanding of basic electronic principles will help to show how electronic control units work. This understanding is a great aid when fault-finding electrical and electronic systems. These range from simple interior light delay circuits to the most complicated engine management systems.

Shown in Figure 7.25 are the symbols for some common electronic components. A simple and brief description follows for some of the components shown. Standards for these symbols vary from manufacturer to manufacturer but most are similar.

Resistors are probably the most widely used component in electronic circuits. Two factors must be considered when choosing a suitable resistor, namely the ohms value and the power rating. Resistors are used to limit current flow and provide fixed voltage drops. Most resistors used in electronic circuits are made from small carbon rods – the size of the rod determines the resistance. A thermistor is a resistor that changes resistance with temperature.

Figure 7.30 Resistors in an electronic control unit

Capacitors are two plates separated by an insulator. The value is determined mostly by the area of the

plates and the distance between them. Capacitors are often constructed out of metal foil sheets insulated by paper, which are rolled up together inside a tin can. The two plates can hold a charge of electricity.

Diodes can be described as one-way valves. For most uses, this is a good description. The diode is made

Figure 7.31 Capacitor in a flasher unit

from two types of silicon (N type and P type). Electrons can flow from negative (N type) to the positive (P type) material, but not the other way round. Zener diodes are very similar in operation except that they are designed to conduct in the reverse direction at a preset voltage. They can be thought of as a type of pressure-relief valve.

Figure 7.32 Diodes are one-way valves for electricity

Transistors are the devices which have allowed the development of today's complex and small electronic systems. The transistor is used either as a switch or

Remember! There is more support on the website that includes additional images and interactive features: www.tomdenton.org

231

as an amplifier. Transistors are constructed from the same materials as diodes but with three terminals. A small voltage (about 0.7V) supplied to the base terminal of a transistor known as an NPN will cause it

Figure 7.33 Transistor in an ignition ECU

to fully switch on, joining the collector and emitter. It is sometimes useful to think of a transistor as a type of relay. However, with a transistor a smaller voltage will partially switch the collector-emitter circuit on and hence the component works as an amplifier.

Inductors are most often used as part of an oscillator or amplifier circuit. In these applications, it is essential for the inductor to be stable and of reasonable size. The basic construction of an inductor is a coil of wire wound on a former. It is the magnetic effect of the changes in current flow which give this device the properties of inductance. The inductor is also used as a filter because it tends to prevent changes in signals.

Figure 7.34 Electronic control unit

Integrated circuits, or ICs, are constructed on a single slice of silicon. Combinations of some

of the components mentioned previously can be combined to carry out various tasks. These tasks can range from a simple switching action to the operation of a microprocessor of a computer. The components required for these circuits can be made directly on to one slice of silicon. The advantage of this is not just the size of the ICs (which can be very small) but the speed at which they can be made to work.

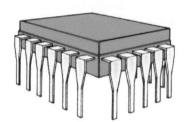

Figure 7.35 IC Package

The simplest form of **amplifier** involves just one resistor and one transistor. A small change on the input terminal will cause a similar change of current through the transistor and an amplified signal will be seen at the output terminal. However, the output will be inverted compared to the input. This very simple circuit has many applications when used as a switch. For example, a very small current flowing to the input can be used to operate a relay winding connected in place of the resistor.

Practical amplifier One of the main problems with the previous transistor amplifier is that the gain can be variable and non-linear. To overcome this, some type of feedback is used to make a circuit with more appropriate characteristics. In the circuit shown online, resistors Rb1 and Rb2 set the base voltage of the transistor and because the base emitter voltage is constant at about 0.7V, this in turn will set the emitter voltage. The standing current through Rc and Re is hence defined and the small signal changes at the input will be reflected in an amplified form at the output, albeit inverted.

Integrated circuit differential amplifiers are very common – one of the most common is known as the '741 op-amp'. This type of amplifier has a gain in the region of 10 000. Operational amplifiers are used in many applications and in particular can be used as signal amplifiers. A major role for this device is to act as a buffer between a sensor and a load such as a display. The internal circuit of these types of device can be very complicated but external connections and components can be kept to a minimum.

Figure 7.36 Operational amplifier on a chip

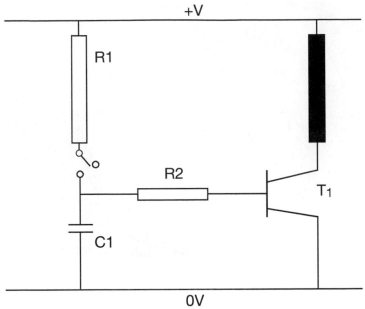

Figure 7.37 Basic timer circuit

Bridge circuits There are many types of bridge circuits but they are all based on the principle of the Wheatstone bridge. A simple calculation will show that the meter will read zero when each side of the bridge is balanced. A bridge and amplifier circuit, which may be typical of a motor vehicle application, is shown online. In this circuit, R1 could be a hot wire airflow sensor or a temperature measurement thermistor. The output of the bridge changes with even small changes in the resistance of R1.

The **Schmitt trigger** is used to change variable signals into crisp, square wave signals for use in digital or switching circuits. For example, a sine wave fed into a Schmitt trigger will emerge as a square wave with the same frequency as the input signal. The output signal from an inductive-type distributor or a crank position sensor on a motor vehicle will be passed through a Schmitt trigger. This will ensure that further processing is easier and the switching action is positive.

Timers In its simplest form, a timer consists of just two components, a resistor and a capacitor. When the capacitor is connected to a supply via the resistor, it is accepted that it will become fully charged in 5CR seconds (where R is the resistor value in ohms and C is the capacitor value in farads). The discharge time is the same if the resistor is connected across the capacitor. Timer circuits similar to this are used in wiper delay units and flasher units.

Filters A filter that prevents large particles of contaminates reaching a fuel injector is an easy concept to grasp. In electronic circuits, the basic idea is just the same except the 'particle size' is the frequency of a signal. Electronic filters come

Figure 7.38 Flasher unit

in two main types: low pass filters, which block high frequencies, and high pass filters, which block low frequencies. The two components of a basic filter work together as a voltage divider, just like two resistors connected to a battery. However, the 'resistance' of the capacitor changes depending on the signal frequency. Filters are often used to block interference signals.

A **Darlington pair** is a simple combination of two transistors, which will give a high current gain – typically several thousand. The transistors are usually mounted on a heat sink and overall the device will

Figure 7.39 Fuel filter

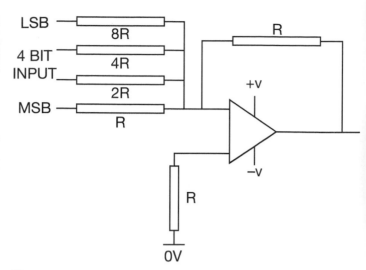

Figure 7.41 DAC operation

have three terminals marked as a single transistor – base, collector and emitter. The Darlington pair configuration is used for many switching applications. A common use is for the switching of coil primary current in the ignition circuit.

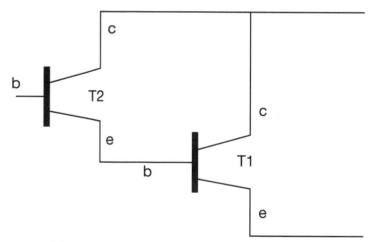

Figure 7.40 Two transistors as a Darlington pair

Digital to analog conversion (DAC) Conversion from digital signal to an analog signal is a relatively simple process. When an operational amplifier is configured with shunt feedback, the input and feedback resistors determine the gain (amplification). The 'weighting' of each input line can be determined by choosing suitable resistor values. In the case of the four bit digital signal, the most significant bit will be amplified with a gain of one. The next bit ½, the next bit a ¼, and in this case, the least significant bit will be amplified with a gain of ⅛. The output signal produced is therefore a voltage proportional to the value of the digital input number.

The purpose of this circuit is to convert an analog signal, such as that received from a temperature thermistor, into a digital signal for use by a computer or a logic system. Most systems work by comparing the output of a digital to analog converter (DAC) with the input voltage. The output of a binary counter is connected to the input of the DAC, the output of which will be an increasing voltage. This voltage is compared with the input voltage and the counter is stopped when the two are equal. The count value is then a digital representation of the input voltage. The operation of the other digital components in this circuit will be explained in the next section. ADCs are available in IC form and can work to very high speeds.

 Look back over the previous section and write out a list of the key bullet points from this section

7.1.5 Digital systems ❸

Introduction With most electronic systems on the car, we don't have to worry about what the electronics do in detail. It is good practice to think of electronic systems as having inputs and outputs. The 'brain' of the system will be the electronic control unit or ECU. Shown online as an example, is an anti-lock brake system. The inputs supply information to the ECU about how the car is operating. The ECU 'decides' what to do and then controls the outputs of the system, which in this case are the brakes.

Figure 7.42 Wheel sensors

Figure 7.43 Modulator

Figure 7.44 ECU

Digital electronic systems When working on a system of this type if it is not working correctly, we have to consider just one part at a time until the fault is found. This type of work is very interesting but you will need to understand the operation of all the basic principles first. It is not necessary to

understand the operation of ECUs in detail, but a basic knowledge of how they work is essential. The basic building blocks of digital systems are known as logic gates.

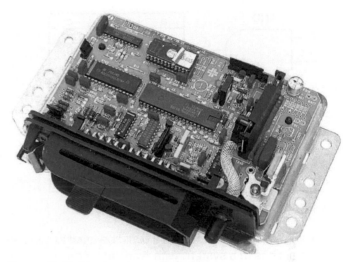

Figure 7.45 Electronic control unit

Logic gates A truth table is used to describe what combination of inputs will produce a particular output. The AND gate will only produce an output of '1' if both inputs (or all as it can have more than two inputs) are also at logic '1'. Output is '1' when inputs A AND B are '1'. The OR gate will produce an output when either A OR B (OR both), are '1'. Again more than two inputs can be used. A NOT gate is a very simple device where the output will always be the opposite logic state from the input. In this case A is NOT B. The AND and OR gates can each be combined with the NOT gate to produce the NAND and NOR gates respectively. These two gates have been found to be the most versatile and are used extensively for the construction of more complicated logic circuits.

Logic gates are made from simple electronic components such as resistors and transistors. The circuit shown here is a simplified NOT gate. This gate simply inverts the input.

Combinational logic Circuits consisting of many logic gates are called combinational logic circuits. They have no memory or counter circuits. These circuits can be represented by a simple block diagram with inputs and outputs. The first stage in the design process of creating a combinational logic circuit is to define the required relationship between the inputs and outputs. Let's consider we need a circuit to compare two sets of three inputs and if they are not the same to provide

235

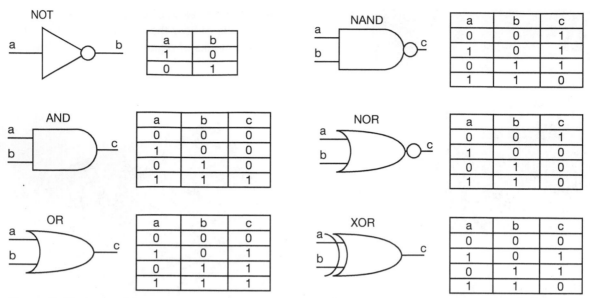

NOT

a	b
1	0
0	1

NAND

a	b	c
0	0	1
1	0	1
0	1	1
1	1	0

AND

a	b	c
0	0	0
1	0	0
0	1	0
1	1	1

NOR

a	b	c
0	0	1
1	0	0
0	1	0
1	1	0

OR

a	b	c
0	0	0
1	0	1
0	1	1
1	1	1

XOR

a	b	c
0	0	0
1	0	1
0	1	1
1	1	0

Figure 7.46 Truth tables

a single logic '1' output. This could be used to compare the actions of a system with twin safety circuits, such as an ABS electronic control unit. The logic circuit could be made to operate a warning light if a discrepancy exists between the two safety circuits. Shown here is one way in which the circuit could be constructed.

Counters are constructed from a series of bistable (two steady states) devices. Bistables are often called flip-flops! A binary counter will count clock pulses at its input. These counters are called 'ripple through' or non-synchronous because the change of state ripples through from the least significant bit and the outputs do not change simultaneously. The counters can be configured to count up or down.

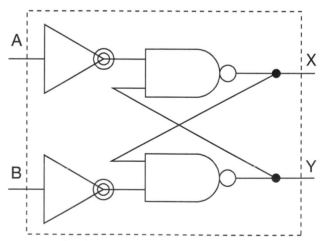

Figure 7.48 Basic memory building block made from gates

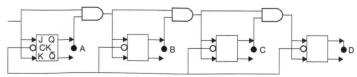

Figure 7.49 Binary counter

Figure 7.47 Safety circuits compare inputs

Sequential logic The combinational logic circuit discussed previously was a combination of various gates. The output of each system was only determined by the present inputs. Circuits which have the ability to memorise previous inputs or logic states are known as sequential logic circuits. In these circuits, the sequence of past inputs determines the current output. Sequential circuits store information after the inputs are removed – they are the basic building blocks of computer memory.

Memory circuits Electronic circuits constructed using flip-flops, as described previously, are a form of memory. If eight flip-flops are connected together, they form a simple eight-bit word memory. This is usually called a register rather than memory but it will store one byte (eight bits) of information. When more than one register is used, an address is required to access or store the data in a particular register. Figure 7.50 shows a block diagram of a four-byte memory. Each area of this memory is allocated a unique address. A control bus is also needed to determine if the memory is to be read from or written to.

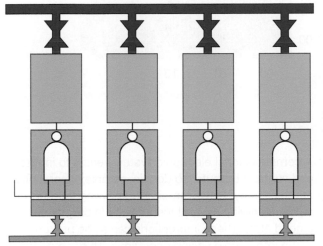

Figure 7.50 Four-byte memory

RAM The memory which has just been described above, together with the techniques used to access the data, is typical of most computer systems. This is the type of memory known as random access memory (RAM). Data can be written to and read from this type of memory but note that the memory is volatile – in other words it will 'forget' all its information when the power is switched off!

Figure 7.51 Random access memory chips

ROM The type of memory that can be 'read from' but not 'written to' is known as read only memory (ROM). This type of memory has data permanently stored which is not lost when power is switched off. There are many types of ROM which hold permanent data, but one in particular is worth a mention – EPROM. This stands for erasable, programmable, read only memory. Its data can be changed with special equipment (some are erased with ultraviolet light) but otherwise its memory is permanent. In an engine management ECU, operating data and a controlling program are stored in ROM whereas instantaneous data (engine speed, load, temperature, etc.) are stored in RAM.

Figure 7.52 Read only memory chip

Microprocessor The advent of the microprocessor has made it possible for tremendous advances in all areas of electronic control, not least of these in the motor industry. Designers have found that the control of vehicle systems, which is now required to meet the customer's needs and the demands of regulations, has made it necessary to use computer control. Figure 7.56 shows a block diagram of a microcomputer containing the four major parts. These are the input and output ports, some form of memory and the CPU or central processing unit (microprocessor). It is likely that some systems will incorporate more memory chips and other specialised components. Three buses carry data, addresses and control signals.

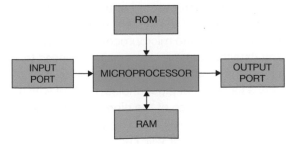

Figure 7.53 Modern microprocessor system

Microprocessor operation A microprocessor is operated at very high speed by the system clock. The microprocessor has a 'simple' task. It has to fetch an instruction from memory, decode the instruction and then carry out or execute the instruction. This cycle, which is carried out relentlessly, even if the instruction is to do nothing, is known as the fetch-decode-execute sequence.

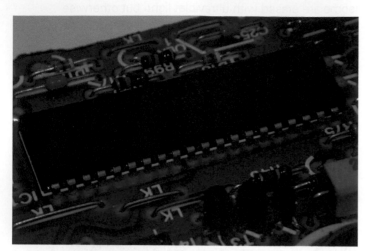

Figure 7.54 Microprocessor (microcontroller in this case)

Fetch-decode-execute sequence The full sequence of events is represented here. The execute phase can be as simple as adding two numbers inside the microprocessor or it may require data to be output to a port. If this is the case then the address of the port will be placed on the address bus and a control bus 'write' signal generated.

► the microprocessor places the address of the next memory location on the address bus
► at the same time, a memory read signal is placed on the control bus
► the data from the addressed memory location is placed on the data bus
► the data from the data bus is temporarily stored in the microprocessor
► the instruction is decoded in the microprocessor internal logic circuits
► the 'execute' phase is now carried out.

Memory The way in which memory actually works was discussed previously. We will now look at how it is used in a microprocessor-controlled system. Memory is the part of the system which stores both the instructions for the microprocessor (the program) and any data that the microprocessor will need to execute the instructions. It is convenient to think of memory as a series of pigeonholes, each of which is able to store data. Each of the pigeonholes must have an address to distinguish them from each other and so that the microprocessor will 'know' where a particular

piece of information is stored. The microprocessor reads the program instructions from sequential memory addresses and then carries out the required actions in turn.

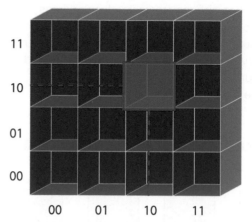

Figure 7.55 The address of the memory location shown is 1010

A complex digital electronic system reacts to inputs and controls outputs. To do this, complex operations are necessary. However, simply knowing that the unit is following a set of instructions (a program) is a good start. These instructions can all be broken down into simple yes/no operations. This is a good way to comprehend a complicated system. Modern computers, 'in car or on desk', contain millions and millions of logic gates! A simulation program is available from the website.

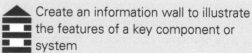
Create an information wall to illustrate the features of a key component or system

7.1.6 Central electrical control ❸

Introduction For many years the trend with automotive electrical systems has been towards some sort of networked central control. This makes sense because many systems can share one source of information and, with the proper equipment, diagnostics can be made easier. Also, centralization allows facilities to be linked and improved. For example, networking and centralization of control units makes it easier to have a system where the engine will not start if a door is open, or where selection of reverse gear can operate a rear wiper when the fronts are switched on.

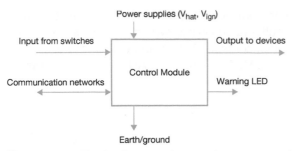

Figure 7.56 The basic central control system can be simplified in a way that is represented here

Central control usage The most common usage of central control is for body systems such as lighting, wipers, doors, seats and windows. In some cases these systems are controlled by slave units via a communication network; in other cases, one unit controls everything. In almost all cases, this central unit is networked to other ECUs.

Some central control modules connect via normal wires to switches that supply normal voltage on/off signals; others use switches that communicate on the CAN or LIN networks. The outputs from the module are sent via relays or solid-state switches on standard wires.

Names Manufacturers have different names for these systems and the control units but most have a similar function. Four example names follow:

▶ Body Control Module (BCM)
▶ General Electronic Module (GEM)
▶ Central Control Unit (CCU)
▶ Central Control Module (CCM)

Figure 7.58 Central control module (Source: Continental)

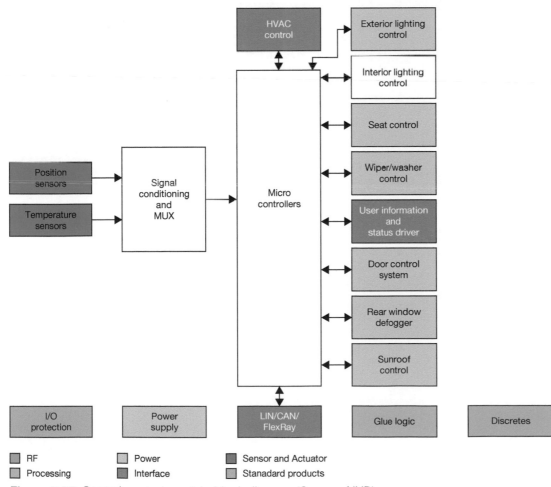

Figure 7.57 Central control module block diagram (Source: NXP)

239

Detailed system and circuit diagrams The images in Figures 7.59 and 7.60 show a full circuit diagram (in two halves) that has been adapted from materials supplied by Ford Motor Company. The circuit shows a general electronic module (GEM) and how it is used to control the wipers (in this case). Note that the multifunction wiper switch contains a series of switch contacts that all connect directly back to the GEM. Also note the CAN connection to the module.

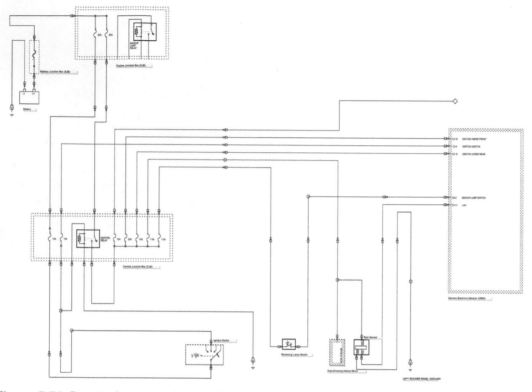

Figure 7.59 Part 1 of a wiper circuit using a central control module (Source: Ford Motor Company)

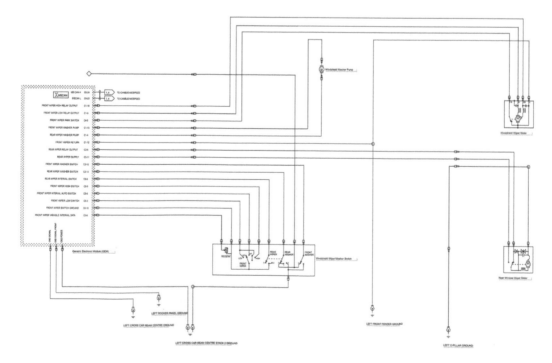

Figure 7.60 Part 2 of a wiper circuit using a central control module (Source: Ford Motor Company)

Lighting system Figure 7.61 is from a Ford vehicle with adaptive front lighting. The light switch in this case has a supply (30) and an earth/ground (31) connection, but all commands to operate the lights are sent via the LIN bus. The GEM supplies outputs to operate the lights and a separate module is used in this case for the adaptive features of the lights.

Control units A central body control module (BCM) is the primary hub that maintains body functions, such as:

▶ internal and external lighting
▶ security and access control
▶ comfort features for doors and seats
▶ other convenience controls.

It will in many cases also link to other systems such as the powertrain.

Control module details A company called FreeScale produces high-quality 16-bit MCU families that target many BCM applications. A single board computer (SBC) combines voltage regulation with a CAN or LIN physical interface in a single package. H-bridge

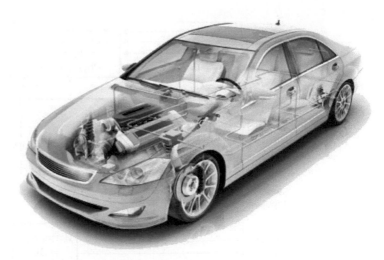

Figure 7.62 Control systems

drivers and a series of high-side switches drive high-current loads and replace relays. The gateway serves as the information bridge between various in-car communication networks, including Ethernet, FlexRay, CAN, LIN and MOST protocols. It also serves as the car's central diagnostic interface.

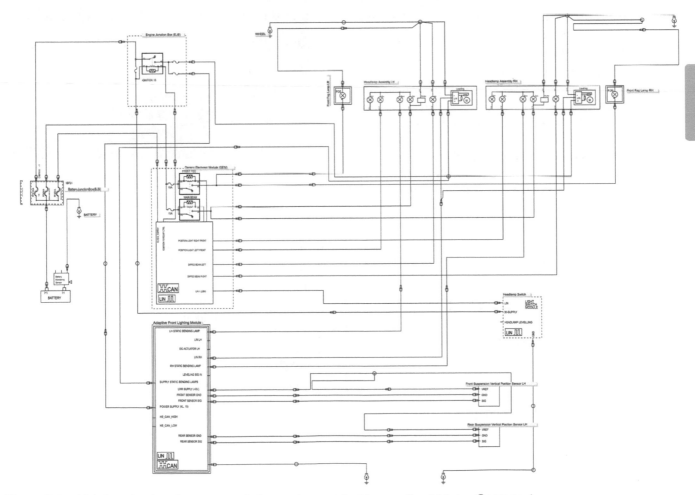

Figure 7.61 Lighting circuit using a general electronic module (Source: Ford Motor Company)

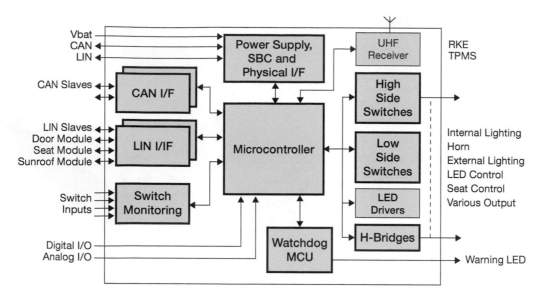

Figure 7.63 Details of the body control module and central gateway (Source: FreeScale Electronics)

Exterior lighting plays an important role in the safety of car passengers and other road users. Different types of lamps (e.g. halogen, xenon or LED) are used in a variety of lighting functions, such as brake lights, turn indicators, low and high beam headlights, daytime running lights and others. More advanced functions include light bending, levelling and shaping to adapt to changing driving conditions.

Lighting control The FreeScale 'eXtreme' switch product family of intelligent high-side switches use performance profiles tailored for different lamp types. They feature extensive diagnostic functionalities to detect faults and malfunctions and provide 'wave-shaping' to improve system-level EMC performance. Figure 7.65 shows the internal configuration of a lighting control module that uses solid-state switching.

Figure 7.64 Modern vehicle high intensity discharge (HID) lights

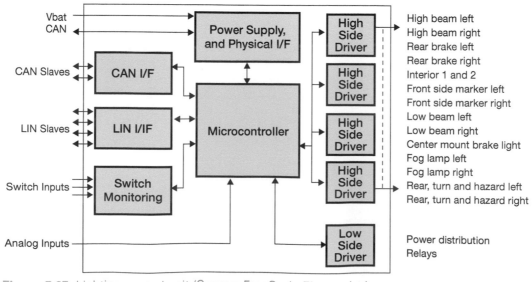

Figure 7.65 Lighting control unit (Source: FreeScale Electronics)

Switching components The system is designed for low-voltage automotive lighting applications. Its four MOSFETs (dual 10mΩ/dual 12mΩ) can control four separate 55W/28W bulbs, and/or Xenon modules, and/or LEDs. Programming, control and diagnostics are accomplished using a 16-bit SPI interface. (Serial to peripheral interface (SPI) is a communications protocol developed by Motorola and later adopted by others in the industry. It is a simple four-wire serial communications interface used by many microprocessor/microcontroller peripheral chips that enables the controllers and peripheral devices to communicate each other.) Its output with selectable slew rate improves electromagnetic compatibility (EMC) behaviour. The device allows the user to program the fault current trip levels and duration of acceptable lamp inrush via the SPI. The device has a fail-safe mode to provide fail-safe functionality of the outputs in case of MCU damage.

Figure 7.66 eXtreme or high side switch devices (Source: FreeScale)

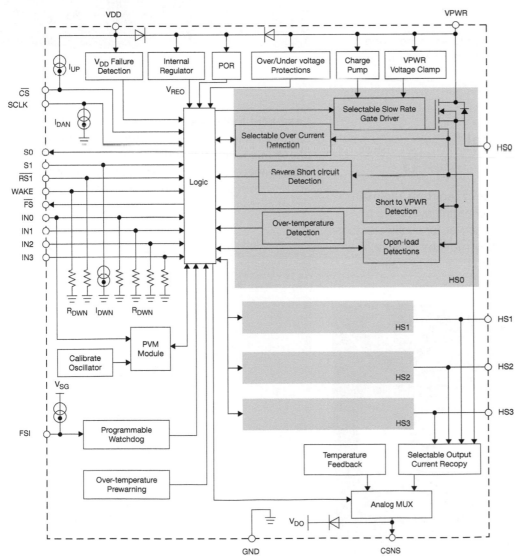

Figure 7.67 Internal configuration of an eXtreme or high side switch (Source: FreeScale)

Summary This section has outlined and showed examples of how central control systems are configured. At first view they can appear complex but, compared to separate switches and wires and relays for every electrical component on the car, centralization actually simplifies the system as well as making it easy to add new features.

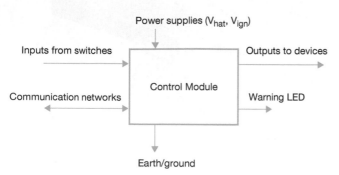

Figure 7.68 Central control system

 Make a simple sketch to show how one of the main components or systems in this section operates

7.2 Engine electrical

7.2.1 Battery introduction ❶

Battery and charging system function The main function of the battery and charging system is to provide a source of electric power for all the electrical systems on the vehicle. They must be capable of providing the electric power under all operating conditions.

Figure 7.69 Charging system

Starting system function The main function of the starting system is to crank the engine at sufficient speed to begin the internal combustion process. This will then allow the engine to run and to be fully controlled by the vehicle driver.

Figure 7.70 Starter system

Lead-acid batteries The majority of vehicle batteries are of conventional design, using lead plates in a dilute sulphuric acid electrolyte. This feature leads to the common description of 'lead-acid' batteries. The output from a lead-acid battery is direct current (DC).

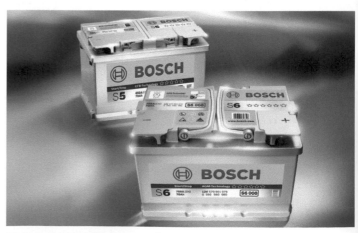

Figure 7.71 Vehicle batteries

Battery chemistry A rechargeable battery is an electrochemical unit that converts an electric current into a modified chemical compound. This chemical reaction can be reversed to release an electric current.

The modified chemical compound in the battery stores energy, which is available as electricity when connected to a circuit.

Routine maintenance Some old batteries have open cells that require routine maintenance to the electrolyte level. This usually consists of topping up with distilled water at regular intervals. Most modern lead-acid battery designs have improved plate construction and case design. This, together with precise alternator charge control, allows low-maintenance and maintenance-free types to be used.

A vehicle 12V battery is made up from six cells. Each lead-acid cell has a nominal voltage of 2.1V, which gives a value of 12.6V for a fully charged battery under no-load conditions. The six cells are connected in series, internally in the battery, with lead bars. The cells are formed in the battery case and are completely separate from each other.

Each cell has a set of interleaved positive and negative plates kept apart by porous separators. The separators prevent contact of the plates, which would give an internal short circuit and affect the chemical reaction in the battery cell. The cell plates are supported above the bottom of the case. This leaves a sediment trap below the plates so that any loose material that falls to the bottom does not cause a short circuit between the plates.

Plate construction The cell plates are formed in a lattice grid of lead-antimony or lead-calcium alloy. The grid carries the active material and acts as the electrical conductor. The active materials are lead peroxide for the positive plate and spongy lead for the negative plate.

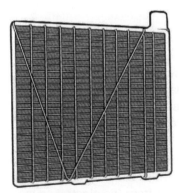

Figure 7.72 Cell plate construction

Charged battery When a battery is in a charged state, the positive plates of lead peroxide (PbO_2) are reddish brown in colour, and the negative plates of spongy lead (Pb) are grey in colour.

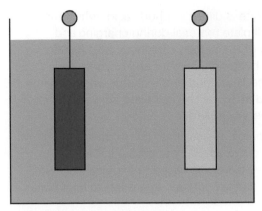

Figure 7.73 Charged state

Discharged battery When the battery is discharging, a chemical reaction with the electrolyte changes both plates to lead sulphate ($PbSO_4$).

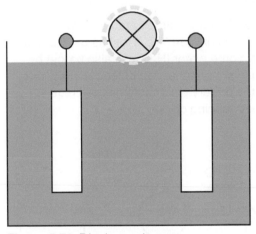

Figure 7.74 Discharged state

Reversible chemical reaction Applying an electric current to the battery reverses the process. The charged battery stores chemical energy. This can be released as electrical energy when the battery is connected into a circuit.

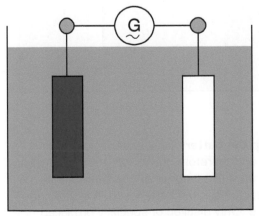

Figure 7.75 Chemical changes reverse when current is applied

245

The electrolyte is dilute sulphuric acid, which reacts with the cell plate material during charging and discharging of the battery. Sulphuric acid (H_2SO_4) consists of hydrogen, sulphur and oxygen. These chemicals separate during the charge and discharge process and attach to the cell plate active material or return to the electrolyte.

During discharge, the sulphate (SO_4) combines with the lead to form lead sulphate ($PbSO_4$). The oxygen in the positive plate is released to the electrolyte and combines with the hydrogen that is left to form water (H_2O).

During charging, the reverse process occurs with the sulphate (SO_4), leaving the cell plates to reform with the hydrogen in the electrolyte to produce sulphuric acid (H_2SO_4). Oxygen in the electrolyte is released to reform with the positive cell plate material as lead peroxide (PbO_2).

Gassing Near the fully charged state, some hydrogen (H_2) and oxygen (O_2) is lost as gas from the battery vent. Some water (H_2O) can also be lost by vaporisation in hot weather. On older batteries, this meant that the battery electrolyte needed regular inspection and topping up.

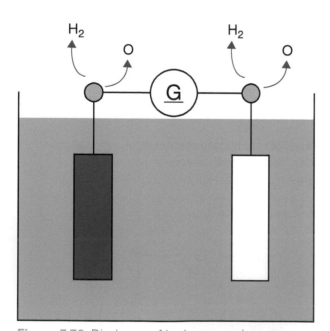

Figure 7.76 Discharge of hydrogen and oxygen

Topping up the battery Only water is lost from the battery and therefore only water should be used for topping up. Any contaminants will affect the chemical reactions in the battery and, therefore, the performance. Only distilled or specially produced topping up water should be used. Tap water is not suitable for topping up a battery. Acid should never be used as this will strengthen the acid solution and alter the chemical reactions.

Note: Not many batteries can be topped up now but there are a few out there!

 Create a word cloud for one or more of the most important screens or blocks of text in this section

7.2.2 Battery types and charging 🔋

Modern batteries Many modern batteries use a modified plate design that has a centralised plate lug and radial grid construction. To reduce gassing, the grid material is a lead-calcium alloy with a small portion of antimony. This plate design gives improved electric current flow and is lighter than the earlier design. This means that lighter-weight batteries are now available with the same performance as the older type.

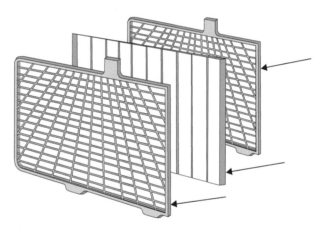

Figure 7.77 Modern cell plate design 1. Plate, 2. Separator, 3. Plate

Low maintenance These batteries are low-maintenance or maintenance-free types. When used with an alternator, with accurate charge control, they require maintenance either at yearly intervals or never!

Maintenance free batteries are completely sealed or have only a very small vent. The maintenance free battery does not lose water from the electrolyte in the same way as conventional lead-acid batteries do. A number of changes in the chemical composition of the plates and in the construction of the battery case reduce gassing to almost zero. Liquid, gas and vapour are also captured and returned to the battery cells.

plates, which are then pressed together, so that there is no free acid in the cells. In a recombination battery, there is slightly more negative plate material than positive plate material. This allows the oxygen released by the positive plates, near to the fully charged position, to combine with the negative plate material rather than be released as gas. As there is no loss of gas from these batteries, they can be fully sealed.

A hybrid battery, or deep cycle battery, produces a high performance cold cranking amperage. The 'hybrid' design of this type of battery refers to the use of a lead-antimony alloy for the positive plate grids and a lead-calcium alloy for the negative plate grids. This allows the battery to provide a high current for cold starting in very cold conditions, without permanent harm to the battery.

Battery chargers There are two types of off-vehicle charger – the 'bench charger', which has a current output of up to about 10A, and the 'fast-charger', that can recharge a battery in about 30 minutes, with a current of about 50A. Not all batteries are suitable for fast charging and reference to the manufacturer's instructions is required before charging any battery.

Bench charger The bench charger should be situated in a well-ventilated area of the workshop. Smoking should also be prohibited. The bench top should be resistant to acid and be made from an electrically insulating material.

Charger operation The charger is connected to the mains electrical system and uses a transformer to reduce to 24, 12 or 6V, depending on the battery voltage. The actual output will be slightly higher than the nominal battery voltage in order to give an effective charge. The charger also includes a rectifier to change the AC (alternating current) mains supply to the DC (direct current) voltage required by the battery.

Voltage control Bench chargers with voltage control have high initial current outputs, which fall as the battery charges. Chargers with current control can be adjusted to suit individual batteries.

Charge rates The usual rule for battery charging is that the charge current should be set to a tenth of the Ah (ampere hour) rating of the battery. Alternatively, about a sixteenth of the reserve capacity, or a fortieth of the cold cranking amps figure, gives a good guide. A fully discharged battery will take about 12 hours to fully recharge. When recharging partially charged batteries, it is recommended that they should be checked at regular intervals.

Figure 7.78 Sealed maintenance-free battery

The **plate grids** are of radial design, made from a lead-calcium alloy, and filled with a high-density active material. The plates are enclosed in chemically inert separator envelopes. At the top of each cell is a liquid and gas separator area with a drain to return any liquid to the cell. The cells are sealed off from each other and the connecting bars are sealed where they pass through the cell partitions.

Recombination battery A further development in maintenance-free battery design is the 'recombination battery'. These batteries have all the electrolyte held in microporous envelope separators around the

Figure 7.79 Fast chargers

Disconnecting the charger Always switch off the charger and leave it for about five minutes before disconnecting the leads and carrying out any tests. This is to allow any hydrogen gas to dissipate into the atmosphere and reduce ignition and fire hazards from accidental sparks. Hydrogen is highly flammable and explosive in an enclosed space such as inside the top of a battery.

Fast chargers are portable items of equipment that will charge a battery in a short space of time. They can often be used for engine starting, depending on design.

Build-up of gas Where possible, the battery tops should be removed during charging to prevent a build-up of gas in the battery case. The area around the battery should be marked as a no smoking area.

Overcharging There is a risk of overcharging and overheating a battery with fast chargers, and, therefore, if a temperature sensor is fitted, it must be used. If a sensor is not fitted, frequent checks should be made to check for gassing and battery temperature. Charging should be stopped when heavy gassing is evident or if the battery feels more than just warm to the touch.

External features The external features of a battery are the type and size of the terminal posts, the dimensions of the battery and the method of fixing to the vehicle. The terminal posts are clearly identifiable as positive or negative, by the positive sign (+) and/or red colour and the negative sign (-) and/or black colour.

Tapered round terminal posts The positive and negative posts, on tapered round terminals, are different sizes. The larger post is the positive one and the smaller is negative. The difference in the sizes is used to minimise the risk of incorrect fitting. There are two size ranges, with a smaller version used by some manufacturers, generally in the far eastern geographical zone, and a larger size used by western European and American manufacturers. These round post terminals have a cast cap-type or clamp-type cable terminal.

Figure 7.81 Round battery posts

'L' terminal posts Some manufacturers use batteries with a flat, or 'L' terminal on the battery, and a flat terminal on the cable, and a nut and bolt to complete the connection. Some American vehicles use a side terminal, which has an internal thread, and the connection is made with a bolt through a flat terminal on the cable.

Battery cables must have sufficient cross-sectional area to carry the starter motor and electrical systems current. The feed to the starter motor is a heavy-duty insulated cable, and the earth or ground cable is of similar construction, or may be a braided strap.

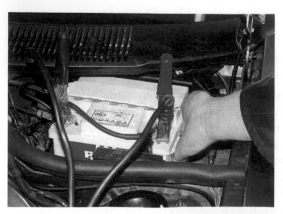

Figure 7.80 Use of temperature sensor OR check the battery by hand

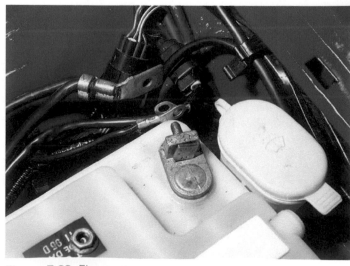

Figure 7.82 Flat posts

248

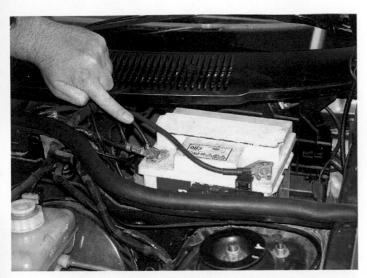

Figure 7.83 Component earth

 Look back over the previous section and write out a list of the key bullet points from this section.

7.2.3 Absorbent glass mat (AGM) batteries ❸

Introduction Absorbent glass mat is an improved lead acid battery with higher performance than the common wet or flooded type. Instead of submerging the plates into liquid electrolyte, the electrolyte is absorbed in a mat of fine glass fibres. This makes the battery spill-proof and therefore it can be transported without hazardous material restrictions. The plates can be made flat like the standard flooded lead acid and placed in a rectangular case or wound into cylinders.

Internal resistance AGM batteries have a very low internal resistance so they can deliver high currents and offer long service even if occasionally deep-cycled. These batteries are lower weight and more reliable than the conventional lead acid type. They also perform at high and low temperatures and have a low self-discharge. Other advantages over regular lead acid are a better specific power rating (high load current) and faster charge times (up to five times faster). The disadvantages are that they have a slightly lower specific energy (capacity) and higher manufacturing costs.

Performance AGM batteries are found in high-end vehicles to run power-hungry accessories such as heated seats, steering wheels, mirrors and wipers.

NASCAR and other auto racing leagues choose AGM products because they are vibration resistant. Start-stop batteries are almost exclusively AGM because the conventional types do not respond well to repeated micro cycling.

Charging All sealed units such as gel and AGM batteries are sensitive to overcharging. They can be charged to 2.40V/cell (and higher) without problem. However, the float charge, used to keep the battery at 100%, should be reduced to between 2.25 and 2.30V/cell. Vehicle charging systems for conventional lead acid batteries often have a fixed voltage setting of 14.40V (2.40V/cell). A direct replacement with a sealed unit could therefore cause the battery to be overcharged. As always, the correct battery is the one specified by the vehicle or battery manufacturer.

Heat AGM and gel electrolyte batteries do not like heat. Manufacturers recommend stopping charge if the battery core reaches 49°C. Many batteries on modern vehicle now include a temperature and current sensor of some sort that is linked to a smart charging system.

 Use a library or the web search tools to further examine the subject in this section

7.2.4 Battery testing ❶

Introduction For testing the state of charge of a non-sealed type of battery, it was traditional to use a hydrometer. The hydrometer is a syringe which draws electrolyte from a cell and a float which will float at a particular depth in the electrolyte according to its density. The relative density or specific gravity is then read from the graduated scale on the float. A fully charged cell should show 1.280, when half charged, 1.200, and if discharged, 1.120.

However, almost all vehicles are now fitted with maintenance free batteries and a hydrometer cannot be used to find the state of charge. This can, however, be determined from the voltage of the battery, as given in the table. An accurate voltmeter is required for this test.

Battery volts at 20°C	State of charge
12.0V	Discharged (20% or less)
12.3V	Half charged (50%)
12.7V	Charged (100%)

7

Figure 7.84 Multimeter

Figure 7.85 A high-rate discharge tester can indicate battery state-of-health, but it still relies on the skill and judgement on the technician; in addition, there are several health and safety-related issues to this approach! Note: The tester in the picture is a fixed load, and not really suitable for the battery shown (Source: www.autoelex.co.uk)

Energy requirements Modern vehicles have sophisticated energy requirements and electronic consumers that need a stable, clean voltage supply. Workshops are seeing obscure faults with electronic systems, including fault code errors, brought on by failing batteries. Traditionally, a failing battery would manifest itself by having insufficient power to crank the engine and start the vehicle. This was often more apparent in winter, when cold starts need more torque to overcome the friction of a cold engine, with thicker, cold lubricating oil. However, with modern vehicles a failing battery is likely to produce a fault, of an unrelated nature, before this 'non start' symptom occurs. Battery technology has progressed in-line with the vehicle systems, but a different method of establishing the serviceability was needed.

Traditional test methods There are two traditional methods of checking a wet, lead-acid, vehicle battery. The first is state of charge (SOC) which can be determined via measuring the specific gravity (SG) of the electrolyte in each cell with a hydrometer (there is also a less accurate option, to measure the battery terminal voltage). Assuming the battery is reasonably well charged (>75%), then a performance test, indicating state-of-health (SOH), could be executed via a discharge test. This test is performed using a high-rate discharge tester, with the appropriate load according to the battery capacity, and it would indicate the battery capability to supply a large current (as would be required under starting conditions). From these measurements, an experienced technician could make a judgement on the battery's fitness for purpose.

Reasons for change There are several reasons that the methods mentioned previously are no longer applicable:

▶ Many modern batteries have no access to the cell electrolyte, thus hydrometer readings are simply not possible. Although the battery may have a built-in hydrometer, this is of limited use; it's just a general indicator.

▶ In order to execute a high-rate discharge test, the battery has to be disconnected from the vehicle – this can present time-consuming problems for the technician, e.g. lost radio codes, ECU memory loss, etc.

▶ There are health and safety issues; wet batteries contain acid and generate volatile gases. High rate loads tests can create sparks and heat – all potential problems in a safety conscious workshop.

▶ The measurements still rely on the knowledge and experience of the technician to make a judgement on the battery SOH. This is subjective and could be the source of inaccurate diagnoses.

Digital battery testers – conductance testing Along with the progress in battery and vehicle technology, technology developments have also provided alternative methods for testing. Battery testers now use a completely different approach to evaluating the battery condition, providing an objective measurement of battery condition and capability along with a more accurate SOH assessment. These testers are intelligent units, with menu-guided test procedures. The technique itself is known as conductance measurement.

Figure 7.86 Intelligent, digital battery testers are much safer and more appropriate for testing modern battery technology (Source: www.autoelex.co.uk)

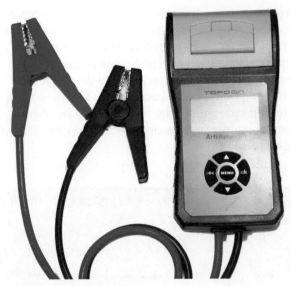

Figure 7.87 Battery tester

The conductance test is a completely different method of establishing the battery condition and performance and is ideally suited for modern vehicle battery test applications. It can also be applied to older, wet lead-acid batteries. The conductance tester applies an AC voltage, of known frequency and amplitude, across the battery terminals, and monitors the subsequent current that flows with respect to phase shift and ripple. The AC voltage is superimposed on the battery's DC voltage and acts as brief charge and discharge pulses. This information is utilized to calculate the impedance (measure of opposition to alternating current) of the battery, and, from this, the conductance value can be established (impedance and conductance have a reciprocal relationship).

Battery Tester Traditionally batteries were tested with a voltmeter and a high rate discharge tester. This monitored the battery voltage while causing a current discharge of 100s of amps. Some batteries exploded during the process! The device shown here will test 12V batteries rated from 100 to 2000 CCA. It uses advanced conductance measurement technology, to show battery health status including voltage, cranking power, state of charge and more. It displays one of five easy-to-understand results that can be printed out for the customer:

▶ Good battery
▶ Good battery, need recharge
▶ Replace, life cycle
▶ Replace, bad cell
▶ Charge, retest.

Built in reverse-connection protection prevents damage to the battery or tester. It can be used for batteries on or off the vehicle.

 Using images and text, create a short presentation to show how a component or system works

7.2.5 Battery charging ❷

Battery charging should only be carried out in a well-ventilated area specially designated for the purpose. A suitable acid-resistant and non-conductive bench is recommended. A face shield, to prevent acid splashes on the face and in the eyes, should be kept close by. Sterile eyewash should also be available for use if acid does splash into the eyes. 'No smoking' signs should be clearly displayed.

Slow and fast charging There are two ways of charging batteries in the workshop – one is a slow or trickle charge and the other a fast charge. These require two different types of charger. Most batteries can be fast charged, but this should only be carried out infrequently. If a high charge is used, this can cause some deterioration of the battery active materials.

Charger operation A slow charger or bench charger uses mains electricity. Inside is fitted a transformer to reduce the voltage to 6, 12 or 24v, to suit the battery or batteries on charge. Also fitted is a rectifier to change the AC volts of the mains supply to the DC volts needed for charging batteries. The charger is connected to the battery terminals with the correct polarity. After setting the control switches, the charger is then turned on at the main switch.

 7

Charge rate There are a number of different types of charger and these should be used in accordance with the manufacturer's instructions. The recommended charge rate for a battery is one-tenth of the ampere hour capacity. A 40Ah battery should be charged at 4A. If the ampere hour capacity is not known, set the rate to one-sixteenth of the reserve capacity. Where the charge current can be adjusted, this facility should be used to set the rate.

Figure 7.88 Correct polarity is important

Figure 7.89 Battery charger

Safety first It is important that the charger is switched off before it is disconnected from the battery. For further safety, leave the batteries for about five minutes before the charger leads are disconnected. This will allow any flammable gas to dissipate to the atmosphere.

Fast charging A fast charger can be connected to a battery on a vehicle in order to give a quick boost when a battery has a low charge. Some of these chargers have an engine start facility. Always follow the equipment manufacturer's instructions when using this type of charger. Some batteries are not suitable for fast charging and, therefore, always refer to the vehicle or battery manufacturer's data for recommendations.

Fast charge rate and time Fast chargers have a time clock for setting the charger for a fixed charge period. Some have a temperature probe included to switch off the charger if the battery becomes overheated. Keep a close watch on the battery temperature if a fast charger does not have a temperature probe. The maximum setting for a fast charge should not exceed one hour at five times the normal charge rate.

> Select a routine from section 1.3 and follow the process to study a component or system

7.2.6 Starting system 1 2

Introduction The engine starting system consists of a heavy-duty motor, with a drive pinion that engages with a gear on the engine flywheel, and an electrical control circuit to operate the motor.

Figure 7.90 Pre-engaged starter

The starter motor power output has to be able to crank a cold engine at sufficient speed to start the engine. A two-litre petrol engine will have a starter motor of about one kilowatt and will spin the engine at about 150 rev/min. A similar sized diesel engine will require double the power and twice the cranking speed to start.

Starter motor and the control circuit The main components of the starter motor are the magnetic fields, armature, drive pinion and solenoid. The circuit consists of a battery supply, earth cables and the starter switch.

The **starter motor** is a direct current (DC) electromagnetic unit that usually has two pairs of magnetic pole shoes arranged at opposite positions inside the motor casing. The casing acts as the yoke for the magnetic poles. The magnetic pole shoes can be strong permanent magnets or electromagnets using a winding.

Figure 7.91 Pre-engaged motor

The armature, which consists of a series of wire conductor loops wound around a laminated iron core, is mounted on the motor spindle. The conductor loops are terminated into segments of a commutator. Carbon or composite brushes conduct the motor electrical supply through the commutator segments to the individual conductor loops.

Figure 7.92 Armatures

Basic motor operation The construction of a simple, direct current motor is shown in Figure 7.129. The magnetic force between the poles is from north to south. A loop conductor inside the magnetic field is provided with a DC electrical supply through a split slip ring, which forms a simple commutator.

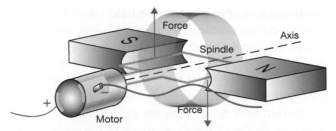

Figure 7.93 Direct current motor operating principle

Magnetic field When an electrical current is passed through a conductor, a magnetic field is formed around that conductor. The magnetic field direction depends on the direction of the current flow. When the conductor is placed inside a fixed magnet, the magnetic field distorts to produce a repelling magnetic force, which pushes on the conductor.

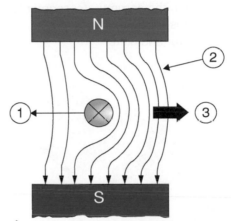

Figure 7.94 Electric current through a conductor in a magnetic field

Loop conductor When an electrical current is passed through a loop conductor, the magnetic field around the conductor is in the opposite direction in each side of the loop. The loop conductor on a motor is fitted to, or forms part of, a spindle so that it is free to rotate.

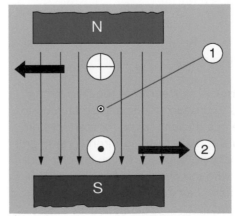

Figure 7.95 Magnetic forces reacting to give rotary motion

Loop conductor in a magnetic field When the loop conductor is placed inside a magnetic field, and an electrical current is passed through the loop, the resulting magnetic forces will cause the loop to rotate until it is out of the magnetic attract and repel positions of the magnetic fields. At this point, the current flow direction in the loop conductor is reversed, by the changed positions of the commutator contacts, and the inertia of the loop brings it again into the effective magnetic field position. The loop will, therefore, continue to rotate.

Armature windings In practice, it requires a large series of loop conductors to provide a motor with continuous rotation and good torque characteristics. In order to supply each loop winding, when it is in alignment with the field magnets, and to maintain the current in the proper direction, a commutator is fitted. Current is passed to the commutator segments through spring-loaded brushes held in position by brush holders on the motor end plate.

Figure 7.96 A series of windings on the armature give a continuous rotary motion

Speed and torque A starter motor requires strong magnetic forces to produce the speed and torque to crank an engine at sufficient speed for starting. For this, the armature is made with soft iron cores in order to make strong electromagnets, which are able to change polarity with the direction of current flow in the loop conductors. Laminations of soft iron are used for the cores in order to reduce magnetisation losses. They are insulated from each other and assembled as a single unit on the armature.

Magnetic fields The magnetic strength of the field magnetic poles is usually produced by using an electrical winding around the pole shoe. The wire coil is wound around one pole shoe and then the other in the opposite direction, so that the opposing field poles are produced opposite each other in the casing.

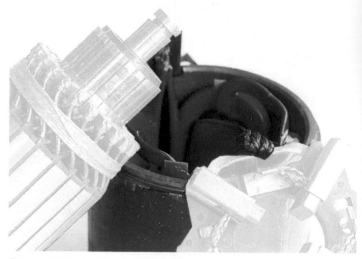

Figure 7.97 Field windings

Motor drive gear The drive from the motor is taken from a pinion gear on the spindle to the large diameter starter ring gear on the engine. The starter ring gear is fitted to the outside of the flywheel on manual transmission vehicles, or the torque converter drive plate on automatic transmission vehicles. The pinion meshes with the ring gear only during starting and is made to slide axially on, or with, the spindle in order to engage the drive when operated.

Figure 7.98 Ring gear on a flywheel

Figure 7.99 Starter pinion

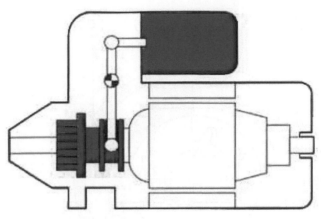

Figure 7.101 Pinion movement

Pre-engaged starter On a pre-engaged starter motor, the drive pinion is brought into mesh by the action of an electromagnetic solenoid mounted on the starter motor casing. The solenoid has a soft iron plunger, which is drawn into the magnetic field that is produced inside the solenoid when an electrical current is passed through the solenoid windings.

Figure 7.100 Starter in position

Pinion engagement Connected to the plunger is a lever, which is pivoted so that as one end is pulled into the solenoid, the opposite end pushes the pinion into mesh with the starter ring gear. The pinion is mounted on a unidirectional clutch, which is fitted to a sleeve with an internal spline to take the drive from the starter spindle. On the outside of the sleeve is a radial groove to take the fork of the engagement lever.

At the other end of the solenoid are the electrical contacts that form the switch to pass the electrical current to the motor. The solenoid on many pre-engaged starter motors has two windings. These are the 'closing' and 'holding' windings.

The closing winding, or pull-in coil, operates as soon as the solenoid is energised. This winding has an earth, or ground, return through the motor windings. This passes a current into the motor so that it rotates slowly during the engagement phase. Once the switch contacts are fully engaged, the holding winding holds the switch in place. The closing winding does not conduct once the motor current has been switched on.

A holding coil is wound around the solenoid. This creates the magnetic field required to hold the solenoid in the engaged position during starting. When the starter switch is released, a spring returns the solenoid plunger to its 'off' position.

If the engine were to start under these conditions, it would drive the motor spindle at an excessive speed. To prevent this occurring on pre-engaged drive starter motors, a unidirectional overrun clutch is fitted on the pinion. This allows the motor to drive the engine but stops the engine driving the motor. A roller-type overrun clutch is a popular method although a few other types are used. These clutch units are sealed for life and require replacement if they fail in service.

Figure 7.102 One-way clutch and pinion

7

255

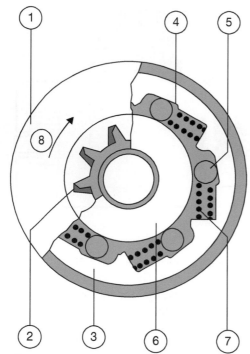

Figure 7.103 Clutch operation: 1-Casing, 2-Pinion, 3-Clutch shell, 4-Roller race, 5-Roller, 6-Pinion shaft, 7-Spring, 8-Dierection of rotation

There are some variations in starter motor design and construction. The conventional field coil construction consists of four coils wound around pole shoes. The direction of current flow in the coil windings produces two north and two south poles opposite each other. The electrical feed from the battery is connected to the field coils and then conducted into the armature windings by brushes that rub on a commutator.

Series wound fields Simple electrical circuits for the field coils and armature windings are shown here. These windings can be shunt-wound (parallel), but are usually series-wound, which is the more commonly used circuit.

Series-parallel fields An alternative arrangement is the series-parallel motor, which has parallel field coil connections and field to armature connections in series. The circuit diagram for this type of motor is shown here. The advantage of this arrangement is in the reduced resistance in the circuit, allowing a higher electrical current to give increased power and torque.

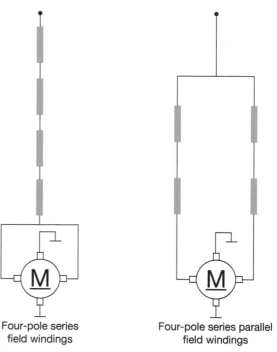

Four-pole series field windings

Four-pole series parallel field windings

Figure 7.105 Series and series-parallel circuit

Permanent magnets (PM) motors Many small, modern starter motors are using permanent magnets for the field poles. These motors have high speed and low torque characteristics and are suitable, without additional gearing, for petrol or gasoline engines up to 1.9 litres.

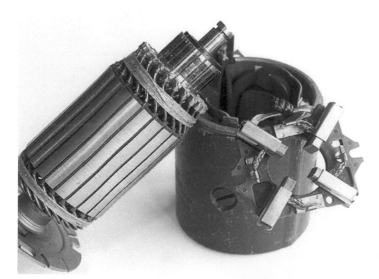

Figure 7.104 Starter case and field windings

Figure 7.106 Bosch starter motor

Intermediate drive gears On permanent magnet starter motors, for light diesel engines and petrol/gasoline engines up to five litres, an intermediate planetary gear set between the motor and drive pinion is used. This intermediate gearing modifies the speed and torque characteristics of the motor and makes it possible to construct starter motors that can be 40% lower in weight. The starter electrical current is passed through the armature only. On a planetary gear motor, the spindle is fitted with a sun wheel, and the motor casing with the annulus. The output to the drive pinion is made from the planetary gear carrier.

Figure 7.107 Intermediate transmission

Starter motor control circuits use a heavy-duty electrical relay, called a solenoid, to switch the large starter current to the motor. The solenoid is an electromagnetic switch and, on modern pre-engaged starter motors, is attached to the top of the motor, where it performs the switching function, and is also used to slide the motor drive pinion into mesh with the starter ring gear on the engine flywheel.

Figure 7.108 Solenoid construction

A **basic starter circuit** is shown here. The main components are the battery, starter switch, which is usually part of the ignition switch, the solenoid and motor, connecting cables and the earth, or ground, return circuit. The battery and starter cables are of heavy-duty construction to carry a large current to the motor. The control cables are standard low current cable sizes. If any of these cables have to be replaced, cables of the same size, or as specified by the vehicle manufacturer, should always be used.

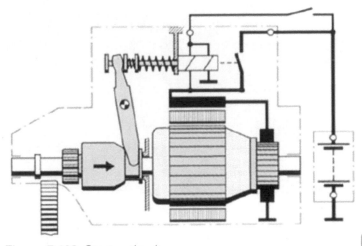

Figure 7.109 Starter circuit

Automatic transmission systems Starter motor circuits may have additional automatic switching to prevent the engine being started in particular situations. Automatic transmission systems incorporate an inhibitor switch on the gear selector, which allows engine starting in the park and neutral positions only. This prevents the engine being started with the transmission in gear, which could result in the vehicle pulling away unexpectedly. The inhibitor switch must be carefully checked and adjusted so that there is no risk of incorrect operation.

Electronic anti-theft systems Many modern anti-theft electronic systems have an engine immobiliser that must be disengaged before the engine can be started.

 Use a library or the web search tools to further examine the subject in this section

7.2.7 **Starting system testing** ❷

Visual checks The basic starter system test procedure, to identify any faults, should start with a visual inspection of the battery terminals and cables. Look closely for signs of corrosion and cable strands that may be broken close to the terminals. Check the terminals on the starter motor and solenoid for the main motor feed and for the control circuit.

Figure 7.112 . . . cranking volts

Starter voltage when cranking Connect a digital voltmeter between the main starter terminal and the starter body. Crank the engine and note the voltage. The reading should be no more than half a volt less than the reading taken at the battery.

Figure 7.110 Checking starter cables and terminals

Battery voltage when cranking Connect a digital voltmeter across the battery terminals and note the voltage. Follow the vehicle manufacturer's instructions to disable the ignition circuit to prevent the engine from starting, and, if necessary, the injectors, so that fuel does not enter the exhaust and catalytic converter. Crank the engine for about 10 seconds and note the voltage reading. The cranking voltage should not drop below 10V.

Figure 7.113 Voltmeter connected to starter earth . . .

Figure 7.111 A digital voltmeter connected to show . . .

Figure 7.114 . . . and starter supply

Earth circuit tests To check the earth, or ground return connect the probes to the motor body and battery earth terminal. A reading of 0.5V during cranking should not be exceeded.

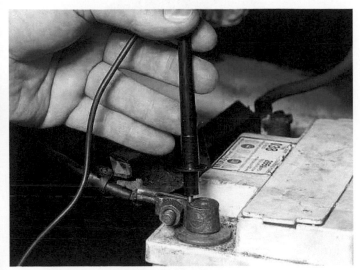

Figure 7.115 Voltmeter connected to check earth circuit

Solenoid feed To check the solenoid feed from the ignition/starter switch, connect the probe to the solenoid switch feed cable at the solenoid. Voltage readings with the terminal connected should be within 0.5V of the battery voltage.

Figure 7.116 Voltmeter connected to check solenoid supply

Main contacts The solenoid main contacts can be checked by connecting the meter probes across the solenoid main terminals. The meter reading during cranking should not exceed 0.5V. However, a reading of 0V is normal.

Summary If the battery, cables and connections are all in good order but the motor is not operating correctly, it will need to be removed for further tests. The main components to look at are the brushes and internal windings. If the motor is failing to engage correctly, the drive pinion and starter ring gear on the engine should be inspected.

Figure 7.117 Starter motor

 Look back over the previous section and write out a list of the key bullet points from this section

7.2.8 Charging system ❶ ❷

Electrical generator The electrical generator on modern vehicles is an alternator. Older vehicles used a dynamo, which gives a direct current without the need for a rectifier.

Figure 7.118 Alternator on an engine

Rotor and stator There are two main parts to an alternator – these are the rotor and the stator. Together they produce an AC voltage output. An electrical current is induced or generated in the stator by the magnetic fields produced in the rotor.

Figure 7.119 Alternator stator and rotor

The **rectifier** changes the AC voltage to a DC voltage, because that is what is needed for battery charging. Diodes in a bridge formation are used to route the electrical current in such a way as to convert the AC voltage to a DC voltage.

Figure 7.120 Rectifier and main output terminals

The voltage **regulator** senses the alternator output voltage. It then controls the rotor magnetic field strength to maintain the voltage at the correct level. The ignition or charge warning light circuit is used to produce an initial magnetic field in the rotor during engine starting.

Figure 7.121 Voltage regulator

On **modern alternators**, all the main components are enclosed in a lightweight aluminium casing. The vehicle engine provides power to the alternator, through a drive belt and pulleys to the rotor, which is mounted on bearings in the end covers of the alternator casing. A typical alternator electrical circuit is also shown in Figure 7.161.

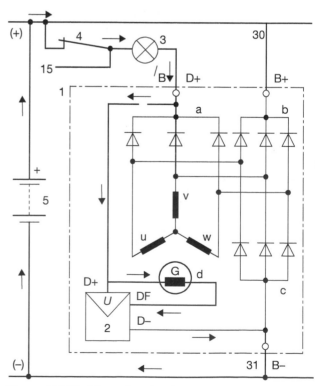

Figure 7.122 Alternator circuit

The generation of an electrical current requires a changing magnetic field around a conductor. This process is known as electromagnetic 'induction' and occurs under a variety of conditions, one of which is shown in the online image. The voltage produced is alternating current (AC).

By increasing the loop conductors to three, at 120° intervals, a three-phase output is produced. This is similar to the method used in an alternator except that a series of magnets (rotor) are made to rotate inside a conductor made up from three inductive coils (stator).

Rotor windings In a light vehicle, alternator magnetic fields are produced around magnetic poles on the rotor by an electrical current passing through coil windings. The poles are made from iron and shaped like claws with six fingers. There are two of these, one for each pole, facing each other, and set at each end of the rotor.

Figure 7.124 Rotor magnetic field

Excitation current The initial electrical current to 'excite' the windings is provided through the ignition or generator warning light circuit. The light acts as an indicator that the generator field is being provided with an initial current to 'excite' the rotor windings, and as a warning when the voltage from the stator is less than battery voltage. Under normal conditions, the light should go out as soon as the engine is running.

Figure 7.123 Rotor, slip rings and brushes

Brushes and slip rings Wound inside the poles is the rotor winding, which is connected to slip rings at one end of the rotor. Carbon brushes are used to conduct an electrical current to the rotor windings through the slip rings. The arrangement of slip rings can be of either cylindrical or face type.

Magnetic field When an electrical current is passed through the rotor windings, they become 'excited' and a magnetic field is produced. The strength of the magnetic field is proportional to the voltage in the windings. The voltage in the windings is provided by the alternator during the charging phase and then controlled to regulate the alternator output voltage.

Figure 7.125 Sectioned alternator

The charge warning light goes out when the engine is running because of the nature of electrical current flow. Electricity always flows by the easiest route to earth or ground. When the alternator is charging, the electrical route is from the stator to the field windings and is of a higher value than the battery. The higher voltage takes precedence and the warning light circuit is bypassed.

The stator, which is fitted inside the alternator casing, is made from soft iron laminations wound with three sets of windings. The three sets of windings give three separate outputs, or phases, of alternating current. The electrical current induced in the alternator flows in the stator because of the changing magnetic fields produced by rotation of the rotor. The speed of rotation and the magnetic strength of the rotor determine the value of the voltage that is produced.

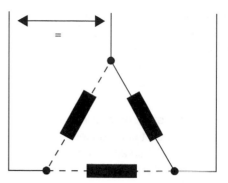

Figure 7.128 Delta stator windings

Semiconductor diodes Modern alternator rectifiers use semiconductor diodes, in a bridge formation, to provide rectification of the alternating current (AC) to the direct current (DC) required to charge the vehicle battery.

Figure 7.126 Stator construction

Star and delta windings The windings are enamel-coated copper wire of a heavy gauge and, for light vehicle applications, are connected in a 'star' formation. The windings can also be connected in a 'delta' formation, and this is often used for larger vehicles. The voltage and current outputs from the two formations are different for the same magnetic field strength and alternator speed. The voltage is higher and current lower for the star formation in comparison with the delta formation.

Figure 7.129 Rectifier

Half wave rectification Rectification of alternating current (AC) to direct current (DC) is achieved in a rectifier by allowing a forward flow of current to pass through a diode and then preventing the reverse flow from passing through. This simple method is called partial, or half wave, rectification.

Full wave rectification A more efficient and effective system is full wave rectification. To achieve this, a series of diodes are connected in a 'bridge' arrangement, so that the current flow is routed through open paths created by the bias of the diodes. There are two open paths in the bridge rectifier – one for each direction of current flow. The output current flow is always in the same direction, and this gives a direct current flow.

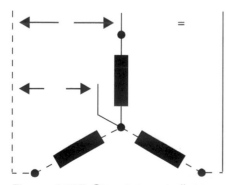

Figure 7.127 Star stator windings

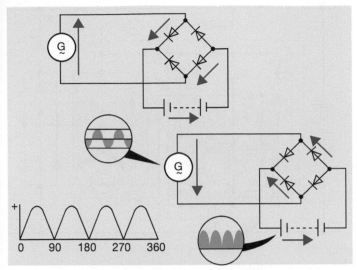

Figure 7.130 Bridge rectifier showing full wave rectification

Three phase rectification The three phases of the alternator output require six diodes arranged in the circuit as seen here. These arrows show the current flow for each phase and direction of flow of the alternating current. The output from the rectifier is connected to the vehicle battery and the circuit completed by an earth, or ground return connection through the alternator casing.

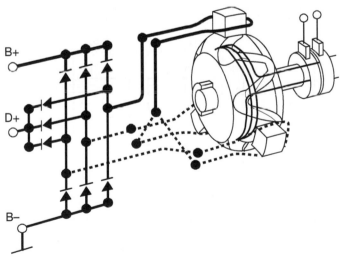

Figure 7.131 Alternator rectifier and current flow paths

Field diodes Three additional diodes are fitted in the circuit so that part of the output from the stator can be passed to the rotor. This is to increase the magnetic field strength when the stator voltage increases as the engine speed rises.

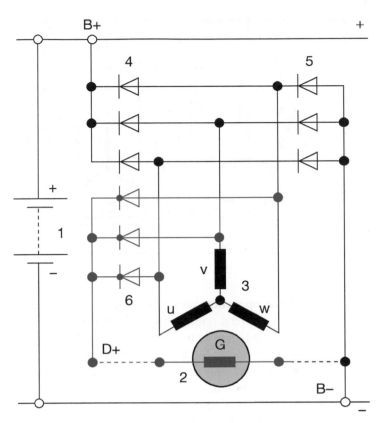

Figure 7.132 Field diodes and current flow to rotor and regulator

The voltage from the stator increases as the engine speed increases. Without a control system, the voltage would rise to high levels and cause extensive damage to the alternator and electrical systems on the vehicle. The regulator is connected into the rotor field circuit to control the rotor winding voltage and therefore the rotor magnetic field strength.

Field current control As the alternator output is dependent on speed and rotor magnetic field strength, it is necessary to reduce the magnetic field strength as the speed increases. The regulator maintains a constant alternator output voltage. This is usually at about 14.2V, which is sufficient to charge the battery without causing excessive gassing, and, for maintenance-free batteries, is the optimum voltage level for correct charging.

Zener diode The regulator consists of a small electronic circuit built around a zener diode. A zener diode conducts an electrical current only when its rated voltage is applied. In the circuit diagram of a regulator (see Figure 7.133) the zener diode, resistors, and switching transistor are used to switch on a Darlington circuit to route an electrical current through the rotor windings, or off to bypass the rotor windings to earth or ground.

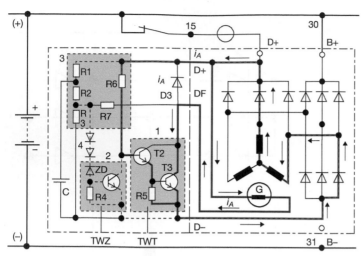

Figure 7.133 Regulator circuit – conducting

Regulator operation When the voltage on the zener diode is below the rated voltage, current passes alongside this route to the rotor windings. The magnetic field strength increases and the voltage induced in the stator increases. This is conducted in the circuit onto the zener diode and, when the rated voltage is reached, the zener diode becomes conductive, so that current is passed to the base of a switching transistor and opens the route to bypass the rotor windings.

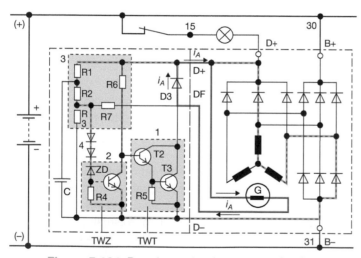

Figure 7.134 Regulator circuit – not conducting

Rotor winding current As the voltage in the rotor windings is switched off, the voltage induced in the stator also reduces until it falls below the rated voltage of the zener diode. The diode becomes non-conductive and stops the current flow to the transistor so that it switches off the bypass route, which, in turn, allows the current to return once again to the rotor windings.

Remember! There is more support on the website that includes additional images and interactive features: www.tomdenton.org

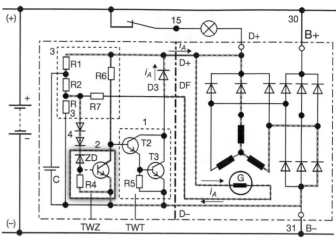

Figure 7.135 Rotor winding is switched on and off as a result of the zener diode conducting

This cycle occurrs repeatedly in the regulator to maintain a constant voltage output from the stator.

Alternator circuit The alternator circuit arrangement, described in the previous sections, is the most common system in use. It is known as a 'machine sensed' type. This type of alternator has two cable terminals and is earthed, or grounded, through the casing. The two terminals are the '+', which is connected directly to the battery positive terminal, and the 'Ind', '61' or 'D', which is connected to the charge warning lamp.

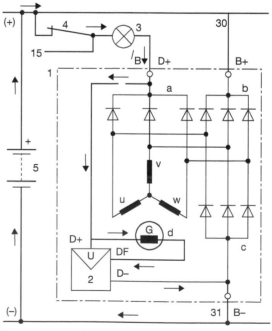

Figure 7.136 Internal and external alternator circuit

Alternator and charging circuit A typical circuit for a modern alternator is shown in figure 7.137. The extra diodes from the centre of the stator help to improve the overall efficiency.

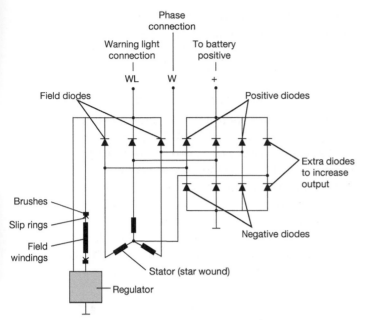

Figure 7.137 Modern alternator circuit

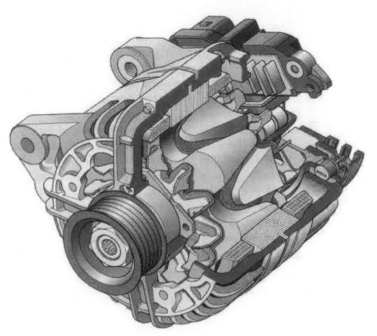

Figure 7.139 Modern alternator

Water-cooled alternator When an alternator operates, the electrical flow produces heat. In this alternator, the heat is used initially to help the engine warm up. Once up to temperature, the engine cooling system then helps to keep the alternator cool. Most types, however, use a simple cooling fan.

Figure 7.138 Water cooling improves efficiency

Modern high output alternator Developments in design have produced very high output machines. Some are now capable of producing 140A or more.

Efficiency The ratio between the power supplied to a machine and the power it actually produces is known as its efficiency. In an alternator there are five key areas where power is lost as it is converted from mechanical (or kinetic) energy to electrical energy:

▶ Copper losses: These are caused by the electrical resistance of the stator and rotor windings. It is proportional to the square of the current ($P = I^2R$)

▶ Iron losses – stator: These occur in the iron of the stator core. They are caused by alternating magnetic fields and hysteresis, which results in eddy currents that produce heat. They can be reduced by laminating the stator core.

▶ Iron losses – rotor: Eddy currents are produced on the surface of the rotor because of the fluctuations of magnetic flux caused by the slits in the stator core.

▶ Rectifier losses: Rectifier diodes cause a voltage drop. This results in heat and hence the need for the rectifier heat sink. High efficiency diodes (HEDs) can reduce but not eliminate this loss

▶ Mechanical losses: These are caused by friction in the bearings and sliding contacts, but mostly by air resistance, sometimes called windage. This is caused by the rotor and fan and is particularly noticed at higher speeds.

Ongoing developments aim to optimise all these losses.

7

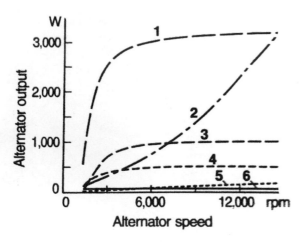

Figure 7.140 Comparison of losses in a 220A alternator. 1. Power output, 2. Iron losses, 3. Copper losses (stator), 4. Rectifier losses, 5. Friction losses, 6. Copper losses (rotor).

Alternator noise is caused by the cutting of magnetic flux. Interestingly on some alternators, it is possible to determine if they are working by listening for the 'whine' they produce. Of course proper measurements are needed too! One area of research that may reduce the noise is the use of a 5-phase, pentagram-connected stator winding as shown here. Another method is to use two 3-phase windings offset by 30 degrees.

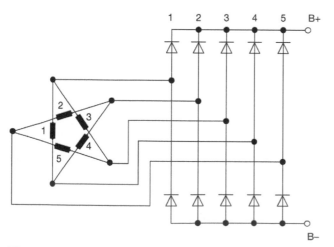

Figure 7.141 Pentagram connected 5-phase alternator with a bridge rectifier.

Bosch 'Efficiency Line' (EL) series alternators are designed for vehicles that feature several safety and comfort functions and therefore require a greater amount of electrical energy. By further improving the electrical design of this alternator and adding new diodes, its efficiency is as much as 77%. This means

that the alternator needs less mechanical energy to generate the same electrical power. In real driving conditions, fuel consumption and CO_2 emissions can be reduced by as much as 2%. These alternators can also withstand temperatures of up to 125°C. Series production of these alternators started in 2010.

The EL alternators are available in three sizes, covering a range between 130 and 210A. They ensure that the battery is well charged even when the engine is idling or running at low rpm. A well charged battery is a condition for start-stop systems. This means that if this fuel-saving system is used, consumption is reduced even further. The combination of a Bosch EL alternator and a start-stop system can reduce fuel consumption (and therefore CO_2 emissions) in urban driving by as much as 10%.

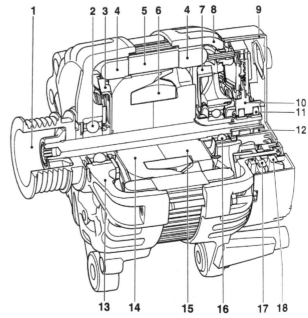

Figure 7.142 Efficiency Line alternator components: 1. Belt pulley, 2. A-side (drive-end) ball bearing, 3. A-side fan, 4. Stator winding heads, 5. Laminated stator core, 6. Rotor winding (excitation winding), 7. B-side fan, 8. B-side end shield, 9. Protective cap, 10. Brush holder, 11. Carbon brush, 12. Collector ring, 13. A-side end shield, 14. A-side claw pole, 15. B-side claw pole, 16. B-side ball bearing, 17. Negative heat sink of rectifier, 18. Positive heat sink of rectifier. (Source: Bosch)

LIN interface and intelligent charging More fuel can be saved if the electrical energy is generated as far as possible during overrun (when rolling or braking). This calls for intelligent alternator control, which is provided by an interface such as the local interconnected network (LIN). The variable control of charging voltage lowers fuel consumption and extends the battery's life cycle.

Figure 7.143 Alternator 28V regulator with a LIN interface

Testing The standard (simplified) method of testing a charging system is to measure battery voltage with the engine off and then again when running at about 3000 rpm with the lights on. If the voltage increases to about 14.2V then the system is almost certainly ok. See the section on testing charging for more details.

Smart charging alternators however are different. The alternator is controlled by the PCM (power control module) which monitors certain parameters such as the engine temperature, battery temperature and electrical demand. If the alternator does not receive a signal from the PCM, the battery light is illuminated on the vehicle. This can sometimes be misdiagnosed as an alternator failure. Many smart charge systems are designed to be used only with a silver calcium battery not the lead acid type. This is due to the voltages used which may damage a lead acid battery and give incorrect readings.

Disconnect the 3 pin PCM plug from the alternator (with the engine switched off) and if it starts charging again, it shows the alternator is being controlled by the PCM – or there is a wiring fault. The 3 pin plug will need to be reconnected immediately after the test to avoid any damage to the vehicle. This test is a temporary check only and is not a resolution to a possible fault.

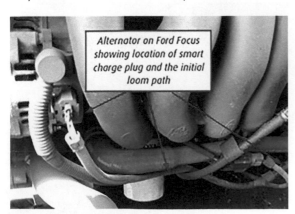

Alternator on Ford Focus showing location of smart charge plug and the initial loom path

Figure 7.144 Smart charge plug on a Ford Focus Valeo alternator (Source: Valeo)

Summary The development of the alternator has been quite amazing over the last 100 years or so. As well as huge increase in output and a reduction in size, this has all been achieved with a component that is able to withstand these operating conditions or requirements:

▶ High rotational speed
▶ Engine vibration
▶ Extreme engine compartment climate
▶ Low acoustic impact

 Use the media search tools to look for pictures and videos relating to the subject in this section

7.2.9 Charging system testing ❷

The most common cause of undercharging is a defective drive belt. Look at the belt very carefully and replace it if there is any doubt about serviceability. The belt is made from a rubber compound reinforced with fabric webbing. It is possible for the webbing to fracture and the belt to look good when it is cold but, when warm, the rubber becomes pliable and the belt will slip.

Figure 7.145 Alternator and drive belt

Belt tension A 'V' belt grips the sides of the pulleys and any slippage will polish the belt and pulleys. If this is found, replace the drive belt. The pulleys may also need replacing if excessively worn. The belt tension is important and, as a rule, the free play on the long side of the belt should be about half an inch or 13mm. Multi-V belt tension may be checked in a similar way. However, it is important to follow the vehicle manufacturer's instructions for this adjustment.

7

Figure 7.146 'V' belt and pulley

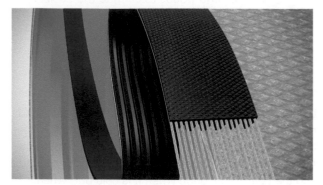

Figure 7.147 Ribbed drive belts are now the most common (Source: Continental)

Alternator output The output from an alternator has two values – the regulated voltage and the current output. The voltage should always be the same, as this is controlled by the regulator at the rated voltage. Modern alternators are usually 14.2V (+/- 0.2), but some older types were higher. Check the manufacturer's data for this value, and for the maximum current output.

Figure 7.148 Alternator output voltage

Voltage and current tests A basic method to carry out a quick check is to connect a digital voltmeter across the battery. A clamp-on ammeter should also be connected on the alternator output cable (not the starter motor feed).

Running tests Start the engine and take voltage and current readings at both idle speed and at about 3000 rpm. The voltage should be as specified in the manufacturer's data (14.2V). The current will vary according to the state of charge of the battery and the load being used by the vehicle electrical systems.

Maximum current output test Check that the alternator can supply all electrical systems at the same time and still charge the battery. Switch on the vehicle headlamps, fog lamps, heated rear window, air conditioning and any other systems. Check that the alternator output current is near its rated output. If necessary the output of the alternator can be made to increase nearer to its maximum. Discharge the battery by leaving the lights on for a few minutes before starting the engine and commencing the test.

Figure 7.149 Turn on vehicle systems such as lights

Figure 7.150 Turn off radio

Regulator bypass test Some alternators can be tested by bypassing the regulator terminals so that the alternator provides the maximum output. Carefully follow the manufacturer's instructions. This test is not often necessary so refer to manufacturer's recommendations.

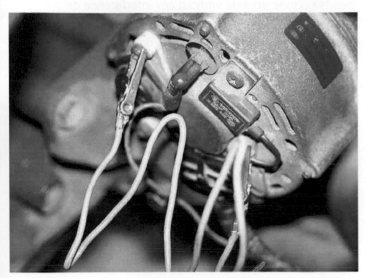

Figure 7.151 Regulator bypass test for alternator maximum output

Charge warning light Check that the charge warning light comes on before the engine is started. The final charging system check is to make sure that the warning light goes out when the engine speed is increased.

Summary It is now common to use combined battery, starter and alternator testers such as the one outlined in the battery test section.

 Using images and text, create a short presentation to show how a component or system works

7.2.10 Stop-start control ❸

Automatic starting This system differs from a normal starting system in that the driver is not in direct control. The driver sends a 'start' request to an ECU, which then in turn can carry out a number of different checks. For example:

▶ Is the driver authorised to start the engine?
▶ Is the engine stationary?

▶ Is battery state of charge adequate at the current temperature?
▶ Is the clutch disengaged or is the automatic transmission in neutral?

The checks only take a fraction of a second after which the ECU initiates the start. Once a set engine speed has been reached, the starter is switched off. This always ensures the shortest possible start time and reduces starter wear. A similar process can be used for start-stop operation.

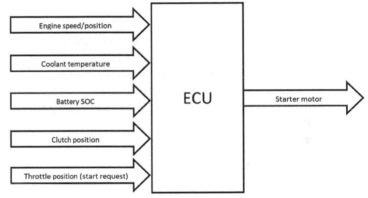

Figure 7.152 Automatic starting system block diagram)

Start-stop systems were mostly used on cars with manual transmission but it is now possible to achieve the same feature to work with automatic transmission. The benefit of this system is 5% or more reduction in CO_2 and fuel consumption in urban traffic. Almost all European automakers are now integrating Bosch start-stop technology into compact cars, premium sedans, light trucks and even in powerful sports cars. Interest in this fuel-saving technology is also on the rise in the US and China.

Reliability To ensure reliable operation, a high degree of control of other systems is necessary; for example, electrical energy management that incorporates battery charge detection. Measures such as a DC/DC converter may also be needed to stabilise the supply to other electrical systems during the voltage drop caused by the starter. Engine systems must also be optimized to ensure a fast start and finally the starter and ring gear components must be enhanced so as to not wear out prematurely.

Start-stop systems, which largely use existing components such as the starter motor and engine control unit, can be quickly and cost-effectively adapted to different engines and vehicles. This is why this starter-motor-based method has now become very popular. Efficient generators that charge the battery as fast as possible are an ideal addition to start-stop systems, since they allow the start/stop function to be used more often.

Automatic transmission In vehicles fitted with automatic transmission, the start-stop function is very easy to use. The driver only has to step on the brake pedal and as soon as the car has stopped the engine stops automatically. It starts again when the brake pedal is released. This is very convenient for the driver, who only has to step on the accelerator and brake.

Sensor information The system works conveniently in the background and evaluates a wide array of sensor information before being activated. For instance, a battery sensor determines the battery's level of charge and the engine will only be switched off if a quick restart is guaranteed. Another example is the vehicle's cabin temperature. If the cabin temperature is too cold or too hot, the engine will continue to run until the occupants' desired temperature is reached. Finally, there is a DC/DC converter for stabilizing the voltage of the electrical system during starts to ensure that the radio, navigation system, or hands-free telephone operate without any interference or interruption.

Developments In order to further reduce consumption, and CO_2 emissions, engineers are continually working to extend engine shutdown times. This will apply initially to the time when the vehicle is coasting to a stop and then it will also apply to periods while driving, when the driver is not accelerating. The effectiveness of all safety and comfort functions will be guaranteed, even during these longer periods with the engine switched off. The enhanced start-stop systems could save up to an additional 10% of fuel.

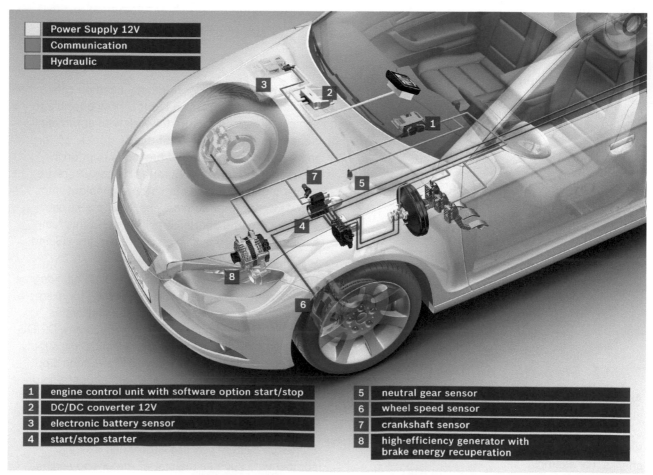

1	engine control unit with software option start/stop
2	DC/DC converter 12V
3	electronic battery sensor
4	start/stop starter

5	neutral gear sensor
6	wheel speed sensor
7	crankshaft sensor
8	high-efficiency generator with brake energy recuperation

Legend: Power Supply 12V · Communication · Hydraulic

Figure 7.153 Start-stop systems can reduce fuel consumption and therefore CO_2 emissions in the New European Driving Cycle (NEDC) by up to 5% (Source: Bosch Media)

The **electronic starter control** system incorporates a static relay on a circuit board integrated into the solenoid switch. This will prevent cranking when the engine is running. Starter control can be supported by an ECU and 'smart' features can be added to improve comfort, safety and service life.

- Starter torque can be evaluated in real time to tell the precise instant of engine start.
- The starter can be simultaneously shut off to reduce wear and noise generated by the free-wheel phase.
- Thermal protection of the starter components allows optimization of the components to save weight and to give short circuit protection.
- Electrical protection also reduces damage from misuse or system failure.
- Modulating the solenoid current allows redesign of the mechanical parts allowing a softer operation and weight reduction.

Summary There are a number of enhancements to the starting system that are designed to improve driver experience and enhance component life but the main benefit is a reduction in fuel consumption and CO_2 emissions.

 Use a library or the web search tools to further examine the subject in this section

7.2.11 **Electronic starter motor control** 3

Introduction An electronically controlled starting system differs from a normal system in that the driver is not in direct control. The driver sends a 'start' request to an ECU which then in turn can carry out several different checks. For example:

- Is the driver authorised to start the engine?
- Is the engine stationary?
- Is battery state of charge adequate at the current temperature?
- Is the clutch disengaged or is the automatic transmission in neutral?

Control The checks only take a fraction of a second after which the ECU initiates the start. Once a set engine speed has been reached the starter is switched off. This always ensures the shortest possible start time and reduces starter wear. A similar process can be used for start-stop operation.

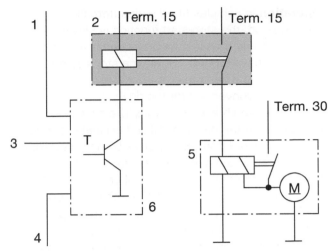

Figure 7.154 Automatic starting circuit: 1. Start signal from driver, 2. Relay, 3. Other input signals. 4. Park/neutral or clutch signal, 5. Starter, 6. ECU (Terminal 15 is ignition live and 30 is battery live)

Stop-start systems The benefit of this system is 5% or more reduction in CO_2 and fuel consumption in urban traffic. Almost all European automakers are now integrating stop-start technology into compact cars, premium sedans, light trucks and even in powerful sports cars. To ensure reliable operation a high degree of control of other systems is necessary. For example, electrical energy management that incorporates battery charge detection. Engine systems must also be optimized to ensure a fast start and finally, the starter and ring gear components must be enhanced to not wear out prematurely.

Components Stop-start systems, which largely use existing components such as the starter motor and engine control unit, can be quickly and cost-effectively adapted to different engines and vehicles. Therefore, this starter motor based method has now become very popular. Efficient generators that charge the battery as fast as possible are an ideal addition to these systems, since they allow the function to be used more often.

Figure 7.155 Starter motor

Operation In vehicles fitted with automatic transmission, the stop-start function is very easy to use. The driver presses the brake pedal, and as soon as the car has stopped, the engine stops automatically. It starts again when the brake pedal is released. This is very convenient for the driver, who uses the accelerator and brake as normal. The system works completely in the background and evaluates a wide array of sensor information before being activated. For instance, a battery sensor determines the battery's level of charge and the engine will only be switched off if a quick restart is guaranteed. Another example is the vehicle's cabin temperature. If the cabin temperature is too cold or too hot, the engine will continue to run until the occupants' desired temperature is reached.

Enhancements To further reduce consumption, and CO_2 emissions, engineers are continually working to extend engine shutdown times. This will apply initially to the time when the vehicle is coasting to a stop and then it will also apply to periods while driving, when the driver is not accelerating. The effectiveness of all safety and comfort functions will be guaranteed, even during these longer periods with the engine switched off. The enhanced stop-start systems could save up to an additional 10% of fuel.

 Select a routine from section 1.3 and follow the process to study a component or system

7.2.12 48V technology ❸

Introduction The benefits of 48V systems have been under discussion for many years but at last they seem to be gaining ground.

Developments It is thought that 48V architectures will transform and enhance just about every major subsystem of conventional vehicles. However, they will also be a bridge technology that will allow current combustions engines to evolve into the all-electric future. One area that will be important for combustion engine power and efficiency is the use of electrically-accelerated turbochargers. The higher power that 48V can deliver also enables use of electromechanically actuated valves.

Electrification of mechanical components also allows additional digital control and all the benefits that brings. For example, Audi are working on an electromechanical rotary-damper system to replace conventional hydraulic dampers. This would have been much more difficult with 12V. New possibilities for adjusting the suspension will now exist as the

response characteristics of a component (or system) will be controlled by software.

48V micro-hybrids This system employs a starter/generator hence it can be used to charge the batteries as well as start the vehicle. It is a 3-phase device and includes two sensors, position and current, which provide information back to the ECU. The controller area network (CAN) and local interconnect network (LIN) connections make data available across the vehicles as well as data from other sensors available to the micro-hybrid system. Infineon is one of the leading companies that offers a micro-hybrid 48V control system.

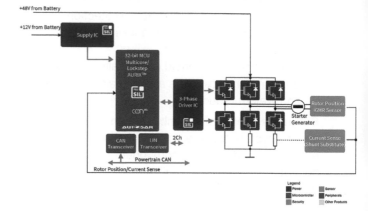

Figure 7.156 Micro-hybrid 48V control system block diagram (Source: Infineon, www.infineon.com)

 Create an information wall to illustrate the features of a key component or system

7.2.13 Jump start without a battery ❸

Introduction A jump starter without a battery is a device that as the names suggests is 'batteryless'. Instead it contains super-capacitors. These components have a higher energy flow density but not a high-power storage capacity. For this reason, it is possible to charge the capacitors with a relatively small power source. Inside the unit, a DC-DC converter acts like a transformer to step up the voltage to charge the capacitors with sufficient energy for a single start operation.

Capacitors This is where the capacitors win over a battery due to their power delivery capability. Another interesting development is that using the power transform capability, the capacitors can be charged from the dead battery itself if it has some voltage and power capacity. This is difficult to comprehend but remember that the normal failure mode of a chemical battery involves its ability to deliver high power for starting. It is less often the case that the battery cannot deliver any power, so even a small amount over a longer period can be utilised for charging the capacitors for a start assistance.

Figure 7.157 Batteryless jump start pack (Source: Jack Sealey, www.sealey.co.uk)

Operation The device shown is sold as a 'batteryless' jump start pack. It is small and light when compared to a battery based device. It needs no maintenance and charges from the dead battery, or from another vehicle or even via USB! It can also be used with the vehicle battery open circuit, for situations where the battery is completely dead.

Figure 7.158 Sealey batteryless unit charging from the 'dead' battery (www.autoelex.co.uk)

Summary In service, this type of jump starter works well. The package is aimed at passenger car users and can supply about 300A, which is more than enough power to start most cars. The device also includes a diesel glow plug support mode to allow pre-heating time.

 Use a library or the web search tools to further examine the subject in this section

7.3 Lighting and indicators

7.3.1 Lighting systems ❶

Introduction Vehicle lighting systems are very important, particularly where road safety is concerned. If headlights were suddenly to fail at night and at high speed, the result could be serious. Remember, that lights are to see with, and to be seen by . . .

Lighting clusters Lights are arranged on a vehicle to meet legal requirements and to look good. Headlights, sidelights and indicators are often combined on the front. Taillights, stoplights, reverse lights, and indicators are often combined at the rear.

Figure 7.159 Lights . . .

Figure 7.160 . . . are positioned . . .

Figure 7.161 . . . in different ways by . . .

Figure 7.162 ... different manufacturers

Bulbs The number, shape and size of bulbs used on vehicles is increasing all the time. A common selection is shown here. Most bulbs used for vehicle lighting are generally either conventional tungsten filament bulbs or tungsten halogen.

Figure 7.163 Selection of bulbs

In the conventional bulb, the tungsten filament is heated to incandescence by an electric current. The temperature reaches about 2300°C. Tungsten, or an alloy of tungsten, is ideal for use as filaments for electric light bulbs. The filament is normally wound into a 'spiralled spiral' to allow a suitable length of thin wire in a small space and to provide some mechanical strength.

Tungsten halogen bulbs Almost all vehicles now use tungsten halogen bulbs for the headlights. The bulb will not blacken and, therefore, has a long life. In normal

Figure 7.164 Headlight bulb

gas bulbs, about 10% of the filament metal evaporates. This is deposited on the bulb wall. Design features of the tungsten halogen bulb prevent deposition. The gas in halogen bulbs is mostly iodine. The glass envelope is made from fused silicon or quartz.

Headlight reflectors The object of the headlight reflector is to direct the random light rays produced by the bulb into a beam of concentrated light, by applying the laws of reflection. Bulb filament position relative to the reflector is important if the desired beam direction and shape are to be obtained.

Figure 7.165 Reflector

Reflector construction A reflector is a layer of silver, chrome or aluminium deposited on a smooth and polished surface such as brass or glass. Consider a mirror reflector that 'caves in' – this is called a concave reflector. The centre point on the reflector is called the pole, and a line drawn perpendicular to the surface from the pole is known as the principal axis.

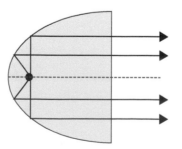

Figure 7.166 Concave reflector

Focused beam If a light source is moved along the principal axis, a point will be found where the radiating light produces a reflected beam parallel to the axis. This point is known as the focal point, and its distance from the pole is known as the focal length.

Divergent and convergent beams If the filament is between the focal point and the reflector, the reflected beam will diverge – that is, spread outwards along the principle axis. If the filament is positioned in front of the focal point, the reflected beam will converge towards the principle axis.

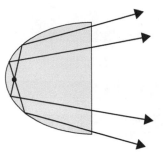

Figure 7.167 Light source behind the focal point

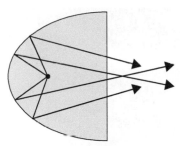

Figure 7.118 Light source in front of the focal point

The intensity of reflected light is strongest near the beam axis, except for the light cut off by the bulb itself. The intensity, therefore, drops off towards the outer edges of the beam. A common type of reflector and bulb arrangement is shown, where the dip filament is shielded. This gives a nice sharp cut-off line when on dip beam. It is used with asymmetric headlights.

Headlight lenses A good headlight should have a powerful, far-reaching central beam, around which the light is distributed both horizontally and vertically in order to illuminate as great an area of the road surface as possible. The beam formation can be considerably improved by passing the reflected light rays through a transparent block of lenses. It is the function of the lenses to partially redistribute the reflected light beam and any stray light rays. This gives better overall road illumination.

Lenses work on the principle of refraction. The headlight front cover is the lens. It is divided up into a large number of small rectangular zones, each zone being formed optically in the shape of a concave flute or a combination of flute and prisms. Each individual lens element will redirect the light rays to obtain an improved overall light projection or beam pattern.

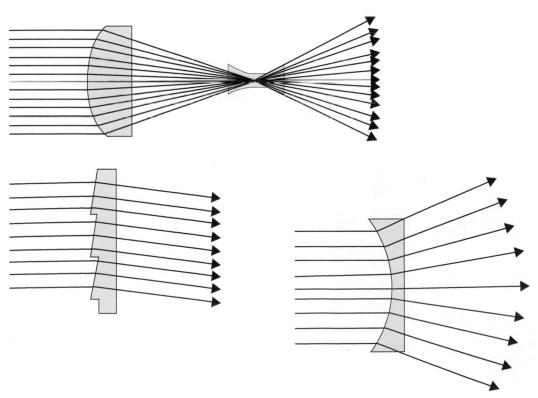

Figure 7.169 Headlight lens details

Complex shape reflectors Many headlights are now made with clear lenses, which means that all the direction of the light is achieved by the reflector. The clear lens does not restrict the light in any way. This makes the headlights more efficient as well as more attractive.

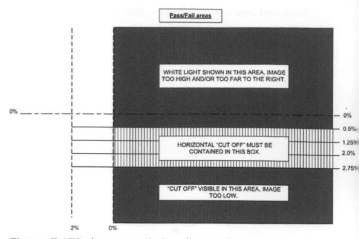

Figure 7.172 Asymmetric headlamp alignment requirements

Figure 7.170 Modern headlights

Other lights Sidelights, taillights, brake lights and others are relatively straightforward. Headlights present the most problems. This is because on dipped beam they must provide adequate light for the driver, but not dazzle other road users.

Headlight alignment In the UK, the required alignment is part of the MOT test. The requirements are shown in Figure 7.172.

The function of a levelling actuator is to adjust the dipped or low beam in accordance with the load carried by the car. This will avoid dazzling oncoming

Figure 7.173 Beam setter in use (Source: Hella)

traffic. Manual electric levelling actuators are connected up to a control on the dashboard. This allows the driver to adjust beam height.

Automatic static actuators adjust beam height to the optimum position in-line with vehicle load conditions. The system includes two sensors (front and rear), which measure the attitude of the vehicle. An electronic module converts data from the sensors and drives two electric gear motors (or actuators) located at the rear of the headlamps, which are mechanically attached to the reflectors.

> Make a simple sketch to show how one of the main components or systems in this section operates

Figure 7.171 Rear lights

7.3.2 Stoplights and reverse lights

Introduction Stoplights, or brake lights, are used to warn drivers behind that you are slowing down or stopping. Reverse lights warn other drivers that you are reversing, or intend to reverse. The circuits are quite simple. One switch in each case operates two or three bulbs. A relay may be used.

The circuits for these two systems are similar. Shown online is a typical stoplight or reverse light circuit. Most incorporate a relay to switch on the lights, which is in turn operated by a spring-loaded switch on the brake pedal or gearbox. Links from the stoplight circuit to the cruise control system may be found. This is to cause the cruise control to switch off as the brakes are operated. A link may also be made to the anti-lock brake system.

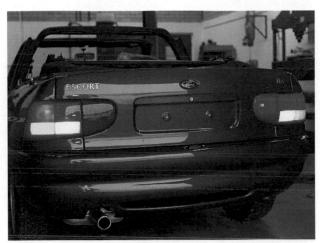

Figure 7.174 Stop and reverse lights form part of the rear light cluster

Switches The circuits are operated by the appropriate switch. The reverse switch is part of the gearbox or

Figure 7.175 Stoplight switch

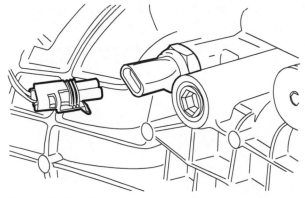

Figure 7.176 Reverse light switch

gear change linkage. The stop switch is usually fitted so it acts on the brake pedal.

Figure 7.183 is the complete lighting circuit of a vehicle. The colour codes used are discussed in the basic electrical learning sections. However, you can follow the circuit by looking for the labels on the wires. 'N' for example, means 'Brown' but this has no effect on how it works! Operation of part of this circuit is as described over the following paragraphs.

The ignition must be on for the reverse lights to operate. The reverse light switch gets its feed from fuse 16 on the GY wire. When the switch is operated, the supply is sent to the rear lamps on a GN wire. The switch is usually mounted on the gear change linkage or screwed into the gearbox.

The ignition must be on for the brake or stoplights to operate. The brake or stoplight switch gets its feed from fuse 16 on the GY wire. When the switch is operated, the supply is sent to the rear lamps on a GP wire. A connection is also made to the centre high-mounted stoplight. The switch is usually mounted on the pedal box above the brake pedal.

Light emitting diodes (LEDs) are more expensive than bulbs. However, the potential savings in design costs due to long life, sealed units being used and greater freedom of design, could outweigh the extra expense. LEDs are ideal for stoplights.

A further advantage is that they illuminate quicker than ordinary bulbs. This time is approximately the difference between 130ms for the LEDs, and 200ms for bulbs. If this is related to a vehicle brake light at motorway speeds, then the increased reaction time equates to about a car length. This is potentially a major contribution to road safety.

Centre high-mounted stop lamps An LED centre high mounted stop lamp (CHMSL) illuminates faster than conventional incandescent lamps, improving driver response time and providing extra braking distance. Due to their low height and reduced depth,

Figure 7.177 CHMSL

LED CHMSLs can be easily harmonised with all vehicle designs. They can be mounted inside or integrated into the exterior body or spoiler.

Summary Reverse lights are operated by a simple on/off gearbox switch. Stoplights are operated by a simple on/off switch on the pedal box. Both circuits operate in much the same way. High-mounted stoplights are now quite common, many of them using LEDs.

 Select a routine from section 1.3 and follow the process to study a component or system

7.3.3 **Interior lighting** 1

Introduction Interior lighting consists of several systems, the main ones being courtesy lights, map lights and panel illumination lights. The circuits are quite simple; however, they are often linked with the central locking system. Features such as delay and fade out are now common. This requires some electronic control.

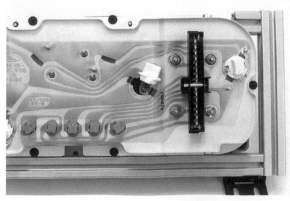

Figure 7.178 Instrument illumination

Map lights are an extra feature to assist with reading a map in the dark! Many types are available. Some are small spotlights, which form part of the interior light assembly. Others are positioned on the centre console of the vehicle.

Interior or courtesy lights are designed to illuminate the vehicle interior when the doors are opened. Most cars have one central interior light above the rear view mirror, or two lights on the sides above the driver's and passenger's shoulders.

Door switches are simple spring-loaded contacts that are made as the door opens. The contacts are broken again as the door closes. Rubber seals are sometimes used to keep water out. The same switches may also be used for the alarm system.

Figure 7.179 Switch positioned in the door pillar

Interior light circuit The circuit shown in Figure 7.219 is typical of many in common use. The sliding switches have three positions, 'off', 'on' and 'door operated'. The control module is to allow delay operation. In this case, it is also used for the central locking system.

Instrument and panel illumination Panel and instrument lights are illuminated when the vehicle sidelights are switched on. Most cars also incorporate a dimmer switch so the level of illumination can be set.

Central control module Ford now fit a general electronic module (GEM) which controls the interior lights as well as a number of other functions. The interior lights are controlled to prevent them being left on and with a delay timer. On this system the lights are illuminated when a door is opened and the ignition is in position 0 or I. They are extinguished 25 seconds after a door is closed, when the car is locked or the ignition switch is moved to position II or III.

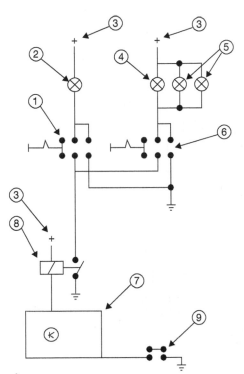

Figure 7.180 Circuit for interior lights

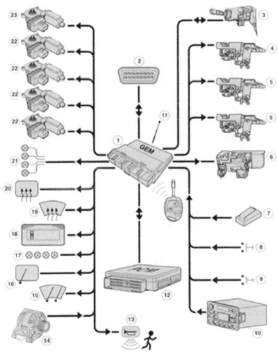

Figure 7.181 General electronic module (GEM) system

Summary Interior lights are important for passenger comfort. Most now operate via some type of electronic control. One enhancement is a switching off delay, after the doors are closed. Some manufacturers are linking functions such as interior lights with other systems through a central control module.

Use a library or the web search tools to further examine the subject in this section

7.3.4 Lighting circuits ❷

Introduction Lighting circuits can appear complex at first view. However, if you concentrate on just one part of the circuit at a time you will find it easier to understand. Relays are often used because they take load off the control switches. They are still simple switches so don't panic! Take your time and you will find electrical circuits an interesting challenge.

Shown online is a simplified lighting circuit. Whilst this representation helps to demonstrate the way in which a lighting circuit operates, it is not often used in this simple form. A full circuit is described later in this section.

The circuit shows how various lights, in and around the vehicle, operate with respect to each other. The headlights for example, cannot be operated without the sidelights first being switched on. The spotlights are wired so that they only work when the headlights are on main beam.

Dim dip circuit Dim dip headlights are an attempt to stop drivers just using sidelights in semi-dark or poor visibility conditions. The circuit is such that when sidelights and ignition are on together, then the headlights will come on automatically at about one-sixth of normal power.

Dim dip lights are achieved in one of two ways. The first uses a simple resistor in series with the headlight bulb. The second uses a 'chopper' module, which switches the power to the headlights on and off rapidly. In either case, the 'dimmer' is bypassed when the driver selects normal headlights.

The diagram shown is the complete lighting circuit of a vehicle. The colour codes used are discussed in Section 7.1.3. However, you can follow the circuit by looking for

Figure 7.182 Headlights on dim dip – and spot lights

279

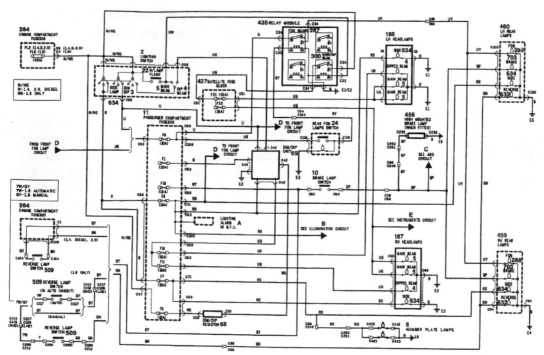

Figure 7.183 Lighting circuit

the labels on the wires. 'R' for example, means 'Red' but this has no effect on how it works! Operation of part of this circuit is described over the following paragraphs.

Operation of the switch allows the supply on the N or N/S wire to pass to fuses 7 and 8 on the R wire. The two fuses then supply left lights on an RB wire, and right lights on a RO wire. The number plate lights are also supplied from here.

When the dip beam is selected, a supply is passed, on a U and UR wire, to the dim dip unit, which is de-energised. This then allows a supply to fuses 10 and 11 on the OU wire. This supply is then passed to the left light on a UK wire and the right light on a UB wire.

Selecting main beam allows a supply on the UW wire to the main/dip relay, thus energising it. A supply is therefore placed on fuses 21 and 22 and hence to each of the headlight main beam bulbs.

When sidelights are on, there is a supply to the dim dip unit on the RB wire. If the ignition supplies a second feed on the G wire from fuse 1, the unit will allow a supply from fuse 5 to the dim dip resistor on the NS wire. This continues on to the dim dip unit on an NG wire. The dim dip unit links this supply to fuses 10 and 11. These are the dip beam fuses. The supply is therefore passed to the left light on a UK wire and the right light on a UB wire.

When the headlights are switched on, a supply is made from the light switch to fuse 9 on a U wire. From this fuse a supply is sent to the fog light relay contacts on a U wire and the rear fog lamp switch on a UR wire. When the fog switch is operated, it sends a supply on

the RY wire to close the relay. The main supply is now fed from the relay on a UY wire to both rear fog lamps.

Summary Following a circuit diagram is easy after a bit of practice. Think of it as a railway map that is used to get from A to B. Electricity will only make the 'journey' if the path is complete.

 Use a library or the web search tools to further examine the subject in this section

7.3.5 Indicators and hazard lights ❷ ❸

Introduction Direction indicators have a number of statutory requirements. The light produced must be amber, but they may be grouped with other lamps.

Figure 7.184 Indicator bulb

The flashing rate must be between one and two per second with a relative 'on' time of between 30 and 57%. If a fault develops this must be apparent to the driver by the operation of a warning light on the dashboard. The fault can be indicated by a distinct change in frequency of operation or the warning light remaining on. If one of the main bulbs fails, then the remaining lights should continue to flash perceptibly.

Legislation exists as to the mounting position of the exterior lamps. The rear indicator lights must be within a set distance of the rear lights, and within a set height. The wattage of indicator bulbs is normally 21W.

Stoplights The wattage of stoplight bulbs is normally 21W. These lights often come under the heading of auxiliaries or signalling. A circuit is examined later in this section. The bulbs are often combined with the rear lights.

Figure 7.185 Stoplight bulb

Flasher units The operation of this unit is based around an integrated circuit. The electronic type shown can operate at least four 21W bulbs (front and rear) and two 5W side repeaters when operating in hazard mode. This will continue for several hours if required. Flasher units are rated by the number of bulbs they are

Figure 7.186 Modern electronic flasher unit

capable of operating. When towing a trailer or caravan, the unit must be able to operate at a higher wattage. Most units use a relay for the actual switching as this provides an audible signal. The thermal-type flasher units are still used but only on older vehicles.

Figure 7.187 Old thermal flasher unit

The **electronic circuit** is constructed together with the relay, on a printed circuit board. Very few components are used as the integrated circuit is specially designed for use as an indicator timer. The resistor and capacitor shown set the flash rate. The unit is designed to give an on-off ratio of 50% and an operating frequency of 1.5Hz (90 per minute).

Figure 7.188 Circuit details

Bulb failure The on-off signal is passed to a transistor circuit, with a diode connected to protect it from back EMF. Bulb failure is recognised when the voltage across a resistor falls. The bulb failure circuit causes the unit to double the speed of operation. Extra capacitors can be used for interference suppression.

Figure 7.189 Indicator warning light

Figure 7.190 shows the complete indicator circuit of a vehicle. The colour codes used are discussed in Section 7.1.3. However, you can follow the circuit by looking for the labels on the wires. 'G' for example, means 'Green' but this has no effect on how it works! Operation of part of this circuit is as described over the following paragraphs.

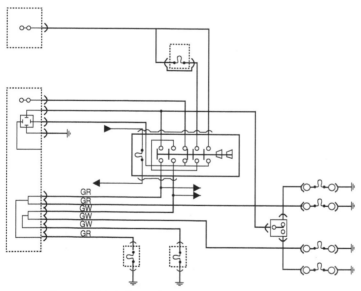

Figure 7.190 Complete indicator circuit

Indicators and hazard circuit An indicator and hazard lights circuit diagram is shown in Figure 7.191. Note how the hazard switch, when operated, disconnects the ignition supply from the flasher unit and replaces it with a constant supply. The hazard system will therefore operate at any time, but the indicators will only work when the ignition is switched on. When the indicator switch is operated left or right, the front, rear and repeater bulbs are connected to the output terminal of the flasher unit. This is what makes it operate and causes the bulbs to flash.

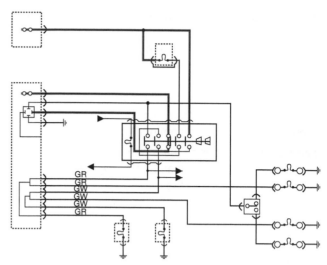

Figure 7.191 Ignition and permanent supplies

When the hazard switch is operated, five sets of contacts move. Two sets connect left and right circuits to the output of the flasher unit. One set disconnects the ignition supply and another set connects the battery supply to the unit. The final set of contacts cause a hazard warning light to be operated. On this, and on most vehicles, the hazard switch is illuminated when the sidelights are switched on.

With the ignition switched on, fuse 1 in the passenger compartment fuse box provides a feed to the hazard warning switch on the G wire. Provided the hazard warning switch is in the off position, the feed crosses the switch and supplies the flasher unit on the LGK wire. When the switch control is moved for a right turn, the switch makes contact with the LGN wire from the flasher unit, which is connected to the GW wire. This allows a supply to pass to the right-hand front and rear indicator lights, and then to earth on the B wire.

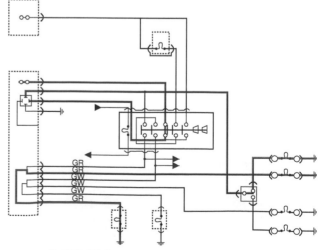

Figure 7.192 Right circuit

When the switch control is moved for a left turn, the switch makes contact with the GR wire, which allows the supply to pass to the left-hand front and rear indicator lights, and then to earth on the B wire. The action of the flasher unit causes the circuit to make and break.

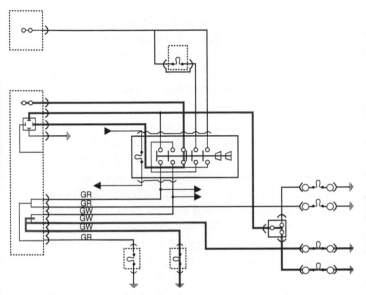

Figure 7.193 Left circuit

By pressing the hazard warning switch a battery, supply on the N0 wire from fuse 3 or 4 in the engine bay fuse box, crosses the switch and supplies the flasher unit on the LGK wire. At the same time, contacts are closed to connect the hazard warning light and the flasher unit to both the GW and GR wires. These are the right-hand and left-hand indicators. The warning light and the main lights flash alternately.

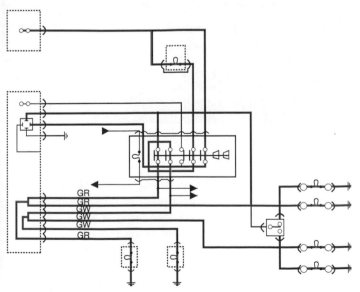

Figure 7.194 Hazard circuit

Summary Indicators and hazard lights are interesting circuits. Hazard lights are intended to show a hazard such as a broken down car.

Using images and text, create a short presentation to show how a component or system works

7.3.6 Other lighting technology

Introduction Manufacturers are constantly working to improve vehicle lighting systems. A number of new technologies have recently been introduced. Light emitting diodes have been used for some time. However, there are still some interesting developments taking place with infrared and ultraviolet lighting. These are covered briefly in this section.

Light emitting diodes The main advantages of light emitting diodes (LEDs), when used for lighting, is that they have a typical rated life of over 50 000 hours. The environment in which vehicle lights have to survive is hostile. Extreme variations in temperature and humidity, as well as serious shocks and vibration, have to be endured. LEDs are being developed in red, green and blue (RGB) groups. This will allow white light as well as other colours. The design possibilities for rear lights are therefore limitless.

Infrared lights Thermal imaging technology promises to make night driving visibly less hazardous. Infrared thermal-imaging systems are going to be fitted to cars. General Motors is now offering a system called 'Night Vision' as an option. After Night Vision is switched

Figure 7.195 LED stoplight

7

on, 'hot' objects, including animals and people, show up as white in the thermal image. The image is projected onto the windscreen. On the vehicle, a camera unit sits in the centre of the car behind the front grille.

Figure 7.196 Night Vision

Gas discharge lamps (GDL) are now being fitted to vehicles. They have the potential to provide illumination that is more effective, and there are new design possibilities for the front of a vehicle. The conflict between aerodynamic styling and suitable lighting positions is an economy/safety trade off, which is undesirable.

The source of light in the gas discharge lamp is an electric arc. The discharge bulb actually used is

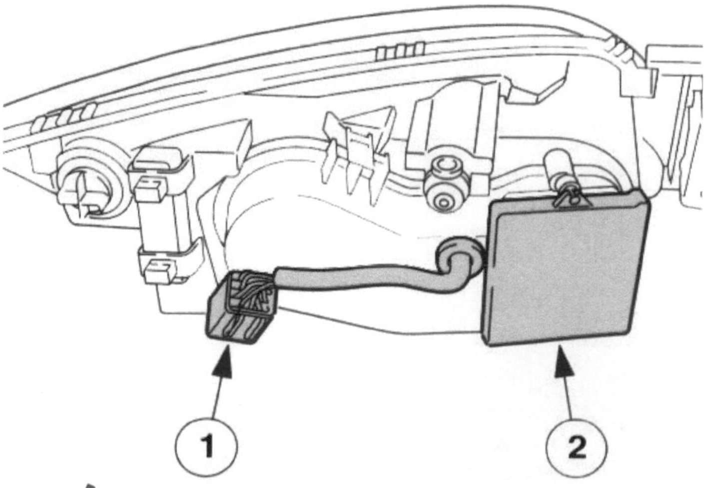

only about 10mm across. Two electrodes extend into the bulb, which is made from quartz glass. The gap between these electrodes is about 4mm. The bulb is filled with xenon gas. These lamps are sometimes described as high intensity discharge (HID).

If the GDL system is used as a dip beam, self-levelling lights are required because of the high intensities. Use as a main beam may be a problem because of the on/off nature. A GDL system for dip beam, which stays on all the time, is supplemented by a conventional main beam.

Ultraviolet headlights The GDL can be used to produce ultraviolet light. Since UV radiation is virtually invisible, it will not dazzle oncoming traffic but will illuminate fluorescent objects such as specially treated road markings and clothing. These glow in the dark much like a white shirt under some disco lights. The UV light will also penetrate fog and mist, as the light reflected by water droplets is invisible. It will even pass through a few centimetres of snow. Cars with UV lights use two conventional halogen main/dip lights and two UV lights. The UV lights come on at the same time as the dipped beams.

Figure 7.198 Road markings glowing at night

Blue lights! Philips produces halogen bulbs called 'BlueVision'. The light stimulates driver concentration and makes driving in the dark less tiring. It also reflects much better on road markings and signs. The bulbs are directly interchangeable with existing bulbs. However, it should be noted that halogen technology is not comparable to the Xenon discharge technology.

Figure 7.197 GDL system headlamp

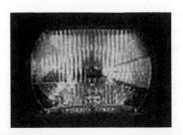

Figure 7.199 Is blue whiter than white?

Jewel aspect lamps are based on the complex shape reflector technology, which is widely used in headlamps. Beam pattern is no longer completely controlled by the lens but by the reflector. Conventional lens optics using prisms are minimised, giving the impression of greater depth and brightness.

Figure 7.200 Great looking lights

Summary Good lights are vital for safe driving at night. Interesting developments are taking place continuously. Many of the sophisticated systems, now being introduced on top of the range cars, will soon be available as standard options. Drive safely.

 Use the media search tools to look for pictures and videos relating to the subject in this section

7.3.7 **BMW laser headlamps** ❸

Introduction All carmakers take safety measures very seriously and the big German manufacturers have

a reputation in this area. Improvements in crumple zones, air bag technology and glass manufacturing have all led to improved safety for drivers and passengers over the last twenty years or so. However, BMW aim to be the new market leader with their headlamp technology which could make night driving safer than ever before.

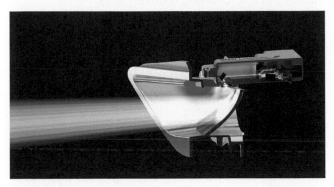

Figure 7.201 BMW laser headlamps where the rapid deflection of a laser creates whatever beam pattern is needed (Source: BMW Media)

Intensity The lights can provide a much-improved intensity over the LED lights which are currently fitted to existing models. It is claimed that the new laser headlamps will function at up to 344 lux when operating in high beam mode. This compares favourably with the current 180 lux which is offered by even the latest LED headlamps. Older headlamp technology, which uses xenon or halogen, rarely tops 120 lux.

Figure 7.202 Headlamp patterns (Source: BMW Media)

Use the media search tools to look for pictures and videos relating to the subject in this section

7.4 Body electrical

7.4.1 Washers, wipers and screens ❶ ❷

The requirements of the wiper system are simple. The windscreen must be clean enough to provide suitable visibility at all times. To do this, the wiper system must meet the following requirements:

▶ efficient removal of water and snow
▶ efficient removal of dirt
▶ operate at temperatures from −30 to 80°C
▶ pass the stall and snow load test
▶ have a service life in the region of 1.5 million wipe cycles
▶ be resistant to corrosion from acid, alkali and ozone.

Wiper blades are made of a rubber compound and are held on to the screen by a spring in the wiper arm. The aerodynamic property of the wiper blades has become increasingly important. The strip on top of the rubber element is often perforated to reduce air drag. A good-quality blade will have a contact width of about 0.1mm. The lip wipes the surface of the screen at an angle of about 45°. The pressure of the blade on the screen is also important.

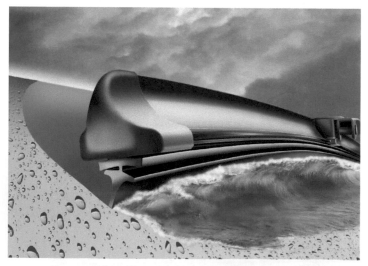

Figure 7.203 Details of a wiper blade

Most wiper linkages consist of a series or parallel mechanism. Some older types use a flexible rack and wheel boxes similar to the operating mechanism of many sunroofs. One of the main considerations for the design of a wiper linkage is the point at which the blades must reverse. This is because of the high forces on the motor and linkage at this time. If the reverse point is set so that the linkage is at its maximum force transmission angle, then the reverse action of the blades puts less strain on the system. This also ensures smoother operation.

Wiper motors All modern wiper motors are permanent magnet types. The drive is taken via a worm gear to increase torque and reduce speed. Three brushes may be used to allow two-speed operation. The normal speed operates through two brushes placed in the usual positions opposite to each other. For a fast speed, the third brush is placed closer to the earth brush. This reduces the number of armature windings between them, which reduces resistance, hence increasing current and therefore speed.

Figure 7.204 Motor brushes and armature

Circuit protection Wiper motors or the associated circuit must have some kind of short circuit protection. This is to protect the motor in the event of stalling – if frozen to the screen, for example. A thermal trip of some type is often used, or a current sensing circuit in the wiper ECU if fitted.

Windscreen washers The windscreen washer system consists of a simple DC permanent magnet motor, which drives a centrifugal water pump. The water, preferably with a cleaning additive, is directed onto an appropriate part of the screen by two or more jets. A non-return valve is often fitted in the line to the jets to prevent water siphoning back to the tank. This also allows 'instant' operation when the washer button is pressed. The washer circuit is normally linked in to the wiper circuit so that when the washers are operated, the wipers start automatically and will continue for several more sweeps after the washers have stopped.

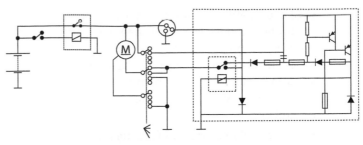

Figure 7.205 Washer pumps

Figure 7.206 Wiper circuit

Washer and wiper circuits Figure 7.206 shows a circuit for fast, slow and intermittent wiper control. The switches are shown in the off position and the motor is stopped and in its park position. Note that the two main brushes of the motor are connected together via the limit switch, delay unit contacts and the wiper switch. This circuit connection causes regenerative braking because of the current, generated by the motor due to its momentum, after the power is switched off. Being connected to a very low resistance loads up the 'motor/generator' and, when the park limit switch closes, it stops instantly.

When either the delay contacts or the main switch contacts are operated, the motor will run at slow speed. When fast speed is selected, the third brush on the motor is used. On switching off, the motor will continue to run until the park limit switch changes over to the position shown. This switch is only in the position shown when the blades are in the parked position.

Central control units Some vehicles use a system with more enhanced facilities. This is regulated by what may be known as a central control unit

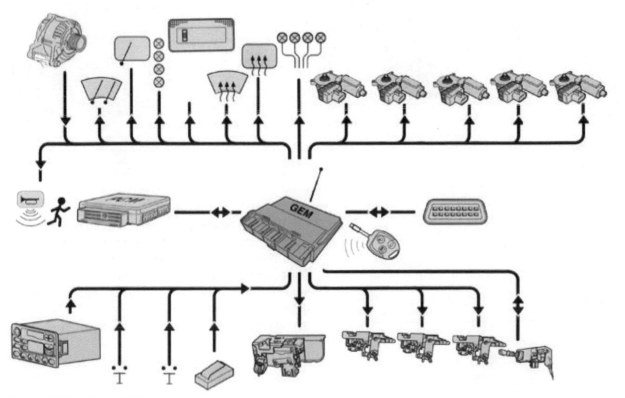

Figure 7.207 Ford GEM and components

(CCU), a multifunction unit (MFU) or a general electronic module (GEM)! These units often control other systems as well as the wipers, thus allowing reduced wiring bulk under the dash area. Electric windows, headlights and heated rear window, to name just a few, are now often controlled by a central unit.

Electronically controlled facilities Using electronic control, a CCU allows the following facilities for the wipers:

- ▶ front and rear wash/wipe
- ▶ intermittent wipe
- ▶ time delay set by the driver
- ▶ reverse gear selection rear wipe operation
- ▶ rear wash/wipe with 'dribble wipe' (an extra wipe several seconds after washing)
- ▶ stall protection.

Wiper blade pressure control A system called wiper pressure control can infinitely vary the pressure of the blade onto the screen, depending on vehicle speed. At high speeds, the air stream can cause the blades to lift and judder. This seriously reduces the cleaning effectiveness. If the original pressure is set to compensate, the pressure at rest could deform the arms and blades. Sensors are used to determine the air stream velocity and intensity of the rain. An ECU then evaluates the data from these sensors and passes an appropriate signal to a servomotor. When the blades are in the rest position, pressure is very low to avoid damage. The pressure rises with increasing vehicle speed and heavy rain.

Linear rear wipers Current wiper systems are based on an alternative rotary movement to cover a wipe area of between 50% and 60% of the total surface area of the rear window. This limit is due to the height/width ratio and the curve of the window. The linear rear wiper concept ensures optimum visual comfort as it covers over 80% of the rear window surface. This increase in the driver's field of vision enhances safety, especially during low-speed manoeuvres such as reversing or parking.

Rear screen heating Heating of the rear screen involves a circuit with a relay, which will usually incorporate a timer. The heating elements are thin metallic strips bonded to, or built inside, the glass. When a current is passed through the elements, heat is generated and the window will defrost or demist.

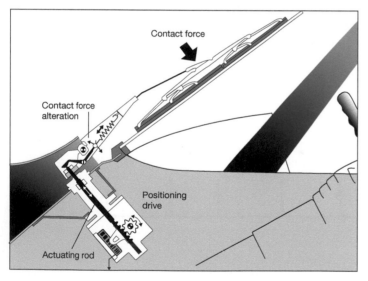

Figure 7.209 Rear screen heater elements

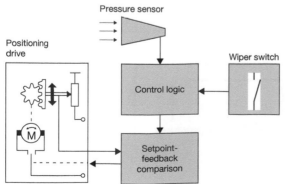

Figure 7.208 Bosch pressure control system

Windscreen heating Front windscreen heating is used on some vehicles. This presents more problems than the rear screen, as vision must not be obscured. The technology used is drawn from the aircraft industry – it involves very thin wires cast in to the glass. As with the heated rear window, this device can consume a large current and uses a timer relay.

Remember! There is more support on the website that includes additional images and interactive features: www.tomdenton.org

High current Screen heaters can draw high current, 10 to 15 amps being typical. Because of this, the circuits often contain timer relays to prevent the heaters being left on too long. The timer will switch off after 10 to 15 minutes.

Figure 7.210 Timer relay

 Create an information wall to illustrate the features of a key component or system

7.4.2 Horns, obstacle avoidance and cruise control ❶ ❷

Regulations in most countries state that the horn (or audible warning device) should produce a uniform sound. This makes sirens and melody-type fanfare horns illegal! Most horns draw a large current and so are switched by a relay.

Horn circuit The standard horn operates by simple electromagnetic switching. Current flow causes an armature, which is attached to a tone disc, to be attracted to a stop. A set of contacts is then opened. This disconnects the current, allowing the armature and disc to return under spring tension. The whole process keeps repeating when the horn switch is on. The frequency of movement, and hence the tone, is arranged to lie between 1.8 and 3.5kHz. This note gives good penetration through traffic noise.

Twin horns systems, which have a high and a low tone horn, are often used. This produces a more pleasing sound but is still very audible in both town and higher speed conditions.

Figure 7.211 Horns removed from vehicle

Obstacle avoidance radar This system, sometimes called collision avoidance radar, can be looked at in two ways. First, it can be an aid to reversing, which gives the driver some indication as to how much space is behind the car. Second, collision avoidance radar can be used as a vision enhancement system. Obstacle avoidance radar, when used as a vision enhancement, is somewhat different. Figure 7.212 is a block diagram to demonstrate the principle of this system. In the future, this may be linked with adaptive cruise control.

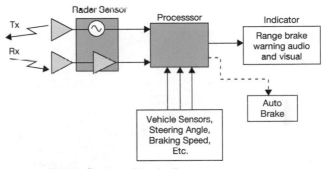

Figure 7.212 System block diagram

Reversing aid The principle of radar as a reversing aid is illustrated in Figure 7.213. This technique is, in effect, a range finding system. The output can

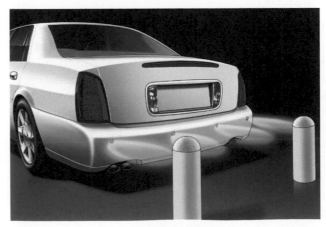

Figure 7.213 Ultrasonic reversing aid

be audio or visual, the latter being perhaps most appropriate, as the driver is likely to be looking backwards. The audible signal is a 'pip pip pip' type sound, the repetition frequency of which increases as the car comes nearer to the obstruction, becoming almost continuous as impact is imminent. The technique is relatively simple as the level of discrimination required is low and the radar only has to operate over short distances. The main problem is to ensure the whole width of the vehicle is protected.

Cruise control is the ideal example of a closed loop control system. The purpose of cruise control is to allow the driver to set the vehicle speed and let the system maintain it automatically.

Speed control The system reacts to the measured speed of the vehicle and adjusts the throttle accordingly. The reaction time is important so that the vehicle's speed does not surge up and down. Other facilities are included, such as allowing the speed to be gradually increased or decreased at the touch of a button. Most systems also remember the last set speed. They will resume to this speed, at the touch of a button.

Figure 7.214 Throttle controller

System description The main switch switches on the cruise control, this in turn is ignition controlled. Most systems do not retain the speed setting in memory when the main switch has been turned off. Operating the 'set' switch programs the memory, but this normally will only work if conditions similar to the following are met:

▶ vehicle speed is greater than 40km/h
▶ vehicle speed is less than 120km/h
▶ change of speed is less than 8km/h/s
▶ automatics must be in 'drive'
▶ brakes or clutch are not being operated
▶ engine speed is stable.

Set and resume Once the system is set, the speed is maintained to within about 3–4 km/h until it is deactivated by pressing the brake or clutch pedal, pressing the resume switch, or turning off the main control switch. The last set speed is retained in memory except when the main switch is turned off. If the cruise control system is required again then either the set button will hold the vehicle at its current speed or the resume button will accelerate the vehicle to the previous set speed. When cruising at a set speed, the driver can press and hold the set button to accelerate the vehicle until the desired speed is reached. If the driver accelerates from the set speed to overtake, for example, then when the throttle is released the vehicle will slow down again.

Control methods A number of methods are used to control the throttle position. Vehicles fitted with drive by wire systems allow the cruise control to operate the same actuator. A motor can be used to control the throttle cable or in many cases, a vacuum-operated diaphragm is used.

In the online diagram, when the speed needs to be increased valve 'x' is opened allowing low pressure from the inlet manifold to one side of the diaphragm. The atmospheric pressure on the other side will move the diaphragm and hence the throttle. To move the other way valve 'x' is closed and valve 'y' is opened allowing atmospheric pressure to enter the chamber. The spring moves the diaphragm back. If both valves are closed then the throttle position is held. Valve 'x' is normally closed and valve 'y' normally open. In the event of electrical failure, cruise will not remain engaged and the manifold vacuum is not disturbed. Valve 'z' provides extra safety – it is controlled by the brake and clutch pedals.

Safety switches The brake switch is very important, as it would be dangerous braking if the cruise control system was still trying to maintain the vehicle speed.

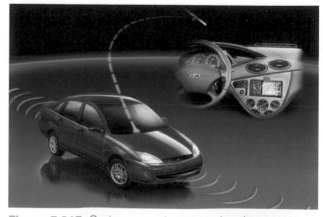

Figure 7.215 Cruise control system development

This switch is normally of superior quality and is fitted in place or as a supplement to the brake light switch activated by the brake pedal. Adjustment of this switch is important. The clutch switch is fitted in a similar manner to the brake switch. It deactivates the cruise system to prevent the engine speed increasing if the clutch is pressed. The automatic gearbox switch will only allow the cruise to be engaged when it is in the 'drive' position. This is to prevent the engine over speeding if the cruise tried to accelerate to a high road speed with the gear selector in position '1' or '2'.

Speed Sensor This will often be the same sensor that is used for the speedometer. If not several types are available, the most common producing a pulsed signal the frequency of which is proportional to the vehicle speed.

Figure 7.216 Road speed sensor in position

Adaptive cruise control Conventional cruise control has now developed to a high degree of quality. It is, however, not always very practical on many roads as the speed of the general traffic is constantly varying and often very heavy. The driver has to take over from the cruise control system on many occasions to speed up or slow down. Adaptive cruise control can automatically adjust the vehicle speed to the current traffic situation. The system has three main aims:

▶ to maintain a speed as set by the driver
▶ to adapt this speed and maintain a safe distance from the vehicles in front
▶ to provide a warning if there is a risk of collision.

System operation The operation of an adaptive cruise system is similar to a conventional system. However, when a signal from the headway sensor detects an obstruction, the vehicle speed is decreased. If the optimum stopping distance cannot be achieved by just backing off the throttle, a warning is given to the driver. The more complex systems can take control of the

vehicle transmission and brakes. It is important to note that adaptive cruise control is designed to relieve the burden on the driver, not take full control of the vehicle!

Figure 7.217 Headway sensor

 Use the media search tools to look for pictures and videos relating to the subject in this section

7.4.3 Mobile multimedia

In Car Multimedia It would be almost unthinkable to not have a quality music system in our vehicles. Shown here is a factory fitted, in car entertainment (ICE) system. It has auxiliary inputs for either a 3.5mm jack plug, USB or Bluetooth. It also has a feature that mirror smartphone apps onto its touch screen, an iPhone in this example.

Figure 7.218 Multimedia options

Linking a phone to the car allows most of the common features to synchronise. Phone address book, music and sat nav for example. This means that some cars do not have features built in because they would only duplicate those that we already have available.

291

Figure 7.219 Satellite navigation on the multimedia screen from an iPhone

Speakers ICE systems usually include at least six speakers: two larger speakers in the rear to produce good low frequency reproduction, two front door speakers for mid-range and two front door tweeters for high frequency notes.

Speaker Construction Speakers are a very important part of a sound system. No matter how good the receiver or CD player is, the sound quality will be reduced if inferior speakers are used. Equally, if the speakers are of a lower power output rating than the set, distortion will result at best and damage to the speakers at worst. Speakers fall into the following categories:

▶ Tweeters – high frequency reproduction
▶ Mid-range – middle range frequency reproduction (treble)
▶ Woofers – low frequency reproduction (bass)
▶ Sub-woofers – very low frequency reproduction.

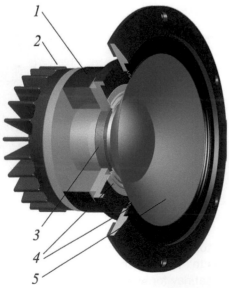

Figure 7.220 Cutaway view of a dynamic midrange speaker. 1. Magnet 2. Cooler (sometimes present) 3. Voicecoil 4.Suspension diaphragm 5. Cone

Radio Data System (RDS) has become standard on many radio sets. It is an extra inaudible digital signal which is sent with FM broadcasts in a similar way to how text is sent with TV signals. RDS provides information so a receiver can appear to act intelligently. The possibilities available when RDS is used are as follows:

▶ The station name can be displayed in place of the frequency
▶ Automatic tuning to the best available signal for the chosen radio station
▶ Traffic information and news broadcasts can be identified, and a setting made so that whatever you are listening to at the time can be interrupted.

Figure 7.221 RDS display on an older set with a CD player

frequency modulation (FM). FM is generally a better source of high-fidelity sound. This is because the quality of AM reception is limited by the narrow bandwidth of the signal. FM does however present problems with reception when mobile. As most vehicles use a rod aerial, which is omni-directional, it will receive signals from all directions. Because of this, reflections from buildings, hills and other vehicles can reach the set all at the same time and distort the signal.

Digital Audio Broadcast (DAB) Digital audio broadcasting is designed to provide high quality digital radio broadcasting for reception by stationary and mobile receivers. It is being designed to operate at any frequency up to 3GHz. The system uses digital techniques to remove redundancy and perceptually irrelevant information from the audio source signal. All transmitted information is then spread in both the frequency and the time domains (multiplexed), so a high-quality signal is obtained in the receiver, even under poor conditions.

Figure 7.222 Digital broadcast radio set

Figure 7.223 Mirrors

Summary Mobile multimedia has changed significantly over the years. On earlier cars it was common for them to not be fitted with a radio. Radio cassette players became standard equipment followed by radio and CD or even multi-CD players. Quality in general of the systems has increased significantly too. Almost all modern cars now have a multimedia unit that does include a radio but allows connection of a smartphone as a minimum to be a music player but now more often full multimedia facilities.

 Use the media search tools to look for pictures and videos relating to the subject in this section

Figure 7.224 Seats

7

7.4.4 Seats, mirrors, roofs and locking ❷

Introduction Electrical movement of seats, mirrors and the sunroof are included in one section as the operation of each system is quite similar. The operation of electric windows and central door locking is also much the same. Fundamentally, all the systems discussed in this section operate using one or several permanent magnet motors, together with a supply reversing circuit.

A typical motor reverse circuit is shown online. When the switch is moved, one of the relays operates and changes the polarity to one side of the motor. If the switch is moved the other way then the polarity of the other side of the motor is changed. When at rest, both sides of the motor are at the same potential. This has the effect of regenerative braking so that when the motor stops it will do so instantly. Further refinements are used to enhance the operation of these systems.

Limit switching, position memory and force limitation are the most common.

Adjustment of the seat is achieved by using a number of motors to allow positioning of different parts of the seat. Movement is possible in the following ways:

▶ front to rear
▶ cushion height rear
▶ cushion height front
▶ backrest tilt
▶ headrest height
▶ lumbar support.

Electrically controlled seat This system uses four positioning motors and one smaller motor to operate a pump, which controls the lumber support bag. Each motor is operated by a simple rocker-type switch that controls two relays as described previously. Nine relays are required for this, two for each motor and one to control the main supply.

Figure 7.225 Seat motors

Seat position When the seat position is set, some vehicles have memories to allow automatic re-positioning if the seat has been moved. This is often combined with electric mirror adjustment. The circuit in Figure 7.226 is constructed to allow position memory. As the seat is moved, a variable resistor, mechanically linked to the motor, is also moved. The resistance value provides feedback to an electronic control unit (ECU). The facility to reposition seats automatically is isolated when the engine is running or the car moving. This is to prevent the seat moving into a dangerous position as the car is being driven. The seats can still be adjusted by operating the switches as normal.

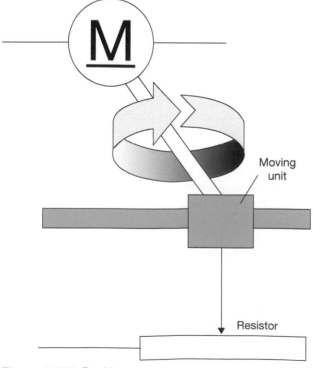

Figure 7.226 Position memory

Electric mirrors Many vehicles have electrical adjustment of mirrors, particularly on the passenger side. The system used is much the same as has been discussed previously in relation to seat movement. Two small motors are used to move the mirror vertically or horizontally. Many mirrors also contain a small heating element on the rear of the glass. This is operated for a few minutes when the ignition is first switched on. The circuit shown in Figure 7.226 includes feedback resistors for position memory.

Figure 7.227 Mirror switch

Figure 7.228 Mirror motors

Electric sunroof operation The operation of an electric sunroof is again based on a motor reverse circuit. However, further components and circuitry are needed to allow the roof to slide, tilt and stop in the closed position. The extra components used are a micro switch and a latching relay. A latching relay works in much the same way as a normal relay except that it locks into position each time it is energised. The mechanism used to achieve this is much like that used in ballpoint pens that have a button on top.

Figure 7.229 Sunroof

Latching relay The micro switch is mechanically positioned such as to operate when the roof is in its closed position. A rocker switch allows the driver to adjust the roof. The switch provides a supply to the motor to run it in the chosen direction. The roof will open or tilt. When the switch is operated, to close the roof, the motor is run in the appropriate direction until the micro switch closes (when the roof is in its closed position). This causes the latching relay to change over, which stops the motor. The control switch has now to be released. If the switch is pressed again, the latching relay will once more change over and the motor will be allowed to run.

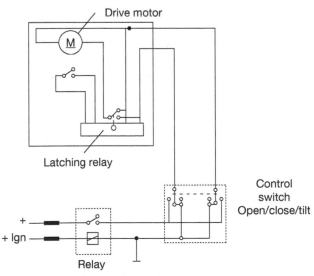

Figure 7.230 Sunroof circuit

Central locking When the key is turned in the driver's door lock, all the other doors on the vehicle should also lock. Motors or solenoids in each door achieve this. If the system can only be operated from the driver's door key, then an actuator is not required in this door. If the system can be operated from

either front door or by remote control then all the doors need an actuator. Vehicles with built-in alarm systems lock all the doors as the alarm is set.

A **door locking circuit** is shown in Figure 7.231. The main control unit contains two change-over relays (reverse circuit), which are actuated by either the door lock switch or, if fitted, the remote key. The motors for each door lock are wired in parallel and all operate at the same time.

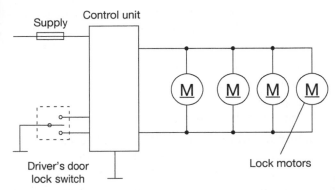

Figure 7.231 Circuit to operate door locks

Actuators Most door actuators are small motors. Via suitable gear reduction, they operate a linear rod in either direction to lock or unlock the doors. A simple motor reverse circuit is used to achieve the required action.

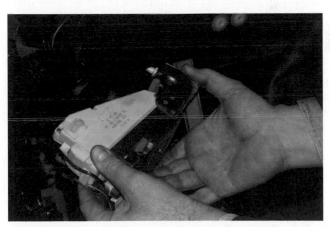

Figure 7.232 Door lock actuator

Remote control Infrared or microwave central door locking is controlled by a small, handheld transmitter and a sensor receiver unit. When the key is operated, by pressing a small switch, a complex code is transmitted. Trillions of different code combinations are used on modern systems. The sensor, in the car, picks up this code and sends it in an electrical form to the main control unit. If the received code is correct, the relays are triggered and the door locks are either locked or unlocked. If an incorrect code is received on three consecutive occasions when attempting to unlock the doors, then some systems will switch off until the door is opened by the key. This technique prevents a scanning-type transmitter unit from being used to open the doors.

Electric window operation The basic form of electric window operation is similar to many of the systems discussed so far in this module – that is, a motor reversing system either by relays or directly by a switch. More sophisticated systems are now popular for reasons of safety as well as improved comfort. The following features are now available from many manufacturers:

▶ one touch up or down
▶ inch up or down
▶ lazy lock
▶ back off or bounce back.

System diagram The complete system consists of an electronic control unit containing the window motor relays, switch packs and a link to the door lock and sunroof circuits. This is represented in Figure 7.233 in the form of a block diagram.

When a window is operated in 'one touch' mode, the window is driven in the chosen direction until the switch position is reversed, the motor stalls or the ECU receives a signal from the door lock circuit. The problem with one touch operation is that if a child, for example, should become trapped in the window there is a serious risk of injury. To prevent this, a bounce back feature is used. An extra commutator is fitted to the motor armature. This produces a signal, via two

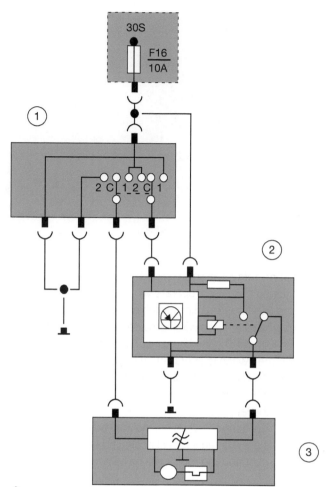

Figure 7.234 Circuit for electric windows

brushes, proportional to the motor speed. Hall sensors are used on some systems. If the rate of change of speed of the motor is detected as being below a certain threshold, the ECU reverses the motor until the window is fully open.

Window position By counting the number of pulses received, the ECU can also determine the window position. This is important, as the window must not reverse when it stalls in the closed position. In order for the ECU to know the window position, it must be initialised. This is often done simply by operating the motor to drive the window first fully open, and then fully closed. If this is not done then the one touch feature and bounce back will not operate.

A 'lazy lock' feature allows the car to be fully secured by one operation of a remote key. This is done by linking the door lock ECU and the window and sunroof ECUs. A signal is supplied and causes all the windows to close in turn, then the sunroof, and finally locks the doors. The alarm will also be set if required. The windows close in turn to prevent the excessive current demand which would occur if they all tried to operate at the same time.

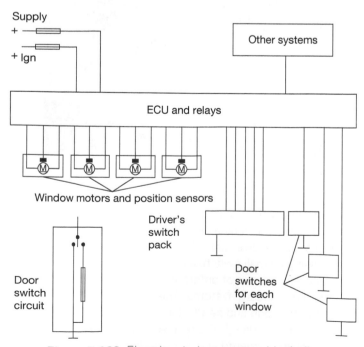

Figure 7.233 Electric window system block diagram

Electric window circuit A circuit for electric windows is shown in Figure 7.234. Note the rear window isolation switch – this is commonly fitted to allow the driver to prevent rear window operation (for child safety, for example).

Motor A typical window lift motor used for cable or arm lift systems is shown in Figure 7.235. Most motors are permanent magnet types and drive through a worm gear – this reduces speed and greatly increases the torque.

Figure 7.236 Alarm switch-type sensor

Figure 7.235 Electric window lift motor

Summary All of the systems examined in this section are based on motor reverse circuits. Door locks, windows, sunroofs, mirrors and seats all operate in this way. Most of the systems are designed to improve driver and passenger comfort.

 Use a library or the web search tools to further examine the subject in this section

7.4.5 Security systems ②

Security Stolen cars and theft from cars account for about a quarter of all reported crime. A huge number of cars are reported missing each year and over 20% are never recovered. Even when returned, many are damaged. Most car thieves are opportunists and even a basic alarm system serves as a deterrent in this case.

Alarm sensors Car and alarm manufacturers are constantly fighting to improve security. Building the alarm system as an integral part of the vehicle electronics has made significant improvements. Even so, retrofit systems can still be very effective. The main types of intruder alarm used are:

▶ switch operated on all entry points
▶ trembler operated
▶ battery voltage sensed
▶ volumetric sensing.

Disabling the vehicle There are four main ways to disable the vehicle:

▶ ignition circuit cut off
▶ fuel system cut off
▶ starter circuit cut off
▶ engine ECU code lock.

A separate switch or transmitter can be used to set an alarm system. Often, they are set automatically when the doors are locked.

Security system The following is an overview of the good alarm systems now available as either a retrofit or factory fit. Most are made for 12V, negative earth vehicles. Some have electronic sirens and give an audible signal when arming and disarming. They are all triggered when the car door opens and will automatically reset after a period of time, often 1 or 2 minutes. The alarms are triggered instantly when an entry point is breached. Most systems are two pieces, with separate control unit and siren – most will have the control unit in the passenger compartment and the siren under the bonnet.

Remote operation Most systems now come with remote 'keys' that use small button-type batteries and have an LED that shows when the signal is being sent; they operate with one vehicle only. Intrusion sensors such as car movement and volumetric sensing can be adjusted for sensitivity. When operating with flashing lights, most systems draw about 5A. Without flashing lights (siren only) the current draw is less than 1A. The sirens produce a sound level of about 95dB, when measured 2m in front of the vehicle.

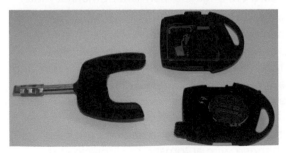

Figure 7.237 Inside a remote key

297

Factory fitted alarms Most factory fitted alarms are combined with the central door locking system. This allows the facility mentioned in a previous section known as 'lazy lock'. Press the button on the remote unit, and as well as setting the alarm, the windows and sunroof close and the doors lock.

Security coded ECUs A security code in the engine electronic control unit is a powerful deterrent. This can only be 'unlocked' to allow the engine to start when it receives a coded signal. Ford, and other manufacturers, use a special ignition key, which is programmed with the required information. Even the correct 'cut' key will not start the engine. Citroën, for example, have used a similar idea but the code has to be entered via a numerical keypad.

 Create a word cloud for one or more of the most important screens or blocks of text in this section

7.4.6 Starting, alarming and locking ❸

Introduction This section is an overview of how starting circuits, alarms, keys and locking systems can be interlinked. It also gives an example of how some keys are programmed and the difference between remote key entry (RKE) and passive key entry (PKE).

Starter circuits The circuit shown in here is from a Ford vehicle fitted with manual or automatic transmission. The inhibitor circuits will only allow the starter to operate when the automatic transmission is in 'park' or 'neutral'. Similarly for the manual version, the starter will only operate if the clutch pedal is depressed. The starter relay coil is supplied with the positive connection by the key switch. The earth path is connected through the appropriate inhibitor switch. To prevent starter operation when the engine is running the power control module (EEC V) controls the final earth path of the relay. A resistor fitted across the relay coil reduces back EMF. The starter in current use is a standard pre-engaged, permanent magnet motor.

Keyless circuit In the diagram shown here, the powertrain control module (PCM) allows the engine to start, only when the passive anti-theft system (PATS) reads a key which transmits a valid code. On a key-free vehicle, the passive key is recognized by the key-free module and if the key is valid the permission to start is issued directly. On vehicles with a manual transmission it is necessary to depress the clutch pedal, on those with automatic transmission the

brake pedal must be pressed. On a key-free system the key-free module switches on the control voltage for the starter relay. The PCM switches the ground in the control circuit of the starter relay which then connects power through to the starter solenoid. As soon as the speed of the engine has reached 750 rpm or the maximum permitted start time of 30 seconds has been exceeded, the PCM switches off the starter relay and therefore the starter motor. If the engine does not turn or turns only slowly, the starting process is aborted by the PCM. These measures help to protect the starter and increase its life.

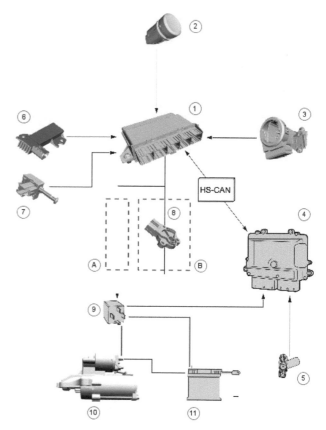

Figure 7.238 Keyless starting system 1 Keyless vehicle module, 2 Start/stop button, 3 Electronic steering lock, 4 Powertrain control module, 5 Crank sensor, 6 Keyless vehicle antenna, 7 Vehicles with manual transmission: Clutch pedal position switch, Vehicles with automatic transmission: Stoplamp switch, 8 The TR sensor, 9 Starter relay, 10 Starter motor, 11 Battery (Source: Ford Motor Company)

Alarm system A modern anti-theft alarm circuit is shown online. As with all complex systems it can be considered as a black box with inputs and outputs. The inputs are signals from key and lock switches as well as monitoring sensors. The outputs are the alarm horn and the hazard lights but also starter inhibitor relays, etc. This system can be operated by remote control or using the key in a door lock. When first activated, the system checks that the doors and tailgate are closed by monitoring the appropriate switches. If all is in order, the anti-theft system is then activated after a 20-second delay. The function indicator LED flashes rapidly during this time and then slowly once the system is fully active. The alarm can be triggered in a number of ways:

▶ opening a door, the tailgate or the bonnet/hood
▶ removal of the radio connector loop

▶ switching on the ignition
▶ movement inside the vehicle.

If the alarm is triggered, the horn operates for 30 seconds and the hazard lights for five minutes. This stops if the remote key or door key is used to unlock the vehicle.

Passive anti-theft system (PATS) This system is a vehicle immobilizer developed by Ford. It is activated directly through the ignition switch by means of an electronic code stored in a special key. Each key has a transponder that stores the code, which does not

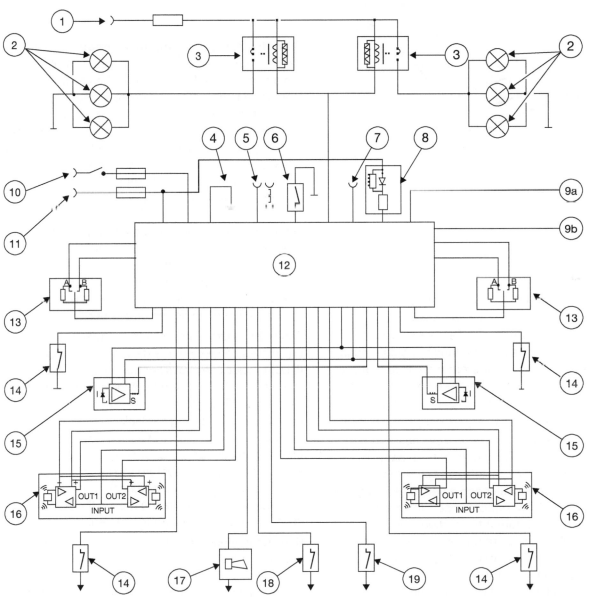

Figure 7.239 Anti-theft alarm system with remote control and interior monitoring: 1 Fused supply, 2 Hazard lights, 3 Hazard lights alarm relay, 4 Earth/Ground, 5 Diagnostic connector, 6 Bonnet/Hood switch, 7 Connector for radio theft protection, 8 Function light, 9 Input signal locked/unlocked, 10 Ignition supply, 11 Battery supply, 12 Anti-theft alarm ECU, 13 Left/Right door key switch, 14 Door switches, 15 Infrared receiver, 16 Ultrasound sensors, 17 Horn, 18 Tailgate switch, 19 Tailgate key switch (Source: Ford Motor Company)

require a battery. The key code is read by the receiver (which is part of the ignition switch) when the key is turned from position 0 to 1 or 2 (usually marked as I or II). If the code matches the one stored in the module, then it allows the engine to start. These systems operate independently of the alarm.

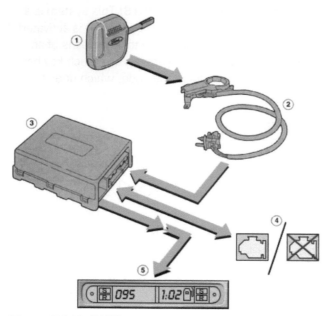

Figure 7.240 PATS components: 1 Key integrated transponder, 2 Transmitter/Receiver, 3 PATS module, 4 Engine start – yes/no, 5 Clock with integrated function indicator (Source: Ford)

Key programming Some keys and/or remotes for later vehicles may need to be reprogrammed if, for example, the battery goes flat or a new key is required. There are several methods of programming remote keys. However, different manufacturers use various methods and it is therefore not possible to cover all of these. An example is given here.

In earlier systems a red key is used as a master – it is exactly the same as the other keys apart from its colour. This key is the only one that can program new keys – if lost the whole system has to be reprogrammed by a dealer and a new master supplied. To program a red key system, insert the master key into the ignition and turn it to position II. When the light on the clock goes out remove the key. The light will come back on if the master key was used. While the light is still on, insert the new key and turn to position II. The light will flash twice and the key is programmed. To program a two key system, both of the original keys are needed. Insert the keys one after the other in the ignition, turn to position II and then remove. After the second key is removed, insert the new un-programmed key, switch to position II and then remove it. The new key is now programmed.

Figure 7.241 PATS components and ECU

Remote keys (programming example) Switch the ignition from I to II quickly four times – this illuminates the alarm warning light. Remove the key from the ignition and point it at the remote sensor (interior mirror usually). Press and hold one of the buttons until the light on the remote flashes. Keep holding the first button, press the other button three times and finally release both buttons. The light on the remote and the warning light will flash five times – the remote key is now programmed. On some vehicles, switching the ignition from I to II quickly four times will activate a chime. Remove the key and press any of the buttons to activate another chime. Finally, replace the key and turn the ignition to position II – the remote key is now programmed. A useful tip is that, on many remotes, changing the batteries within 15 seconds will mean they do not need to be reprogrammed.

Fault diagnosis Many vehicle manufacturers use equipment connected to a diagnostic link connector (DLC) to check several systems, including alarms. This is the same DLC as used for engine management diagnostics. See Chapter 10 for more details. Test equipment is becoming available that can be used by independent repairers. However, it is not often cost-effective to purchase this for specific vehicles. As with others, an alarm system can be treated as a black box system. In other words, checking the inputs and outputs for correct operation means the complexity inside the ECU can be largely ignored.

Remote keyless entry (RKE) Remote keyless entry has been a feature on many cars for a number of years. Remote keys work by transmitting either radio frequency or infrared signals. Door locking is controlled by a small handheld transmitter and a receiver unit, as well as a decoder in the main control unit. This layout varies slightly between different manufacturers. When the remote key is operated (by pressing a small switch), a complex code is transmitted. The number of codes used is in excess of 50 000. The receiver sensor picks up this code and sends it in an electrical form to the main control unit. If the received code is correct, the relays are triggered and the doors are either locked or unlocked. On some systems, if an incorrect code is received on three consecutive occasions when attempting to unlock the doors, the system will switch itself off until the door is opened by the key. This action resets the system and allows the correct code to operate the locks again. This technique prevents a scanning-type transmitter unit from being used to open the doors.

Figure 7.242 Standard key and remote transmitter combined

Passive keyless entry (PKE) Passive keyless entry systems mean the driver doesn't even need to press a button to unlock the vehicle! The electronic key is simply carried in a pocket, on a belt clip or in a bag. The controllers in the doors communicate with the key using radio frequency (RF). This action determines if the correct key is present and, if it is, the doors are unlocked. This communication event is triggered by lifting the door handle, or in some cases the vehicle will even unlock as the key holder approaches it. PKE systems need the same level of security as any other remote locking method. Conventional RKE is

a unidirectional process. In other words, signals are only sent from the key to the receiver. With PKE the communication is two-way. This is because the PKE system carries out an 'identity friend or foe' (IFF) operation for security purposes. The vehicle sends a random challenge to the key, the key encrypts this value and sends it back to the vehicle. The vehicle then performs the same encryption, compares the result with that sent by the key, and unlocks the doors if the values match.

Battery life is a critical issue for PKE. To obtain the required range of operation, 1.5m (5ft), the detection circuit in the key needs to be sensitive enough to detect just a few mV; this consumes significant power. There is also an issue with power consumption for the base station (vehicle) if the doors are designed to unlock as the key approaches. To achieve this, the base station must poll continuously. In other words, it must keep looking for the key. This consumes battery power, which could be an issue if the vehicle was left for a long period. However, this method does have the advantage that the doors will always be locked unless a key is present.

If the method of lifting a handle is used as a trigger, then no power is consumed until needed. The downside of this method is that the user will want to feel the door unlock as the handle is lifted. Texas Instruments has developed a low-frequency RF chip. With a standby current of 5mA and less than 10mV peak-to-peak sensitivity, the chip therefore provides a long battery life. It comes in an industry-standard package small enough to fit into a key fob or credit

Figure 7.243 Nissan Altima keyless stop/start system

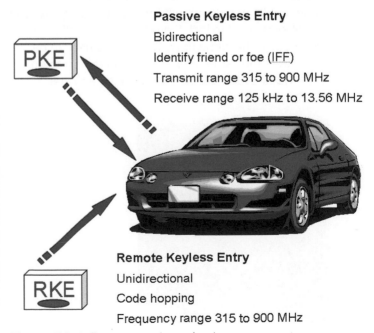

Passive Keyless Entry
Bidirectional
Identify friend or foe (IFF)
Transmit range 315 to 900 MHz
Receive range 125 kHz to 13.56 MHz

Remote Keyless Entry
Unidirectional
Code hopping
Frequency range 315 to 900 MHz

Figure 7.244 Remote and passive key entry systems

card device. This type of system is likely to become very common. Some PKE systems can be set up to recognize multiple keys.

Passive keyless go and exit When the driver enters a car, the key remains in a 'pocket' or at least it will be inside the vehicle. This means, assuming that the key is being recognized, engine starting can be by a simple start button. As the button is pressed the same authentication process that takes place for the door locks starts. The engine can only be started if the key is inside the car, which is a technical challenge for the designers. For example, the key could be in a jacket hanging above the back seat, or it could be in the jacket outside on the roof. After the occupants have left the vehicle, the doors can be locked by pressing a handle or as the driver leaves the vicinity. 'Inside/outside' detection is also necessary for this scenario so the key cannot be locked in the car.

Keypad entry In vehicles equipped with a keypad entry system, the vehicle doors and the boot can be locked and unlocked without using a key. Before unlocking the boot or a passenger door, the driver's door must be unlocked. Usually, if more than five seconds pass between pressing numbers on the keypad, the system will shut down and the code has to be entered again. To unlock the driver's door, the factory code or a personal code is entered. All codes have five numbers. After the fifth number is pressed, the driver's door unlocks. The passenger doors can then be unlocked by pressing the 3/4 button within five seconds of unlocking the driver's door. To unlock the boot, the 5/6 button must also be pressed within five seconds. If this time is exceeded, the code to open the driver's door must be re-entered. The keypad can also be used to lock the doors. To lock all of the car doors at the same time, 7/8 and 9/0 need to be pressed at the same time. It is not necessary to enter the keypad code. This will also arm the anti-theft system if fitted.

Figure 7.245 Keypad entry system

Summary Starting, alarming, keys and locking systems are often part of the same thing because for security reasons they are linked together. Manufacturers are working on new and innovative ways to keep cars secure and make the starting process easy, so I would just add the usual important advice here – always refer to manufacturers' data when working on these systems.

> Look back over the previous section and write out a list of the key bullet points from this section

7.4.7 Safety systems

Passive and active safety Active safety relates to any development designed to actively avoid accidents. It can be considered under four general headings: handling safety, physiological safety, perceptual safety and operational safety. Passive safety relates to developments that protect the occupants of the vehicle in the event of an accident. Airbags are a good example of this.

Airbags and belt tensioners A seat belt, seat belt tensioner and an airbag are at present the most effective restraint system in the event of a serious accident. At speeds in excess of 40km/h, the seat belt alone is no longer adequate. Research after a number of accidents has determined that in 68% of cases an airbag provides a significant improvement. It is suggested that if all cars in the world were fitted with an airbag then the number of fatalities annually would be reduced by well over 50 000. Some airbag safety issues have been apparent in the USA where airbags are larger and more powerful. This is because in many areas the wearing of seatbelts is less frequent.

Figure 7.246 Airbags

Steering wheel unit The method becoming most popular for an airbag system is that of building most of the required components into one unit. This reduces the amount of wiring and connections, thus improving reliability. An important aspect is that some form of system monitoring must be built in, as the operation cannot be tested – it only ever works once.

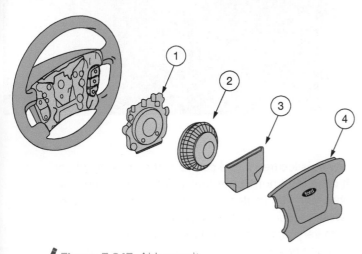

Figure 7.247 Airbag unit

The sequence of events in the case of a frontal impact at about 35km/h, is as follows:

1 The driver is seated in the normal seating position prior to impact. About 15ms after the impact the vehicle is strongly decelerated and the threshold for triggering the airbag is reached. The igniter ignites the fuel tablets in the inflater.

2 After about 30ms, the airbag unfolds and the driver will have moved forwards as the vehicle's crumple zones collapse. The seat belt will have locked or been tensioned depending on the system.

3 At 40ms after impact, the airbag will be fully inflated and the driver's momentum will be absorbed by the airbag.

4 About 120ms after impact, the driver will be moved back into the seat and the airbag will have almost deflated through the side vents allowing driver visibility.

Passenger airbags Passenger airbag deployment events are similar to the previous description. The position is different but the basic principle of operation is the same.

Components and circuit The main components of a basic airbag system are as follows:

▶ driver and passenger airbags
▶ warning light
▶ passenger seat switches
▶ pyrotechnic inflater

Figure 7.248 Side airbag position

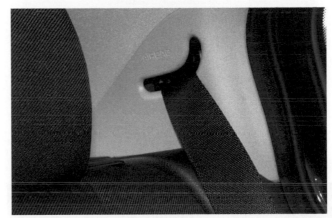

Figure 7.249 Rear airbag position

▶ igniter
▶ crash sensor(s)
▶ electronic control unit.

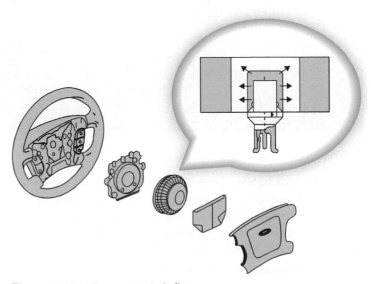

Figure 7.250 Pyrotechnic inflater

303

Figure 7.251 Electronic control unit

Figure 7.253 Warning light

The **airbag** is made of a nylon fabric with a coating on the inside. Prior to inflation, the airbag is folded up under suitable padding, which has specially designed break lines built in. Holes are provided in the side of the airbag to allow rapid deflation after deployment. The driver's air has a volume of about 60 litres and the passenger airbag about 160 litres.

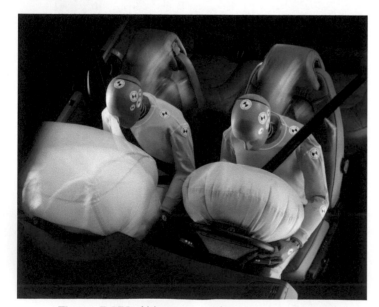

Figure 7.252 Airbags operating

Seat switches A warning light is used as part of the system monitoring circuit. This gives an indication of a potential malfunction and is an important part of the circuit. Seat switches are used, on the passenger side, to prevent deployment when not occupied.

The **pyrotechnic inflater** and the igniter can be considered together. The inflater, in the case of the

driver, is located in the centre of the steering wheel. It contains a number of fuel tablets in a combustion chamber. The igniter consists of charged capacitors, which produce the ignition spark. The fuel tablets burn very rapidly and produce a given quantity of nitrogen gas at a given pressure. This gas is forced into the airbag through a filter and the bag inflates, breaking through the padding in the wheel centre. After deployment, a small amount of sodium hydroxide will be present in the airbag and vehicle interior. Personal protection equipment must be used when removing the old system and cleaning the vehicle interior.

Mechanical-type crash sensor The crash sensor can take a number of forms – these can be described as mechanical or electronic. The mechanical system works by a spring holding a roller in a set position until an impact, above a predetermined limit, provides enough force to overcome the spring and the roller moves, triggering a micro switch. The switch is normally open with

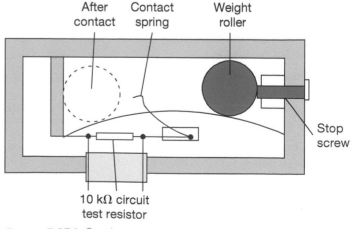

Figure 7.254 Crash sensor

a resistor in parallel to allow the system to be monitored. Two switches similar to this may be used to ensure the airbag is deployed only in the case of sufficient frontal impact. Note the airbag is not deployed in the event of a roll over.

Electronic-type crash sensor The other main type of crash sensor can be described as an accelerometer. This will sense deceleration, which is negative acceleration. A piezoelectric crystal accelerometer, much like an engine knock sensor, is used. Severe change in speed of the vehicle will cause an output from these sensors as the seismic mass moves or the springs bend. Suitable electronic circuits can monitor this and be pre-programmed to react further when a signal beyond a set threshold is reached. The advantage of this technique is that the sensors do not have to be designed for specific vehicles, as the changes can be software based.

Figure 7.256 ECU

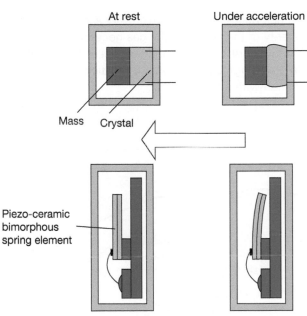

Figure 7.255 Crash sensor

Electronic control unit The final component to be considered is the electronic control unit or diagnostic control unit. When a mechanical-type crash sensor is used, in theory, no electronic unit would be required. A simple circuit could be used to deploy the airbag when the sensor switch operated. However, it is the system monitoring, or diagnostic part of the ECU, which is most important. If a failure is detected in any part of the circuit then the warning light will be operated. Up to five or more faults can be stored in the ECU memory, which can be accessed by blink code or serial fault readers. Conventional testing of the system with a multimeter and jump wires is not

to be recommended as it might cause the airbag to deploy!

Airbag system A block diagram of an airbag circuit is shown online. A digital-based system, using electronic sensors, has about 10ms at a vehicle speed of 50km/h to decide if the restraint systems should be activated. In this time, about 10 000 computing operations are necessary. Data for the development of these algorithms is based on computer simulations but digital systems can also remember the events during a crash allowing real data to be collected.

Seat belt tensioners Taking the 'slack' out of a seat belt in the event of an impact is a good contribution to vehicle passenger safety. The decision to take this action is the same as for the airbag. The two main types are:

▶ spring tension
▶ pyrotechnic.

When the explosive charge is fired, the cable pulls a lever on the seat belt reel, which in turn tightens the belt. The unit must be replaced once deployed. This feature is sometimes described as 'anti-submarining'.

Remember! There is more support on the website that includes additional images and interactive features: www.tomdenton.org

305

Figure 7.257 Belt tensioners

Side airbags Airbags working on the same techniques to those described previously are being used to protect against side impacts. In some cases, bags are stowed in the door pillars or the edge of the roof.

Intelligent Airbag Sensing System Bosch has developed an 'Intelligent Airbag Sensing System', which can determine the right reaction for a specific accident situation. The system can control a one or two-stage airbag inflation process via a two-stage gas generator. Acting on signals from vehicle acceleration

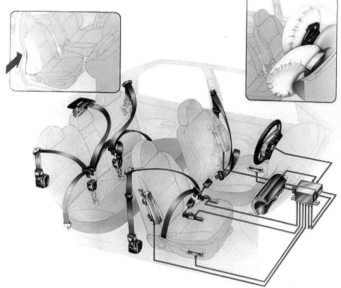

Figure 7.258 Ford seat belt and airbag system

and belt buckle sensors, which vary according to the severity of the accident, the gas generator receives different control pulses, firing off one airbag stage (de-powering), both stages (full inflation), or staged inflation with a time interval.

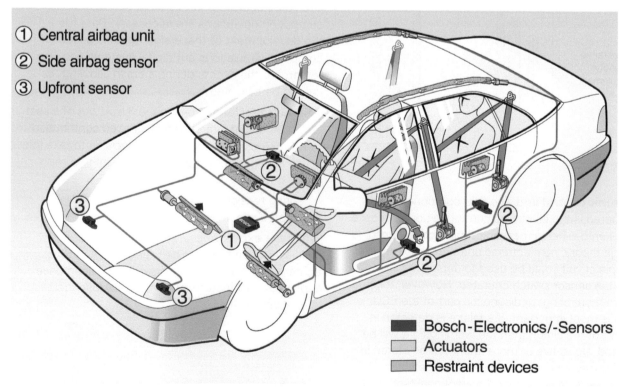

① Central airbag unit
② Side airbag sensor
③ Upfront sensor

■ Bosch-Electronics/-Sensors
☐ Actuators
☐ Restraint devices

Figure 7.259 Bosch safety system

Multistage inflation Future developments will lead to capabilities for multistage inflation, or a controllable sequence of inflation following a pattern determined by the type of accident and the position of the vehicle occupants. The introduction of an automotive occupancy-sensing (AOS) unit that uses ultrasonic and infrared sensors will provide further enhancements. This additional module will detect seat and child occupancy, and be capable of assessing whether a passenger is in a particular position, such as feet on the dashboard!

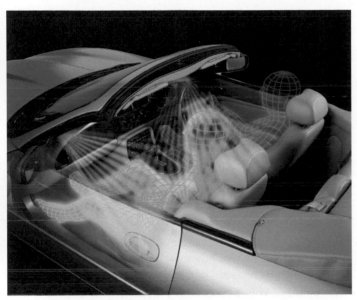

Figure 7.260 Jaguar adaptive restraints

Pre-crash sensor Bosch hopes that the latest radar technology will assist the design of a pre-crash sensor capable of detecting an estimated impact speed prior to collision and activating individual restraint systems, such as seatbelt pre-tensioners. Alternatively, if necessary, all available restraint systems can be operated.

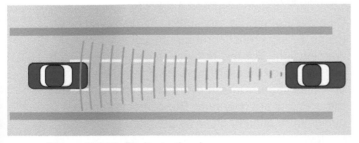

Figure 7.261 Radar technology

Create a word cloud for one or more of the most important screens or blocks of text in this section

7.5 Monitoring and instrumentation

7.5.1 Sensors ❶ ❷

Introduction Sensors are used on vehicles for many purposes. For example, the coolant temperature thermistor is used to provide data to the engine management system as well as for the driver. The information is provided to the driver by a display or gauge.

Sensors The following sections list some of the things on a car that are sensed or measured, together with typical sensors. The sensors convert what is being measured into an electrical signal. This signal can then be used to operate a display, such as a gauge or warning light, on the instrument panel.

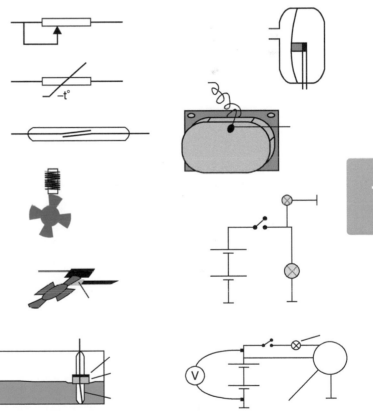

Figure 7.262 Sensors used for instrumentation

Fuel level Fuel level is measured by using a variable resistor that is moved by a float. The position of the float is determined by how much fuel is in the tank. The resistance value is varied by a contact sliding over a resistor.

Temperature The most common temperature measurement is that of the engine coolant. However, outside air, cabin, air intake and many other temperatures are also measured. A thermistor is used for most applications. A thermistor is a special material that changes its resistance with temperature. Most types are described as negative temperature coefficient (NTC) – this means that as temperature increases, their resistance decreases.

Figure 7.265 Temperature sensor

A reed switch consists of two small strips of steel. When these become magnetised, they join and make a circuit. Bulb failure circuits often use a reed relay to monitor the circuit. In the online circuit, the contacts of the reed switch will only close when electricity is flowing to the bulb being monitored.

Road speed is often sensed using an inductive pulse generator. This sensor produces an AC output with a frequency which is proportional to speed. It is like a small generator that is driven by a gear on the gearbox output shaft. This type of sensor is also used to sense engine speed from the flywheel or crankshaft.

Engine speed can be sensed in a number of ways. The Hall effect sensor however, is a very popular choice, as it is accurate and produces a square wave output with a frequency proportional to engine speed. The Hall IC produces a voltage when it is in a magnetic field. The rotating plate shown here alternately prevents and allows the magnetism to reach the IC.

Fluid levels, such as washer fluid or radiator coolant, are often measured or sensed using a float and reed switch assembly. The float has a magnet attached that causes the contacts to join when it is in close proximity. The float moves up or down depending on the fluid level.

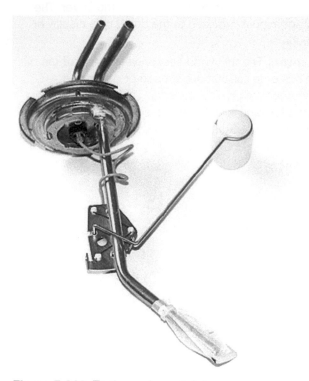

Figure 7.263 Tank sender unit full . . .

Oil pressure may be measured and displayed on a gauge or, as is most common, by using a simple warning light. For this purpose, a diaphragm switch is used. As oil pressure increases, it is made to act on a diaphragm. Once it overcomes spring pressure, the contacts are operated. The contacts can be designed to open or close as pressure reaches a set level.

Brake pad wear is sensed by using a simple embedded contact wire. When the friction material wears down the embedded contact makes contact with the disc to complete a circuit. Some systems use a loop of wire that is broken when the pad wears out.

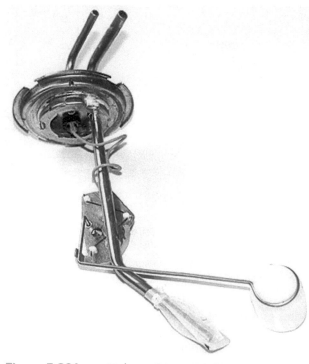

Figure 7.264 . . . and empty

Lights in operation can be monitored by a bulb and simple circuit. However, note that this circuit will only indicate that the switch is on, it will not confirm that the circuit is working. A good example of this is the main beam warning light.

Battery charge rate can be sensed by a simple bulb circuit. The charge warning light is caused to go out when the alternator produces an output on one side of the bulb, which is the same as that supplied by the battery to the other side. If an equal voltage is supplied to both sides, the voltage across the warning light will be zero, and hence it will not be lit!

A wide range of sensors is used to operate instrument displays. Sensors convert what is being measured into an electrical signal. This may be by a simple on/off operation, a changing voltage output or a change in resistance.

 Create a mind map to illustrate the features of a key component or system

7.5.2 Gauges ❶❷

Introduction By definition, an instrumentation system can be said to convert a 'variable' into a readable or usable display. For example, a fuel level instrument system will display a representation of the fuel in the tank using an analog gauge.

Figure 7.266 Instrument panel

Instrumentation is not always associated with a gauge or a read-out-type display. In many cases, a system can be used just to operate a warning light. However, it must still work to certain standards. For example, if a low outside temperature warning light did not illuminate at the correct time, a dangerous situation could develop.

Thermal-type gauges, which are ideal for fuel and engine temperature indication, have been in use for many years. This will continue because of their simple design and inherent 'thermal' damping. The gauge works by utilising the heating effect of electricity and the widely adopted benefit of the bimetal strip.

Figure 7.267 Fuel gauge display

As a current flows through a simple heating coil wound on a bimetal strip, heat causes the strip to bend. The bimetal strip is connected to a pointer on a suitable scale. The amount of bend is proportional to the heat, which in turn is proportional to the current flowing. Providing the sensor can vary its resistance in proportion to the fuel level or temperature, the gauge will indicate a suitable representation.

Damping The inherent damping is due to the slow thermal effect on the bimetal strip. This causes the needle to move very slowly to its final position. This is a particular advantage for displaying fuel level, as the variable resistor in the tank will move, as the fuel moves, due to vehicle movement. If the gauge reacted quickly, it would be constantly moving. The

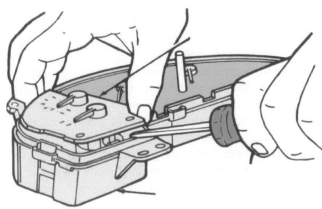

Figure 7.268 Thermal gauge being removed for repair

movement of the fuel, however, is in effect averaged out and an accurate display can be obtained.

Variable resistance Thermal-type gauges are used with a variable resistor. This is either a float in the fuel tank or a thermistor in the engine water jacket. The sender resistance is usually at a maximum when the tank is empty or the engine is cold.

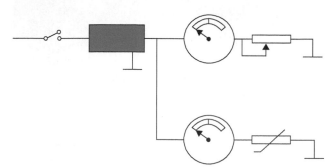

Figure 7.269 Fuel and temperature gauge circuit

A constant voltage supply is required to prevent changes in the system voltage affecting the reading. This is because if system voltage increased, the current flowing would increase, and the gauges would read higher. Most voltage stabilisers are simple zener diode circuits as shown in Figure 7.270.

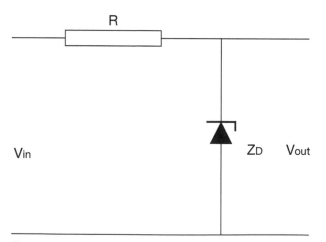

Figure 7.270 Voltage stabiliser

Moving iron gauges The moving iron gauge was in use earlier than the thermal type but is again gaining popularity on some cars. Two small electromagnets are used, which act upon a small soft iron armature. This armature is then connected to a pointer. The armature will position itself in between the cores of the electromagnets depending on the magnetic strength of each. The ratio of magnetism in each core is changed as the variable resistance sender changes.

Figure 7.271 Gauge on an older vehicle

This type of gauge reacts very quickly and is prone to swing about with movement of the vehicle. Some form of external damping can be used to improve this problem. Resistor R1 is used to balance out the resistance of the tank sender. A good way to visualise the operation of the circuit is to note that when the tank is half full, the resistance of the sender will be the same as the resistance of R1. This makes the circuit balanced and the gauge will read half full. The sender resistance is at a maximum when the tank is full.

Air cored gauges work on the same principle as a compass needle lining up with a magnetic field. The needle of the display is attached to a very small permanent magnet. Three or more coils of wire are used and each produces a magnetic field. The magnet, and therefore the needle, will line up with the resultant of the three fields. The current, therefore, flowing through each coil is the key to moving the needle position.

Figure 7.272 Gauge details

The online animation shows the principle of the air-cored gauge, together with the circuit for use as a temperature indicator. The resistor on the left is used to limit maximum current and the calibration resistor is used for calibration! The thermistor is the temperature sender. As the thermistor resistance increases, the current in all three coils will change. Current through C, which is one coil but wound in two parts, will be increased but the current in coils A and B will decrease. As the resistance decreases, the opposite will occur, moving the needle from cold to hot.

The air-cored gauge has a number of advantages. It has almost instant response and, as the needle is held in a magnetic field, it will not move as the vehicle changes position. The gauge can be arranged to continue to register the last position even when switched off. If a small 'pull off' magnet is used, it will return to its zero position. A change in system voltage would affect the current flowing in all three coils. Variations are therefore cancelled out so a voltage stabiliser is not needed. The operation is similar to the moving iron gauge.

Other gauges A variation of any of the above types of gauge can be used to display other required outputs, such as voltage or oil pressure. Gauges to display road or engine speed, however, need to react very quickly to changes. Many systems now use a stepper motor or other type of electrical gauge for this purpose.

Figure 7.273 Stepper motor tachometer

Cable speedometer Some cars still use conventional cable driven speedometers. The head units usually work by either friction or magnetism. The frictional or magnetic 'drag' increases as speed increases and this is used to move a needle. The flexible cable is driven from the gearbox output. It has square ends to transfer the rotation.

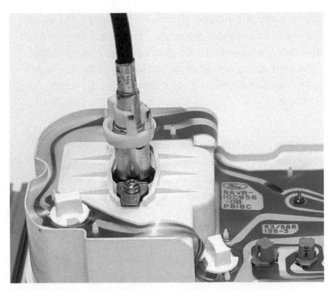

Figure 7.274 Speedometer cable and head

Shown here is a block diagram of a speedometer, which uses an ammeter as the gauge. This system uses a quenched oscillator sensor. This sensor produces a constant signal, even at very low speed. The frequency of the signal is proportional to road speed. The sensor is driven from the gearbox or a final drive output. The gauge will read an average of the pulses from the sensor. This average value is dependent on the frequency of the input signal, which in turn is dependent on vehicle speed. The odometer is driven by a stepper motor, which is driven by the output of a divider and an amplifier.

Rev counter or tachometer The system for driving most rev counters is similar to the electronic speedometer system. Pulses from the ignition primary circuit are often used to drive the gauge. The rev counter needle response is damped to give a steady reading.

Figure 7.275 Tachometer display

Summary A number of different gauges are used for instrumentation displays. The most common for fuel and temperature display are thermal, moving iron and air-cored. Speedometers and tachometers use stepper motors, electrical gauges or mechanical systems.

 Use the media search tools to look for pictures and videos relating to the subject in this section

7.5.3 Instrument displays ❶❷

Visual displays The function of any visual display is to communicate information. Most displays used in a vehicle must provide instant data but the accuracy is not necessarily important. Analog displays provide almost instant feedback from one short glance. For example, if the needle of the temperature gauge is positioned in the middle, the driver can assume that the engine temperature is within suitable limits. A digital read-out of temperature such as 98°C would not be as easy to interpret. This is why when digital processing and display techniques are used, the read-out may still be in analog form.

Figure 7.276 Analog display

Display types Numerical and other forms of display are used for many applications. Some of these are:

▶ vehicle map
▶ trip computer
▶ clock
▶ radio panel
▶ route finding screen
▶ general instruments.

These displays can be created in a number of ways – the most common use light emitting diodes, vacuum fluorescence or liquid crystals. The following pages examine each of these in detail.

Figure 7.277 Digital display

Light emitting diodes If the junction of a diode is manufactured from a special material, light will be emitted from it when a current is made to pass in the forward-biased direction. This is described as a light-emitting diode (LED). With slight changes in the manufacturing process, it will produce red, yellow or green light. LEDs are used extensively as indicators on electronic equipment and in digital displays. They last for a very long time and draw only a small current.

LED arrangements LED displays are tending to be replaced for automobile use by the liquid crystal type, which can be backlit to make it easier to read in the daylight. However, LEDs are still popular for many applications. The display will normally consist of a number of LEDs arranged into a suitable pattern for the required output. This can range from the standard seven-segment display to show numbers, to a custom designed display as shown in Figure 7.277.

Liquid crystals are substances that do not melt directly from the solid to the liquid phase. They first pass through a 'para-crystalline' stage in which the molecules are partially ordered. In this stage, a liquid crystal is a cloudy or translucent fluid. However, it still has some of the optical properties of a solid crystal.

Detailed display capability Mechanical stress, electric and magnetic fields, pressure, and temperature can alter the molecular structure of liquid crystals. A liquid crystal also scatters light that shines on it. Because of these properties, liquid crystals can be used to display letters and numbers on automobile instrument and other displays. LCDs are now used for portable computers and even television screens.

Figure 7.278 Liquid crystal display

An **LCD display** is achieved by only allowing polarised light to enter the liquid crystal, which as it passes through the crystal, is rotated by 90°. The light then passes through a second polariser, which is set at 90° to the first. A mirror at the back of the arrangement reflects the light so that it returns through the polariser, the crystal and the front polariser again. The net result is that light is simply reflected, but only when the liquid crystal is in this one particular state.

Display operation When an AC voltage of about 10V at 50Hz is applied to the crystal, it becomes disorganised. The light passing through it is therefore, no longer twisted by 90°. This means that the light polarised by the first polariser will not pass through the second and will therefore not be reflected. This shows as a dark area on the display. These areas are constructed into suitable segments to provide whatever type of display is required. The size of each individual area can be very small, for example to form one pixel of a computer screen.

Figure 7.279 Liquid crystal radio/CD display

LCDs are very low power but do require a source of light to operate. To be able to read the display in the dark some form of lighting is required. Instead of using a reflecting mirror at the back of the display, a source of light known as back lighting can be used. A condition known as DC electroluminescence is an ideal phenomenon. This uses a compound which is placed between two electrodes, in much the same way as the liquid crystal, but it emits light when a voltage is applied.

A **vacuum fluorescent display** (VFD) works in much the same way as a television tube and screen. It is becoming increasingly popular for vehicle use because it produces a bright light, which is adjustable. It produces a wider choice of colours than LED or LCD displays. The VFD system consists of three main components – the filaments, the grid and the screen. Segments are placed appropriately for the intended use of the display. The filament forms the cathode and the segments the anode of the main circuit. The control grid is used to control brightness.

Figure 7.280 VFD in close up

Display operation When a current is passed through the tungsten filaments, they become red hot and emit electrons. The whole unit is made to contain a vacuum so that the electrons are not affected by any outside influence. The segments are coated with a fluorescent substance and connected to a positive polarity control wire. When electrons strike the segments, they fluoresce, emitting light. If the potential of the grid is changed, the number of electrons striking the segments changes, thus affecting the brightness.

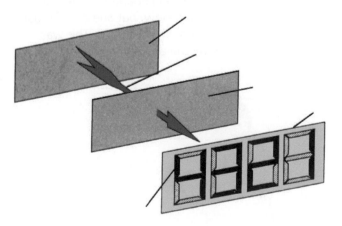

Figure 7.281 Principle of a VFD display

VFD control The circuit shown here is used to control a VFD. Note how the voltage of the segments, when activated, is above that of the grid. The electronic

313

display controller connects one or more of the appropriate segments to a supply to produce the desired output. The glass front of the display can be coloured to improve the readability and aesthetic value.

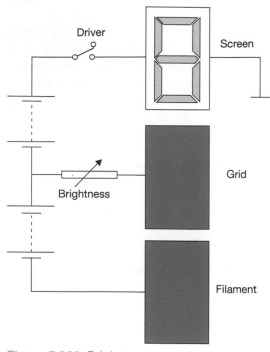

Figure 7.282 Brightness control

Head up displays One of the main problems with any automobile instrument or monitoring display is that the driver has to look away from the road to see the information. Because of this, if the driver does not look at the display, an important warning such as low oil pressure could be missed. Many techniques can be used, such as warning beepers or placing the instruments almost in view, but one of the most innovative is the head up display (HUD). This was originally developed by the aircraft industry for fighter pilots.

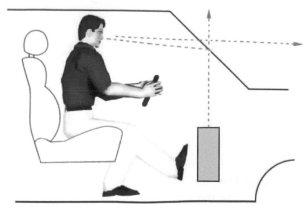

Figure 7.283 HUD in use

Information from a display device is directed onto a partially reflecting windscreen. A great deal of data could be presented in this way. However, under normal circumstances, the driver would be able to see the road through the screen. The brightness of the display has to be adjusted to suit ambient lighting conditions. The main problem, however, is what information to provide in this way. The speedometer could form part of a lower level display and a low oil pressure could cause a flash right in front of the driver.

Display techniques Most of the discussion in previous sections has been related to the activation of an individual display device. The techniques used for, and the layout of, dashboard or display panels are very important. Largely this comes back to readability. When so many techniques are available to the designer, it is tempting to use the most technologically advanced. However, this is not always the best!

Summary The layout and the way that instruments are combined is an area in which much research has been carried out. This relates to the time it takes the driver to gain the information required when looking away from the road to glance at the instrument pack. Figure 7.284 shows the instrument panel and other read-out displays on the 'S-type' Jaguar. Note how compact it is, so that the information can be absorbed almost without the driver having to scan to each read-out in turn. The aesthetic looks of the dashboard are an important selling point for a vehicle.

Figure 7.284 Jaguar dashboard and instruments

 Use the media search tools to look for pictures and videos relating to the subject in this section

7.5.4 Vehicle condition monitoring ❷

Introduction Vehicle condition monitoring (VCM) is a form of instrumentation. It has now become difficult to separate it from the normal instrumentation system discussed in the first section. The complete VCM system can include driver information relating to a wide range of systems. Some of these are discussed on the following pages. Figure 7.285 is a typical display unit with a vehicle map.

Figure 7.286 High temperature warning light

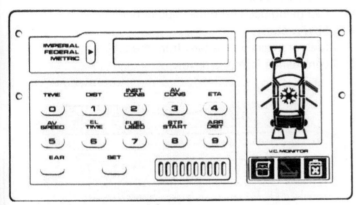

Figure 7.285 Display unit and vehicle map

Monitored systems include:

▶ bulb operation, by monitoring the current drawn by the lights
▶ doors, bonnet or boot position, by signals from switches
▶ brake pad wear by contact wires embedded in the friction material.

Systems which can be monitored for fluid level include fuel, brake fluid, coolant and the screen washers. Many of these systems work by the action of a magnet on a reed switch. However, some manufacturers use different techniques so check their data.

The **temperatures** monitored are the engine coolant and outside air. When a higher than normal engine coolant temperature is sensed, a warning light is illuminated. When a low outside temperature is sensed, a snowflake-shaped warning light is lit! The sensors used are thermistors.

Figure 7.287 is of a circuit which can be used to operate a bulb failure warning light. The principle is that the reed relay is only operated when the bulb being monitored draws current. The reed switch and coil may be described as a current relay.

Oil level can be monitored by measuring the resistance of a heated wire on the end of a dipstick.

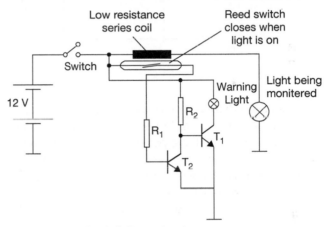

Figure 7.287 Bulb failure circuit

A small current is passed through the wire to heat it. How much of the wire is covered by oil will determine its temperature and therefore its resistance.

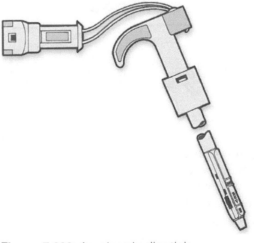

Figure 7.288 An electric dipstick

315

Dual resistance system Many of the circuits monitored use a dual resistance system so that the circuit itself is also checked. Shown here is the equivalent circuit for this technique. The circuit will produce one of three possible outputs – high resistance, low resistance or an out of range reading. The high or low resistance readings are used to indicate, say, correct fluid level and low fluid level. A figure outside these limits would indicate a circuit fault of either a short or an open circuit.

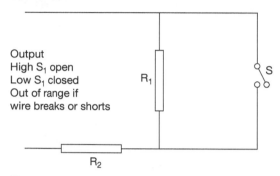

Output
High S_1 open
Low S_1 closed
Out of range if
wire breaks or shorts

Figure 7.289 High/low resistance circuit

Vehicle map display The display for a vehicle map is often just a collection of LEDs or a back lit LCD. These are arranged into suitable patterns and shapes such as to represent the circuit or system being monitored. A door open will illuminate a symbol which looks like the door of the vehicle map is open. A low outside temperature or ice warning is often a large snowflake.

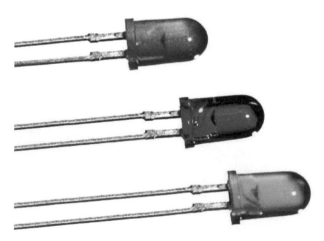

Figure 7.290 LEDs

Trip computer outputs The trip computer used on many top-range vehicles is a popular accessory. The display of a typical trip computer is shown here. The basic functions available on most systems are:

▶ time and date
▶ elapsed time or a stop watch

▶ estimated time of arrival
▶ average fuel consumption
▶ range on remaining fuel
▶ trip distance.

Trip computer inputs The information can usually be displayed in Imperial, US or metric units. In order to calculate the different outputs, the following inputs to the system are required:

▶ clock signal from an oscillator
▶ vehicle speed signal from a speed sensor or the instrumentation ECU
▶ fuel being used from the injector open time or a flow meter
▶ fuel remaining in the tank from the tank sender unit
▶ commands from the driver.

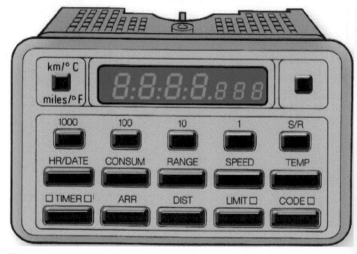

Figure 7.291 Trip keypad

Other systems use the same inputs and many of these systems 'communicate' with each other. This can make the overall wiring very bulky and complicated. This type of interaction and commonality between systems has been the main reasons for the development of multiplexed wiring techniques.

Summary Vehicle condition monitoring systems and trip computers are becoming common on lower range cars. Monitoring a system, so that early warning of a problem can be given, is a significant contribution to safety. It is also important, therefore, to ensure that the monitoring systems are working correctly.

 Use a library or the web search tools to further examine the subject in this section

7.5.5 Digital instrumentation system ❸

Introduction The components shown in Figure 7.292 form part of a typical digital instrumentation system. All signal conditioning and logic functions are carried out in the ECU. This will often form part of the dashboard assembly.

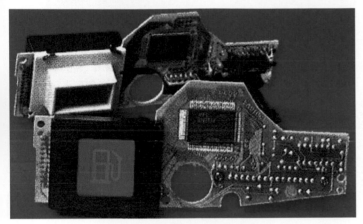

Figure 7.292 Instrument electronics

Standard sensors provide information to the ECU via the multiplexer (MUX). The ECU contains a memory, which allows it to be programmed to a specific vehicle. The gauges used may be digital units or analog, as described in the previous section. The gauges are driven in turn via the de-multiplexer (DeMUX).

Other system functions Digital systems allow extra functions to be incorporated. For example, a low fuel warning light can be made to illuminate at a particular fuel tank sender unit resistance reading. A high engine temperature warning light can be made to operate at a set resistance of the thermistor.

Figure 7.293 Low fuel warning light

Readability To prevent the temperature gauge fluctuating as the cooling system thermostat operates, the gauge can be made to read only at, say, five set figures. For example, even if the input resistance varies from 240 to 200 ohms as the thermostat operates, the ECU will output just one reading corresponding to 'normal' on the gauge. If the resistance is much higher or lower, the gauge will be made to read at one of the five higher or lower positions. This technique gives a low-resolution reading but it is quick and easy for the driver to read.

Oil pressure or other warning lights can be made to flash. This is more likely to catch the driver's attention when a fault is serious. In general, only the oil pressure and high temperature warning lights would be important enough to flash. The lights could be distracting but, in these cases, the consequence of not stopping the engine could be serious.

Service or inspection interval warning lights can be used. These warning lights are progressively lit over a period of time, to show when a service is due. This time period is known as the service interval. However, the service interval is reduced if the engine often experiences high speeds or high temperatures. Oil condition sensors are also now used to help determine service intervals.

Trip Computer Signals from the instrument ECU are supplied to the trip computer to provide information on fuel quantity and road speed. Alternatively, a signal from the trip computer may provide information on fuel usage to the instrument ECU!

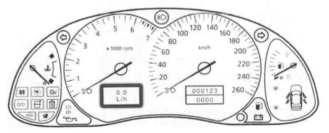

Figure 7.294 Trip display as part of the instrument panel

Summary Many vehicles now use digital Instrument systems. However, the final displays may still be made up of conventional gauges. The main advantage of digital systems is that extra functions and facilities can be included. A good example would be the ability to switch between miles per hour and kilometres per hour readings.

 Use the media search tools to look for pictures and videos relating to the subject in this section

7.5.6 Global positioning system ❸

History From 1974 to 1979, a trial using six satellites allowed navigation in North America for just four hours per day. This trial was extended worldwide by using 11 satellites until 1982 at which time it was decided that the system would be extended to 24 satellites, in six orbits, with four operating in each orbit. There are now some 31 satellites in use. They are set at a height of about 21 000km (13 000 miles), inclined 55° to the equator and take approximately 12 hours to orbit the Earth. The orbits are designed so that there are always six satellites in view from most places on the Earth.

Figure 7.295 Display in a Jaguar

Accuracy The system was developed by the American Department of Defence. Use of an encrypted code allows a ground location to be positioned to within a few centimetres. The signal employed for civilian use is artificially reduced in quality so that positioning accuracy is in the region of 50m. Some systems, however, now improve on this and can work down to about 15m.

Figure 7.296 If you look really hard you can see the satellites . . .

Triangulation GPS satellites send out synchronized information 50 times a second. Orbit position, time and identification signals are transmitted. A modern GPS receiver will typically track all of the available satellites, but only a selection of them will be used to calculate position. The times taken for the signals to reach the vehicle are calculated and from this information the computer can determine the distance from each satellite.

The current vehicle position can then be worked out using three coordinates. Imagine the three satellites forming a triangle (represented in Figures 7.334, 7.335 and 7.336 as A, B, C), the position of a vehicle within that triangle can be determined if the distance from each fixed point (satellite) is known. This is called triangulation.

Satellites The GPS receiver gets a signal from each GPS satellite. The satellites transmit the exact time that the signals are sent. By subtracting the time the signal was transmitted from the time it was received, the GPS can tell how far it is from each satellite. The GPS receiver also knows the exact position in the sky of the satellites at the moment they sent their signals.

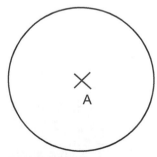

Figure 7.297 At a known distance from a fixed point 'A' you could be anywhere on a circle

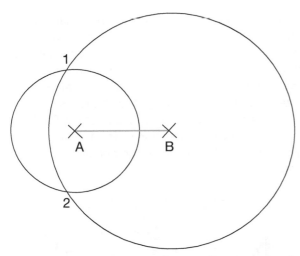

Figure 7.298 At a known distance from two fixed points 'A and B' you must be at position 1 or 2

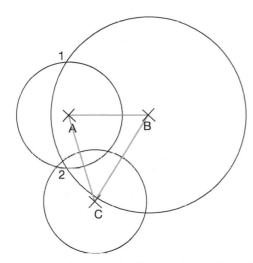

Figure 7.299 At a known distance from three fixed points 'A, B and C' then you must, in this case, be at position 2

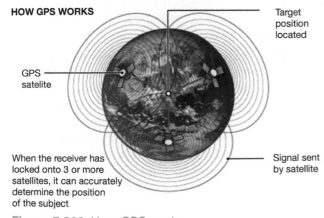

HOW GPS WORKS

Target position located

GPS satelite

When the receiver has locked onto 3 or more satellites, it can accurately determine the position of the subject

Signal sent by satellite

Figure 7.300 How GPS works

So, given the travel time of the GPS signals from three satellites and their exact position in the sky, the GPS receiver can determine their positions in three dimensions – east/west, north/south and altitude.

Calculations To calculate the time the GPS signals took to arrive, the GPS receiver needs to know the time very accurately. The GPS satellites have atomic clocks that keep very precise time, but it is not feasible to equip a GPS receiver with such a device. However, if the GPS receiver uses the signal from a fourth satellite it can solve an equation that lets it determine the exact time, without needing an atomic clock.

Figure 7.301 Four satellites used to determine vehicle position (Source: Ford)

Number of satellites If the GPS receiver is only able to get signals from three satellites, position can still be calculated, but less accurately. If only three satellites are available, the GPS receiver can get an approximate position by making the assumption that you are at mean sea level. If you really are at sea level, the position will be reasonably accurate, but if you are driving in the mountains the two dimensional fix could be several hundreds of metres out.

Additional inputs As well as the satellite data, some in car systems also process the following input signals:

▶ wheel speed sensors
▶ reverse light switch
▶ magnetic field sensor
▶ turn angle sensor.

It should also be noted that many GPS units now work without any additional inputs and accurate positioning can even be achieved by GPS receivers built into smartphones.

319

Wheel speed sensors The wheel speed sensors provide information on distance covered. The sensors on the non-driven wheels are used because the driven wheels slip when accelerating. ABS wheel speed sensors have become smaller and more efficient. Recent models not only measure the speed and direction of rotation but can be integrated into the wheel bearing. On some systems turn angle is calculated by comparing left- and right-hand signals. This is not necessary when a turn angle sensor is used.

Figure 7.302 ABS wheel speed sensors (Source: Photo Bosch)

The **reverse light switch** is used on some systems. This is because the signals from the wheel speed sensors do not indicate if the vehicle is travelling forwards or in reverse.

Figure 7.303 Reverse light switch

The **magnetic field sensor** determines direction of travel in relation to the Earth's magnetic field. It also senses the changes in direction when driving round a corner or a bend.

The two crossed measuring coils sense changes in the Earth's magnetic field because it has a different effect in each of them. The direction of the Earth's field can be calculated from the polarity and voltage produced by these two coils. The smaller excitation coil produces a signal that causes the ferrite core to oscillate. The direction of the Earth's magnetic field causes the signals from the measuring coils to change depending on the direction of the vehicle.

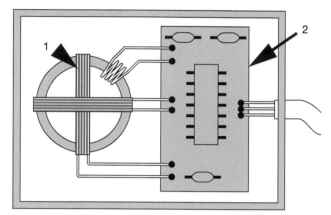

Figure 7.304 Field sensor: 1 crossed coils, 2 control circuit (Source: Ford)

All GPS units must have some sort of compass

The **turn angle sensor** allows the navigation computer to follow a digital map, in conjunction with other sensor signals, because it provides accurate information about the turning of the vehicle around its vertical axis. It is mounted in the main unit and supersedes the magnetic compass. The sensor is like a tiny tuning fork that is made to vibrate, in the kilohertz range, by the two lower piezoelectric elements. The upper elements sense the acceleration when the vehicle changes direction. This is because the twisting of the piezo elements causes an electrical charge. This signal is processed, converted into a voltage that corresponds to vehicle turning movement, and sent on to the main computer.

System use To use most satellite navigation systems, the destination address is entered using a joystick control, cursor keys or something similar. The systems usually 'predict' the possible destination as letters are entered so it is not usually necessary to enter the complete address. Once the destination is set, the unit will calculate the journey. Options may be given for the shortest or quickest routes at this stage. Driving instructions, relating to the route to be followed, are given visually on the display and audibly through speakers.

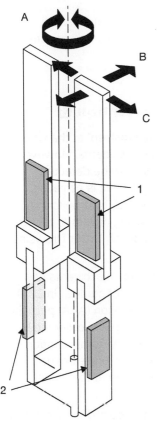

Figure 7.305 Turn angle sensor: 1 Piezoelectric element (picks up acceleration in the twisting direction B around the vertical axis of the vehicle A), 2 Piezoelectric element (causes vibration in direction C) (Source: Ford)

Figure 7.306 Directions

Dead reckoning Even though the satellite information only provides a positional accuracy of about 50m, using dead reckoning intelligent software can still get the driver to their destination with an accuracy of about 5m in some cases. Dead reckoning means that the vehicle position is determined from speed sensor and turn angle signals.

The computer can update the vehicle position from the GPS data by using the possible positions on the stored digital map. This is because in many places on the map only one particular position is possible – it is assumed that short cuts across fields are not taken! Dead reckoning even allows navigation when satellite signals are disrupted.

Figure 7.307 A digital map

As the driver follows the instructions for the first right turn here, the system will 'know' the location to within a metre or so if steering angle is used as in input

New developments In July 2008, the Pentagon announced that Boeing would work on a High Integrity global positioning system demonstration contract that runs until January 2011.

The European Space Agency (ESA) has a program called the European Geostationary Navigation Overlay Service (EGNOS), which is an interim step until its competing Galileo GPS constellation can be built and deployed. EGNOS uses three satellites in geostationary orbit, correlating their information with GPS to improve civilian positioning accuracy from 15m to 2m.

The Office of Naval Research in the USA aims to work with an existing commercial constellation, the low-bandwidth Iridium constellation of satellites. This could create a GPS service that provides quicker positioning fixes and improved accuracy for military users. It would also be more resistant to jamming and other forms of damage.

Summary Vehicle global positioning systems use a combination of information from satellites and sensors to accurately determine the vehicle position on a digital map. A route can then be calculated to a given destination. Like all vehicle systems, GPS

Figure 7.308 GPS IIF satellite

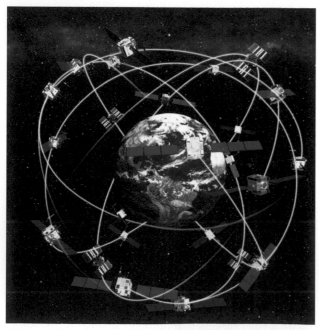

Figure 7.309 NAVSTAR constellation (Source: http://www.defenseindustrydaily.com.)

continues to develop and will do for some time yet as more features are added to the software. Already it is possible to 'ask' many systems for the nearest fuel station or restaurant, for example.

 Using images and text, create a short presentation to show how a component or system works

7.5.7 Drowsiness detection ❸

Introduction Fatigue and 'microsleep' at the wheel are often the cause of serious accidents. However, the initial signs of fatigue can now be detected before a critical situation arises. This is achieved, in a system Bosch has developed, by monitoring steering movements and advising drivers to take a break in time. It is called the Bosch Driver Drowsiness Detection system.

Studies The influence of fatigue on accidents has been demonstrated in a number of studies. In 2010, the American Automobile Association (AAA) published an analysis based on the accident data collected by the National Highway Traffic Safety Administration (NHTSA) in the United States. The assessment showed that overtired drivers were at the wheel in 17% of all fatal accidents in the US.

Steering angle sensor The required information is provided either by the car's electric power steering system, or by the steering angle sensor which is part of the car's ESP® anti-skid system. The feature can therefore be installed cost-effectively and helps further increase road safety. The system can be used in passenger cars and light commercial vehicles, and can also be integrated into various control units in vehicles. It was first introduced as a standard feature in 2010, in the Volkswagen Passat.

Figure 7.310 Steering angle sensor (Source: Bosch Media)

Behaviour and response time Lack of concentration and fatigue result in changes to the driver's steering behaviour and response time. Fine motor skills deteriorate, and steering action becomes less precise. The driver corrects small steering mistakes more often. The driver drowsiness detection function is based on an algorithm which begins recording the driver's steering behaviour the moment the trip begins. It then recognizes changes over the course of long trips. Typical signs of waning concentration are phases during which the driver is barely steering, combined with slight, yet quick and abrupt steering movements to keep the car on track. Based on the frequency of these movements and other parameters (among them the length of a trip, use of turn signals, and the time of day) the function calculates the driver's level of fatigue. If that level exceeds a certain value, an icon such as a coffee cup flashes on the instrument panel to warn drivers that they need a rest.

Figure 7.311 Bosch drowsiness detection – and a recommendation

Summary Drowsiness is often the cause of serious accidents. The Bosch Driver Drowsiness Detection system is able to monitor steering movements (using the steering angle sensor) and advises drivers to take a break if necessary.

 Look back over the previous section and write out a list of the key bullet points from this section

7.5.8 Laser head-up displays (HUDs) ❸

Introduction HUDs are now available as standard or optional equipment on many high-end cars. They are also being incorporated into a range of advanced driver assistance systems (ADAS). This is done to help drivers stay focused on the road ahead rather than looking down at their dashboard. The HUD places road speed, warning signals and indicator arrows on the driver's windscreen directly in their line of sight.

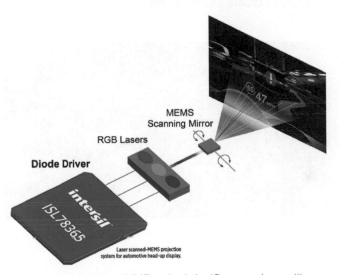

Figure 7.312 Laser HUD principle (Source: Intersil)

Laser diodes One of the newest HUDs uses laser diode drivers to pulse high-intensity red, green and blue (RGB) lasers. These project high-definition (HD) video onto the windshield. These are described as augmented-reality HUDs and can, for example, paint a transparent arrow directly onto the road in front of the car, to make GPS directions easy to follow. Another key feature is that they can highlight other objects, like pedestrians, animals or vehicles that might present a hazard.

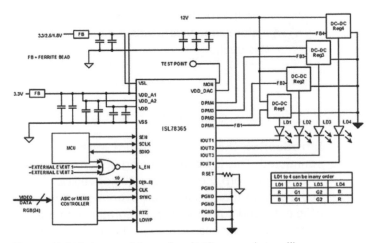

Figure 7.313 Laser driver circuit (Source: Intersil)

Components The laser scanned projection system enables the use of lower cost optics. The main components are the laser diode driver, laser diodes, shaping optics, the oscillating mirror and control electronics. The RGB colour laser diodes are pulsed synchronously as the mirror is scanned across the display field. The image is then drawn pixel-by-pixel across the display field, which is shown on the screen. Of course, this happens very quickly so the driver just sees the completed images. Each pixel is pulsed very rapidly to create HD resolution.

Figure 7.314 Pedestrians can be 'enhanced' (Source: Nvidia)

Efficiency Laser systems offer better electrical efficiency than earlier LCD frame-based projection systems. This is because the instrument information does not fill the entire HUD field display area as it only presents time-sensitive information for a short duration of time. Augmented reality information can have over 70% of the display pixels turned off. A laser scanned HUD only consumes electrical power when there are relevant pixels to be projected, thus reducing power consumption.

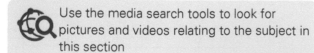
Use the media search tools to look for pictures and videos relating to the subject in this section

7.5.9 Connected vehicles 🔞

Introduction The car of the future will be connected. This is because using up-to-the-minute information from the internet will get vehicle occupants to their destination even more safely, efficiently and conveniently. Integration into the internet of things also unlocks a host of vehicle-related services.

Figure 7.315 Bosch technology puts the car online (Source: Bosch Media)

Information on traffic jams, black ice and wrong-way drivers is available in the cloud. When combined with infrastructure data from parking garages and charge spots, this provides a broader perspective known as the connected horizon. In the connected vehicle, the driver can 'see' over the top of the next hill, around the next bend and beyond. Because future cars will warn drivers in plenty of time about sudden fog or about a line of cars stopped behind the next bend, driving will be safer.

Figure 7.316 A connected car drives more proactively than a human! (Source: Bosch Media)

5G networks 5G follows previous mobile networks generations 2G, 3G and 4G. Compared to current networks (mostly 4G and 3G technology), 5G is designed to be faster and more reliable. It also has greater capacity and lower response times. The key benefit of 5G is speed, which can be in excess of 1Gb/s (1000Mbit/s) and could reach ten times this figure. The speed we experience in the real world depends on many factors, such as how far we are from a base station and how many other people are using the network at the same time. However, a typical user experienced data rate for downloads is expected to be a minimum of 100Mbit/s (which is much faster than existing systems).

Network	Max download speeds	Download an HD film
3G	384Kbps	More than a day
4G	100Mbps	7 minutes+
4G+	300Mbps	2.5 minutes
5G	1–10Gbps (theoretical)	4 to 40 seconds

Comparing 4G and 5G

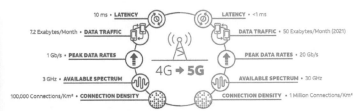

Figure 7.317 Comparing 4 and 5G (Source: https://policyforum.att.com)

Bandwidth There is potentially an even bigger benefit with 5G and that is low latency. 4G and other network standards have a long delay between a command being issued and a response being received. This is about 65ms for 3G and 40ms for advanced 4G. Fixed 'landline' broadband connections in the UK have a latency of 10ms to 20ms. 4G has a target of 4ms but for mission critical applications (like ADVs) this can be 1ms.

5G will also has access to more spectrum and at higher frequencies, in particular millimetre wave, which is the band of between 30Ghz and 300Ghz. This means that networks will be able to handle many high-demand applications all at once. Because of these improved speeds and short latency, automakers will be able to download more data and use the cellular network for some safety-related vehicle-to-everything (V2X) features.

Figure 7.318 Vehicle occupants will exchange lots of different types of data so security of the link will be essential (Source: IBM)

Features like tele-operated driving and cooperative manoeuvres will become possible. Data collected from vehicle cameras, meanwhile, can be collated to create maps that include real-time updates such as accidents and temporary roadworks. Because of the higher bandwidth, large high resolution map tiles could be uploaded and downloaded as needed. Cybersecurity will be a major issue, and most experts believe communications will have to be stored using techniques such as blockchain. This will ensure that information in messages cannot be tampered with, and that IDs are trusted.

Figure 7.319 V2X can prevent accidents (Source: Continental)

Connectivity also enhances vehicle efficiency. For example, precise data about traffic jams and the road ahead makes it possible to optimize charging management in hybrid and electric vehicles along the selected route. And because the car thinks ahead, the diesel particulate filter can be regenerated just before the car exits the motorway, and not in the subsequent stop-and-go traffic. Connectivity improves convenience as well, as it is a prerequisite for automated driving. It is the only way to provide unhurried braking in advance of construction zones, traffic jams and accident scenes.

Figure 7.320 Cloud connection (Source: Bosch Media)

325

Vehicle data Along with driving data and information on the vehicle's surroundings, the connected car also captures data on the operation of individual components. Running this data through sophisticated algorithms permits preventive diagnostics. For example, the data collected from an injection nozzle can be put through distributed algorithms in the cloud and in the vehicle to predict the part's remaining service life. The driver or fleet operator can be notified immediately, and an appointment made with the workshop in good time. In this way, it is often possible to avoid expensive repair and down times, especially for large commercial vehicles.

Driver's view Buildings, hedges or a truck can quickly obscure drivers' view, especially at intersections. If a road user is driving carelessly, it is often a matter of milliseconds that decide whether there is a collision or not. However, vehicle connectivity can greatly reduce the number of resulting accidents by promptly providing information that is outside the driver's and the vehicle's field of vision.

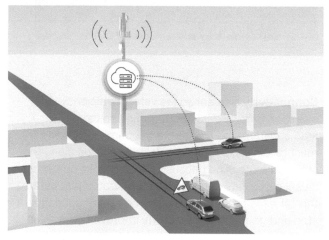

Figure 7.321 Local 'cell based' cloud intersection assistant (Source: Bosch Media)

Local clouds Outside of cities, where vehicles travel at higher speeds, there is a definite advantage if data takes the short route via the local cloud. Compared to solutions that exchange information via a central cloud, local cloud approaches are at least three times faster and they have much lower variances in the case of vehicle-to-vehicle latencies under 2ms. In some situations, this can make the difference as to whether the information reaches the car on time and the driver or the safety function can react quickly enough.

Vehicle-to-everything (V2X) To enable connected and automated driving in the future, vehicles must be able to easily communicate with one another as well as with their surroundings. There is currently no globally standardized technical basis for this exchange of data, which is known as vehicle-to-everything communication, or V2X. Instead, vehicles will in future communicate using the wide variety of different standards implemented by countries and vehicle manufacturers around the world.

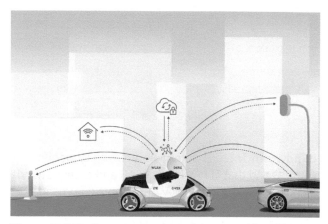

Figure 7.322 Communication with everything (Source: Bosch Media)

Making the connection Cars can use the Wi-Fi networks available in cities, while elsewhere they can communicate using, for instance, cellular networks. The complex task of managing these diverse communication options is handled by software. It continuously searches for the best transmission technology that suits the particular requirements and switches automatically between the available alternatives. The software therefore maintains continuous and seamless vehicle connectivity, ensuring cars can, for example, reliably alert one another to accidents and passengers can enjoy uninterrupted music streaming.

Figure 7.323 Connectivity unit (Source: Bosch media)

Cloud connection Initially, most vehicles will connect directly to the cloud. In the near future, increasing numbers of vehicles will also be able to communicate directly with one another (V2V) as well as with traffic signals, road construction sites, pedestrian crossings and buildings, etc. They will then be able to alert one another to potential hazards like the approaching tail end of a traffic jam, accidents, and icy conditions.

Figure 7.324 Software continuously searches for the best transmission technology (Source: Bosch media)

Critical information As well as keeping an eye on which V2X communication technologies are currently available for use, the software also closely monitors the costs and data transmission latency of each alternative connection option, since not every technology is suitable in every situation.

For example, when it comes to alerting a driver to another vehicle that is about to pull out in front of them from a side street, every millisecond counts. This kind of critical information must be communicated in real time using highly reliable technology that is always ready for use, even if that means the resulting data transmission costs are greater. Large volumes of data can be transmitted via Wi-Fi in a short space of time, though a downside is that public or home Wi-Fi hotspots are not always available.

Figure 7.325 V2X connectivity

Use a library or the web search tools to further examine the subject in this section

7.5.10 Hacking and cybersecurity ❸

Introduction The interconnectivity of current and future vehicles makes them potential targets for attack. Connectivity opens vehicle systems to the dark side of the Internet, forcing automakers to quickly develop strategies to ensure that they don't join the litany of corporations hit by hack attacks. SAE Recommended Practice J3061, 'Cybersecurity Guidebook for Cyber-Physical Vehicle Systems', is the first document tailored for vehicle cybersecurity. As more systems on vehicles are connected to the outside world by radio waves of some sort, or they scan the world outside of the car, then more opportunities are presented to hackers. Manufacturers are working very hard to reduce the chances of this happening and are helped in this process by what can be described as ethical hackers. There have been several interesting examples in the news recently; presented below are two examples that illustrate why this is an important area. In addition, there is currently a debate about the legality of hacking your own car as well as ways to stop other hackers.

Challenges Securing vehicles of the future is a cyber security challenge, to say the least. When something goes wrong with your home computer, 'crash' is only a metaphor. A recent survey found that nearly 100% of today's cars include wireless technologies that could be insecure and most manufacturers may not be able to easily determine if their vehicles have been hacked. Physical attacks via onboard diagnostic devices have shown it could be possible to manipulate some systems, steering for example, even while cars are moving.

Security, and specifically cybersecurity, is an increasingly urgent issue for the automotive industry. Systems are becoming more complex and the threat environment is also becoming more capable and sophisticated. This issue will only be made worse for automated driving vehicles (ADVs) because of vehicle to everything (V2X) communication. A range of best practices exist ranging from management focus down to technical measures, which can help to control the risk.

7

Figure 7.326 Cyber security is essential

Connected environment The shift from independent, closed vehicle systems to that of a connected environment is a massive change for the industry.

All vehicle systems much therefore have three mutually reinforcing properties:

▶ **Secure**: Prevention is better than cure and effective risk management begins by preventing system breaches in the first place
▶ **Vigilant**: Hardware and software can degrade, and the nature and type of attacks can change. No level of security is perfect. Security must therefore be monitored to ensure it is still secure or to see if it has been compromised
▶ **Resilient**: When a breach occurs, there must be a system in place to limit the damage and re-establishing normal operations. The system should also neutralise threats and prevent further spread.

Securing sensors The importance of securing individual sensors is critical in connected cars. Keep in mind these vehicles are a kind of internet connected data centre on wheels! A typical car can contain:

▶ About 70 computational systems running up to 100 million lines of code
▶ GPS devices that aid navigation and report on real-time traffic
▶ Diagnostic systems that check maintenance needs and send an alert in the event of an accident or breakdown.

As infrastructure evolves, cars will also be able to communicate with roadside devices such as traffic lights. Security must be part of the design and development, not bolted on at the end!

Key principles As vehicles get smarter, cyber security in the automotive industry is becoming an increasing concern. Whether we're turning cars into Wi-Fi

connected hotspots or equipping them with millions of lines of code to create fully autonomous vehicles, cars are more vulnerable than ever to hacking and data theft.

It's essential that all parties involved in the manufacturing supply chain, from designers and engineers to retailers and senior level executives, are provided with a consistent set of guidelines that support this global industry.

Figure 7.327 There are many aspects to cyber security

Hacking The interconnectivity of current and future vehicles makes them potential targets for attack. Connectivity opens vehicle systems to the dark side of the Internet, forcing automakers to quickly develop strategies to ensure that they don't join the litany of corporations hit by hack attacks. SAE Recommended Practice J3061, 'Cybersecurity Guidebook for Cyber-Physical Vehicle Systems', is the first document tailored for vehicle cybersecurity.

Summary Increasing use of connected system means the cybersecurity risk is greater. Manufacturers are working very hard to reduce this, but it continues to be a challenge.

 Use a library or the web search tools to further examine the subject in this section

7.6 Heating, ventilation and AC

7.6.1 Ventilation systems

Introduction Fresh air helps to keep the driver of a vehicle alert. Most cars now allow a wide range of settings for ventilation.

To allow fresh air from outside the vehicle to be circulated inside the cabin, a pressure difference must be created. This is achieved by using a plenum chamber. A plenum chamber by definition holds a gas, in this case air, at a pressure higher than the ambient pressure.

Airflow The plenum chamber on a vehicle is usually situated just below the windscreen, behind the bonnet. When the vehicle is moving the airflow over the vehicle will cause a higher pressure in this area. Suitable flaps and drains are utilised to prevent water entering the car through this opening.

Recirculated air Many vehicles allow a choice between fresh or recirculated air. The main reason for this is to decrease the time taken to heat or cool the car interior. The other reason is that, for example, in heavily congested traffic the outside air may not be very clean.

Figure 7.328 Ventilation controls

Air distribution By means of distribution trunking, control flaps and suitable nozzles, the air can be directed as required. This system is enhanced with the addition of a variable speed blower motor.

Air outlets When extra air is forced into a vehicle cabin, the interior pressure would increase if no outlets were available – most passenger cars have outlet grills on each side of the vehicle above the rear quarter panel.

Blower motors The motors used to increase airflow are simple permanent magnet two brush motors.

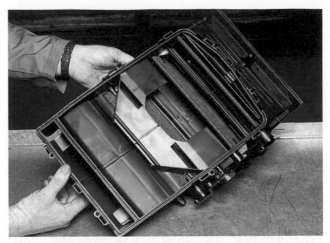

Figure 7.329 Control flaps

The blower fan is often of the centrifugal type and, in many cases, the blades are positioned asymmetrically to reduce resonant noise.

Figure 7.330 Typical motor and fan arrangement

Speed control The motor speed is controlled by varying the voltage supplied. This is achieved by using dropping resistors. The speed can be made infinitely variable by the use of a variable resistor. In most cases, the motor is controlled to three or four set speeds.

Three-speed control system Shown here is a circuit diagram typical of a three-speed control system. The resistors are usually wire wound and are placed in the air stream to prevent overheating. These resistors will have low values in the region of 1Ω or less.

Summary Ventilation, as well as the obvious need for fresh air, contributes to road safety by helping to keep the driver alert. Most systems are a simple arrangement of flaps, trunking and vents.

Make a simple sketch to show how one of the main components or systems in this section operates

7.6.2 Vehicle heating 1 2

Introduction Any heating and ventilation system has a simple set of requirements. These are summarised as follows:

- ▶ an adjustable temperature in the vehicle cabin
- ▶ heat must be available as soon as possible
- ▶ heat can be distributed to various parts of the vehicle
- ▶ fresh air ventilate possible but with minimum noise
- ▶ all windows can be demisted
- ▶ easy control operation.

Figure 7.332 Heater radiator

Figure 7.331 Heater controls

Heat from the engine can be used to increase the temperature of the car interior. This is achieved by use of a heat exchanger, often called the heater matrix. Due to the action of the thermostat in the engine cooling system, the water temperature remains reasonably constant. The air being passed over the heater matrix is therefore heated to a set level.

The **heater matrix** is like a small radiator. It consists of many tubes surrounded by fins to increase the surface area. Copper was used at one time but most modern heater matrixes are aluminium with plastic header tanks.

Hot air A source of hot air is now available for heating the vehicle interior. However, some form of control is required over how much heat is needed.

Heat control The control method used on most modern vehicles is blending. There is a control flap, which determines how much of the air being passed into the vehicle is directed over the heater matrix. Some systems use a valve to control the hot coolant flowing to the heater matrix.

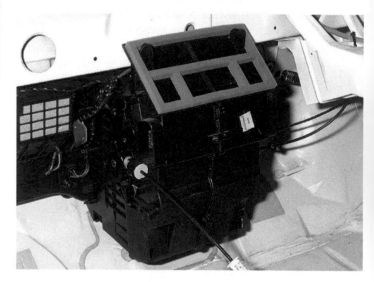

Figure 7.333 Heater box and flaps

Direction control Through the appropriate arrangement of flaps, it is possible to direct air of the chosen temperature to selected areas of the vehicle interior. Basic systems allow the warm air to be adjusted between the inside of the windscreen and the driver and passenger foot wells. Fresh cool air outlets with directional nozzles are also fitted.

Heater blower motor A blower motor in the air intake duct boosts airflow through the heater. The motor is usually fitted with a series of resistors in order to provide a range of speeds. The motor switch routes the electric current through the appropriate resistor for the speed selected on the switch.

Demisting Outlets are positioned for directing air onto the inside of the screen. Some cars also have outlets to direct air onto the side windows. All the output air from the heater can be directed in this way if required.

Figure 7.334 Double fan

Figure 7.335 Wide fan

Figure 7.336 Single fan

Figure 7.337 Blower box

Figure 7.338 Demister outlets

Air-cooled systems have the air stream passing directly over the engine cylinders and cylinder heads to remove the heat at the source. Fins are cast into the cylinders and cylinder heads to increase the surface area of the components and therefore ensure that sufficient heat is lost.

Because of how air cooling works, it is difficult to collect heat for use in the vehicle. Some systems use a heat exchanger as part of the exhaust system. The danger of this is that if the exhaust corrodes gases can be taken into the vehicle. Flaps are used in the same way as for water-cooled systems to control temperature and direction. However, controlling the heat output from these systems is a problem.

Summary Heating is an important passenger comfort system. The demisting function is an indispensable feature. Most cars use heat from the cooling system to heat the interior. A heater matrix is used for this purpose. A blower motor, together with distribution and blending flaps, provides the control.

 Create a word cloud for one or more of the most important screens or blocks of text in this section

7.6.3 Air conditioning fundamentals ❷

Introduction A vehicle fitted with air conditioning allows the temperature of the cabin to be controlled to the ideal or most comfortable value. This is usually determined by the ambient conditions. Air conditioning can be manually controlled or, as is often the case, combined with some form of electronic control. The system as a whole can be thought of as a type of refrigerator or heat exchanger. Heat is removed from the car interior and dispersed to the outside air.

Principle of refrigeration To understand the principles of air conditioning or refrigeration, the following terms and definitions will be useful:

▶ Heat is a form of energy.
▶ Temperature is the degree or intensity of heat of a body, and the condition that determines whether or not it will transfer heat to, or receive heat from, another body.
▶ Heat will only flow from a higher to a lower temperature.
▶ Change of state describes the changing of a solid to a liquid, a liquid to a gas, a gas to a liquid or a liquid to a solid.
▶ Evaporation describes the change of state from a liquid to a gas.
▶ Condensation describes the change of state from gas to liquid.
▶ Latent heat describes the energy required to evaporate a liquid without changing its temperature.

Latent heat, in the change of state of a refrigerant, is the key to air conditioning. As an example of this, when you put a liquid such as methylated spirits on your hand it feels cold. This is because it evaporates and the change of state, liquid to

gas, uses heat from your body. This is why the process is often thought of as 'unheating' rather than cooling. Remember, however, that methylated spirits is flammable. The refrigerant used in many air conditioning systems changes state from liquid to gas at −26.3°C.

Refrigerant The refrigerant used in many current air conditioning systems is known as R134a. This substance changes state from liquid to gas at −26.3°C. R134a is HFC based. Earlier types were CFC based and caused problems with atmospheric ozone depletion. While R134a has been one of the global standard automotive air-conditioning refrigerants, it does not meet the EU F-gas legislation (517/2014) which came into force in January 2015. A new HFO gas known as R1234yf has a GWP of 4 and will therefore meet the EU regulation. It has physical properties like R134a so has the potential to be used in current R134a systems with limited system modifications. However, different types of refrigerant are NOT compatible and must NEVER be mixed.

A key to understanding refrigeration is to remember that low-pressure refrigerant will have low temperature, and high-pressure refrigerant will have a high temperature.

Figure 7.340 R134a refrigerant

Air conditioning system The layout of an air conditioning or refrigeration system is shown in Figure 7.341. The main components are the evaporator, the condenser and the pump or compressor. The evaporator is situated in the car, the condenser outside the car, in the air stream, and the compressor is driven by the engine.

As the compressor operates it causes the pressure on its intake side to fall. This allows the refrigerant in the evaporator to evaporate and draw heat from

Figure 7.339 A liquid evaporating uses heat

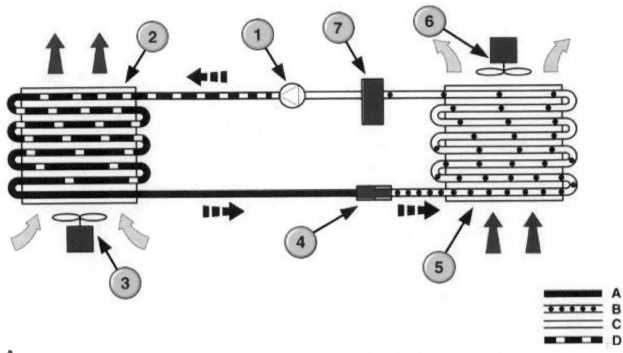

	A
	B
	C
	D

Figure 7.341 System layout

the vehicle interior. The high-pressure or output side of the pump is connected to the condenser. The pressure causes the refrigerant to condense, in the condenser, thus giving off heat outside the vehicle as it changes state.

The compressor pumps low pressure but heat laden vapour from the evaporator, compresses it and pumps it as a superheated vapour under high pressure to the condenser. The temperature of the refrigerant at this stage is much higher than the outside air temperature. It therefore gives up its heat, via the fins on the condenser, as it changes state back to a liquid. This high-pressure liquid is then passed to a receiver drier, which stores any vapour that has not yet turned back to a liquid. Alternatively, a suction accumulator is used on the low-pressure side.

Drying agent A desiccant, which is a drying agent, removes any moisture that is contaminating the refrigerant. Refrigerant, like brake fluid, is hygroscopic, which means it absorbs water. The high-pressure liquid is now passed through the thermostatic expansion valve, or a fixed orifice, and is converted back to a low-pressure liquid as it passes through a restriction into the evaporator.

As the liquid changes state to a gas in the evaporator, it takes up heat from its surroundings, thus cooling or 'unheating' the air which is forced over the fins. The low-pressure vapour leaves the evaporator returning to the pump, thus completing the cycle.

Temperature of the refrigerant If the temperature of the refrigerant increases beyond certain limits, the condenser cooling fans can be switched in to supplement the ram air effect.

Figure 7.342 Condenser cooling fans

Summary Changing a liquid into a gas uses energy. This energy, in the form of heat, is taken from inside the vehicle. When the gas is compressed, it gets hotter and the heat can be given off outside the vehicle. This turns the gas back into a liquid and the cycle starts again.

 Use a library or the web search tools to further examine the subject in this section

7.6.4 Air conditioning components ❸

The main components of an air conditioning system are:

▶ a compressor
▶ a condenser
▶ an evaporator
▶ a control valve
▶ a drier.

Each part is examined over the following pages.

Compressor An air conditioning system compressor is shown in Figure 7.343. It is belt driven from the engine crankshaft and it causes refrigerant to circulate through the system. The compressor is controlled by an electromagnetic clutch, which may be under either manual control or electronic control depending on the type of system.

Figure 7.343 Compressor in place

The shaft is driven by the engine via a multi-V belt. Five double pistons are arranged around the driving shaft. The swash plate, which is mounted on the mainshaft, causes the pistons to move backwards and forwards.

This compressor consists of two helices, one within the other. One is fixed, the other moves as the shaft rotates. This causes chambers to expand and contract. The refrigerant is drawn in as the chambers expand, and compressed as they contract.

Vane compressors are similar in operation to oil pumps of this design. As the central shaft rotates, the vanes are thrown out against the stator. This causes the pumping action because the volumes are increased and decreased as the shaft rotates.

The **condenser** is fitted in front of the vehicle radiator. It is very similar in construction to the radiator and fulfils a similar role. The heat is conducted through the aluminium pipes and fins to the surrounding air and then by a process of radiation and convection is dispersed by the air movement.

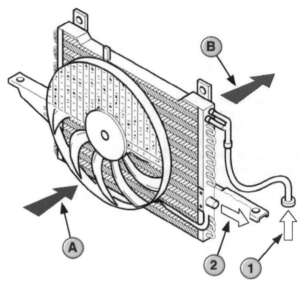

Figure 7.344 The condenser is very similar to the radiator

Receiver/drier Shown here is a typical receiver/drier assembly. It is connected in the high-pressure line between the condenser and the thermostatic expansion valve. This component carries out four tasks:

▶ to hold refrigerant in a reservoir until a greater flow is required
▶ to prevent contaminants circulating through the system by using a filter
▶ to retain vapour until it converts back to a liquid
▶ the removal of moisture from the system using a drying agent.

Low-pressure accumulator/drier Systems that use a fixed orifice control system usually use a low-pressure accumulator instead of a receiver/drier. This component carries out the same tasks as a receiver/drier.

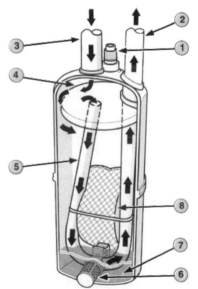

Figure 7.345 Ford low-pressure accumulator assembly

A sight glass is fitted to some receiver/driers. This gives an indication of refrigerant condition and system operation. The refrigerant generally appears clear if all is in order.

A **thermostatic expansion valve** is shown in Figures 7.346 and 7.347. The main function of this valve is to control the flow of refrigerant as demanded by the system. This in turn controls the temperature of the evaporator. A temperature sensor is fitted in the evaporator on some systems.

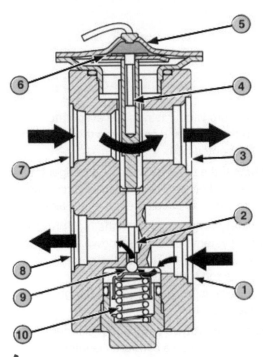

Figure 7.346 Valve body in section

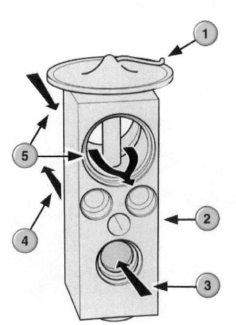

Figure 7.347 Complete valve body

Some systems use a fixed orifice control valve. The operation is quite simple. A fixed orifice, which is a small hole, only allows a certain flow rate! Filters are included to prevent contamination. The fixed orifice is the connection between the low- and high-pressure systems.

The **evaporator** is similar in construction to the condenser, consisting of fins to maximise heat transfer. It is mounted in the car under the dash panel forming part of the heating and ventilation system. As well as cooling the air passed over it, the evaporator also removes moisture from the air. This is because the moisture in the air condenses on the fins. The action is much like breathing onto a cold pane of glass. A drain is fitted to remove water.

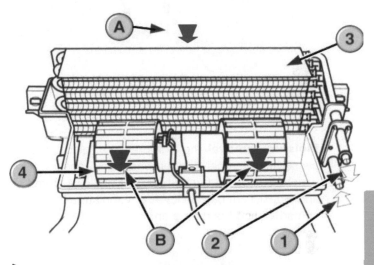

Figure 7.348 Evaporator housing and blower motor

The key components of an air conditioning system are the condenser, evaporator and compressor. Refrigerant takes heat from the car as it evaporates. It is then compressed, and condenses in the condenser. It gives off heat to the atmosphere.

 Make a simple sketch to show how one of the main components or systems in this section operates

7.6.5 Air conditioning systems 3

Introduction The operation of an air conditioning system is a continuous cycle.

High-pressure vapour The compressor pumps low pressure but heat laden vapour from the evaporator, compresses it and pumps it as a superheated vapour under high pressure to the condenser. The temperature of the refrigerant at this stage is much higher than the outside air temperature; hence, it gives up its heat via the fins on the condenser, as it changes state back to a liquid.

Compressor clutch The compressor is connected to its drive pulley by a magnetic clutch. When the field coil is energised, the driven plate is made to connect with the drive pulley. When the current is switched off, the magnetism falls and the driven plate moves away under spring pressure.

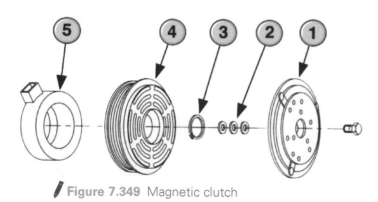

Figure 7.349 Magnetic clutch

Moisture removal The high-pressure liquid is then passed to the receiver drier where any vapour, which has not yet turned back to a liquid, is stored, and a desiccant bag removes any moisture that is contaminating the refrigerant.

Figure 7.350 Receiver drier

The **high-pressure liquid** is now passed through the thermostatic expansion valve (or a fixed orifice) and converted back to a low-pressure liquid as it passes through a restriction into the evaporator. This valve is the element of the system that controls the refrigerant flow and hence the amount of cooling provided. As the liquid changes state to a gas in the

evaporator, it takes up heat from its surroundings, thus cooling or 'unheating' the air that is forced over the fins.

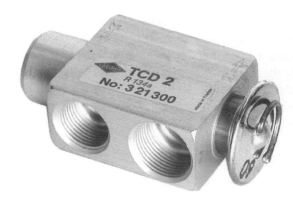

Figure 7.351 Thermostatic expansion valve

The low-pressure vapour leaves the evaporator returning to the pump and completes the cycle. The cycle is represented here.

Cooling fans If the temperature of the refrigerant increases beyond certain limits, condenser cooling fans can be switched in, to supplement the ram air effect.

Safety A safety switch is fitted in the high-pressure side of most systems. It is often known as a high-low pressure switch. It will switch off the compressor if the pressure is too high due to a component fault, or if the pressure is too low due to a leakage. Some systems use two switches.

Automatic temperature control Full temperature control systems provide a comfortable interior temperature in line with the driver's requirements. An electronic control unit (ECU) has control of:

- ▶ fan speed
- ▶ air distribution
- ▶ air temperature flaps
- ▶ fresh or recirculated air flaps
- ▶ air conditioning pump magnetic clutch.

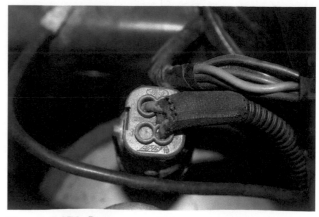

Figure 7.352 Pressure switch connection

Figure 7.353 Pressure switch

Figure 7.355 Detecting solar light improves operation

Figure 7.354 Compressor pump clutch

in the most appropriate manner. Control of the flaps can be either by solenoid controlled vacuum actuators or by small motors. The main blower motor is often controlled by a heavy-duty power transistor, which gives a constantly variable speed. These systems can provide a comfortable interior temperature in exterior conditions ranging from −10 to +35°C and in bright sunlight.

Summary The key components of an air conditioning system are the condenser, evaporator and compressor. When working together correctly, the temperature in a vehicle can be held at a comfortable level. Essential in some places . . .

 Create a word cloud for one or more of the most important screens or blocks of text in this section

Sensors A number of sensors can be used to provide input to the ECU:

▶ An ambient temperature sensor mounted outside the vehicle to allow compensation for extreme temperature variations. This device is usually a thermistor.
▶ A solar light sensor mounted on the fascia panel.
▶ The in car temperature sensor is a thermistor. However, to allow for an accurate reading a small motor and fan can be used. This takes a sample of interior air and directs it over the sensor.
▶ A coolant temperature sensor is used to monitor the temperature of the coolant supplied to the heater matrix.

Electronic control The ECU takes information from all of the above sources and can adjust the system

7.6.6 **Other heating systems** ❸

Introduction Electrical heating is used for screens, windows, seats and mirrors. Some heavy vehicles also incorporate cab heaters, which use fuel from the tank. As far back as the 1920s, when vehicle heaters were not fitted, electrically heated gloves were available. Beware of short circuits!

Rear screen heating involves a circuit with a relay and usually a timer. The heating elements are thin metallic strips bonded to, or built inside the glass. When a current is passed through the elements, heat is generated and the window will defrost or demist.

Remember! There is more support on the website that includes additional images and interactive features: www.tomdenton.org

Figure 7.356 Rear screen heater elements

High current This circuit can draw high current, 10 to 15 amps being typical. Because of this, the circuit often contains a timer relay to prevent the heater being left on for too long. The timer will switch off after 10 to 15 minutes. The rear screen elements are usually shaped to defrost in the rest position of the rear wiper blade, if fitted.

Figure 7.357 Timer relay

Windscreen heating Front windscreen heating is used on some vehicles. This presents more problems than the rear screen, as vision must not be obscured. The technology used, drawn from the aircraft industry, involves very thin wires cast in to the glass. As with the heated rear window, this device can consume a large current and uses a timer relay.

Seat heaters The concept of seat heating is simple. A heating element is placed in the seat, together with an on-off switch and a control to regulate the heat. However, the design of these heaters is more complex than first appears.

Seat heating requirements:

▶ the heater must only supply the heat loss experienced by the person's body
▶ heat must only be supplied at the major contact points
▶ heating elements must fit the design of the seat
▶ The elements must pass the same rigorous tests as the seat, such as squirm, bounce and bump tests.

The main method of control is a thermostat switch. Recent developments, however, tend to favour electronic control combined with a thermistor. These seat heaters will heat up to provide an initial sensation in one minute and to full-regulated temperature in three minutes. Very nice on a cold morning!

Mirror heaters Some vehicles are fitted with heated mirrors. These may come on with the screen heaters or be on all the time with the ignition. Small elements are fitted behind the glass. This is a particularly useful system for defrosting.

Figure 7.358 Heated mirror

Summary All forms of heating system improve driver and passenger comfort. This can be a major contribution to road safety.

 Use a library or the web search tools to further examine the subject in this section

7.6.7 Heat pumps ❸

Introduction A heat pump is a device that transfers heat energy from a source of heat to a destination called a heat sink. Heat pumps are designed to move thermal energy in the opposite direction of natural heat flow by absorbing heat from a cold space and releasing it to a warmer one. A heat pump uses a small amount of external power to do this,

Operating principle When a heat pump is used for heating, it uses the same basic refrigeration-type cycle used by an air conditioner or a refrigerator but in the opposite direction. This therefore releases heat into the conditioned space rather than the surrounding environment. Imagine a car AC system working in reverse so it takes heat from outside and releases it in the cabin rather than the other way around. In heating mode, heat pumps are three to four times more efficient in their use of electric power than simple electrical resistance heaters.

Performance Heat naturally flows from warm to cold. A heat pump can absorb heat from a cold space and release it to a warmer one. Heat (energy) is not conserved in this process, which requires some level of external energy, such as electricity. Heat pumps are used to transfer heat because less high-grade energy is required for their operation than is released as heat. In electrically powered heat pumps, the heat transferred can be three or four times larger than the electrical power consumed, giving the system a coefficient of performance (COP) of 3 or 4, as opposed to a COP of 1 for a conventional electrical resistance heater, in which all heat is produced from input electrical energy.

Refrigerant Heat pumps use a refrigerant as an intermediate fluid to absorb heat where it vaporizes, in the evaporator, and then to release heat where the refrigerant condenses, in the condenser. The refrigerant flows through insulated pipes between the evaporator and the condenser, allowing for efficient thermal energy transfer. Reversible heat pumps work in either direction to provide heating or cooling to the internal space. They use simple valves to reverse the flow of refrigerant from the compressor through the condenser and evaporator.

The refrigerant needs to reach a sufficiently high temperature, when compressed, to release heat through the condenser. Similarly, the fluid must reach a sufficiently low temperature when allowed to expand in the evaporator. The pressure difference must be great enough for the fluid to condense at the hot side and still evaporate in the lower pressure region at the cold side. The greater the temperature difference, the greater the required pressure difference and consequently, the more energy is needed to compress the fluid. Therefore, the efficiency decreases with increasing temperature difference.

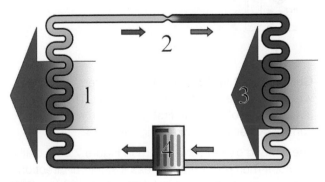

Figure 7.359 A simplified diagram of a heat pump's vapor-compression refrigeration cycle: 1) condenser, 2) expansion valve, 3) evaporator, 4) compressor.

Operation in a vehicle The following three diagrams show the three modes of operation:

▶ Heating
▶ Cooling
▶ Mixed

In cooling mode, the system acts like normal air conditioning where excess heat is taken from the cabin via the evaporator and is lost through the external condenser.

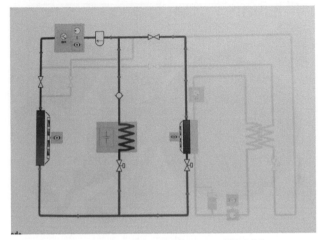

Figure 7.360 Cooling mode (Source: BMW)

In heating mode, refrigerant shut off valves are closed and the heating ones opened. The hot (pressurised) refrigerant is now circulated through a heat pump heat exchanger, which causes the heat to be transferred to the coolant for the heater circuit. To ensure this transfer takes place, an expansion valve at the outlet of the pump is used to further increase refrigerant pressure and hence its heat. The normal expansion valve used for the AC is also activated, further increasing the temperature. The return circuit for the now cooled refrigerant is the opposite way through the condenser.

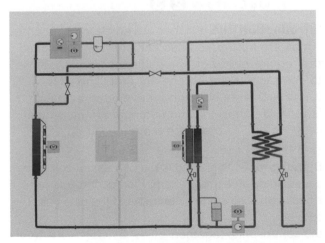

Figure 7.361 Heating mode (Source: BMW)

In mixed operating mode, the expansion valves are opened and closed as needed to achieve the desired temperature, and there is no reverse flow. The evaporator is cold to ensure dehumidification and the high voltage battery (in this example) can also be cooled. Refrigerant heat is used to ensure the desired cabin temperature is reached.

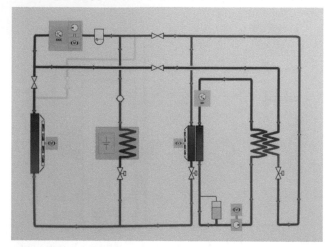

Figure 7.362 Mixed mode (Source: BMW)

Summary Heat pumps are starting to be used on some electric vehicle because the improve overall efficiency. They can be thought of as an AC system working in reverse.

 Select a routine from section 1.3 and follow the process to study a component or system

7.7 Multiplexing

7.7.1 Overview ❷ ❸

Introduction The number of vehicle components which are networked has considerably increased the requirements for the vehicle control systems to communicate with one another. The CAN (Controller Area Network), developed by Bosch, is today's communication standard in passenger cars. However, there are a number of other systems.

Signal path Multiplexing is the process of combining several messages for transmission over the same signal path. The signal path is called the data bus. The data bus is basically just a couple of wires connecting the control units together.

A data bus consists of a communication or signal wire and a ground return, serving all multiplex system nodes. The term node is given to any sub-assembly of a multiplex system (such as a control unit) that communicates on the data bus.

Benefits Not all vehicle systems are networked. However, the application of partial multiplex has achieved the following:

▶ a reduction of wiring harness weight and bulk
▶ an improvement of system functions and responsiveness with integrated control
▶ a reduction in production methods and cost
▶ improvement in troubleshooting as the system can self-diagnose any faults at the final point of assembly and at any time during service.

Early multiplexing systems On some vehicles, early multiplex systems used three control units. These were the door control unit, the driver's side control unit and the passenger's side control unit. These three units replaced the following:

▶ integrated unit
▶ interlock control unit
▶ door lock control unit
▶ illumination light control
▶ power window control unit
▶ security alarm control unit.

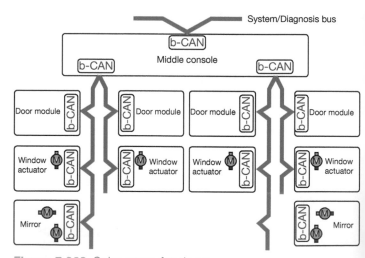

Figure 7.363 Subsystem for doors

Generation 1 and 2 Most manufacturers' systems went through a number of generations. The construction of a typical first generation multiplex control system is shown in Figure 7.364.

The second generation system in Figure 7.365, unlike previous multiplex systems, has a single bi-directional bus-type communication line between the driver ECU and passenger ECU.

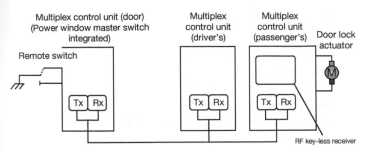

Bus-type (bi-directional) communication system using microcomputer-based communication ports

Figure 7.364 Generation 1

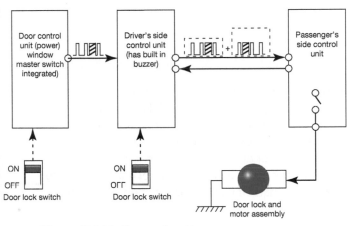

Figure 7.365 Generation 2

Generation 3 The third generation added more features:

▶ communication transfer speed of up to 16kBbit/s
▶ faster system shutdown
▶ diagnostic modes
▶ gauge module self-diagnosis function.

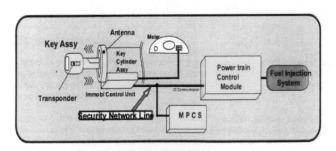

Figure 7.366 Gauge control module

Generation 4 The fourth generation is a controller area network system (CAN) which has many control units operating over two networks:

▶ Basic CAN (B-CAN), also known sometimes as Body CAN
▶ Fast CAN (F-CAN)

There will be more details on CAN later.

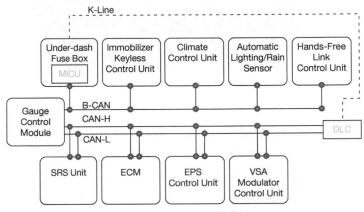

Figure 7.367 Combined B-CAN and F-CAN

Multiplex operation When a switch is operated, a coded digital signal is generated and communicated, according to its priority, via the data bus. All control units receive the signal but only the control unit for which the signal is intended will activate the desired response.

Data bus Only one signal can be sent on the bus at any one time. Therefore each signal has an identifier that is unique throughout the network. The identifier defines not only the content but also the priority of the message. Some systems make changes or adjustments to their operation much faster than other systems. Therefore, when two signals are sent at the same time, it is the system which requires the message most urgently whose signal takes priority.

Control functions Multiplex control systems have 'wake-up' and 'sleep' functions to decrease parasitic draw on the battery.

The multiplex control unit stops the functions (communication and CPU control) when the system is not required to operate. For example when the ignition switch is turned off, the control units will go into sleep mode ten seconds later, assuming that all the doors are closed. As soon as any operation happens (for example a door is opened) the related control unit receives a wake-up call.

Each control unit also has a hardware fail-safe function that fixes the output signal when there is any CPU malfunction. It also has a software fail-safe function, which ignores the signal from the malfunctioning control unit, and allows the system to operate normally.

7

Self-diagnosis A multiplex control system has the advantage of self-diagnosis. This allows quick and easy troubleshooting and verification using diagnostic trouble codes (DTCs)

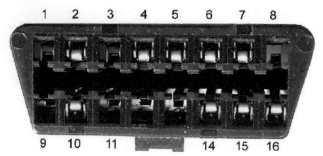

Figure 7.368 16-pin data link connector

Summary Many vehicles contain over a kilometre of wiring to supply all their electrical components. Luxury models may contain considerably more because of elaborate driver's aids. The use of multiplexing means that considerably less wiring is used in a vehicle along with fewer multi-plugs and connectors, etc.

An additional advantage is that existing systems can be upgraded or added to without modification of the original system.

The advantages are therefore:

▶ reduction in vehicle mass
▶ fewer raw materials are used
▶ there are fewer components to dispose of
▶ there is a reduction in build time
▶ there is less need for research and development – a system can be used in many models
▶ reliability is increased
▶ the system can be expanded.

 Use a library or the web search tools to further examine the subject in this section

7.7.2 Controller area network (CAN) ❸

Introduction CAN is a serial bus system especially suited for networking 'intelligent' devices, as well as sensors and actuators, within a system or subsystem. It operates in a broadly similar way to a wired computer network. CAN stands for controller area network and means that control units are able to interchange data. It was first developed by Bosch.

■ Surround sensors (radar, video)
■ Brake control system
■ Occupant safety
■ Electric power steering
■ CAN bus

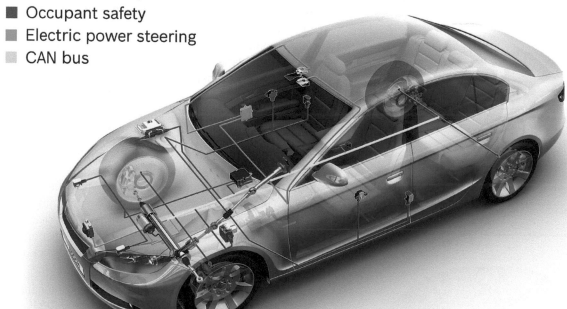

Figure 7.369 Bosch technologies for driver assistance (Source: Bosch Media)

High-integrity CAN is a high-integrity serial data communications bus for real-time applications. It operates at data rates of up to 1Mbit/s. It also has excellent error detection and confinement capabilities.

CAN was originally developed by Bosch for use in cars but is now used in many other industrial automation and control applications.

Multi-master capabilities CAN is a serial bus system with multi-master capabilities. This means that all CAN nodes are able to transmit data and several CAN nodes can request use of the bus simultaneously. In CAN networks there is no addressing of subscribers or stations, like on a computer network, but, instead, prioritized messages are transmitted. A transmitter sends a message to all CAN nodes (broadcasting). Each node decides on the basis of the identifier received whether it should process the message or not. The identifier also determines the priority that the message enjoys in competition for bus access.

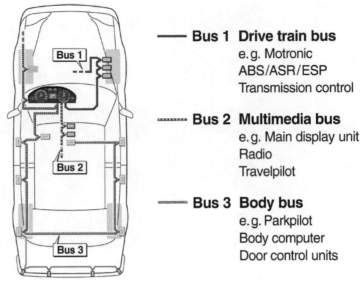

Figure 7.371 Three different speed buses in use (Source: Bosch Media)

Circuits The ECUs on the B-CAN and F-CAN transmit and receive information in the form of structured messages that may be received by several different ECUs on the network at one time. These messages are transmitted and received across a communication circuit that consists of a single wire that is shared by all the ECUs. However, as messages on the F-CAN network are typically of higher importance, a second wire is used for communication circuit integrity monitoring. This CAN-H and CAN-L circuit forms the CAN-bus.

Figure 7.370 Much simplified CAN message protocol flowchart

F-CAN and B-CAN Fast controller area networks (F-CAN) and basic (or body) controller area networks (B-CAN) share information between multiple electronic control units (ECUs). B-CAN communication is transmitted at a slower speed for convenience-related items such as electric windows. F-CAN information moves at a faster speed for real time functions such as fuel and emissions systems. To allow both systems to share information, a control module translates information between B-CAN and F-CAN.

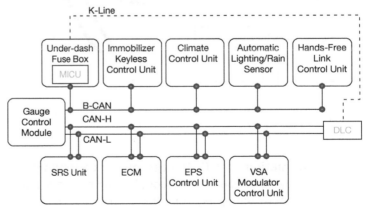

Figure 7.372 F-CAN uses CAN-H (high) and CAN-L (low) wires

A **multiplex control unit** is usually combined with the under-dash fuse/relay box. It controls many of the vehicle systems related to body electrics and the B-CAN. It also carries out much of the remote switching of various hardwired and CAN controlled systems.

Figure 7.373 The multiplex control unit is incorporated in this fuse box

Error correction One of the outstanding features of the CAN protocol is its high transmission reliability. The CAN controller registers a station's error and evaluates it statistically in order to take appropriate measures. These may extend to disconnecting the CAN node producing the errors.

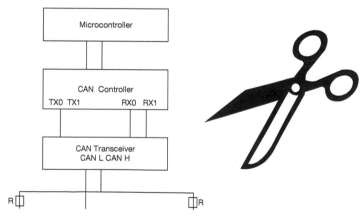

Figure 7.374 CAN nodes can be disconnected by the control program

Information Each CAN message can transmit from 0 to 8 bytes of user information. Longer messages can be sent by using segmentation, which means slicing a longer message into smaller parts. The maximum transmission rate is specified as 1Mbit/s. This value applies to networks up to 40m which is more than enough for normal cars and trucks.

Summary CAN is a serial bus system designed for networking ECUs as well as sensors and actuators. CAN, originally developed by Bosch, stands for controller area network and means that control units are able to share and exchange data.

 Construct a crossword or wordsearch puzzle using important words from this section

7.7.3 CAN data signal ❸

CAN message signal The CAN message signal consists of a sequence of binary digits or bits. A high voltage present indicates the value 1, a low or no voltage indicates 0. The actual message can vary between 44 and 108 bits in length. This is made up of a start bit, name, control bits, the data itself, a cyclic redundancy check (CRC) for error detection, a confirmation signal and finally a number of stop bits.

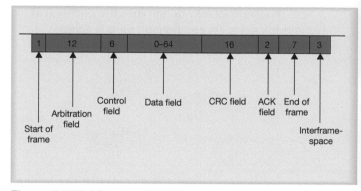

Figure 7.375 Message format

An example of binary format

1000101010001010111100001110101111101010101 0100011111010111100110011000001111110101010 0001111111111000000001

The message identifier or name This portion of the signal identifies the message destination and also its priority. As the transmitter puts a message on the data bus, it also reads the name back from the bus. If the name is not the same as the one it sent, then another transmitter must be in operation, which has a higher priority. If this is the case it will stop transmission of its own message. This is very important in the case of motor vehicle data transmission.

Errors in a message are recognised by what is known as a cyclic redundancy check (CRC). This is an error detection scheme in which all the bits in a block of data are divided by a predetermined binary number. A check character, known to the transmitter and receiver, is determined by the remainder. If an error is recognised, the message on the bus is destroyed. This in turn is recognised by the transmitter, which then sends the message again. This technique, when combined with additional tests, makes it possible to discover all faulty messages.

The basic idea behind CRCs is to treat the message string as a single binary word M, and divide it by a key word k that is known to both the transmitter and the receiver. The remainder r left after dividing M by k constitutes the 'check word' for the given message. The transmitter sends both the message string M and the check word r, and the receiver can then check the data by repeating the calculation, dividing M by the key word k, and verifying that the remainder is r.

Self-monitoring Because each node, in effect, monitors its own output, interrupts disturbed transmissions, and acknowledges correct transmissions, faulty stations can be recognised and uncoupled (electronically) from the bus. This prevents other transmissions from being disturbed.

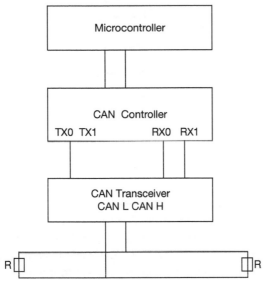

Figure 7.376 CAN Node

Summary A CAN message may vary between 44 and 108 bits in length. This is made up of a start bit, name, control bits, the data itself, CRC error detection, a confirmation signal and finally a number of stop bits.

 Use the media search tools to look for pictures and videos relating to the subject in this section

7.7.4 Local interconnect network (LIN) ❸

Introduction A local interconnect network (LIN) is a serial bus system especially suited for networking 'intelligent' devices, sensors and actuators within a

subsystem. It is a concept for low-cost automotive networks, which complements existing automotive multiplex networks such as CAN.

Hierarchical vehicle network LIN enables the implementation of a hierarchical vehicle network. This allows further quality enhancement and cost reduction of vehicles.

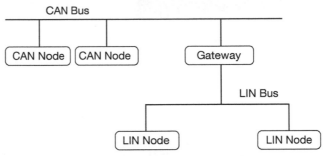

Figure 7.377 Structure using CAN and LIN

Standard The LIN standard includes the specification of the transmission protocol, the transmission medium, the interface between development tools, and the interfaces for software programming. LIN guarantees the interoperability of network nodes from the viewpoint of hardware and software, and predictable electromagnetic compatibility (EMC) behaviour.

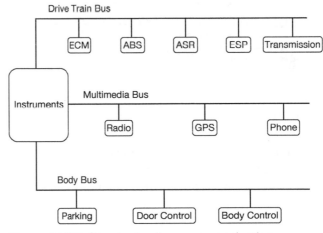

Figure 7.378 Standards allow communication between different systems

Single master, multiple slave network concept LIN is a time-triggered single master, multiple slave network concept. It is based on common interface hardware, which makes it a low-cost solution. Additional attributes of LIN are:

▶ multicast reception with self-synchronization
▶ selectable length of message frames
▶ data checksum security and error detection
▶ single-wire implementation
▶ speed up to 20kBit/s.

345

Figure 7.379 LIN package (Source: Freescale Semiconductor)

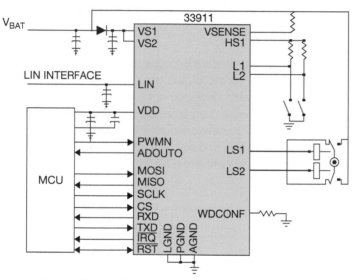

Figure 7.380 System basis chip (SBC) with a LIN transceiver. (Source: Freescale Semiconductor)

Summary LIN provides a cost efficient bus communication where the bandwidth and versatility of CAN are not required. It is used for non-critical systems.

 Make a simple sketch to show how one of the main components or systems in this section operates

7.7.5 FlexRay ❸

Introduction FlexRay is a fast and fault-tolerant bus system for automotive use. It was developed using the experience of well-known OEMs. It is designed to meet the needs of current and future in-car control applications that require a high bandwidth. The bit rate for FlexRay can be programmed to values up to 10MBit/s.

Data exchange The data exchange between the control devices, sensors and actuators in automobiles is mainly carried out via CAN systems. However, the introduction of X-by-wire systems has resulted in increased requirements. This is especially so with regard to error tolerance and speed of message transmission. FlexRay meets these requirements by message transmission in fixed time slots, and by fault-tolerant and redundant message transmission on two channels.

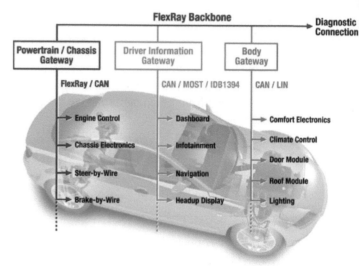

Figure 7.381 FlexRay backbone (Source: Fujitsu Media)

The **physical layer** means the hardware, that is, the actual components and wires. FlexRay works on the principle of time division multiple access (TDMA). This means that components or messages have fixed time slots in which they have exclusive access to the data bus. These time slots are repeated in a cycle and are just a few milliseconds long.

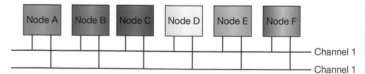

Figure 7.382 FlexRay topology with two channels (Source: Eberspacher)

Bandwidth The fixed allocation of the bus bandwidth to the components or messages by means of fixed time slots has the disadvantage that the bandwidth is not fully used – for example, if a component is simply not in use at its slot-time. To get over this, FlexRay subdivides the cycle into static and dynamic segments. The fixed time slots are situated in the static segment at the beginning of a bus cycle. In the dynamic segment, the time slots are assigned dynamically – in other words, as they are needed. Exclusive bus access is only enabled for a short time in 'mini-slots'. This mini-slot is then only extended if a bus access occurs. Bandwidth is therefore used up only when it is actually needed.

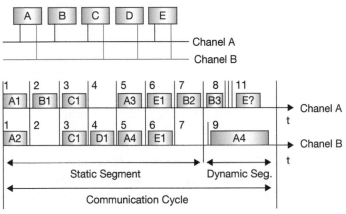

Figure 7.383 Bandwidth timeslots

Data rate FlexRay communicates via two physically separated lines with a data rate of up to 10Mbit/s (1.25 Mbps) on each. The two lines are mainly used for redundant and therefore fault-tolerant message transmission, but they can also transmit different messages.

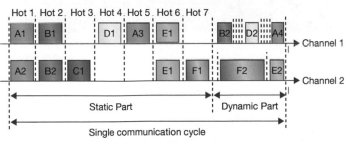

Figure 7.384 FlexRay communication cycle (Source: Eberspacher)

Synchronization In order to implement synchronous functions and use all available bandwidth, the distributed nodes on the network require a common time base. Clock synchronization messages are therefore transmitted in the static segment of each cycle.

FlexRay ECU A FlexRay ECU consists of a host processor, a FlexRay communication controller (CC), a bus guardian (BG) and a bus driver (BD). The host processor supplies and processes the data, which

are transmitted via the controller. The process is as follows:

1 Bus guardian monitors access to the bus.
2 Host processor informs the bus guardian which time slots the communication controller has allocated.
3 Bus guardian allows the communication controller to transmit data only in these time slots.
4 Bus driver is enabled.

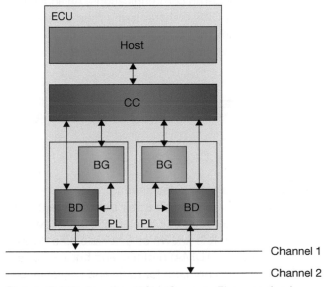

Figure 7.385 FlexRay ECU (Source: Eberspacher)

Summary FlexRay is a fast and fault-tolerant bus system that was developed to meet the needs of high bandwidth applications such as X-by-wire systems. Error tolerance and speed of message transmission in these systems is essential.

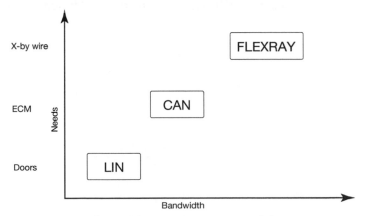

Figure 7.386 Comparing requirements and data rates of the three systems

 Make a simple sketch to show how one of the main components or systems in this section operates

347

7.7.6 **Other networking methods** ❸

Time triggered CAN (TT-CAN) Network applications will soon need bit rates of 5 to 10 Mbps. CAN is limited to 1Mbps so new protocols are under development by Bosch and others. One answer instead of using event triggered transmission is to trigger the communication at a precise time. Because of this, speeds close to 'real time' can be achieved. TT-CAN is an extension of the existing CAN protocol so the system will still be able to use event triggered communications. An additional advantage of TT-CAN is that missing messages are detected immediately, and there is much better protection against unauthorised bus access.

Figure 7.387 TT-CAN signal with the beginning and end joined to make a continuous cycle

Byteflight is a protocol designed for use with safety-critical systems such as air bags and seat belt tensioners. It has a high fault tolerance and allows high speed data. It is known as a flexible time division multiple access (FTDMA) protocol and because of the flexibility can also be used for body and convenience functions such as locking, windows and seat control.

Star topography is used but the protocol also works on linear systems. Communication is by plastic optic fibre (POF) and the data rate, depending on loads is 5 to 10 Mbps. One example of its use is the BMW 1 series where it links the airbag crash sensor satellite ECUs. The protocol was not designed for X-by-wire but it is high performance and has many of the required features.

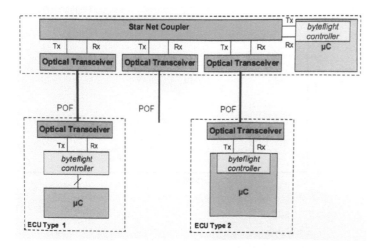

Figure 7.388 Byteflight uses a star topography as this ensures that a node failure cannot bring down the network

MOST protocol MOST (Media Oriented System Transport) generally uses fibre optic connections as represented in this image.

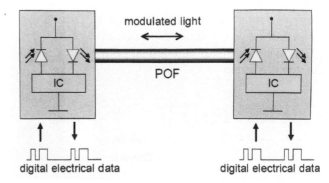

Figure 7.389 Representation of MOST communication using plastic optic fibre (POF). The normal transfer rate is 21.2 Mbps

Fibre optics The cable consists of a fibre core, an inner sheath and a protective casing. It is very flexible but should not be bent, crushed or twisted to excess!

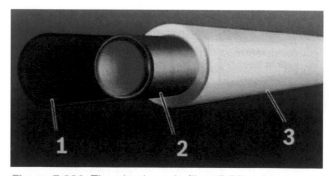

Figure 7.390 The plastic optic fibre (POF) cable. 1. Fibre core, 2. Inner sheath, 3. Protective casing

Figure 7.391 POF cable and connector

Attenuation Some attenuation can occur in the transmission of light waves, particularly over long distances and if multiple connectors are used. However, a key feature of this system is that the signal received by each control unit is re-transmitted so returns to full power.

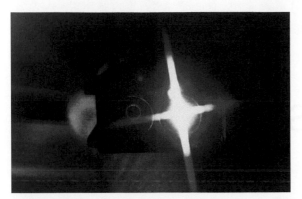

Figure 7.392 There is light at the end of the... cable!

Ethernet is a family of computer networking technologies commonly used in local area networks, metropolitan area networks and wide area networks. It was commercially introduced in 1980 and first standardized in 1983 as IEEE 802.3. Automotive Ethernet does more than support mobile connectivity and complex, high-bandwidth automotive applications; it is a key enabler for the fully connected autonomous vehicle (CAV). NXP, for example, has designed a broad portfolio of robust, flexible and cost-effective automotive Ethernet products to connect vehicle systems faster and more efficiently.

Figure 7.393 Ethernet cable (CAT 5)

Ethernet bandwidth In an automotive Ethernet setup, multiple vehicle systems simultaneously access high bandwidth over a single cable (a twisted pair). The Ethernet is the backbone of the vehicle network and supports higher levels of data processing and more communication types. Each port receives dedicated bandwidth and the entire backbone is capable of IP connectivity. Instead of supporting individual high-bandwidth functions, the architecture supports all the high-bandwidth functions that reside on the same physical network but use logically separated virtual networks.

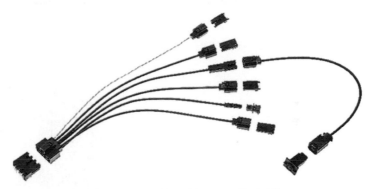

Figure 7.394 Ethernet cables (Source: Delphi)

Automotive Ethernet will connect the vehicle's internal digital devices, connect the vehicle with other vehicles and even make the car a productive part of the IoT. It supports the high bandwidth and real-time processing required for today's infotainment and ADAS systems and provides a platform for the development of truly autonomous driving.

 Use the media search tools to look for pictures and videos relating to the subject in this section

7

CHAPTER 8

Chassis systems

After successful completion of this chapter you will be able to show you have achieved these objectives:

- Understand how light vehicle steering systems operate.
- Understand how light vehicle suspension systems operate.
- Understand how light vehicle braking systems operate.
- Understand how light vehicle wheel and tyres systems operate.
- Understand how to check, replace and test light vehicle chassis units and components.
- Understand how to diagnose and rectify faults in light vehicle chassis systems.

DOI: 10.1201/9781003173236-8

8.1 Suspension

8.1.1 Reasons for suspension ❶

Introduction The suspension system is the link between the vehicle body and the wheels. Its purpose is to:

▶ locate the wheels, whilst allowing them to move up and down, and steer

▶ maintain the wheels in contact with the road and minimise road noise

▶ distribute the weight of the vehicle to the wheels

▶ reduce vehicle weight as much as possible – in particular the unsprung mass

▶ resist the effects of steering, braking and acceleration

▶ work in conjunction with the tyres and seat springs to give acceptable ride comfort.

Figure 8.1 Suspension plays a key role

Compromise The previous list is difficult to achieve completely, so some sort of compromise has to be reached. Because of this, many different methods have been tried, and many are still in use. Keep these requirements in mind, and it will help you to understand why some systems are constructed in different ways.

Sprung and unsprung mass Unsprung mass is usually the mass of the suspension component, the wheels and the springs. However, only 50% of the spring mass and the moving suspension arms are included. This is because they form part of the link between the sprung and unsprung masses. It is beneficial to have the unsprung mass as small as possible in comparison with the sprung mass (main vehicle mass). This is so that when the vehicle hits a bump the movement of the suspension will have only a small effect on the main part of the vehicle. The overall result is therefore improved ride comfort.

Further in suspension A vehicle needs a suspension system to cushion and damp out road shocks. This provides comfort to the passengers and prevents damage to the load and vehicle components. A spring between the wheel and the vehicle body allows the wheel to follow the road surface. The tyre plays an important role in absorbing small road shocks. It is often described as the primary form of suspension. The vehicle body is supported by springs located between the body and the wheel axles. Together with the damper, these components are referred to as the suspension system.

Figure 8.2 Suspension system

The effect of suspension As a wheel hits a bump in the road, it is moved upwards with quite some force. An unsprung wheel is affected only by gravity, which will try to return the wheel to the road surface. However, most of the energy will be transferred to the body. When a spring is used between the wheel and the vehicle body, most of the energy in the bouncing wheel is stored in the spring and not passed to the vehicle body. The vehicle body will only move upwards through a very small distance compared to the movement of the wheel.

Springs These parts of the suspension system take up the movement or shock from the road. The energy of the movement is stored in the spring. The actual spring itself can be in many different forms, ranging from a steel coil to a pressurised chamber of nitrogen. Soft springs provide the best comfort, but stiff springs can be better for high performance. Vehicle springs and suspension therefore are made to provide a compromise between good handling and comfort.

Figure 8.3 Coil spring

Figure 8.4 Leaf spring

Figure 8.5 Gas spring

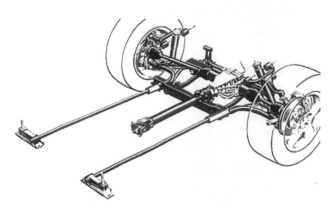

Figure 8.6 Torsion bar spring

Dampers or shock absorbers The energy stored in the spring after a bump has to be got rid of or else the spring will oscillate (bounce up and down). The damper damps down these oscillations by converting the energy from the spring into heat. If working correctly, the spring should stop moving after just one bounce and rebound. Shock absorber is a term which is often used to describe a damper.

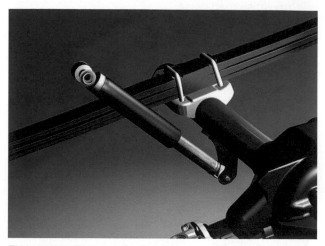

Figure 8.7 Telescopic damper

Strut The combination of a coil spring with a damper inside it, between the wheel stub axle and the inner wing, is often referred to as a strut. This is a very popular type of suspension.

A **wishbone** is a triangular shaped component with two corners hinged in a straight line on the vehicle body. The third corner is hinged to the moving part of the suspension.

Figure 8.8 MacPherson

Figure 8.9 Front suspension wishbone

Figure 8.11 Heavy vehicle axle

Bump stop When a vehicle hits a particularly large bump, or if it is carrying a heavy load, the suspension system may bottom out (reach the end of its travel). The bump stop, usually made of rubber, prevents metal-to-metal contact, which would cause damage.

Figure 8.10 Rubber stop

A **link** is a very general term, which is used to describe a bar or other similar component that holds or controls the position of another component. Other terms may be used such as tie-bar or tie-rod.

Beam axle This is a solid axle from one wheel to the other. It is not now used on the majority of light vehicles. However, as it makes a very strong construction, it is still common on heavy vehicles.

Gas/fluid suspension The most common types of spring are made from steel. However, some vehicles use pressurised gas as the spring (think of a balloon or a football). On some vehicles, a connection between wheels is made using fluid running through pipes from one suspension unit to another.

Figure 8.12 Gas suspension unit

Independent suspension Independent front and rear suspension (IFS/IRS) was developed to meet the demand for improved ride quality and handling. The main advantages of independent suspension are as follows:

▶ when one wheel is lifted or drops, it does not affect the opposite wheel
▶ the unsprung mass is lower and therefore the road wheel stays in better contact with the road
▶ problems with changing steering geometry are reduced
▶ there is more space for the engine at the front
▶ softer springing with larger wheel movement is possible.

Anti-roll bar The main purpose of an anti-roll bar is to reduce body roll on corners. The anti-roll bar can be thought of as a torsion bar. The centre is pivoted on the body and each end bends to make a connection with the suspension/wheel assembly. When the suspension

8

is compressed on both sides, the anti-roll bar has no effect because it pivots on its mountings. As the suspension is compressed on just one side, a twisting force is exerted on the anti-roll bar. Part of this load is transmitted to the opposite wheel, pulling it upwards. This reduces the amount of body roll on corners.

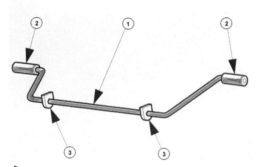

Figure 8.13 Shape of an anti-roll bar

The **Panhard rod** was named after a French engineer. Its purpose is to link a rear axle to the body. The rod is pivoted at each end to allow movement. It takes up lateral forces between the axle and body thus removing load from the radius arms. The radius arms now only have to transmit longitudinal forces.

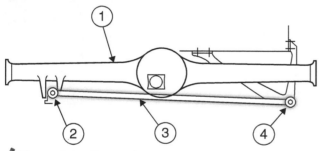

Figure 8.14 Rear axle with Panhard rod

Summary A wide variety of suspension systems and components are used. Engineers strive to achieve optimum comfort and handling. However, these two main requirements are often at odds with each other. As is common with all vehicle systems, electronic control is one way in which developments are now being made.

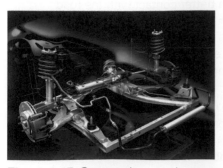

Figure 8.15 Suspension continues to develop

Use a library or the web search tools to further examine the subject in this section

8.1.2 Springs ❶

Introduction The requirements of the springs can be summarised as follows:

▶ to absorb road shocks from uneven surfaces
▶ to control ground clearance and ride height
▶ to ensure good tyre adhesion
▶ to support the weight of the vehicle
▶ to transmit gravity forces to the wheels.

There are a number of different types of spring in use on modern vehicles.

Coil springs Although modern vehicles use a number of different types of spring medium, the most popular is the coil (or helical) spring. Coil or helical springs used in vehicle suspension systems, are made from round spring steel bars. The heated bar is wound on a special former and then heat treated to obtain the correct elasticity (springiness). The spring can withstand any compression load but not side thrust. It is also difficult for a coil spring to resist braking or driving thrust. Suspension arms are used to resist these loads.

Figure 8.16 Coil spring in position

Independent suspension systems Coil springs are generally used with independent suspension systems – the springs are usually fitted on each side of the vehicle, between the stub axle assembly and the body. The spring remains in the correct position because recesses are made in both the stub axle assembly and body. The spring is always under compression due to the weight of the vehicle and hence holds itself in place.

Figure 8.17 Coil spring upper fitting

Figure 8.18 Coil spring lower fitting

Coil spring features The coil spring is a torsion bar wound into a spiral. It can be progressive if the diameter of the spring is tapered conically. A coil spring cannot transmit lateral or longitudinal forces, hence the need for links or arms. It produces little internal damping. No maintenance is required and high travel is possible.

Figure 8.19 Details of a coil spring

Leaf springs The leaf spring can provide all the control for the wheels during acceleration, braking, cornering, and general movement caused by the road surface.

They are used with fixed axles. Leaf springs can be described as:

▶ laminated or multi-leaf springs
▶ single leaf or mono-leaf springs.

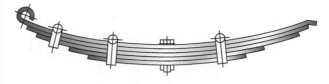
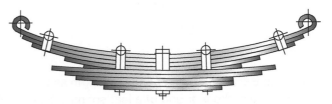

Figure 8.20 Laminated springs

Multi-leaf springs The multi-leaf spring was widely used at the rear of cars and light vehicles, and is still used in commercial vehicle suspension systems. It consists of a number of steel strips or leaves placed on top of each other and then clamped together. The length, cross section, and number of leaves are determined by the loads carried.

Figure 8.21 Commercial vehicle leaf spring

Leaf spring fixings The top leaf is called the main leaf and each end of this leaf is rolled to form an eye. This is for attachment to the vehicle chassis or body. The leaves of the spring are clamped together by a bolt or pin known as the centre bolt. The spring eye allows movement about a shackle and pin at the rear, allowing the spring to flex. The vehicle is pushed along by the rear axle through the front section of the spring, which is anchored firmly to the fixed shackle on the

8

355

vehicle chassis or body. The curve of leaf springs straightens out when a load is applied to it, and its length changes.

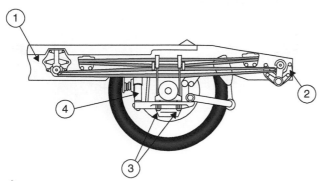

Figure 8.22 Details of a leaf spring

Shackles Because of the change in length as the spring moves, the rear end of a leaf spring is fixed by a shackle bolt to a swinging shackle. As the road wheel passes over a bump, the spring is compressed and the leaves slide over each other. As it returns to its original shape, the spring forces the wheel back into contact with the road. The leaf spring is usually secured to the axle by means of U bolts. As the leaves of the spring move, they rub together. This produces interleaf friction, which has a damping effect.

Single leaf spring A single leaf spring, as the name implies, consists of one uniformly stressed leaf. The spring varies in thickness from a maximum at the centre to a minimum at the spring eyes. This type of leaf spring is made to work in the same way as a multi-leaf spring. Advantages of this type of spring are:

▶ simplified construction
▶ constant performance over a period, because interleaf friction is eliminated
▶ reduction in unsprung mass.

Torsion bars This type of suspension uses a metal bar, which provides the springing effect as it is twisted. It has the advantage that the components do not take up too much room. The torsion bar can be round or square section, solid or hollow. The surface must be finished accurately to eliminate pressure points, which may cause cracking and fatigue failure. They can be fitted longitudinally or laterally.

Torsion bar features Torsion bars are maintenance free but can be adjusted. They transmit longitudinal and lateral forces and have low mass. However, they have limited self-damping. Their spring rate is linear and life span may be limited due to fatigue.

Figure 8.23 Tapered single leaf

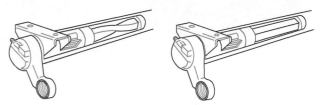

Figure 8.24 Torsion bar in a guide tube

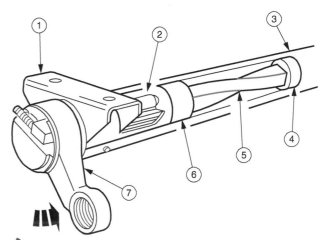

Figure 8.25 An anti-roll bar is a torsion spring

Pneumatic suspension Steel springs must be stiff enough to carry a vehicle's maximum load. However, this can result in the springs being too stiff to provide consistent ride control and comfort when the vehicle is empty. Pneumatic suspension can be made self-compensating. It is fitted to many heavy goods vehicles and buses, but is also becoming popular on some off-road light vehicles.

Air spring system The pneumatic or air spring is a reinforced rubber bellow fitted between the axle and the chassis or vehicle body. An air compressor is used to increase or decrease the pressure depending on

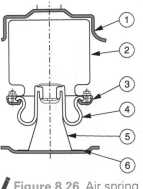

Figure 8.26 Air spring

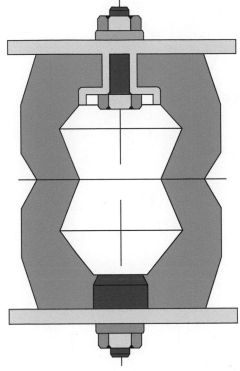

Figure 8.28 Hollow rubber spring

the load in the vehicle. This is done automatically but some manual control can be retained for adjusting the height of the vehicle or stiffness of the suspension. Air springs can be thought of as being like a balloon or football on which the car is supported. The system involves compressors and air tanks. The system is not normally used on light vehicles.

Rubber springs This is now a very old system, but often old ideas come back! The suspension medium, or spring, is simply a specially shaped piece of rubber. This technique was used on early Minis, for example. The rubber did not require damping in most cases. Nowadays, rubber springs are only used as a supplement to other forms of spring. They are however popular on trailers and caravans.

Hydrolastic suspension The suspension unit is supported by a rubber spring. Under the spring, a chamber of fluid is connected by a pipe to the corresponding front or rear unit. This system was the forerunner to the hydragas system.

Hydragas suspension In the hydragas suspension system, each wheel has a sealed displacer unit. This contains nitrogen gas under very high pressure, which works in much the same way as the steel spring in a conventional system. A damper is also incorporated within the displacer unit. The lower part of the displacer unit is filled with a suspension fluid (a type of wood alcohol usually). The units can be joined by pipes or used individually.

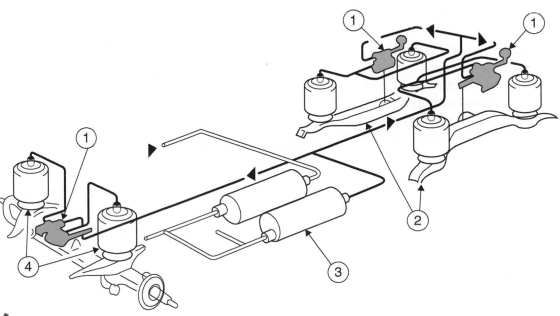

Figure 8.27 Air spring suspension system

357

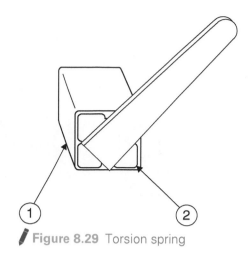

Figure 8.29 Torsion spring

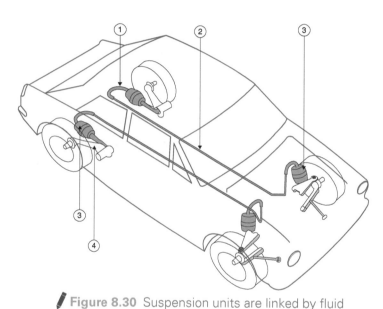

Figure 8.30 Suspension units are linked by fluid

Hydragas connections Connecting suspension units, using fluid-filled pipes, helps to improve the ride quality. Linking front to rear makes the rear unit rise as the front unit is compressed by a bump. This tends to keep the vehicle level and reduce pitch. Ride height control can be achieved by pumping oil into or out of the working chamber.

Summary Suspension springs can be made from a variety of materials and in many different ways. The most common is the coil spring. This is because it has many advantages and is reasonably inexpensive.

 Use the media search tools to look for pictures and videos relating to the subject in this section

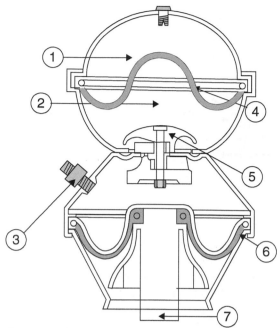

Figure 8.31 Suspension unit

8.1.3 Dampers/shock absorbers 🚹 🔢

Introduction As a spring is deflected, energy is stored in it. If the spring is free to move, the energy is released in the form of oscillations, for a short time, before it comes to rest. This principle can be demonstrated by flicking the end of a ruler placed on the edge of a desk. The function of the damper is to absorb the stored energy, which reduces the rebound oscillation. A spring without a damper would build up dangerous and uncomfortable bouncing of the vehicle.

Hydraulic dampers are the most common type used on modern vehicles. They work on the principle of forcing fluid through small holes. In a hydraulic damper, the energy in the spring is converted into heat. This is caused as the fluid (a type of oil) is forced rapidly through small holes (orifices). The oil temperature in a damper can reach over 150°C during normal operation. As an example, think of using a hand oil pump and how hard it is to make the oil flow quickly.

Damper functions The functions of a damper can be summarised as follows:

▶ to ensure directional stability
▶ to ensure good contact between the tyres and the road
▶ to prevent build-up of vertical movements
▶ to reduce oscillations
▶ to reduce wear on tyres and chassis components.

Friction damper This is not used on cars today, but you will find this system used as part of caravan or trailer stabilisers.

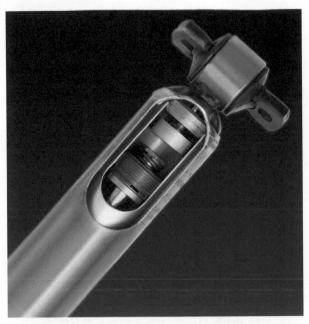

Figure 8.32 Damper

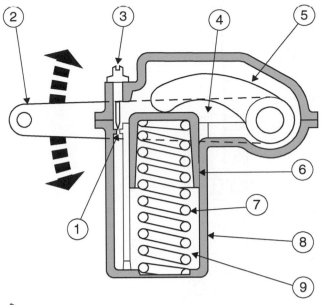

Figure 8.34 Lever-arm damper

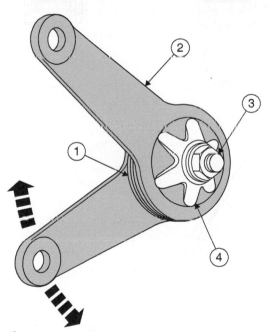

Figure 8.33 Principle of friction damping

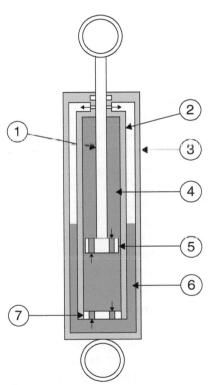

Figure 8.35 Twin tube system

Lever-arm damper A lever-arm damper, which was used on earlier vehicles, works on the same principle as the telescopic type. The lever operates a piston which forces oil into a chamber.

Twin tube telescopic damper This is the most commonly used type of telescopic damper – it consists of two tubes. An outer tube forms a reservoir space and contains the oil displaced from an inner tube. Oil is forced through a valve by the action of a piston as the damper moves up or down. The reservoir space is essential to make up for the

changes in volume as the piston rod moves in and out.

Single tube telescopic damper This is often referred to as a gas damper. However, the damping action is still achieved by forcing oil through a restriction. The gas space behind a separator piston is to compensate for the changes in cylinder volume, which is caused as the piston rod moves. The gas is at a pressure of about 25 bar.

8

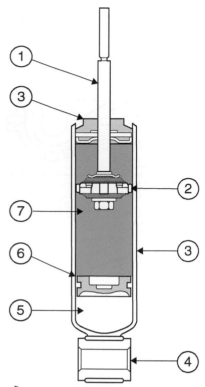

Figure 8.36 Single tube system

The **twin tube gas damper** is an improvement on the well-used twin tube system. The gas cushion is used in this case to prevent oil foaming. The gas pressure on the oil prevents foaming, which in turn ensures constant operation under all operating conditions. The gas pressure is set at about 5 bar. If bypass grooves are machined in the upper half of the working chamber, the damping rate can be made variable. With light loads the damper works in this area with a soft damping effect. When the load is increased the piston moves lower down the working chamber away from the grooves resulting in full damping effect.

Electronically controlled dampers These are dampers where the damping rate can be controlled by solenoid valves inside the units. With suitable electronic control, the characteristics can be changed within milliseconds to react to driving and/or load conditions. When it is activated, the solenoid allows some of the oil to be diverted.

Electronic system In Figure 8.39 you can see the sensors and other components necessary for electronic damper control. Electronic control allows a combination of high comfort and performance. Adjustments can be made automatically or preset by the driver.

Summary Dampers (or shock absorbers) are used to prevent the suspension springs oscillating. This improves handling, comfort and safety.

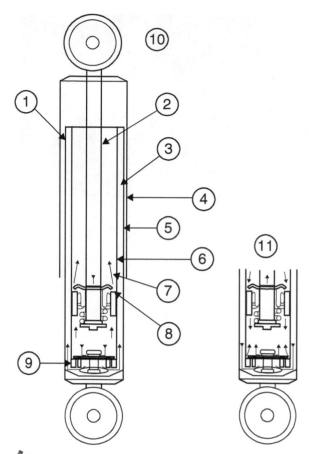

Figure 8.37 Twin tube gas system

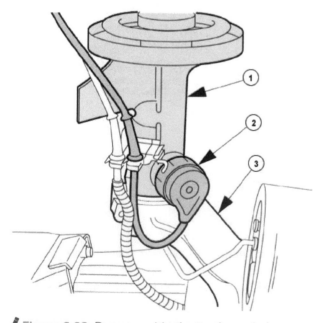

Figure 8.38 Damper with electronic control

Use the media search tools to look for pictures and videos relating to the subject in this section

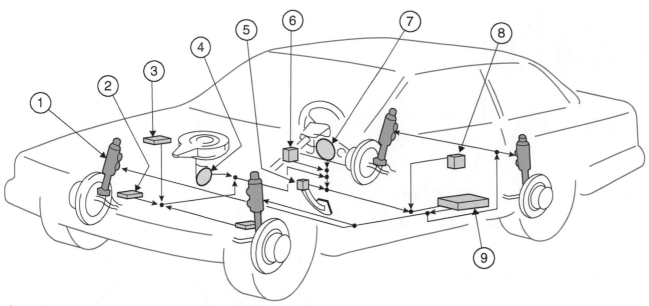

Figure 8.39 Electronic suspension system

Figure 8.40 Suspension in action

8.1.4 Front suspension layouts ❷

Introduction As with most design aspects of the vehicle, compromise often has to be reached between performance, body styling and cost. The following pages compare the main front axle suspension systems.

Wishbone suspension Twin, unequal length wishbone suspension, is widely used on light vehicles. A coil spring is fitted between two suspension arms. The suspension arms are 'wishbone' shaped and the bottom end of the spring fits in a plate in the lower wishbone assembly. The top end of the spring is located in a section of the body. The top and bottom wishbones are attached to the chassis by rubber

bushes. A damper is fitted inside the spring. The stub axle and swivel pins are connected to the outer ends of the upper and lower wishbones by ball or swivel joints.

Figure 8.41 Twin, unequal length wishbone system

Strut-type suspension This type of suspension system has been used now for many years. It is often referred to as the MacPherson system. With this system, the stub axle is combined with the bottom section of a telescopic tube, which incorporates a damper. The bottom end of the strut is connected to the outer part of a transverse link by means of a ball joint. The inner part of the link is secured to the body by rubber bushes. The top of the strut is fixed to the vehicle body by a bearing, which allows the complete strut to swivel. A coil spring is located between the upper and lower sections of the strut.

Figure 8.42 Suspension strut in position

Double transverse arms This type of suspension system results in independently suspended wheels, which are located by two arms perpendicular to the direction of travel. The arms support stub axles. This system allows a low bonnet line and there are only slight changes of track and camber with suspension movements. However, a large number of pivot points are required and production costs are high.

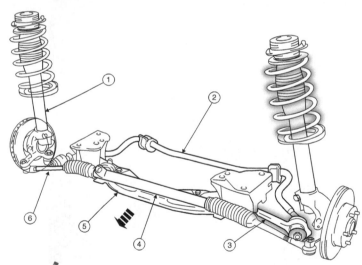

Figure 8.43 Transverse arms and coil spring

Transverse arms with leaf spring With this system, a transverse arm and a leaf spring locate the wheel. The spring can act as an anti-roll bar, hence the low cost. However, it has a harsh response when lightly loaded. There are major changes of camber as the vehicle is loaded. This system is not used on modern cars.

Transverse arm with MacPherson This system is a combination of the spring, damper, wheel hub, steering arm and axle joints in one unit. It is the most popular method in current use. There are only slight changes in track and camber with suspension movement. Forces on the joints are reduced because

of the long strut. However, the body must be strengthened around the upper mounting and a low bonnet line is difficult.

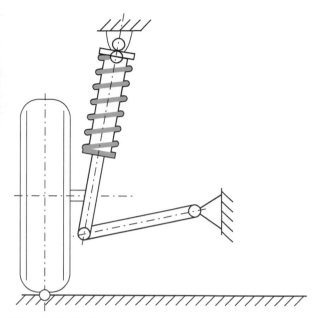

Figure 8.44 Strut system

Double trailing arms In this system, two trailing arms support the stub axle. These can act on torsion bars and are often formed as a single assembly. There is no change in castor, camber or track with suspension movement. It can be assembled and adjusted off the vehicle. However, lots of space is required at the front of the vehicle and it is expensive to produce. Acceleration and braking cause up and down movements which, in turn, change the wheelbase.

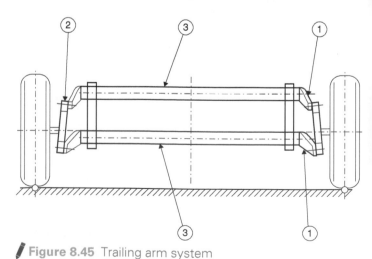

Figure 8.45 Trailing arm system

Remember! There is more support on the website that includes additional images and interactive features: www.tomdenton.org

Summary There are just four main types of front suspension layouts. However, the two most popular are the MacPherson and the unequal length wishbone systems.

 Use the media search tools to look for pictures and videos relating to the subject in this section

8.1.5 Rear suspension layouts 2

Introduction The systems used for the rear suspension of light vehicles vary depending on the requirements of the vehicle. In addition, the systems are different if the vehicle is front- or rear-wheel drive. Older and heavy vehicles use leaf-type springs. The two main types using independent rear suspension are the:

▶ strut type for front-wheel drive
▶ trailing and semi-trailing arm with coil springs for rear-wheel drive.

Strut system The strut type is very much the same as used at the front of the vehicle. Note that suitable links are used to allow up and down movement but to prevent the wheel moving in any other direction. Some change in the wheel geometry is designed in, to improve handling on corners.

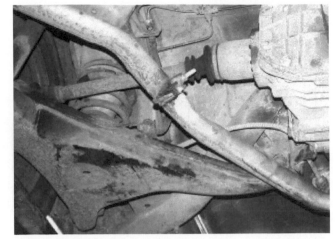

Figure 8.47 Rear semi-trailing arm

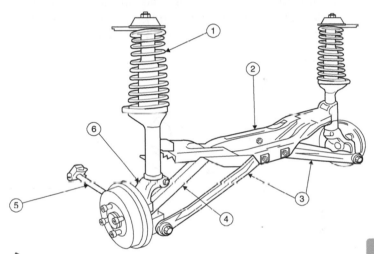

Figure 8.48 Rear suspension struts

Trailing arms and semi-trailing arms Trailing arm suspension and semi-trailing arm suspension both use wishbone-shaped arms hinged on the body. Trailing arms are at right angles to the vehicle centreline and semi-trailing arms are at an angle. This changes the geometry of the wheels as the suspension moves. The final drive and differential unit are fixed with rubber mountings to the vehicle body. Drive shafts must therefore be used to allow drive to be passed from the fixed final drive to the moveable wheels. The coil springs and dampers are mounted between the trailing arms and the vehicle body.

Rigid axle with leaf springs The final drive, differential and axle shafts are all one unit. With this system, the rear track remains constant, reducing tyre wear. It has good directional stability because no camber change causes body roll on corners. This is a

Figure 8.46 Rear struts

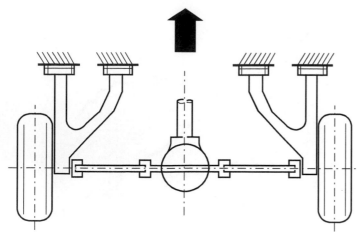

Figure 8.49 Trailing arms

strong design for load carrying. However, it has a high unsprung mass. The interaction of the wheels causes lateral movement, reducing tyre adhesion when the suspension is compressed on one side.

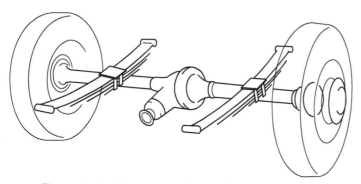

Figure 8.50 This method is used on older and heavy vehicles

Rigid axle with A-bracket This system has a solid axle with coil springs and a central joint supports the axle on the body. It tends to make the rear of the vehicle pull down on braking, which stabilises the vehicle. It results in a high unsprung mass.

Rigid axle with compression/tension struts Coil springs provide the springing and the axle on this system is located by struts. Suspension extension is reduced when braking or accelerating and the springs are isolated from these forces. However, there are high loads on the welded joints, and it has a high overall weight and a large unsprung mass.

Torsion beam trailing arm axle Two links are used, connected by a 'U' section that has low torsional stiffness but a high resistance to bending. Track and camber does not change as the suspension moves. It has a low unsprung mass and is simple to produce. It is a space-saving design but torsion bar springing on this system can be more expensive than coil springs.

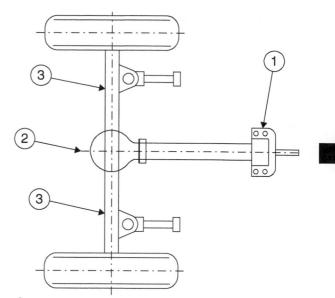

Figure 8.51 'A' bracket system

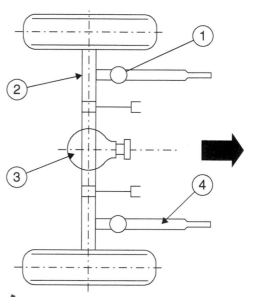

Figure 8.52 Welded compression/tension struts

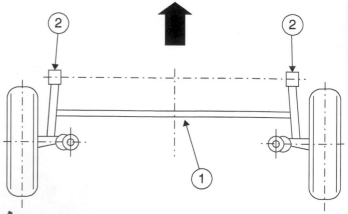

Figure 8.53 Torsion beams twist to provide the spring action

Torsion beam axle with Panhard rod Two links are welded to an axle tube or 'U' section. The lateral forces are taken by a Panhard rod. Track and camber does not change as the suspension moves and simple flexible joints connect it to the bodywork. Torsion bar springing on this system can be more expensive than coil springs.

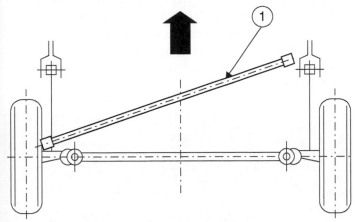

Figure 8.54 Panhard rod and torsion beam system

Trailing arms Trailing arms are always mounted with the pivots at 90° to the direction of travel. When braking, the rear of the vehicle pulls down, giving stable handling. Track and camber does not change and the design is space saving. There is a slight change of wheelbase when the suspension is compressed. Variable length driveshafts with two UJs are used.

Semi-trailing arms – fixed length drive shafts
The semi-trailing arms are pivoted at an angle to the direction of travel. Only one UJ is required. This is because the radius of the suspension arm is the same as the driveshaft when the suspension is compressed. There is only a very small dive when braking. This system has a lower cost than when variable length driveshafts are used. There are however, sharp changes in track when the suspension is compressed. This results in tyre wear and there is a slight tendency to oversteer.

Semi-trailing arms – variable length drive shafts
With this method, the final drive assembly is mounted to the body and two UJs are used on each shaft. The two arms are independent of each other and only slight track changes occur. However, there are large camber changes. The system has a relatively high cost because of the drive shafts and joints.

Summary There are a number of rear suspension systems. Each has advantages and disadvantages. Engineers strive to achieve the optimum design. The system shown in Figure 8.57 is known as 'quadralink'. It is very similar to front strut systems.

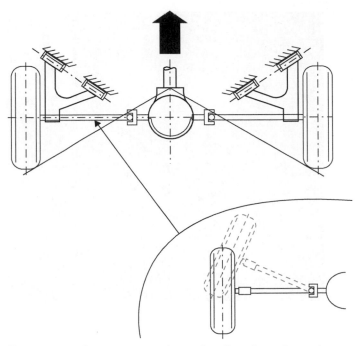

Figure 8.55 Arms at an angle to the direction of travel

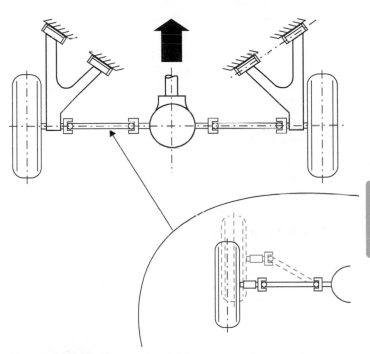

Figure 8.56 Track remains fairly constant but camber changes are large

Electronically Controlled Suspension Electronic control of suspension or active suspension, like many other innovations, was born in the Grand Prix world. It is now slowly becoming more popular on production vehicles.

Conventional suspension systems are always a compromise between soft springs for comfort and harder springing for better cornering ability. Active systems have the ability to switch between the two extremes.

8

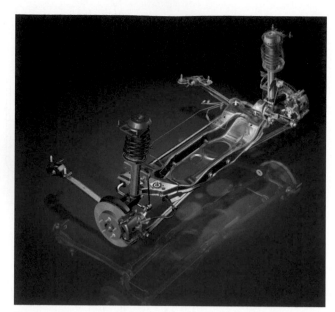

Figure 8.57 Quadralink suspension (Source: Ford Media)

Electronic control One way of achieving the ideal springing is by replacing the conventional springs with double-acting hydraulic units. These are controlled by an ECU, which receives signals from various sensors. Oil at a pressure in excess of 150 bar (that's 150 times atmospheric pressure) is supplied to the hydraulic units from a pump.

Benefits of active suspension The main benefits of active suspension are as follows:

▶ improvements in ride comfort, handling and safety
▶ predictable control of the vehicle under different conditions
▶ no change in handling between laden and unladen.

Delphi MagneRide Delphi have developed an active damper as part of their 'Unified Chassis Control' system. This system uses a type of magnetic fluid that changes the damping characteristics almost instantly, as solenoids are activated.

Figure 8.58 Variable damping technology

Revolution in damper technology The presentation of the new BMW 7 Series is helping to create a breakthrough for electronically controlled vehicle damping. Four gas-pressure twin tube dampers form the core of the system in the BMW 7 Series. Their interior valves are continuously adjustable. A control valve is integrated in each of the damper's pistons. The electrical wiring runs through the hollow piston rod. Three sensors located in the front and rear axle areas detect vertical acceleration and pass this information on to a microprocessor, which has infinitely variable control over the damping characteristics.

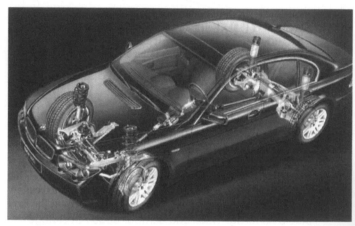

Figure 8.59 Sachs damper system on the BMW

Summary The benefits are considerable and as component prices fall, the system will become available on more vehicles. It is expected that even off-road vehicles may be fitted with active suspension in the near future.

 Create a mind map to illustrate the features of a key component or system

8.1.6 Active suspension ❸

A traditional or a conventional suspension system, consisting of springs and dampers, is passive. In other words, once it has been installed in the car, its characteristics do not change.

Advantages and disadvantages The main advantage of a conventional suspension system is its predictability. Over time the driver will become familiar with a car's suspension and understand its capabilities and limitations. The disadvantage is that the system has no way of compensating for situations beyond its original design.

Components Active suspension systems consist of the following components:

- ▶ electronic control unit (ECU)
- ▶ adjustable dampers and springs
- ▶ sensors at each wheel and throughout the car
- ▶ levelling compressor (some systems).

Components vary between manufacturers, but the principles are the same.

Operation Active suspension works by constantly sensing changes in the road surface and feeding that information to the ECU, which in turn controls the suspension springs and dampers. These components then act upon the system to modify the overall suspension characteristics by adjusting damper stiffness, ride height (in some cases) and spring rate.

Figure 8.60 Jaguar suspension system (Source: Jaguar Media)

Active suspension An active suspension system (also known as computerized ride control) has the ability to adjust itself continuously. It monitors and adjusts its characteristics to suit the current road conditions. As with all electronic control systems, sensors supply information to an ECU which in turn outputs to actuators. By changing its characteristics in response to changing road conditions, active suspension offers improved handling, comfort, responsiveness and safety.

Figure 8.62 Audi system components (Source: www. robson.m3Rlin.org/cars)

Example situation Assume that a car with conventional suspension is cruising down the road and then, after turning left, hits a series of potholes on the right-hand side, each one larger than the next. This would present a serious challenge to a conventional suspension system because the increasing size of the holes could set up an oscillation loop and bottom out the system. An active system would react very differently.

Active system reaction Sensors send information to the ECU about yaw and lateral acceleration. Other sensors measure excessive vertical travel, particularly in the right-front region of the car, and a steering angle sensor provides information on steering position. The ECU analyses this information in about

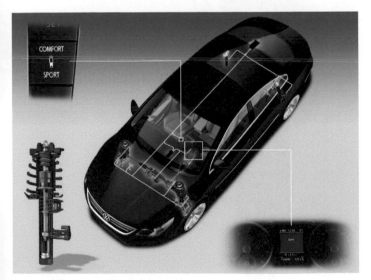

Figure 8.61 Active suspension also allows adjustments, in this case, between sport and comfort settings (Source: Volkswagen Media)

10 milliseconds. It then sends a signal to the right-front spring to stiffen up. A similar signal is sent to the right-rear spring, but this will not be stiffened as much. The rigidity of the suspension dampers on the right-hand side of the vehicle is therefore increased. Because of these actions, the vehicle will drive through the corner, with little impact on driveability and comfort.

Sensors One of the latest types of sensor is produced by Bosch. The sensor simultaneously monitors three of a vehicle's movement axes – two acceleration or inclination axes (ax, ay), and one axis of rotation (Ωz). Previously, at least two separate sensors were required for this. The integration of the sensors for lateral-acceleration and yaw rate reduces space requirements in the vehicle and the assembly work for the complete system.

Actuators There are a number of ways of controlling the suspension. However, in most cases it is done by controlling the oil restriction in the damper. On some systems ride height is controlled by opening a valve and supplying pressurised fluid from an engine driven compressor.

Later systems are starting to use special fluid in the dampers that reacts to a magnetic field, which is applied from a simple electromagnetic coil. The case

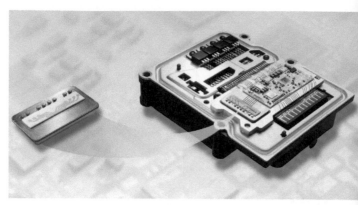

Figure 8.64 Sensor in the ECA (Source: Photo Bosch)

study of a Delphi system in the next section looks at this method in detail.

Figure 8.63 Integrated sensor (Source: Photo Bosch)

Figure 8.65 Suspension strut and actuator connection (Source: Delphi Media)

Summary The improvements in ride comfort are considerable, which is why active suspension technology is becoming more popular.

In simple terms, sensors provide the input to a control system that in turn actuates the suspension dampers in a way that improves stability and comfort.

 Create a mind map to illustrate the features of a key component or system

8.1.7 Delphi MagneRide ❸

Description MagneRide was the industry's first semi-active suspension technology that employed no electro-mechanical valves and small moving parts. The MagneRide Magneto-Rheological (MR) fluid-based system consists of MR fluid-based single tube struts, shock absorbers (dampers), a sensor set and an on-board controller.

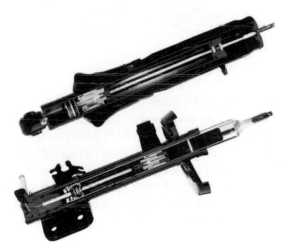

Figure 8.66 MagneRide suspension components (Source: Delphi Media)

System operation Magneto-Rheological (MR) fluid is a suspension of magnetically soft particles such as iron microspheres in a synthetic hydrocarbon base fluid. When MR fluid is in the 'off' state, it is not magnetized and the particles exhibit a random pattern. But in the 'on' or magnetized state, the applied magnetic field aligns the metal particles into fibrous structures, changing the fluid rheology to a near plastic state.

Rheology is the study of friction between liquids.

Performance advantages By controlling the current to an electromagnetic coil inside the piston of the damper, the MR fluid's shear strength is changed, varying the resistance to fluid flow. Fine tuning of the

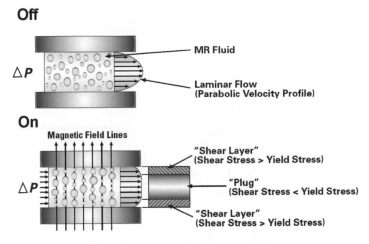

Figure 8.67 Fluid in the on and off states (Source: Delphi Media)

magnetic current allows for any state between the low forces of 'off' to the high forces of 'on' to be achieved in the damper. The result is continuously variable, real time damping.

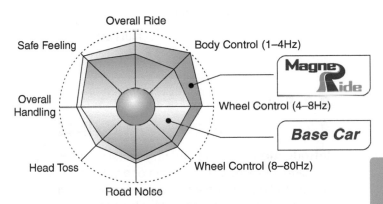

Figure 8.68 Representation of improvements when suspension is controlled (Source: Delphi Media)

Control system The layout here shows the inputs and outputs of the MagneRide system. Note the connections with the ESP system and how the information is shared over the controller area network (CAN).

Summary The MagneRide system, produced by Delphi, uses a special fluid in the dampers. The properties of this fluid are changed by a magnetic field. This allows for very close control of the damping characteristics and a significant improvement in ride comfort and quality.

 Use a library or the web search tools to further examine the subject in this section

369

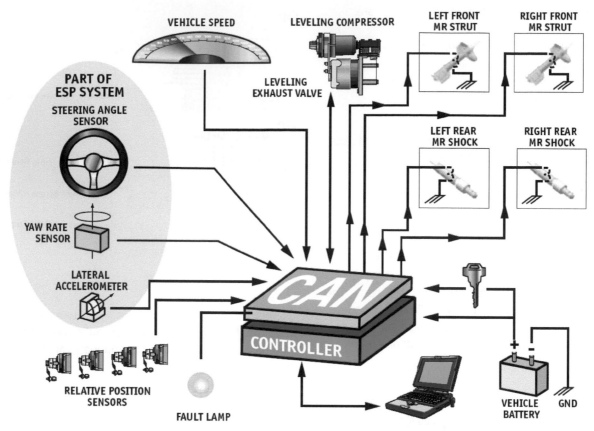

Figure 8.69 Control system (Source: Delphi Media)

8.2 Steering

8.2.1 Steering introduction 🄴 🄵

Introduction Accurate steering systems are critical for the safe and efficient operation of a vehicle and were developed before the internal combustion engine. As technology progresses into different powertrains, the steering system will follow the same principles. Steering systems work in partnership with suspension, with many cars incorporating passive steering or even four-wheel steering into their design. Knowledge and understanding of these areas will remain essential to the technician, regardless of what provides power.

Swinging-beam system On early vehicles, a swinging-beam axle was steered by way of a single pivot (fifth wheel) situated in the centre of the vehicle. The steering was inaccurate, there was a serious risk of overturning and the tyre wear was significant. This inaccuracy is a problem for vehicles that travel at high speeds, making this method unusable for modern vehicles.

Ackermann In 1817, Rudolf Ackermann patented the first stub axle steering system in which each front wheel was fixed to the front axle by a pivot point. An idea developed by Munich carriage builder,

Lankensperger. This made it possible to cover a larger curve radius with the wheel on the outside of the curve than the curve formed by the front wheel on the inside. The two pivot points at the wheels can be achieved with a kingpin or ball joint arrangement to provide the steering action from the track rods. Methods used to translate the work from the steering wheel to the stub axles may vary, but this principle remains the same.

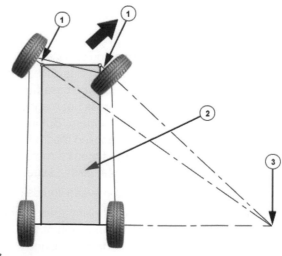

Figure 8.70 Swinging beam axle steering

The advantage of this principle is that it is accurate and reduces tyre wear, even at higher speeds.

The Necessity for Steering Systems Motor vehicles are generally steered via the front wheels, the rear wheels following the front wheels on a smaller radius. With motor vehicles, two factors must be considered:

▶ Dead weight or axle loading
▶ Steered wheels' contact area.

Figure 8.71 Steering components

Steering and Suspension must always be regarded as a unit. If the suspension system is not working correctly, it will have a considerable influence on the vehicle's steering characteristics. For example, defective shock absorbers or dampers reduce the wheel contact with the road, limiting the ability to steer the vehicle. The driving safety of a motor vehicle depends largely on the steering. Reliable steering at high speeds is required, together with easy manoeuvrability.

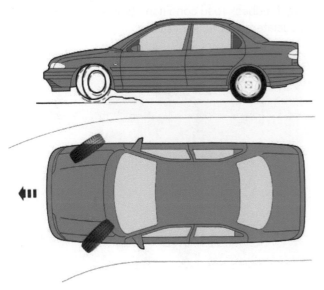

Figure 8.72 Suspension and steering interact

Manoeuvrability Crucial to the manoeuvrability of a motor vehicle is the turning circle, which in turn is directly dependent on the track circle. Designers strive for the smallest possible track and turning circle. The wheel housing should enclose the wheels as tightly as possible; however, sufficient clearance must be left so that the tyres do not rub when the wheels are turned.

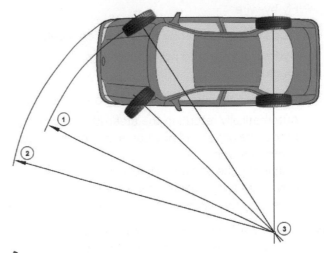

Figure 8.73 Front axle geometry

In this type of steering, the stub axle of the steered front wheel is swivelled about the steering axis. When steering, the wheelbase remains constant. The space between the steered wheels can be used for the installation of deep-seated components such as the power plant. The low centre of gravity contributes to road handling characteristics by reducing body roll. Even at large steering angles, the stability of the vehicle is maintained since the area of support is only slightly reduced.

 Look back over the previous section and write out a list of the key bullet points from this section

8.2.2 Steering control ❶ ❷

Introduction To overcome the friction forces more easily, many different types of steering gear have been developed. Powered steering reduces the effort required and increases safety and comfort. Transmitting the steering movements of the driver to the wheels requires several components. The steering movement is transmitted by way of the steering wheel, shaft, gear and linkage to the front wheels. The rotational movement of the steering wheel is transmitted via the steering shaft to the steering pinion in the steering gear. The steering shaft is supported in the steering column tube, which is fixed to the vehicle body.

Methods The methods used by the driver to translate power from the steering wheel to the wheels mainly fall into the following categories:

▶ Steering Box
▶ Rack and pinion
▶ Fly-by-wire

Methods to assist steering effort are commonly:

▶ Hydraulic (known as power steering)
▶ Electric

Capabilities Steering systems must be capable of:

▶ Automatically returning the steered front wheels to the straight-ahead position after cornering (self-centring action)
▶ Translating the steering wheel rotation so that only about two rotations of the steering wheel are necessary for a steering angle of about forty degrees.

Engineers came up with an innovative idea that requires no mechanical effort for the steering to self-centre, this is achieved through the wheel alignment principles of castor and toe and will be explored in the wheel alignment section.

The forces that can affect steering are considered next.

Bump steer is the term used to describe the tendency of a wheel to steer as it moves through the suspension stroke when a wheel hits a pothole or object. As the wheels rise together after the impact and during the leaning motion of cornering, one wheel toes in, the other out causing alignment principles to come into play and steering the vehicle. Bump steer is overcome with an effective suspension arrangement. A vehicle suffering from severe bump steer may have worn suspension components, be heavily loaded, or have been modified.

Roll steer is the term used to describe a similar effect to bump steer, whereas bump steer exists when two wheels rise together, a live rear back axle will only cause the impacted wheel to be affected.

Torque steer generally affects high powered, front wheel drive cars and describes the influence of torque on steering during heavy acceleration. Honda overcame torque steer with the Civic Type R by arranging the suspension to accommodate the steering axis within the wheel, reducing the scrub radius and this significantly reduces the tug the motorist feels at the steering wheel.

Steering box is the term used to describe a gear arrangement that transmits driver input to the steering components. This is a popular method to steer heavy or rear wheel drive vehicles due to diversity of gearing

ratios available in comparison to a rack. There is less feedback through a steering box than a rack and pinion, making it less preferable for precision steering.

Rack and pinion steering developed in the history of the car, became more popular when front-wheel drive was used more, as it requires little space and production costs are lower. The first hydraulic power steering was produced in 1928. There was no great demand for this until the fifties caused the development of power steering systems to stagnate.

Figure 8.74 Steering rack

Power Steering Systems Increasing standards of comfort stimulated the demand for power steering systems. Speed-sensitive or variable-assistance power steering (VAPS) systems were developed using electronic controls. These represent the latest major innovation to the steering system in production vehicles. The demand for safety and comfort will lead to further improvements in steering systems and how they feedback road conditions to the driver and any associated automatic systems.

Figure 8.75 Power steering pump

Fly by wire steering has the advantage of removing the lag of mechanical steering, providing a crisp steering response to the motorist. The other advantage is the removal of unwanted steering feedback, such as poor road surface, whilst communicating information the motorists wants, like a slippery surface. Fly by wire integrates with advanced driver assistance systems, such as lane assist to improve safety. Nissan's development of this system has a mechanical link to the wheels and uses a clutch system. Patents made by Tesla suggest that the fly by wire system also has power efficiency gains.

Create a mind map to illustrate the features of a key component or system

8.2.3 Steering components ❶ ❷

Construction of the steering system In order to transmit the steering movements of the driver to the wheels, several components are required. The steering movement is transmitted by way of the steering wheel, shaft, gear and linkage to the front wheels. The rotational movement of the steering wheel is transmitted via the steering shaft to the steering pinion in the steering gear. The steering shaft is supported in the steering column tube, which is fixed to the vehicle body.

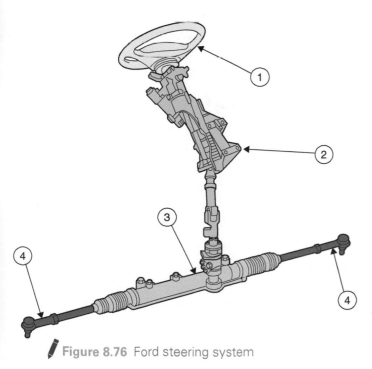

Figure 8.76 Ford steering system

The **steering gear** translates (reduces) the steering force applied by the driver. It also converts the rotational movement of the steering wheel into push or pull movements of the track rods. The converted movement is transmitted to the linkage, which in turn moves the wheels in the desired steering direction. Track rods are required to transmit the steering movement from the steering gear to the front wheels. Different track rods are used depending on the type of front axle.

One-piece track rod moved by drop arm This is the simplest design of steering linkage needing only three joints. One-piece track rods are found only with rigid axles since the distance of the steering swivel pins or joints cannot vary.

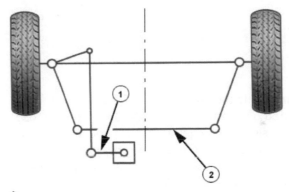

Figure 8.77 One-piece track rod

Two-piece track rod moved by drop arm Two-piece track rods may be split centrally or to one side. They are necessary on vehicles with independent suspension, since the suspensions of the steered wheels are compressed independently of one another. The split reduces the effect of bump steering.

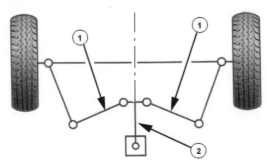

Figure 8.78 Two-piece track rod

Two-piece track rod moved by rack In this type of steering, frequently fitted to light vehicles, the steering linkage is operated by the rack of a rack and pinion gear. Two designs are encountered. The rack either forms part of the track rod or acts directly on the split track rod.

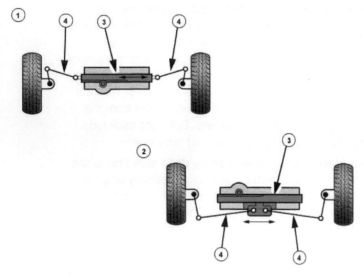

Figure 8.79 Two-piece track rod

The **kingpin** is the predecessor of the ball joint. It is only fitted on commercial vehicles and a few off-road vehicles, since these generally have rigid front axles in which the distances of the track rods do not vary. The kingpin is not maintenance free; it must be supplied with grease via a grease nipple.

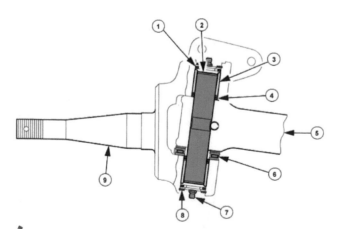

Figure 8.80 Kingpin on a steered front axle

Ball joints allow parts of the steering linkage to rotate about the longitudinal axis of the ball joint. They also allow limited swivel movements transversely to the longitudinal axis. The lubricated ball pivot is supported in steel cups or between preloaded plastic cups. A gaiter prevents lubricant losses. Ball joints are generally maintenance-free and must always be renewed if the gaiter is damaged.

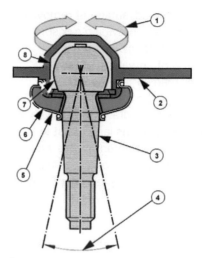

Figure 8.81 Ball joints are used instead of kingpins

Worm and nut steering gear This system consists of a steering screw on which the steering nut is displaced axially as the steering wheel is moved. Slide rings on the circumference of the steering nut transmit the movement to the steering fork and thereby to the drop arm. The drop arm performs a movement of up to 90°. In this type of steering, the wear is relatively high. The steering nut play cannot be adjusted and this is a disadvantage. With this type of steering gear, the steering is linear.

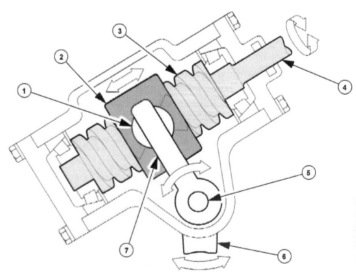

Figure 8.82 Screw and nut steering box

Recirculating ball steering gear Owing to the high friction in the screw and nut steering gear, gears with roller friction have become more common. In the recirculating ball steering gear, the steering screw and steering nut have ball groove threads. The threads do not touch one another because they form channels for the balls. When the steering screw is turned, the balls roll in the ball groove thread in two closed recirculating ball races. The balls are returned by two tubes. The drop arm is moved by means of a gear sector. The advantage of the recirculating ball steering gear is that it functions virtually free of wear.

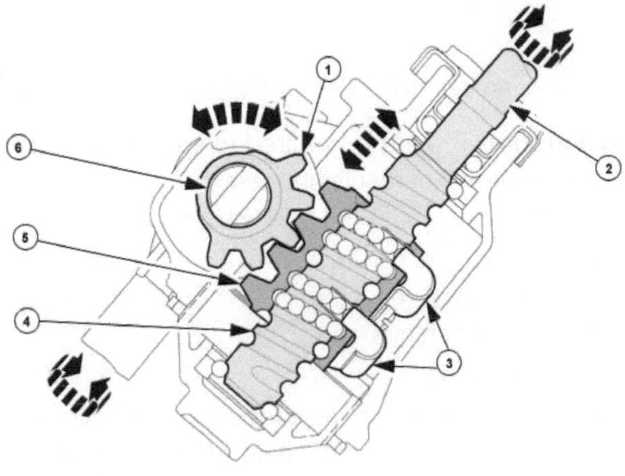

movement about its centre when the steering wheel is turned. The drop arm can perform a swivel movement of up to 90°. Advantages include low wear, ease of steering and only a small space is required. The steering play can be adjusted and the steering is free of play when running in a straight line. With this type of steering gear, the steering is linear.

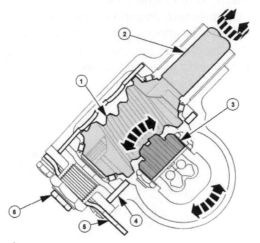

Figure 8.83 Recirculating ball system

Figure 8.85 The roller moves with the worm

The **worm and sector steering gear** has a cylindrical worm, which due to its screw motion turns a steering sector back and forth. The drop arm is fixed to the sector. It can perform a swivel movement of up to about 70°. Worm steering gears are characterised by high transmission ratios, for example 22:1. One disadvantage is the high wear due to the sliding friction between the sector and the cylindrical worm. In addition, it requires large steering forces. In this type of steering gear, the steering is linear.

The **worm and rolling finger steering gear** has a cylindrical screw with an uneven thread pitch. When the worm is rotated, the tapered rolling finger rolls on the flanks of the worm. The rolling finger is displaced. This movement is converted by the shaft into a swivel movement of the drop arm. This system has low wear and ease of steering. The longitudinal play of the worm and the shaft, and the play between rolling finger and worm thread, are adjustable. In this type of steering gear, the steering is progressive due to the uneven thread pitch on the worm.

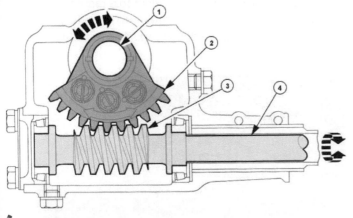

Figure 8.84 The sector moves with the worm

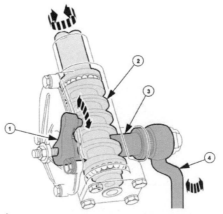

Figure 8.86 Progressive steering boxx

The **worm and roller steering gear** has a roller instead of the sector. The steering worm is not cylindrical but tapers towards the middle like an hourglass. The roller, driven by the worm, can thus perform a steering

Remember! There is more support on the website that includes additional images and interactive features: www.tomdenton.org

8

Rack and pinion steering The steering rack housing generally contains a helically toothed pinion, which meshes with the rack. By turning the steering wheel and hence the pinion, the rack is displaced transversely to the direction of travel. A spring-loaded pressure pad presses the rack against the pinion. For this reason the steering gear always functions without backlash. At the same time, the sliding friction between pressure pad and rack acts as a damper to absorb road shocks. Advantages of rack and pinion steering include the shallow construction, the very direct steering, the good steering return and the low cost of manufacture.

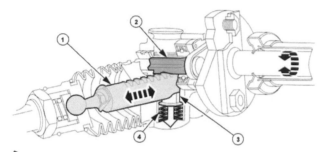

Figure 8.87 Linear steering rack

Variable pitch rack The basic construction and the advantages are similar to those of a rack and pinion steering gear with constant pitch. In a rack and pinion steering gear with variable pitch, a rack is used which has teeth that diminish in size towards the ends. This makes it possible to increase the transmission ratio constantly. This means, in practice, that more steering wheel turns are required but that less effort is required in order to turn the wheels. As a result, the steering moves more easily when applying lock than when moving in a straight line. This makes parking considerably easier.

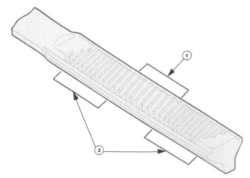

Figure 8.88 Progressive steering rack

Summary There is a wide range of steering boxes and steering layouts. On light vehicles, the most common, by far, is the steering rack. This

is because it has a shallow construction, is very direct, has good steering return and the cost of manufacture is low.

> Make a simple sketch to show how one of the main components or systems in this section operates

8.2.4 Power steering introduction **1** **2**

Introduction The effort required to steer the front wheels depends primarily on the axle load. This is particularly apparent in the following situations:

▶ low speed
▶ low tyre pressures
▶ large tyre contact area
▶ tight cornering.

Steering ratio cannot be increased too much, because a large number of steering wheel turns would be necessary for the steering movement. Generally, a steering force of 250N should not be exceeded. Therefore, the need arises for power steering in heavy cars, trucks and buses. Power assistance is generally produced by hydraulic pressures – however, electric systems are now becoming popular.

The requirements of a power steering system are:

▶ precise onset of power assistance
▶ maintenance of driver feel
▶ continued ability to steer should the power system fail.

Hydraulic power assisted steering (PAS) systems use an engine driven pump to supply pressurised fluid. A control valve directs the fluid to a ram that assists with movement of the steering. If the fluid supply or ram fail, the steering works like a manual system.

Figure 8.89 PAS rack

Figure 8.90 PAS pump

Electric Power Steering Early electric power assisted steering systems used a motor to drive a hydraulic pump. It is now becoming common for the electric motor to act directly on to the steering rack, or the steering shaft. An advantage of electric over hydraulic is that it draws less horsepower, making the vehicle more fuel efficient. Hydraulic systems can use a relatively small force to move a large load with few components, making a heavy vehicle easy to manoeuvre.

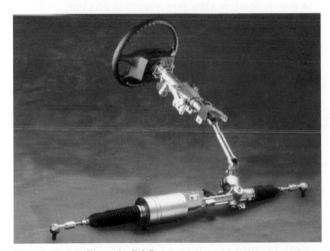

Figure 8.91 Electric PAS

Four-wheel steering is easier to understand if you recall or imagine the effects of rear-wheel steering. If you have ever driven, or watched the movement of a forklift truck, you will realise the different effect moving the rear wheels has on vehicle position. This is the same effect on a normal car when reversing – it is why some drivers have trouble reversing into a parking slot or out of a garage! The key point is that the trailing end of the vehicle tends to slew in the direction that the wheels are turned.

Direction of movement When all four wheels are turned, the overall effect on the vehicle changes again. The effect varies depending on which way the rear wheels are moved. The effects could be described as a turn or a drift. At low speeds, the wheels are turned in opposite directions to improve the drag or slip on the tyres as well as reducing the turning circle. At high speeds, the wheels are turned in the same direction, such as for changing lanes on a motorway. The amount of turn on the rear wheels is much less than the front.

Four-wheel steering system Figure 8.92 shows the layout of the components on one system in current use. As is common with many, if not all aspects of the vehicle, electronic control is now playing a role in four-wheel steering systems. This is used to determine the amount and direction of rear-wheel movement.

Summary Power assistance is used to make steering operation easier. This also improves safety. Two sources of 'power assistance' are hydraulic and electric. Hydraulic systems were common but the use of electric systems is becoming more popular. The more complex four-wheel steering systems improve vehicle handling but add significant cost to the manufacture of the vehicle. Despite this, pressure is increasing to develop fuel efficient manoeuvrability, especially as emissions become more tightly scrutinised.

8

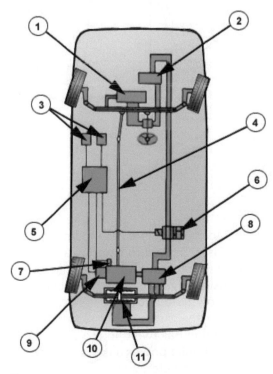

Figure 8.92 System Components

Using images and text, create a short presentation to show how a component or system works

8.2.5 Hydraulic power steering ☑

Introduction Hydraulic power assisted steering (PAS) systems use an engine driven pump to supply pressurised fluid. A control valve directs the fluid to a ram that assists with movement of the steering. If the fluid supply or ram cylinder fail, the steering works like a manual system.

In power-assisted steering systems of modular design, parts of the steering gear take the form of a hydraulic piston and cylinder. This gives a compact construction. In power steering systems of semi-modular design, an external hydraulic cylinder is fitted to the steering gear, which exerts its force on the steering linkage by way of connecting rods.

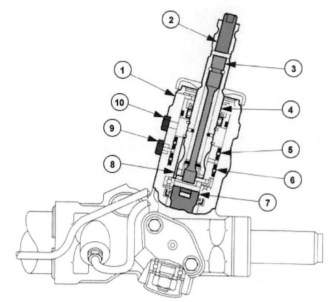

Figure 8.94 Steering rack control unit

When the steering wheel is turned, the control valve is actuated, admitting hydraulic fluid into the ram cylinder. Hydraulic fluid under pressure in the ram cylinder assists with the steering force exerted by the driver. Return hydraulic fluid flows through the outlet at the other end of the ram cylinder into the reservoir. When the steering movement is interrupted, the control piston assumes a neutral position. In this neutral position, the pressure in the ram cylinder is reduced.

With fully hydraulic power steering, the wheel is no longer mechanically connected to the road wheels. Turning the steering wheel actuates the control pump. The hydraulic fluid flowing through the control pump acts on the control valve. As a result hydraulic fluid flows to one side of the ram cylinder, the return fluid flows from the other side of the ram cylinder to the reservoir, the piston of the ram cylinder is displaced, thereby moving the steering linkage. When it is running in a straight line, the hydraulic fluid is delivered directly into the reservoir.

Safety feature Fully hydraulic power steering is only fitted in slow speed tractors and large construction machines. If the power steering pump fails, the ability to steer the vehicle must be ensured by an emergency steering pump, which is connected to the final drive assembly.

Variable assistance power steering (VAPS) is controlled electronically. Variable power steering (sometimes called progressive power steering) makes steering easier at low speeds and provides good driver feel at higher speeds. The main components are shown in Figure 8.95.

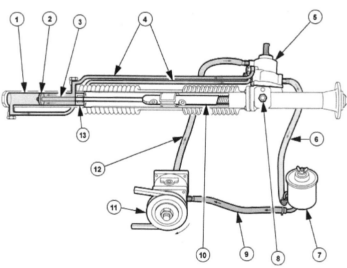

Figure 8.93 PAS system

Control valve Most control valves incorporate a torsion bar. This is designed to twist by a small amount as steering force is exerted. As the torsion bar twists, it allows valves to open and close. These valves supply fluid under pressure to the appropriate side or the ram cylinder. Splines limit the amount of torsion bar twist. In the event of a failure of the hydraulic power assistance, the driver can steer the vehicle by purely mechanical means.

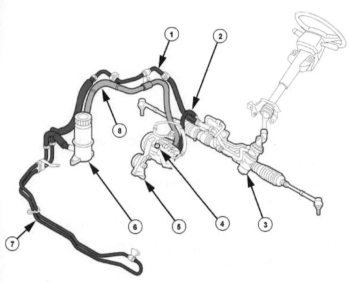

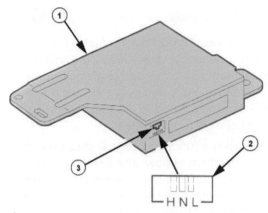

Figure 8.97 Control unit and switch

Figure 8.95 Steering system

Steering angle sensor The electronic control unit monitors the signals from the vehicle speed sensor and the steering position sensor. From this data, it can then work out the power assistance required. The solenoid valve controls the amount of assistance, because the valve in turn controls fluid pressure. Maximum power assistance occurs at speeds less than 10km/h (6mph) or when the steering wheel is rotated more than 45°.

Summary Hydraulic power assisted steering (PAS) systems use an engine driven pump to supply pressurised fluid. A control valve directs the fluid to a ram that assists with the movement of the steering. Variable assistance systems are also used. These usually involve some electronic control. Progressive PAS is controlled in this way or by a restrictor valve that changes with road speed. Pressure switches, when used, often inform engine management systems that PAS is in use. This allows idle speed to be increased if necessary.

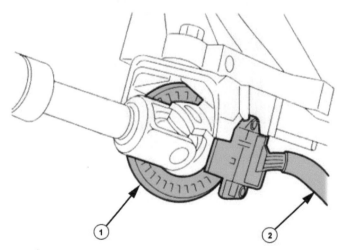

Figure 8.96 This sensor monitors steering position

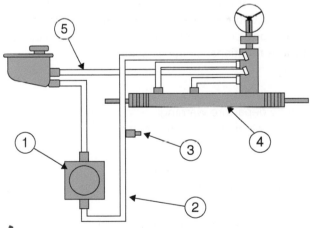

Figure 8.98 Hydraulic circuit

Control module The VAPS electronic control unit has an interesting feature. A slide switch on the side of the unit allows the following settings for the power steering:

▶ Switch position 'H': 10% greater effort required when steering
▶ Switch position 'N': normal adjustment
▶ Switch position 'L': 10% less effort required when steering.

Use a library or the web search tools to further examine the subject in this section

Remember! There is more support on the website that includes additional images and interactive features: www.tomdenton.org

379

8.2.6 Electric power steering ❷❸

Introduction Honda introduced the first 'all electric' power-steering system on the 1993 Acura NSX. Since then, development has continued and electric power assisted steering (EPAS), also called electric power steering (EPS), is now fitted to the majority of new cars and light commercial vehicles.

Figure 8.99 shows a functional control diagram with electrical and mechanical flow. The basic principle is that as the driver turns the steering wheel, sensors register the corresponding steering torque and steering speed with absolute precision. The signals are used by an ECU to calculate the required steering assistance and, based on the calculated results, to control the servomotor. In the final step, the motor transmits the optimum servo torque, via a worm gear or recirculating ball gear. This is applied to the steering column, to a second pinion, or directly to the steering rack.

If the vehicle power supply fails, the driver can continue to steer the vehicle through the mechanical connection between the steering wheel and the steered wheels.

Hydraulic pump Eliminating the hydraulic power steering pump can reduce weight and improve fuel economy. EPS also offers greater handling and steering feel while improving vehicle safety by adapting the steering torque to the vehicle's speed. In addition, it can provide active torque in critical driving situations. There are many advantages to using electric power steering on a vehicle when compared to more traditional hydraulic systems:

Safety

▶ stabilization functions
▶ lane departure warning
▶ obstacle avoidance assistance.

Comfort

▶ straight line running correction
▶ parking assistant

▶ lane keeping assistant
▶ variable assistance.

Steering

▶ steering feel and performance adapted to all conditions
▶ reduced noise
▶ maintenance free
▶ a modular design means less space is used.

Emissions

▶ CO_2 saving: 10g/km* or 20g/km**

Consumption

▶ fuel saving: 0.4 litres/100km* or 0.8litres/100km.**

*New European Driving Cycle (NEDC) with 2-litre spark ignition engine

**Only city traffic

Power consumption Earlier electrical PAS systems used hydraulics but with an electric pump. However, this has now been completely replaced by full-electric systems, for the reasons stated above. One of the key advantages of electric PAS is the reduction in power consumption of about 80%, shown on the chart in Figure 8.100.

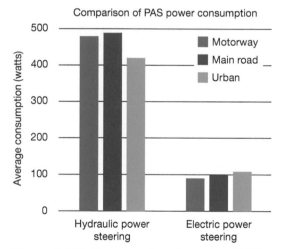

Figure 8.100 Power assisted steering (PAS) power consumption

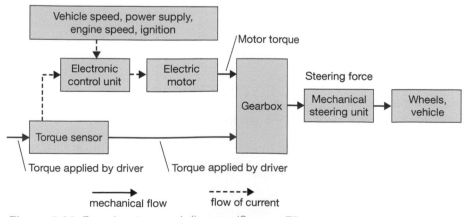

Figure 8.99 Functional control diagram (Source: ZF Servoelectric)

Modular design The modular design of modern systems where all components are combined to one unit, as shown in Figure 8.101, means huge savings in space as well as faster development times. Additional functions such as variable assistance were available on earlier systems but not to the level that is now possible. Steering has become another electronically controlled system where an ECU receives inputs such as steering angle, driver input and road speed. It then outputs to a motor to achieve the best outcome for the situation.

Methods There are two common ways of using electric power for steering assistance (the second is now almost universal):

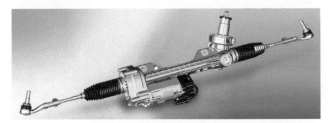

Figure 8.101 Modular electric power steering unit (Source: ZF Servoelectric)

1 Replacing the conventional system pump with an electric motor whilst the ram remains much the same.
2 Using a drive motor, which directly assists with the steering and has no hydraulic components.

The EPAS system is further divided in to three different types, all using rack and pinion steering, and tend to be fitted on different sized vehicles with the axle loads roughly as stated:

1 Servo unit on the steering column (for steering axle loads up to 1 000kg).
2 Servo unit on a second pinion (for steering axle loads up to 1 200kg).
3 Paraxial servo unit (for steering axle loads up to 1 600kg).

Figure 8.102 Electric PAS (Source: Ford Media)

Motor currents in the region of 80A are typical and therefore suitable wiring is needed, as well as powerful control electronics.

Sensors On many systems, an optical torque sensor is used to measure driver effort on the steering wheel (all systems use a sensor of some sort). This sensor works by measuring light from an LED which is shining through holes. These are aligned in discs at either end of a torsion bar fitted into the steering column. An optical sensor element identifies the twist of two discs on the steering axis with respect to each other. From this information the system calculates the torque as well as the absolute steering angle.

One common type of sensor is the Gray code angle-encoder. As a disc is rotated, the digital signal produced by the detectors changes. A 5-bit device is shown in Figure 8.103. The Gray code, named after Frank Gray, is a binary system where two successive values differ in one bit. The reflected binary code was originally designed to prevent spurious output from electromechanical switches but is now also used for error correction. The position (rotation) error due to misalignment is at most one bit.

Figure 8.103 Gray code optical sensor: 1 Light sources, 2 Capture plate, 3 Detectors, 4 Coded disc

Hall effect Another type of sensor uses the Hall effect to measure angle. The sensor shown in Figure 8.104, measures relatively over an unlimited range. Its typical steering-angle signal resolution is 1.5°. When using a Hall effect sensor, a multi-pole magnet is fixed to the steering column. Hall elements detect changes in the sensor's magnetic field without contacts and without gear wheels. As two or more Hall elements are used, any rotary motion generates square-wave signals, which show a certain phase shift relative to each other. These square-wave signals are transmitted directly to the control unit.

Processing of the sensor signals is then done by the electronic control unit, which calculates the position, rotation direction, and rotation speed of the steering wheel. The control unit also validates the sensor output signals and detects short circuits. Because it uses the incremental measuring principle, this particular sensor does not have to be calibrated by the automakers.

Figure 8.104 Steering angle sensor that measures relatively over an unlimited measuring range (Source: Bosch Media)

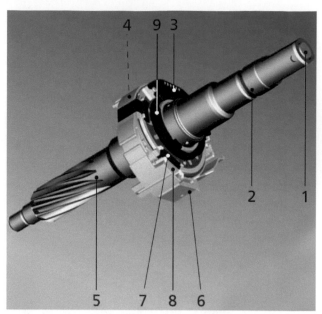

Figure 8.105 Configuration of the torque sensor on the steering pinion: 1 Torsion bar, 2 Input shaft, 3 Sensor module, 4 Clock spring, 5 Steering pinion, 6 Plug, 7 Index magnet (optional), 8 Index sensor (optional), 9 Pole wheel (Source: ZF Servoelectric)

ZF torque sensor This sensor measures the torque applied by the driver at the steering wheel. Based on this, the control unit calculates the steering assistance for the motor. In Figure 8.105 the torque sensor sits on the steering pinion (5). A pole wheel (9) is fitted on the input shaft (2) which is connected to the steering pinion by means of the torsion bar (1). The measuring range covered by the sensor is between +/−8 and +/−10 Nm.

Steering torque If the steering torques are higher, a mechanical angle limiter prevents overload of the torsion bar. When the driver applies a torque on the steering wheel, the torsion bar is rotated, as is the magnet relative to the sensor. The sensor consists of magnetoresistive elements which change their

resistance when the field direction changes. In the process, the voltage follows a sine and cosine curve when the magnet is rotated. The direct rotation angle of the torsion bar is then calculated by means of an inverse tangent function.

The high safety demands on electric steering systems require detection of all faults occurring on the sensor and the creation of a safe condition for the steering system. The sensor data are transmitted to the electronic control unit via a very rugged digital interface. Optionally, the torque sensor can also accommodate an index magnet (7) and an index sensor (8). The index sensor delivers a signal to the ECU for each full steering wheel turn. In combination

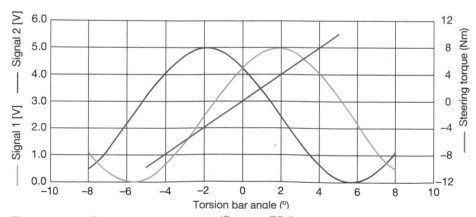

Figure 8.106 Torsion sensor output (Source: ZF Servoelectric)

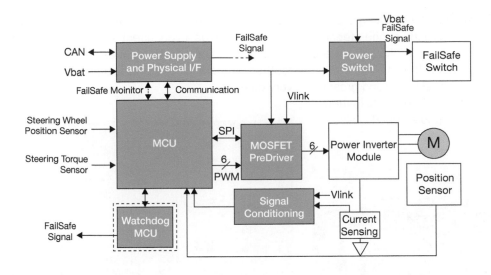

Recommended Products

MCU	Qortwa 32-bit, S12 16-bit, S08 8-bit
Analog	SMARTMOS SBC, physical Interfaces, MOSFET pre-driver

■ Freesacle Technology   ⌐⌐ Optional

Figure 8.107 Electric power steering electrical and electronic systems (Source: Freescale Semiconductor)

with the data from the rotor position sensor and the wheel speeds, the electronic control unit is able to calculate the steering angle with a resolution < 0.05°.

Electronic control The central electronic elements of today's electric power steering systems are 16- and 32-bit MCUs designed for safety-critical applications.

Freescale's 16-bit and Qorivva 32-bit single- and dual-core MCUs provide enhanced computing power and specialized peripherals for complex electric motor control functions. Integrated power supply solutions are also important elements of a power steering control unit. They provide connectivity to automotive buses, such as CAN and LIN.

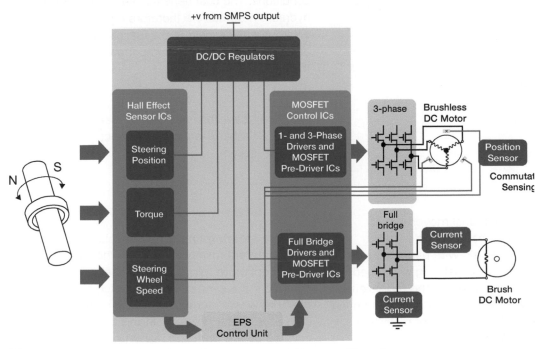

Figure 8.108 Options for AC or DC motor control (Source: Allegro)

Motors and drive mechanism A direct acting system uses an electric motor to work directly on the steering via an epicyclic gear train. This completely replaces the hydraulic pump and servo cylinder. It also eliminates the fuel penalty of the conventional pump and greatly simplifies the drive arrangements. Engine stall when the power steering is operated at idle speed is also prevented.

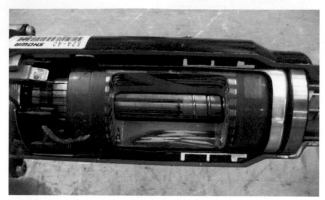

Figure 8.109 Direct acting motor

Figure 8.110 Electric power steering motor (Source: Bosch)

Linear movement The rotational movement of an electric motor must be changed into a linear movement. The ZF Servolectric with paraxial drive system uses a drive concept consisting of toothed belt drive and recirculating ball gear. The recirculating ball gear is a system similar to the non-power steering box technology that has been in use for many years. The noise performance of this system is very good and as such allows the recirculating ball gear to be

rigidly connected to the sub-frame of the vehicle. This in turn gives a very direct steering feel. Despite its ability to transmit high torques safely, the slip-free toothed belt (similar to a timing belt in construction) is very quiet.

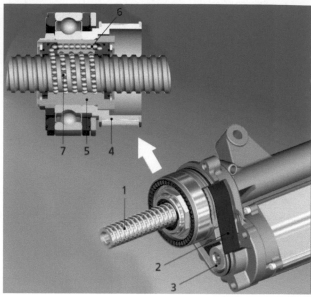

Figure 8.111 Belt drive system: 1 Steering rack, 2 Toothed belt, 3 Toothed disc, small, 4 Toothed disc, big, 5 Ball recirculation nut, 6 Ball return channel, 7 Balls/ball chain (Source: ZF Servolectric)

Summary Electrical PAS occupies little space, something that is at a premium these days, and the motor only averages about 2A under urban driving conditions. The cost benefits over conventional hydraulic methods are therefore considerable. The environmental aspect of the fuel savings of the electrical PAS are another important advantage over the hydraulic system.

 Use a library or the web search tools to further examine the subject in this section

8.2.7 Steering geometry 2

Relative steering angle The wheels of a vehicle cover different distances when cornering. At low speed, optimum rolling of the wheels is only possible if the centrelines of the stub axles, with the front wheels turned, meet the extended centreline of the rear axle. In this case, the paths covered by the front and rear wheels have a common centre. The inside front wheel must, therefore, be turned more than the outside front wheel. This is usually measured with the inside front wheel at a steering angle of 20°.

Wheelbase and track The wheelbase is the distance between the wheel centres of the front and rear wheels. The track is the distance between the wheels, measured from tyre centre to tyre centre on the wheel contact plane. The greater the track and wheelbase, the greater the driving safety, especially when cornering.

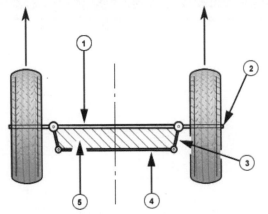

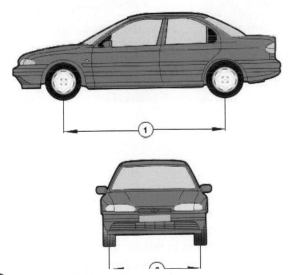

Figure 8.112 The Ackermann principle straight

Figure 8.113 Wheelbase and track

The **wheel toe** is the difference in the distance between the rim flanges in front of and behind the axle in the straight-ahead position. If the distance in front of and behind the axle is the same, the vehicle has zero wheel toe. There is generally some toe-in or toe-out. The wheel toe is given in millimetres or angular degrees and minutes. It is often referred to as tracking. Toe-in occurs when the distance between the rim flanges in the direction of travel is smaller in front of the axle than behind the axle. Toe-out occurs when

the distance between the rim flanges in the direction of travel is greater in front of the axle than behind the axle.

Figure 8.114 The wheel toe is a small angle

Toe-in, toe-out The ideal running direction of the wheels is parallel to the vehicle's longitudinal axis.

Due to deformations in the suspension elements, however, the front wheels are diverted from their ideal line. In the case of front-wheel drive, they are forced inwards in the toe-in direction, and in the case of rear-wheel drive outwards in the toe-out direction. Undesirable toe-out is counteracted by toe-in and undesirable toe-in, by toe-out.

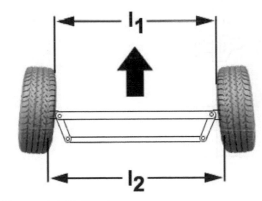

Figure 8.115 Toe-in

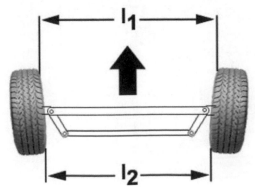

Figure 8.116 Toe-out

385

Camber is the angle between the wheel plane and a line perpendicular to the road. The wheels must be straight ahead. Camber is described as positive when the wheel is inclined out at the top. It has the effect of reducing scrub radius and influences the wheel forces when cornering. Camber is described as negative when the wheel is inclined in at the top. It produces a good cornering force and allows a low vehicle centre of gravity.

Figure 8.120 Drift cars use positive camber

Figure 8.117 Positive camber

Figure 8.118 Negative camber

Figure 8.121 Camber is a small angle

Figure 8.119 F1 cars use negative camber

The **scrub radius** is the distance between the contact point of the steering axis with the road surface plane and the wheel centre contact point. The function of the scrub radius is to reduce the steering force, prevent shimmy and stabilise the straight-ahead position.

Negative scrub radius When the point of contact of the steering axis with the road surface is between the wheel centre and the outside of the wheel, it is termed negative scrub radius. The result of negative scrub radius is that the brake forces acting on the wheel produce a torque, which tends to turn the wheel inwards. As a result, the wheel with the greater braking action is turned inwards, i.e. steered away from the more heavily braked side. This produces automatic countersteer, stabilising the vehicle.

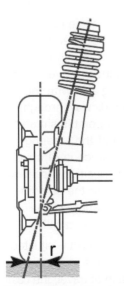

Figure 8.122 Scrub radius – negative

Positive scrub radius When the contact point of the steering axis with the road surface is between the wheel centre and the inside of the wheel, this is termed positive scrub radius. The greater the positive scrub radius, the more easily the wheels can be turned. The result of positive scrub radius is that the brake forces acting on the wheel produce a torque, which tends to turn the wheel outwards. With a large positive scrub radius, the vehicle can be steered very easily. 'Disturbing' forces, such as different road surfaces, act on a long lever and can produce an unwanted steering angle.

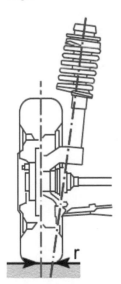

Figure 8.123 Scrub radius – positive

Zero scrub radius When the point of contact of the steering axis with the road surface is in the wheel centre the semi radius is zero. With a zero scrub radius, the wheel swivels on the spot. Steering is

heavy when the vehicle is stationary, since the wheel cannot roll at the steering angle. In this case, no separate torques occur.

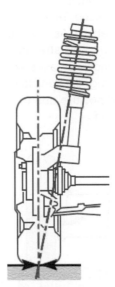

Figure 8.124 Scrub radius – zero

Kingpin inclination The kingpin angle is the angle between the steering axis and the perpendicular to the road surface, viewed in the direction of travel. Scrub radius, wheel camber and kingpin inclination all influence one another. The kingpin inclination mainly affects the aligning torque, which brings the wheels back into the straight-ahead position. Due to the inclination of the steering axes, the vehicle is raised slightly at the front when the steering is turned. The weight of the vehicle, therefore, forces the wheels back into the straight-ahead position.

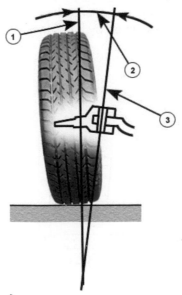

Figure 8.125 Kingpin or swivel axis inclination

Effect of kingpin inclination If the wheels are turned, they move downwards on the inclined plane. On the road, the wheels obviously cannot penetrate the road surface. Therefore, for the load condition on the road, this means that the vehicle is raised. This is counteracted by the weight of the vehicle. As a result, the steered wheels attempt to return to the straight-ahead position. This self-centring action increases with a greater kingpin angle. After cornering, the vehicle is steered back into the straight-ahead position due to this effect and the axle stabilises itself.

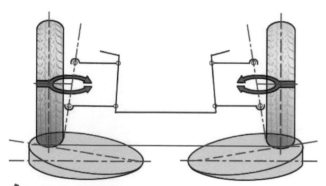

Figure 8.126 Inclined plane effect

Uneven road surfaces Turning out of the straight-ahead position is made more difficult. This is an advantage when travelling over uneven road surfaces. The restoring forces of the kingpin inclination counteract the disturbing forces. They help the driver to hold a course without heavy counter steering. The steering, therefore, becomes smoother. This principle cannot operate if the scrub radius is equal to zero. In this case, the wheel turns on its contact point and does not raise the vehicle body. No steering return forces of any kind are generated. In such cases, the steering return forces are obtained by a positive castor.

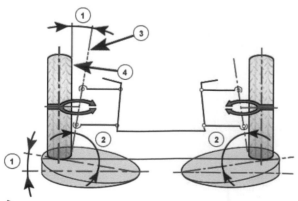

Figure 8.127 Theoretical ramps when the wheels are turned

Positive castor Castor angle is the angle in the vehicle's longitudinal direction between the steering axis and the perpendicular through the wheel centre. The castor trail is the distance between the point of intersection of the steering axis with the road surface plane and the perpendicular through the wheel centre. If the wheel contact point is situated between the point of intersection of the steering axis, with the road surface in the direction of travel, the castor angle and castor trail is positive. Positive castor causes the wheels to return to the straight-ahead position. It influences the steering torque when cornering and the straight-ahead stability.

Negative castor If the wheel contact point is situated in front of the point of intersection of the steering axis, with the road surface in the direction of travel, the castor angle and castor trail are negative. Negative castor, or at least only slight positive castor, is frequently present in front-wheel drive vehicles. This is used in order to reduce the return forces when cornering.

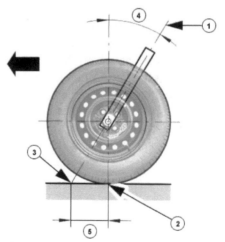

Figure 8.128 Castor angle – positive

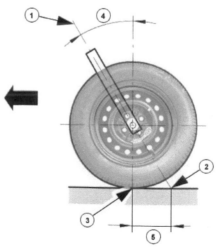

Figure 8.129 Castor angle – negative

Tyres In order to allow the specially tuned suspension systems in today's cars to operate, there must be good contact between vehicle and road surface. The tyre is therefore designed to:

▶ support the weight of the vehicle
▶ ensure good road adhesion
▶ transmit the drive, braking and cornering forces
▶ improve the ride comfort through good suspension
▶ achieve a high mileage.

When drive force is transmitted at the contact area of a rolling wheel, a relative movement occurs between the tyre and road surface. In this case, the distance covered by the vehicle is shorter than that corresponding to the rolling circumference, in other words, slip occurs. The slip percentage represents the difference between the distance covered by a wheel rolling without power transmission, and the distance actually covered with power transmission. When braking with locked wheels the slip is 100%. The slip varies as a function of drive, braking and cornering forces as well as the friction of tyre and road surface.

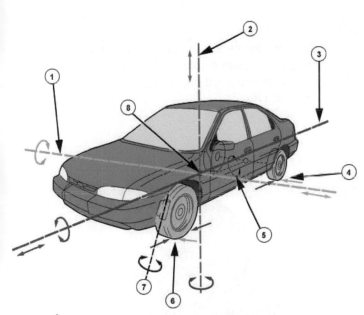

Figure 8.130 Forces transmitted by the tyres

Lateral forces and tyre side deflection A tyre can only transmit lateral forces when it is rolling at an angle to the direction of travel. For this reason the tyre does not roll straight ahead when cornering, but flexes laterally. Due to the flexing, the tyre develops a resistance, or a side force, which keeps the vehicle on course. The side deflection of the

tyre is introduced by the camber and the toe-in of the wheels. It is necessary to transmit the lateral forces in order to absorb disturbing forces such as side winds or negative lift force. When cornering, the centrifugal force represents an additional disturbing force.

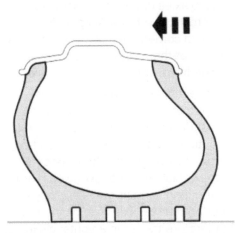

Figure 8.131 Tyre under lateral force loading

Slip angle At higher cornering speeds, the centrifugal force drives the vehicle mass towards the outside of the curve. So that the vehicle can be kept on track, the tyres must transmit cornering forces which counteract the centrifugal force. This is only possible, however, if the tyre flexes laterally. In so doing, the wheels no longer move in their turned direction but drift off at a certain angle from this direction. This means that the tyre is running at an angle to the direction of travel. This angle, which occurs between the tyre's longitudinal axis and the actual direction, is the slip angle.

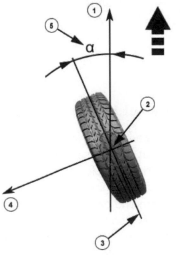

Figure 8.132 Representation of slip angle

Tyres corner best at a slip angle of 15 to 20°. The lateral adhesion depends on the slip angle, the wheel load and the type of road surface. Steering systems are generally designed so that on bends with radii of more than 20 metres, the two steered front wheels lie virtually parallel (that is not in accordance with the Ackermann principle). On bends with smaller radii, the angles of the stub axles differ significantly from one another, in accordance with the Ackermann principle. In high-speed corners, this adjustment leads to improved cornering of the wheel due to the greater turning of the outside front wheel.

The centrifugal force acting at the vehicle's centre of gravity is distributed to the front and rear wheels, according to the position of the centre of gravity. This may result in a direction of travel which deviates from the desired direction of travel. A vehicle oversteers when the rear of the vehicle tends to swing outward more than the front during cornering. The slip angle is significantly greater on the rear axle than on the front axle. The vehicle therefore travels in a tighter circle. If the steering angle is not reduced, the vehicle may break away.

A vehicle understeers when the front of the vehicle tends to swing outward more than the rear during cornering. The slip angle is greater on the front axle than on the rear axle. The vehicle therefore travels in a greater circle. It must be forced into the bend with a greater steering angle. Vehicles with understeer can be carried out of the bend. Front-engined vehicles have a tendency to understeer, since the centre of gravity is situated in front of the vehicle centre.

 Make a simple sketch to show how one of the main components or systems in this section operates

8.2.8 Advanced wheel alignment ❸

Introduction: This section is an overview of how wheel alignment is carried out using laser and camera-based equipment. The principle of the laser system is the same as the earlier 'mirror and scope' methods but is more accurate and can measure more angles. The camera method is also known as 3d measurement. Please refer to the 'Steering Geometry' section for more details on the angles related to steering and suspension.

Figure 8.133 Positive camber angle on an F1 car – but mostly an excuse to show off my signed Ayrton Senna photo!

Wheel alignment Checking and adjusting wheel alignment is routine maintenance that consists of adjusting the angles of the wheels so that they are set to the car manufacturer's specification. The purpose of these adjustments is to reduce tyre wear, and to ensure that vehicle travel is straight and true. For motorsport and off-road applications, the angles can also be altered beyond the maker's specifications to obtain a specific handling characteristic or improved performance, but this can be at a cost of tyre life.

Figure 8.134 Checking the front toe (Source: Tecalemit)

Main angles The main angles are the basic alignment of the wheels relative to each other and to the car body. On some cars, not all of these can be adjusted on every wheel. The main angles are shown in this table:

Angle	Description
Front- and rear-wheel toe	Toe is the angle of each wheel in relation to the centreline (a line straight down the centre of the vehicle) when viewed from above.
Total toe	This is the sum of two individual toe angle readings across an axle added together.
Toe-out on turns (also known as the Ackermann angle/effect)	When a vehicle is turned, the inner front wheel must toe-out more than the outer wheel to avoid scrubbing. Looking from above, the front wheels turn to two different angles.
Set back	Difference between right side and left side wheelbase length.
Rear-wheel thrust line	A line that starts where the centreline of the vehicle meets the rear axle, and ends at the intersection of the toe lines from each rear wheel.
Camber	This angle is measured from a true vertical line, i.e. one perpendicular to the ground. A wheel that is tilted outward at the top has positive camber, if tilted in it has negative camber
Caster	Caster is the forward or rearward tilt of the projected steering axis from true vertical, when viewed from the side. This line is formed by extending a line through the upper and lower steering pivots.
Steering or swivel axis inclination (SAI)	Known also as kingpin inclination. The angle between the centreline of the steering axis and vertical line from the centre contact area of the tyre (viewed from the front). SAI is not normally adjustable, but deviations from a manufacturer's specification may indicate damage.

Example laser equipment The Tecalemit GTR 410/W is a compact easy to use high precision four-wheel laser alignment system that includes wireless PC interface and an integral camber caster gauge. It is manufactured to exacting standards and developed by Tecalemit to provide rapid measurement of key vehicle geometry angles mentioned previously. The system features precision engineered components including a high resolution class B laser. It is easy and quick to set up and very simple to use. It can accommodate wheel sizes from 12″ to 23″ with a particular feature of low clearance measuring heads to clear front spoilers. The Tecalemit system comes with a wall mounted storage unit or a trolley and has a built-in battery charger. The kit includes a pedal applicator, steering wheel clamp and graduated stainless steel turning plates.

Figure 8.135 Trolley mounted equipment with a PC used to record 'before and after' alignment data (Source: Tecalemit)

Bosch video The Bosch FWA 4630 wheel aligner is suitable for use in car workshops and tyre service centres where wheel alignment is a regular service. **3D wheel alignment system** The FWA 4630 wheel aligner from Bosch is a completely new 3D wheel alignment system. The device is particularly suitable for use in car workshops and tyre service centres where wheel alignment is a regular service. With this system, a technician only requires around seven minutes to carry out precise wheel alignment on a car. This represents a time saving of 50% compared with conventional devices. A further cost-related advantage is the fact that the entire measurement procedure, including runout compensation, can be performed by just one person. This means, for example, that wheel alignment can be carried out at the same time as tyre changing.

Measurements are performed using two high-precision cameras per wheel. An integrated reference system with camber and inclination pendulums in each sensor head ensures reliable, reproducible measurement results without complicated, error-prone calibration. The innovative combination of high-tech cameras and an integrated reference system allows the sensors to be positioned wherever convenient

8

Figure 8.136 Results from the system are displayed on a standard monitor and can be saved or printed for the customer (Source: Bosch Media)

in the workshop without time-consuming calibration. The built-in reference system means the sensors do not have to be permanently fixed to a lift. The lift only needs to be fitted with special support brackets so the sensors can be easily fitted or detached, freeing up the lift for other work. This Bosch measuring device is also ideal for use in a pit, where the sensors are simply placed on the floor next to the vehicle.

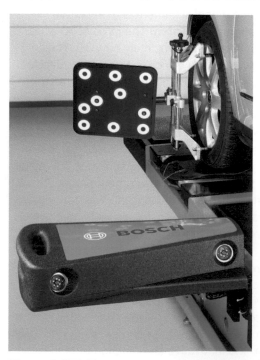

Figure 8.137 Two cameras are used for each wheel (Source: Bosch Media)

Rolling runout compensation, coupled with a high degree of accuracy, is the key to rapid measurement. Unlike conventional runout compensation, the vehicle does not have to be pushed, it can simply be driven. The vibrations caused by the engine do not affect the measurement because the measuring cameras operate at a high frame frequency of 29 frames per second. The wheel alignment process itself is then performed on all wheels simultaneously. This wheel aligner is suitable for aligning all types of cars, through to light transporters and commercial vehicles. The standard boards enable vehicles with a wheelbase of up to 340cm to be aligned. With the larger rear boards available as optional equipment, vehicles with a wheelbase of up to 430cm can be aligned.

Summary: All modern vehicles, not just high-performance cars, require the wheels to be correctly aligned to a high degree of accuracy. As other maintenance costs reduce, this is becoming a more commonplace service within the busy workshop to

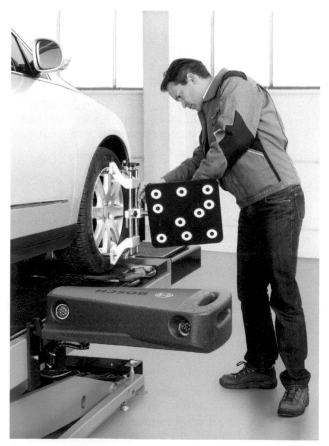

Figure 8.138 Fitting the wheel components (Source: Bosch Media)

improve fuel efficiency, tyre wear and the correct operation of the vehicle, especially those fitted with advanced driver assistance systems.

Use the media search tools to look for pictures and videos relating to the subject in this section

8.3 Brakes

8.3.1 Brakes introduction ❶

Energy conversion The main purpose of the braking system is simple – it is to slow down or stop a vehicle. To do this, the energy in the vehicle movement must be taken away – or converted. This is achieved by creating friction. The resulting heat takes energy away from the movement. In other words, kinetic energy is converted into heat energy.

Braking system The main braking system of a car works by hydraulics. This means that when the driver presses the brake pedal, liquid pressure forces pistons to apply brakes on each wheel. Disc brakes are used on the front wheels of some cars and on all wheels of sports and performance cars. Braking pressure forces brake pads against both sides of a steel disc. Drum brakes are fitted on the rear wheels of some cars and on all wheels of older vehicles. Braking pressure forces shoes to expand outwards into contact with a drum. The important part of brake pads and shoes is the friction lining.

Brake pads are steel backed blocks of friction material, which are pressed onto both sides of the disc. Older types were asbestos based so you must not inhale the dust. Follow manufacturers' recommended procedures. Pads should be changed when the friction material wears down to 2 or 3mm. The circular steel disc rotates with the wheel. Some are solid but many have ventilation holes.

Brake shoes are steel crescent shapes with a friction material lining. They are pressed inside a steel drum, which rotates with the wheel. The rotating action of the brake drum tends to pull one brake shoe harder into contact. This is known as self-servo action. It occurs on the brake shoe, which is after the wheel cylinder, in the direction of wheel rotation. This brake shoe is described as the leading shoe. The brake shoe before the wheel cylinder in the direction of wheel rotation is described as the trailing shoe.

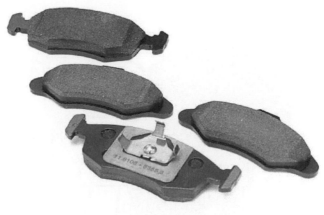

Figure 8.139 Brake pads

Figure 8.140 Brake shoes . . .

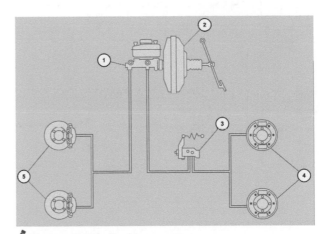

Figure 8.141 Braking system

Hydraulic cylinders The master cylinder piston is moved by the brake pedal. In its basic form, it is like a pump, which forces brake fluid through the pipes. Pressure in the pipes causes a small movement to operate either brake shoes or pads. The wheel cylinders work like a pump only in reverse.

The **brake servo** increases the force applied by the driver on the pedal. It makes the brakes more effective. Vacuum, from the engine inlet manifold, is used to work most brake servos.

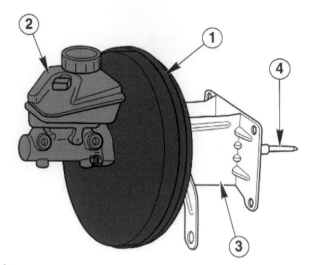

Figure 8.142 Servo unit

Anti-lock Brake System If the brakes cause the wheels to lock and make them skid, steering control is lost. In addition, the brakes will not stop the car as quickly. ABS uses electronic control to prevent this happening.

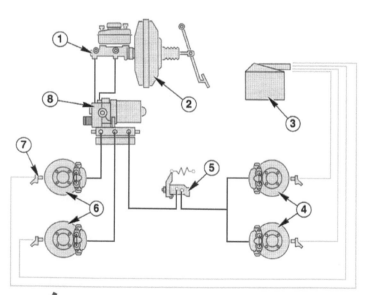

Figure 8.143 ABS layout

Brake pipes Strong, high-quality pipes are used to connect the master cylinder to the wheel cylinders. Fluid connection, from the vehicle body to the wheels, has to be through flexible pipes to allow suspension and steering movement. As a safety precaution (because brakes are quite important!), brake systems are split into two sections. If one section fails, say through a pipe breaking, the other will continue to operate.

Load compensation On most car braking systems, about 70% (or more) of the braking force is directed to

Figure 8.144 Flexible pipes

Figure 8.145 Metal pipes

the front wheels. This is because, under braking, the weight of the vehicle transfers to the front wheels. Load compensation, however, allows the braking pressure to the rear wheels to increase as load in the vehicle increases.

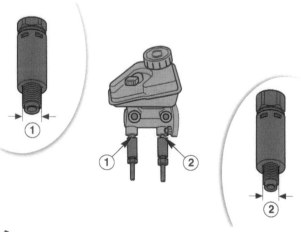

Figure 8.146 Pressure conscious regulator

If brakes become so hot that they cannot convert energy fast enough, they become much less efficient, or in other words, fade away! This is described as brake fade. A more serious form of brake fade can also be caused if the heat generated is enough to melt the bonding resin in the friction material. This reduces the frictional value of the linings or pads.

Test requirements All components of the braking system must be in good working order, in line with most other vehicle systems. Braking efficiency means the braking force compared to the weight of the vehicle. For example, the brakes on a vehicle with a weight of 10kN (1000kg x 10ms^{-2} [g]) will provide a braking force of, say, 7kN. This is said to be 70% efficiency. During an annual test, this is measured on brake rollers.

 Use a library or the web search tools to further examine the subject in this section

8.3.2 Disc, drum and parking brakes ❶ ❷

Disc brakes The caliper shown is known as a single-acting, sliding caliper. This is because only one cylinder is used but the pads are still pressed equally on both sides of the disc by the sliding action. Disc brakes are less prone to brake fade than drum brakes. This is because they are more exposed and can get rid of heat more easily. They also throw off water better than drum brakes. Brake fade occurs when the brakes become so hot they cannot transfer any more energy – and they stop working!

Disc brakes are self-adjusting. When the pedal is depressed, the rubber seal is preloaded. When the pedal is released, the piston is pulled back due to the elasticity of the rubber sealing ring.

Drum brakes Brake shoes are mounted inside a cast iron drum. They are mounted on a steel backplate, which is rigidly fixed to a stationary part of the axle. The two curved shoes have friction material on their outer faces. One end of each shoe bears on a pivot point. The other end of each shoe is pushed out by the action of a wheel cylinder when the brake pedal is pressed. This puts the brake linings in contact with the drum inner surface. When the brake pedal is released, the return spring pulls the shoes back to their rest position.

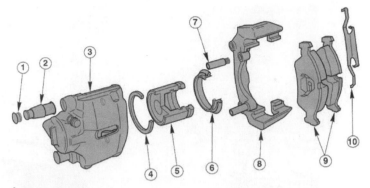

Figure 8.147 Sliding disc brake caliper components

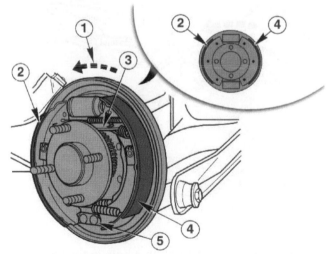

Figure 8.148 Sliding disc brake caliper

Figure 8.149 Rear drum brake

Drum brake features Drum brakes are more adversely affected by wet and heat than disc brakes, because both water and heat are trapped inside the drum. However, they are easier to fit with a mechanical hand brake linkage.

Figure 8.150 Brake drum

Brake adjustments Brakes must be adjusted so that the minimum movement of the pedal starts to apply the brakes. The adjustment in question is the gap between the pads and disc and the shoes and drum. Disc brakes are self-adjusting, because as pressure is released it moves the pads just away from the disc. Drum brakes are different because the shoes are moved away from the drum to a set position by a pull off spring. Self-adjusting drum brakes are almost universal now on light vehicles. A common type uses an offset ratchet, which clicks to a wider position if the shoes move beyond a certain amount when operated.

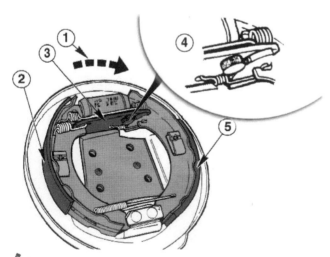

Figure 8.151 Self-adjusting device

Manual adjustment through a hole in the back plate is often used. This involves moving a type of nut on a threaded bar, which pushes the shoes out as it is screwed along the thread. This method is similar to the automatic adjusters. An adjustment screw on the back plate is now quite an old method. A screw or square head protruding from the back plate moves the shoes by a snail cam. As a guide,

tighten the adjuster until the wheels lock, and then move it back until the wheel is just released. You must ensure that the brakes are not rubbing as this would build up heat and wear the friction material very quickly.

Figure 8.152 Brake adjustment hole

Figure 8.153 Square type adjuster (old method)

The precise way in which the shoes move into contact with the drum affects the power of the brakes. If the shoes are both hinged at the same point then the system is said to have one leading and one trailing shoe. As the shoes are pushed into contact with the drum, the leading shoe is dragged by the drum rotation harder into contact, whereas the rotation tends to push the trailing shoe away. This 'self-servo' action on the leading shoe can be used to increase the power of drum brakes. This is required on the front wheels of all-round drum brake vehicles.

Twin leading shoe brakes The shoes are arranged so that they both experience the self-servo action. The shoes are pivoted at opposite points on the backplate and two wheel cylinders are used.

The arrangement is known as twin leading shoe brakes. It is not suitable for use on the rear brakes because if the car is travelling in reverse then it would become a twin trailing shoe arrangement, which means the efficiency of the brakes would be seriously reduced. The leading and trailing layout is therefore used on rear brakes, as one shoe will always be leading no matter which direction the vehicle is moving in.

Figure 8.155 Leading and trailing system

▶ two cables, one to each wheel
▶ equaliser on a single cable pulling a 'U' section to balance effort through the rear cable
▶ single cable to a small linkage on the rear axle.

Disc-type handbrake Some sliding caliper disc brakes incorporate a handbrake mechanism. The footbrake operates as normal and handbrake operation is by a moving lever. The lever acts through a shaft and cam, which works on the adjusting screw of the piston. The piston presses one pad against the disc and because of the sliding action, the other pad also moves.

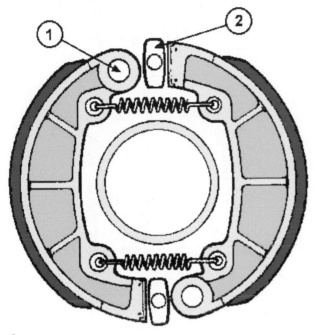

Figure 8.154 Twin leading shoe system

Leading and trailing shoe brakes The standard layout of drum brake systems is normally:

▶ twin leading shoe brakes on the front wheels
▶ leading and trailing shoe brakes on the rear wheels.

Disc brakes are now used on the front wheels of all light vehicles but many retain leading and trailing shoe brakes on the rear. In most cases, it is easier to attach a handbrake linkage to the system with shoes on the rear. This method also provides the braking performance required when the vehicle is reversing.

Inside a brake drum, the hand brake linkage is usually a lever mechanism as shown here. This lever pushes the shoes against the drum and locks the wheel. The hand brake lever pulls on one or more cables and has a ratchet to allow it to be locked in the 'on' position. There are a number of ways in which the hand brake linkage can be laid out to provide equal force, or compensation, for both wheels:

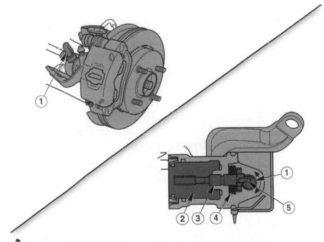

Figure 8.156 Sliding caliper parking brake

Some manufacturers use a set of small brake shoes inside a small drum, which is built in to the brake disc. The caliper is operated as normal by the footbrake. The small shoes are moved by a cable and lever.

Summary Remember that the purpose of the braking system is to slow down or stop a vehicle. This is achieved by converting the vehicle's movement energy into heat. Friction is used to do this. Braking system developments have improved efficiency, reliability and ease of servicing.

8

Look back over the previous section and write out a list of the key bullet points from this section

8.3.3 Hydraulic components ❶ ❷

Introduction Shown here, is the principle of hydraulic brakes. The movement of the piston (2) causes an equal force in all parts of the system. The pistons (1) move a shorter distance. If larger area pistons are used, the force at the brakes can be increased. This is called a liquid lever and acts in addition to the leverage of the brake pedal.

Braking system A complete braking system includes a master cylinder, which operates several wheel cylinders. This system is designed to give the power amplification needed for braking the particular vehicle. On any braking vehicle, a lot of the weight is transferred to the front wheels. Most braking effort is therefore designed to work on the front brakes. Some cars have special hydraulic valves to limit rear wheel braking. This reduces the chance of the rear wheels locking and skidding.

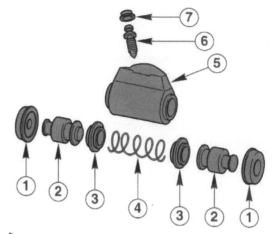

Figure 8.158 Slave cylinder components

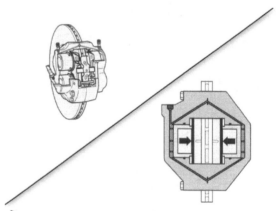

Figure 8.159 Fixed caliper

Disc brake calipers are known as fixed, floating or sliding types. The pistons are moved by hydraulic pressure created in the master cylinder. A number of different calipers are used. Some high-performance calipers include up to four pistons. However, the operating principle remains the same.

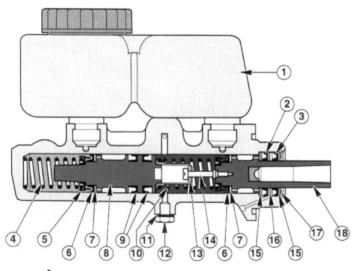

Figure 8.157 Master cylinder

Wheel cylinders Brake shoes can be moved by double- or single-acting wheel cylinders. A common layout is to use one double acting cylinder and brake shoes on each rear wheel of the vehicle and disc brakes on the front wheels. A double-acting cylinder simply means that as fluid pressure acts through a centre inlet, pistons are forced out of both ends.

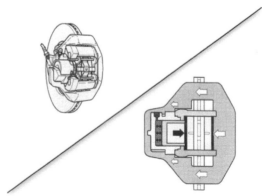

Figure 8.160 Floating caliper

Remember! There is more support on the website that includes additional images and interactive features: www.tomdenton.org

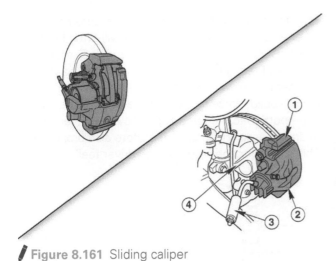

Figure 8.163 Braking and other components

Figure 8.161 Sliding caliper

Brake fluid Always use new and approved brake fluid when topping up or refilling the system. Manufacturers' recommendations must always be followed. Brake fluid is hygroscopic which means that, over time, it absorbs water. This increases the risk of the fluid boiling due to the heat from the brakes. Pockets of steam in the system would not allow full braking pressure to be applied. Many manufacturers recommend that the fluid be changed at regular intervals. Make sure the correct grade of fluid is used. The current recommended types are known as DOT4 and DOT5.

Brake system The online animation shows the main parts of a typical modern braking system. A separate mechanical system is a good safety feature. Most vehicles have the mechanical parking brake working on the rear wheels but a few have it working on the front – take care. Note the importance of flexible connections to allow for suspension and steering movement. These flexible pipes are made of high-quality rubber and are covered in layers of strong mesh to prevent expansion when under pressure.

Figure 8.162 A common type of brake fluid (Source: Castrol)

Safety is built into braking systems by using a double-acting master cylinder. This is often described as tandem and can be thought of as two master cylinders inside one housing. The pressure from the pedal acts on both cylinders but fluid cannot pass from one to the other. Each cylinder is then connected to a separate circuit. These split lines can be connected in a number of ways. Under normal operating conditions, the pressure developed in the first part of the master cylinder is transmitted to the second. This is because the fluid in the first chamber acts directly on the second piston.

Line failure If one line fails, the first piston meets no restriction and closes up to the second piston. Further movement will now provide pressure for the second circuit. The driver will notice that pedal travel increases, but some braking performance will remain. If the fluid leak is from the second circuit, then the second piston will meet no restriction and close up the gap. Braking will now be just from the first circuit. Diagonal split brakes are the most common and are used on vehicles with a negative scrub radius. Steering control is maintained under brake failure conditions.

Multi-circuit systems There are three common 'splits' used on modern braking systems. The first two types listed are the most common:

▶ Diagonal split type, where if a fault occurs, the driver loses half of the front and half of the rear brakes.

▶ Separate front and rear, where if a fault occurs, the driver loses all of the front or all of the rear brakes.

▶ Duplicated front, where if a fault occurs, the driver loses the rear and part of the front or part of the front brakes only. Special front calipers are required when using this method.

399

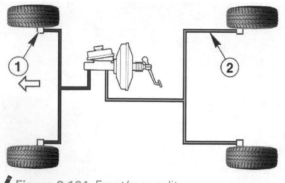

Figure 8.164 Front/rear split

Figure 8.165 Diagonal split

 Make a simple sketch to show how one of the main components or systems in this section operates

8.3.4 Brake servo operation 2

Introduction The brakes of a vehicle must perform well, whilst the effort required by the driver is kept to a reasonable level. This is achieved by the use of a brake servo. It is also called a brake booster. Vacuum-operated systems are commonly used on light vehicles.

Figure 8.166 Vacuum servo

Hydraulic power brakes use the pressure from an engine driven pump. The pump will often be the same as the one used to supply the power assisted steering. Pressure from the pump is made to act on a plunger in line with the normal master cylinder. As the driver applies force to the pedal, a servo valve opens in proportion to the force applied by the driver. The hydraulic assisting force is therefore also proportional. This maintains the all-important 'driver feel'.

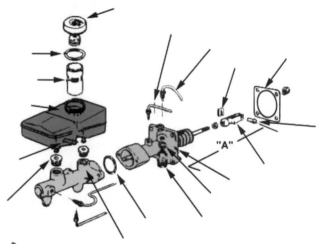

Figure 8.167 Hydraulic brake servo

A **hydraulic accumulator** (a reservoir for fluid under pressure) is incorporated into many systems. This is because the pressure supplied by the pump varies with engine speed. The pressure in the accumulator is kept between set pressures in the region of 70 bar. A warning therefore – if you have to disconnect any components from the braking system on a vehicle fitted with an accumulator, you must follow the manufacturer's recommendations on releasing the pressure first.

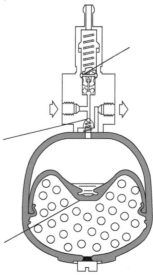

Figure 8.168 Accumulator

Vacuum servo A common servo system uses low pressure (vacuum) from the manifold on one side, and the higher atmospheric pressure on the other side of a diaphragm. The low pressure is taken via a non-return safety valve from the engine inlet manifold. A pump is often used on diesel engined vehicles as most do not have a throttle butterfly and hence do not develop any significant manifold vacuum. The pressure difference, however created, causes a force which is made to act on the master cylinder.

The vacuum servo is fitted in between the brake pedal and the master cylinder. The main part of the servo is the diaphragm. The larger this diaphragm, the greater the servo assistance provided. A vacuum is allowed to act on both sides of the diaphragm when the brake pedal is in its rest position. When pedal force is applied to the piston a valve cuts the vacuum connection to the rear chamber and allows air at atmospheric pressure to enter. This causes a force to act on the diaphragm, assisting with the application of the brakes.

Once the master cylinder piston moves, the valve closes again to hold the applied pressure. Further effort by the driver on the brake pedal will open the valve again and apply further vacuum assistance. In this way, the driver can 'feel' the amount of braking effort being applied. The cycle continues until the driver effort reaches a point where the servo assistance remains fully on.

Vacuum supply On petrol engines the vacuum is obtained from the inlet manifold. On diesel engines a vacuum pump is used. A non-return valve is fitted in the line to keep the vacuum in the servo chamber. This means that it is possible to carry out three or four braking operations, with servo assistance, without the engine running. The valve also prevents fuel vapours getting in the servo and damaging the diaphragm.

Fail safe mode If the vacuum servo stops working, the brakes will still operate but extra force will be required from the driver. The connection to the inlet manifold will normally be via a check valve as an extra safety feature.

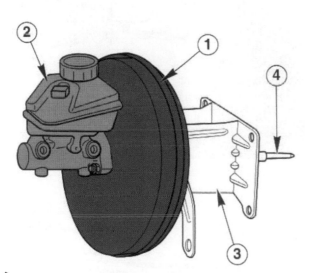

Figure 8.169 Servo unit

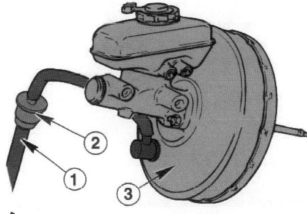

Figure 8.171 A check valve is fitted in the vacuum supply

8

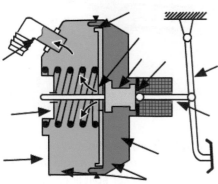

Figure 8.170 Servo construction

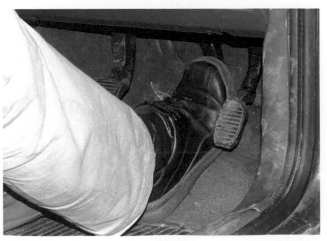

Figure 8.172 Safety is important

iBooster Vehicles without an engine need some other form of servo assistance and many manufacturers now use the iBooster or similar technology. The control principle behind the iBooster is similar to that of vacuum brake boosters: in vacuum brake systems, a valve controls the air supply to provide a boost to the force applied from the driver's foot. With the iBooster, the actuation of the brake pedal is detected via an integrated differential travel sensor and this information is sent to the control unit. The control unit determines the control signals for the electric motor, while a three-stage gear unit converts the torque of the motor into the necessary boost power. The power supplied by the booster is converted into hydraulic pressure in a standard master brake cylinder.

Figure 8.173 iBooster (Source: Bosch Media)

Summary A brake servo assists the driver when the brakes are applied. The 'feel' must be maintained during this operation. Most servos are vacuum operated.

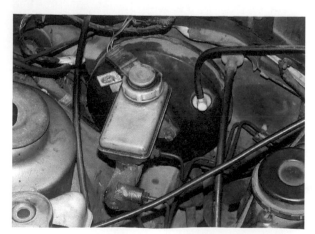

Figure 8.174 Brake servo

 Create a mind map to illustrate the features of a key component or system

8.3.5 Braking force control ❷

There are three main types of braking force control devices:

1 Load apportioning valve
2 Pressure conscious regulator
3 Deceleration-sensing brake pressure reducer.

Braking force distribution The purpose of these devices is to ensure braking force is distributed so that most of the force goes to the front brakes. This improves performance and stability.

Figure 8.175 Pressure conscious regulator

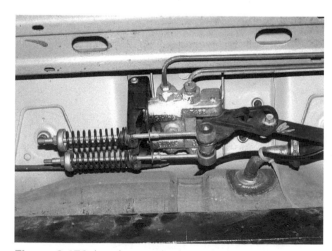

Figure 8.176 Load apportioning valve

Load apportioning valves are fitted between the rear axle and vehicle floor assembly. A single valve is used for vehicles with front to rear split lines, and two valves are used when the split is diagonal.

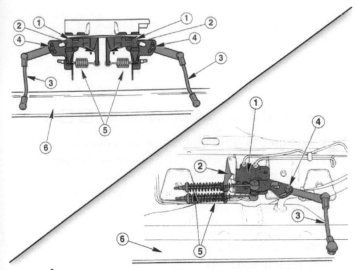

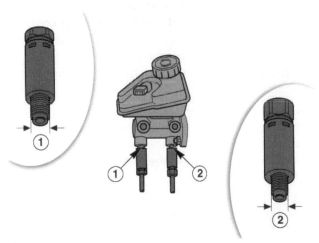

Figure 8.177 Separate and combined units

A lever and tension spring changes the force necessary to make a plunger move. The lever and spring adjust position depending on the vehicle load. Fluid pressure moves the plunger; however, the position of the lever limits the movement. Load in the vehicle sets the valve position. Pressure, and therefore braking force, is controlled by the valve.

The **pressure conscious regulator** is simply fitted in the line, or lines, to the rear brakes. It reduces braking pressure by a fixed amount. An internal control spring is used to set the operating pressure.

Figure 8.178 Regulators in position

The key component in Figure 8.177 is the valve labelled number 2. This valve is held on its seat by the spring. At the start of braking operation, fluid can pass the valve as shown by the arrows.

When the control pressure is reached, the 'control piston' moves against the spring. This closes the passage between the valve seat and the control piston. A further increase in pressure

causes a continuous opening and closing of the valve.

Deceleration sensing One deceleration sensor is used in each brake circuit. The sensors are mounted on the vehicle floor at a set angle to the horizontal. When the deceleration is greater than about 0.5g, the valves allow the pressure to the rear brakes to rise more slowly than the front.

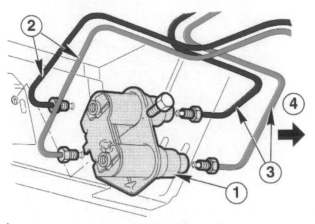

Figure 8.179 Pressure reducer location

Sensor operation In the online animation, when the brakes are applied, fluid is forced through the inlet (2), past the washer (3) and ball (4), through the piston chamber (8) and finally to the outlet (6). At the triggering point of about 0.5g, the ball moves on the angled surface, and closes off the piston bore. This reduces the pressure to the rear brakes. When deceleration reduces, the ball rolls back against the washer.

Controlling brake pressure ensures braking force is distributed so that most of the force goes to the front brakes. As a guide, more than 70% of the braking takes place on the front wheels. This improves performance, control and stability.

> Using images and text, create a short presentation to show how a component or system works

8.3.6 Carbon ceramic brakes

Introduction The Brembo Group has been manufacturing carbon ceramic discs for automotive applications since 2002, when it first supplied these components for the Ferrari Enzo. This high-performance material, made from a special mixture of powders, resins and fibres in a complex manufacturing process, has been used since the 1970s in braking systems for aerospace applications and since the

8

1980s in motorsport. Carbon ceramic materials are ceramics greatly strengthened by the inclusion of carbon fibres. They are therefore very light, extremely strong, and capable of operating at temperatures well in excess of 1500°C.

Use the media search tools to look for pictures and videos relating to the subject in this section

Figure 8.180 Since 2000, carbon ceramic has also been used in the production of braking systems for sports cars

Benefits Carbon ceramic offers substantial benefits in terms of performance, in wet and dry conditions, weight, comfort, corrosion resistance, durability and high-tech appeal. Brembo designed six-piston front and four-piston rear calipers for the Lamborghini Aventador. The design of the Brembo calipers integrates seamlessly with the overall aesthetics of the most exclusive models.

Figure 8.181 Lamborghini brake calipers

8.3.7 Anti-lock brake systems ❷ ❸

Introduction The reason for the development of the anti-lock braking system (ABS) is simple. Under braking conditions if one or more of the vehicle wheels locks (begins to skid) then this has serious consequences:

▶ braking distance increases
▶ steering control is lost
▶ tyre wear is abnormal.

The maximum deceleration of a vehicle is achieved when maximum energy conversion is taking place in the brake system. This is the conversion of kinetic energy to heat energy at the discs and brake drums. The potential for this conversion process between a tyre skidding, even on a dry road, is far less. A good driver can pump the brakes on and off to prevent locking but electronic control can achieve even better results.

ABS is becoming more common on lower price vehicles, which should be a contribution to safety. It is important to remember however, that for normal use, the system is not intended to allow faster driving and shorter braking distances. It should be viewed as operating in an emergency only. Good steering and roadholding must continue when the ABS system is operating. This is arguably the key issue as being able to swerve round a hazard, whilst still braking hard, is often the best course of action.

Fail safe mode In the event of the ABS failing, then conventional brakes must still operate to their full potential. In addition, a warning must be given to the driver. This is normally in the form of a simple warning light.

The system must operate under all speed conditions above walking pace. At this very slow speed, even when the wheels lock, the vehicle will come to rest quickly. If the wheels did not lock, then in theory the vehicle would never stop!

Other operating conditions ABS must be able to recognise aquaplaning and react accordingly. It must also operate on an uneven road surface. The one operating condition still not perfected is braking from slow speed on snow. ABS can actually increase stopping distance in snow; however, steering control will be maintained and this is considered a suitable trade-off.

ABS can be considered as a central control unit with a series of inputs and outputs. An ABS system is represented by the closed loop system block diagram shown online. The most important of the inputs are the wheel speed sensors. The main output is some form of brake system pressure control. The task of the electronic control unit (ECU) is to compare signals from each wheel sensor. From these signals, it can determine the acceleration or deceleration of an individual wheel. Brake pressure can be reduced, held constant or allowed to increase. The maximum pressure is determined by the driver's pressure on the brake pedal.

Wheel acceleration or deceleration A vehicle reference speed is determined from the combination of two diagonal wheel sensor signals. After the start of braking, the ECU uses this value as its reference. The acceleration and deceleration values are live measurements which are constantly changing.

Figure 8.182 ABS warning light

Figure 8.183 Brake disc

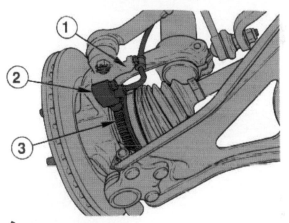

Figure 8.185 Front wheel sensor

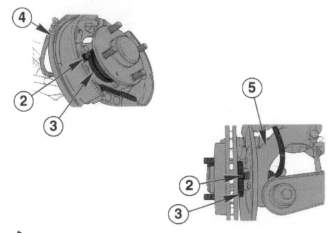

Figure 8.186 Rear wheel sensor

Figure 8.184 Snow conditions

Although brake slip cannot be measured directly, a value can be calculated from the vehicle reference speed. This figure is then used to determine if, and when, ABS should take control of the brake pressure.

ABS components There are variations between manufacturers involving a number of different components. However, for the majority of systems, there are three main components:

▶ wheel speed sensors
▶ electronic control unit
▶ hydraulic modulator.

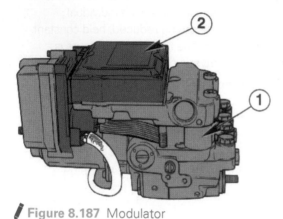

Figure 8.187 Modulator

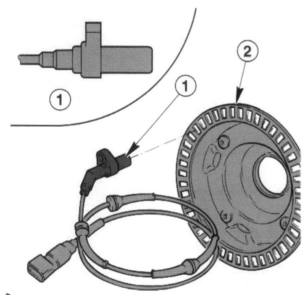

Figure 8.188 Speed sensor

Most of these devices are inductance sensors and work in conjunction with a toothed wheel. They consist of a permanent magnet and a soft iron rod around which is wound a coil of wire. As the toothed wheel rotates, the changes in inductance of the magnetic circuit generate a signal. The frequency and voltage of the signal are proportional to wheel speed. The frequency is the signal used by the electronic control unit.

Sensor operation Some systems now use Hall effect sensors. The Hall sensors are more accurate at lower speed. The main parts of the sensor are a magnet and an integrated circuit containing the sensing element.

Figure 8.189 Hall effect sensor

Electronic Control Unit The ECU takes in information from the wheel sensors and calculates the best course of action for the hydraulic modulator. At the heart of an ABS ECU are two microprocessors which run the same programme independently of each other. This ensures greater security against any fault which could adversely affect braking performance. If a fault is detected, the ABS disconnects itself and operates a warning light. Both processors have non-volatile memory into which fault codes can be written for later service and diagnostic access. The ECU performs a self-test after the ignition is switched on. A failure results in disconnection of the system.

Figure 8.190 Internal circuit

The hydraulic modulator has three operating positions:

▶ pressure release, where the brake line is open to the reservoir
▶ pressure holding, where the brake line is closed
▶ pressure build-up, where the brake line is open to the pump.

The valves are controlled by electrical solenoids which react very quickly.

Modulator operation The start of ABS engagement is known as 'first control cycle smoothing'. This smoothing stage is necessary in order not to react to minor disturbances, such as an uneven road surface, which can cause changes in the wheel sensor signals. The threshold of engagement is critical. If it started too soon, it would be distracting to the driver and cause unnecessary component wear. If too late, steering and stability could be lost on the first control cycle.

Vehicle yaw When braking on a road surface with different adhesion under the left and right wheels, the vehicle will yaw or start to twist. The driver can control this with the steering if time is available. This can be achieved if the pressure to the other front wheel is reduced when the front wheel with poor adhesion becomes unstable. This acts to reduce vehicle yaw, which is particularly important when the vehicle is cornering.

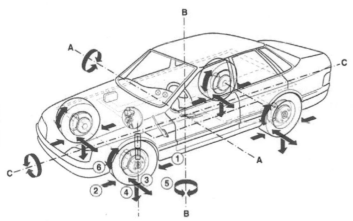

Figure 8.191 Forces acting on a vehicle

Wheel speed instability occurs frequently, and at random, because of axle vibration on rough roads. Due to this instability, brake pressure during ABS operation tends to be reduced more than it is increased. This could lead to loss of braking under certain conditions. A slight delay in the reaction of the ABS due to a delay in signal smoothing, the time taken to move control valves and a time lag in the brake lines, helps to reduce the effect of axle vibration.

The control strategy of the anti-lock brake system can be summarised as follows:

▶ Rapid brake pressure reduction during wheel speed instability. The wheel will, therefore, re-accelerate without too much pressure reduction and avoid under braking.

▶ Rapid rise in brake pressure during and after a re-acceleration to a value just less than the instability pressure.

▶ Discreet increase in brake pressure in the event of increased adhesion.

▶ Sensitivity suited to the prevalent conditions.

▶ Anti-lock braking is not initiated during axle vibration.

Control summary The application of these five main requirements leads to the need for compromise. Optimum programming and prototype testing can reduce the level of compromise but some disadvantages have to be accepted. The best example of this is braking on uneven ground in deep snow, because deceleration is less effective unless the wheels are locked up. In this example, priority is given to stability rather than stopping distance, as directional control is favoured under these circumstances.

Case study The Mercedes SL Class has an impressive package of cutting edge dynamic handling control systems, which includes a new electrohydraulic brake system. Mercedes Benz calls this Sensotronic Brake

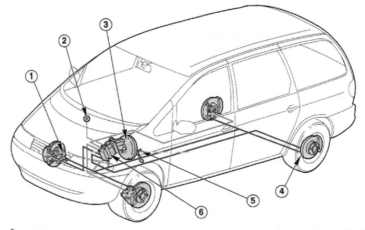

Figure 8.192 Anti-lock brake system

Control (SBC) along with Active Body Control (ABC) and the Electronic Stability Program (ESP). Among the most important performance features of the SBC electrohydraulic braking system are the dynamic building up of brake pressure and the precise monitoring of driver and vehicle behaviour using sensors.

Sensotronic Brake Control In emergencies, SBC instantly increases the pressure in the brake lines and applies the pads to the brake discs so that they can grip instantly with full force when the brake pedal is pressed. Furthermore, thanks to variable brake proportioning, SBC offers enhanced safety when braking on bends. Bosch helped to develop the system. Its own version is called an electrohydraulic brake (EHB) system. The system provides the brakes with a fluid supply from a hydraulic high-pressure reservoir sufficient for several braking events.

Electrohydraulic brake (EHB) system When the brakes are activated, the EHB control unit calculates the desired target brake pressures at the individual wheels. Braking pressure for each of the four

8

Figure 8.193 Mercedes SL

Figure 8.194 EHB components (Source: Continental Corporation)

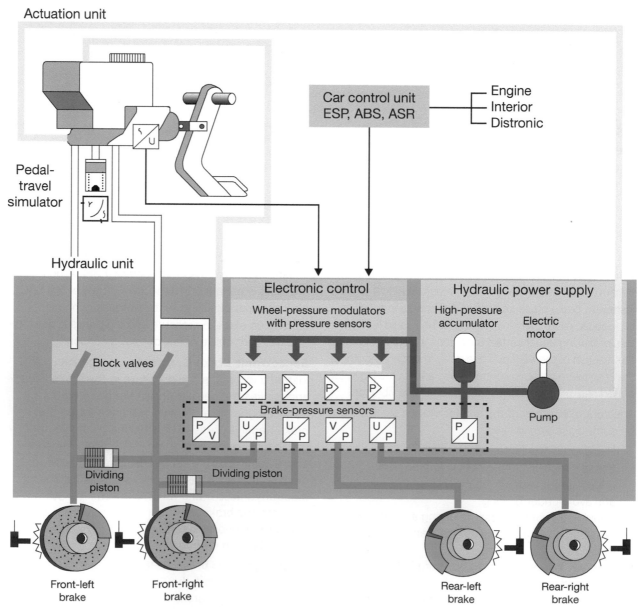

Figure 8.195 EHB system

wheels is regulated individually via a wheel pressure modulator, which consists of one inlet and one outlet valve, controlled electronically. Normally, the brake master cylinder is detached from the brake circuit, with a pedal travel simulator creating normal pedal feedback. If ESP intervenes, the high-pressure reservoir supplies the required brake pressure quickly and precisely to the wheel brakes.

Some of the variations in ABS are shown below. However, the principle of operation of all systems is the same. These discrete operating phases have to be achieved:

▶ pressure reduction
▶ pressure holding
▶ pressure decrease.

Many developments are taking place, however. The main area of development is the integration of ABS with other systems, such as stability control.

 Using images and text, create a short presentation to show how a component or system works

8.3.8 Bosch ABS ❸

Introduction Bosch has taken another important step in the ongoing development of ABS (versions 8 and 9). The key characteristics of this system are:

▶ Scalable product concept with modular software – this allows, amongst other things, optional integration of inertial sensors and pressure sensors to give precise control even at very low brake pressures.
▶ Optional FlexRay interface – allowing links to other systems and further developments.
▶ Optimized microprocessor design and a control unit based on printed circuit technology (running at 20MHz, it has 256kB of memory for ABS and with ESP up to 2MB).
▶ Rare earth magnet motors – to improve power to weight ratio (up to 30% lower weight).
▶ Good noise behaviour and pedal feel with reduced application effort.

The video shown online is an overview of ABS operation (a slightly earlier system but the basic principles are the same).

ABS overview In an emergency braking situation the braking force applied by the driver may be greater than the tyre can handle and the wheel locks. The tyre can now no longer transfer any lateral traction forces. The vehicle becomes unstable and uncontrollable, since the vehicle no longer reacts to the steering input of the

driver. In a vehicle equipped with an Anti-lock Braking System, wheel-speed sensors measure the speed of rotation of the wheels and pass this information to the ABS control unit. If the ABS control unit detects that one or more wheels are tending to lock, it intervenes within milliseconds by modulating the braking pressure at each individual wheel. In doing so, ABS prevents the wheels from locking and ensures safe braking – the vehicle remains steerable and stable. Generally, the stopping distance is also reduced.

Generation 9 of the Bosch ABS is part of the continuous development of this active safety system. The scalable product concept, with modular software architecture, characterizes this latest technical evolution. Compared to the previous generation, weight and size have been reduced by up to 30%. The most compact ABS unit now weighs only 1.1kg.

Figure 8.196 Driving tracks covered on one side with ice or suddenly emerging animals are typical traffic risks. Brake control systems such as the anti-locking system ABS and the Electronic Stability Program increase the safety in such critical situations

Figure 8.197 Bosch brake control systems ABS 8 (left) and ESP 8 (right)

The **ECU** uses the signals from the wheel-speed sensors to compute the speeds of the wheels. Two different operating principles are used – passive

Figure 8.198 Generation 9 ABS and ESP hydraulic modulator and valve block with integrated ECU. It is also the basis for many high-performance safety and assistance functions, such as the ACC adaptive cruise control with stop-and-go function

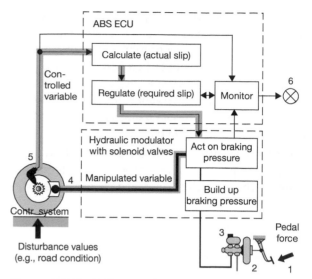

Figure 8.200 ABS control loop: 1 Brake pedal, 2 Brake servo/booster, 3 Master cylinder, 4 Wheel cylinder/ caliper, 5 Wheel-speed sensor, 6 Warning lamp

(inductive) and active (Hall) speed sensors. Active sensors are becoming more and more widespread. They use a magnetic field for the contactless detection of the wheel speed and are capable of recognizing the direction of rotation as well as a standstill.

ABS ECUs contain two microcontrollers for safety and they operate at 20MHz. A CAN or FlexRay connection is also available from the ECU.

Figure 8.199 Bosch ABS generation 8.1, with wheel speed sensors

The **hydraulic unit** puts the ECU's commands into effect and regulates the pressure in the individual wheel brake cylinders by means of solenoid valves. It is located in the engine compartment, between the brake master cylinder and the wheel brake cylinders, so that the hydraulic lines to the brake master cylinder and the wheel brake cylinders can be kept short. The hydraulic unit has input and output solenoid valves for controlling the pressure in the individual wheel brakes. The ECU takes over all electrical and electronic tasks as well as the control functions of the system.

Figure 8.201 ABS 8 exploded view showing the ECU and eight 2-way solenoid valves (two hydraulic positions and two hydraulic connections)

Motorcycle ABS is now possible because of the reduced weight of the system components. 'ABS-9 enhanced' is the most powerful variant of the latest ABS generation 9 made by Bosch. It offers what is known as an eCBS function, an electronic combined brake system.

The system is modified from the usual eight two-way valves used on a split line car brake system as motorcycles only require four such valves. The control algorithm is also different from a car ABS. Most

Remember! There is more support on the website that includes additional images and interactive features: www.tomdenton.org

motorcycle brakes are operated by a hand lever for the front and by pedal for the rear brakes. However, some systems are combined where, for example, the foot pedal operates both front and rear. This means that the ABS system variants are very model specific.

Ongoing developments The further enhancement of driving safety is an important goal for all manufacturers and suppliers. Bosch is no exception and indeed is a key player in this respect.

Networking is what will provide the basis for further developments. ABS is already linked with (well, more accurately, integrated with) stability control but is likely to now be linked with airbag control and other driver assistance systems. New functions then become possible that will improve accident avoidance but also protect car occupants, pedestrians and other road users.

Summary The basic operating principle of ABS-9 has not changed (i.e. pressure holding, pressure

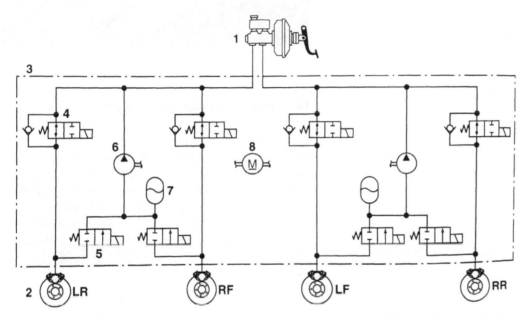

Figure 8.202 ABS hydraulic system diagram: 1 Master cylinder, 2 Wheel cylinders, 3 Hydraulic unit, 4 Inlet valves, 5 Outlet valves, 6 Return pump, 7 Accumulator, 8 Pump motor

The components of Bosch motorcycle ABS including electronic Combined Brake System (eCBS)

1 Hydraulic unit with attached control unit
2 Wheel-speed sensor
– Sensor signal

Figure 8.203 Bosch motorcycle ABS-9 enhanced

411

Figure 8.204 ABS Hydraulic modulator

Figure 8.205 Electric parking brake switch

reduction and pressure increase). However, version 9 has added several new options as well as incremental improvements meaning it will continue to be a popular choice for many manufacturers.

 Use the media search tools to look for pictures and videos relating to the subject in this section

8.3.9 Electric parking brakes ❷ ❸

Introduction The electric (or electronic) parking brake (EPB), also known as an automatic parking brake (APB), is a function offering the driver increased comfort and convenience. In addition, as the hand lever is not used, car manufacturers have more freedom of choice as to where they site the operating parts within the car. Features such as hill- or auto-hold are also possible with an EPB. Hill-hold stops the car rolling away accidentally when standing still or setting off. Auto-hold keeps the brake pressure applied after the driver releases the pedal. If the ABS sensors detect any movement the pressure is increased. If the accelerator is pressed (or the clutch released on a manual) the brakes are let off.

Types of EPB There are two main systems in use, namely the cable-pull and electric-hydraulic type. Both methods include a visual warning light on the dashboard. The second of these is now the most common. A third, fully electric type will start to be used in the near future.

The cable pull system is simply a development of the traditional lever and cable method. As the switch is operated, a motor, or motors, pull the cable by either rolling it on a drum or using an internally threaded gear on a spiral attached to the cable. The electronic parking brake module shown here, also known as the EPB actuator, is fitted to some Range Rover and Land Rover models. The parking brake can be released manually on most vehicles. After removing a plastic cover or similar, pulling a wire cable loop will let off the brake.

Figure 8.206 Cable pull system (Source: Land Rover)

Electric-hydraulic caliper systems These types are usually employed as part of a larger control system, such as an electronic stability program (ESP). When the driver presses the switch to activate the parking brake, the ESP unit automatically generates pressure in the braking system and presses the brake pads against the disc. The calipers are then locked in position by an electrically controlled solenoid valve. The caliper remains locked without any need for hydraulic pressure. To release the brake, the ESP briefly generates pressure again, slightly more than was needed to lock the caliper, and the valve is released.

Electric-hydraulic brake systems also allow other functions such as brake assist. This means shorter braking distances because the system will sense a heavy braking situation and actually apply the brakes harder than the driver would normally do because the system can sense the speed of each wheel.

Figure 8.207 Electric-hydraulic parking brake caliper (Source: Bosch Press)

Testing the parking brake on rollers is possible on both systems. Cable pull types can be tested much like ordinary hand or parking brakes. But there is a risk of locking the wheels. Some manufacturers have test modes – so double check! The types with caliper motors can also be tested on rollers but the procedure is slightly different. Most of the caliper-motor systems have special software incorporated in the ECU for brake testing. When the car is put on a rolling road and the rear wheels are driven by the test equipment, the ECU detects this as a test situation because the rear wheels are moving and the fronts are not. It therefore puts the system into test mode. If a multi-function display is used on the vehicle dashboard, it will display an appropriate message. The technician can then operate the handbrake-switch. The ECU applies the electric parking brake with enough force to obtain a reading on the roller brake

tester. The wheels should not lock. After the test is complete the rollers are stopped and the switch is released. When the switch is activated again, the brakes are re-applied in the normal way and the wheels will be locked.

 Create a word cloud for one or more of the most important screens or blocks of text in this section

8.3.10 Traction control ❸

Introduction The 'steerability' of a vehicle is lost if the wheels spin when driving off under severe acceleration. Electronic traction control has been developed as a supplement to anti-lock brake systems (ABS). This control system prevents the wheels from spinning when moving off or when accelerating sharply while on the move. In this way, an individual wheel which is spinning is braked in a controlled manner. If both or all of the wheels are spinning, the drive torque is reduced by means of an engine control function. Traction control has become known as ASR or TCR.

Traction control is not normally available as an independent system, but in combination with ABS. This is because many of the components required are the same for each. Shown online is a block diagram of a traction control system. Note the links with ABS and the engine control system.

Reasons for traction control Traction control will intervene to achieve the following:

▶ driving stability
▶ reduction of yawing moment reactions
▶ optimum propulsion at all speeds
▶ reduced driver workload.

Intervention An automatic control system can intervene more quickly and precisely than the driver of the vehicle. This allows stability to be maintained at a time when the driver might not have been able to cope with the situation.

Control methods Control of tractive force can be by a number of methods:

▶ throttle control
▶ ignition control
▶ braking effect.

8

413

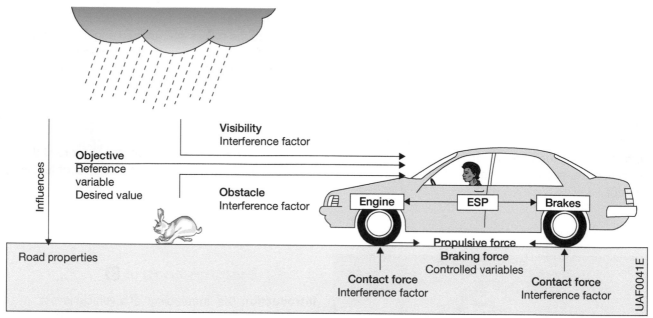

Figure 8.208 'Conditions' acting on the driver and car

Each of these methods is examined further over the next pages.

Throttle control can be via an actuator, which simply moves the throttle cable. If the vehicle employs a 'drive by wire' accelerator, then control will be in conjunction with the engine management system. This throttle control will be independent of the driver's pedal position. This method works alone, but it is relatively slow to control engine torque.

If ignition is retarded, the engine torque can be reduced by up to 50% in a very short space of time. The timing is adjusted by a set ramp value from the actual ignition value.

If the spinning wheel is restricted by brake pressure the reduction in torque at the effected wheel is very fast. Maximum brake pressure is not used to ensure passenger comfort is maintained.

Traction control system A sensor determines the position of the accelerator and taking into account other variables such as engine temperature and speed, the throttle is set at the optimum position by a drive motor. When accelerating, the increase in engine torque leads to an increase in driving torque at the wheels. In order for optimum acceleration, the maximum possible driving torque must be transferred to the road. If driving torque exceeds beyond that which can be transferred, then wheel slip will occur.

Figure 8.209 Throttle actuator

When **wheel spin** is detected, the throttle position and ignition timing are adjusted. However, better results are gained when the brakes are applied to the spinning wheel. When the brakes are applied, a valve in the hydraulic modulator assembly moves over to allow traction control operation. This allows pressure from the pump to be applied to the brakes on the offending wheel. The valves, in the same way as for ABS, can provide pressure build-up, pressure hold and pressure reduction. This all takes place without the driver touching the brake pedal.

Figure 8.210 Hydraulic modulator assembly

Figure 8.211 Stability control can help in a situation like this

ESP systems intervene to ensure stability under a wide range of situations. Shown online is the difference between a vehicle with and without a stability control system. Sensors supply an electronic control unit with information on vehicle movement such as rotation about a vertical axis, which is known as yaw. By controlling the driving force from the engine and the braking force to individual wheels, the vehicle can be kept in a stable condition. This occurs even if the driver is not fully in control!

Traction control is designed to prevent wheel spin when a vehicle is accelerating. This improves traction and ensures vehicle stability. Anti-lock brakes and traction control have now developed into complex stability control systems.

 Create a mind map to illustrate the features of a key component or system

8.3.11 Electronic stability program ❸

Introduction Vehicle stability systems are designed to take over from the driver in extreme conditions to reduce the chances or severity of an accident. Manufacturers describe this in a number of ways – for example, electronic stability control (ESC), dynamic stability control (DSC) and a few other variations on a theme such as vehicle dynamic control (VDC).

Bosch ESP Bosch was the first supplier worldwide in 1995 to start the series production of the Electronic Stability Program (ESP) – so we will stick with this name. The system of the Generation 5.0, as it was called at that time, consisted of eleven components as shown in Figure 8.212.

More modern systems have a reduced number of components and some are integrated together.

Figure 8.212 The first generation of the Bosch Electronic Stability Program

Components The active safety system comprises the following components:

▶ hydraulic unit with attached electronic control unit
▶ wheel speed sensors
▶ steering angle sensor
▶ yaw rate sensor with integrated acceleration sensor.

8

This ESP system monitors vehicle stability 25 times per second.

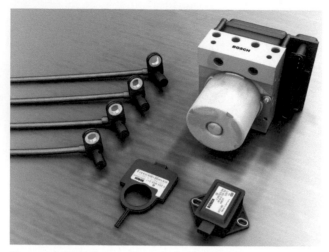

Figure 8.213 Bosch Electronic Stability Program (generation 8)

Figure 8.214 Hydraulic unit and ECU

Integrated ECU The hydraulics component of an ESP system consists of the hydraulics unit and electronics. The unit also forms a part of the anti-lock brake system (ABS).

ESP in action Examine the layout and the components in Figure 8.215 and then watch the

short online video to see how ESP operates in more detail.

Handling ESP improves the safety of a vehicle's handling by detecting and preventing skids. Using a range of sensors, it detects whether the car is following the steering movements. When ESP detects loss of steering control, it automatically applies individual

1 ESP-Hydraulic unit with integrated ECU
2 Wheel speed sensors
3 Steering angle sensor
4 Yaw rate sensor with integrated acceleration sensor
5 Engine-management ECU for communication

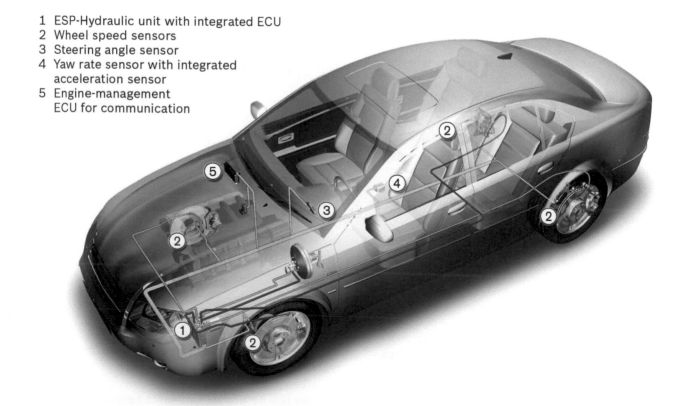

Figure 8.215 ESP Components

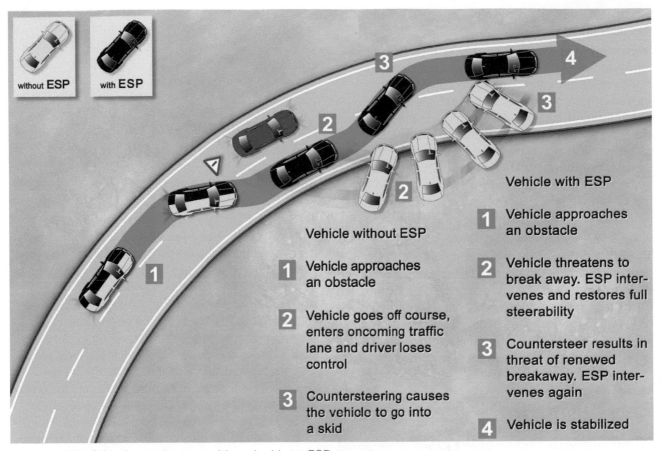

Figure 8.216 Critical manoeuvre – with and without ESP

Vehicle without ESP

1 Vehicle approaches an obstacle

2 Vehicle goes off course, enters oncoming traffic lane and driver loses control

3 Countersteering causes the vehicle to go into a skid

Vehicle with ESP

1 Vehicle approaches an obstacle

2 Vehicle threatens to break away. ESP intervenes and restores full steerability

3 Countersteer results in threat of renewed breakaway. ESP intervenes again

4 Vehicle is stabilized

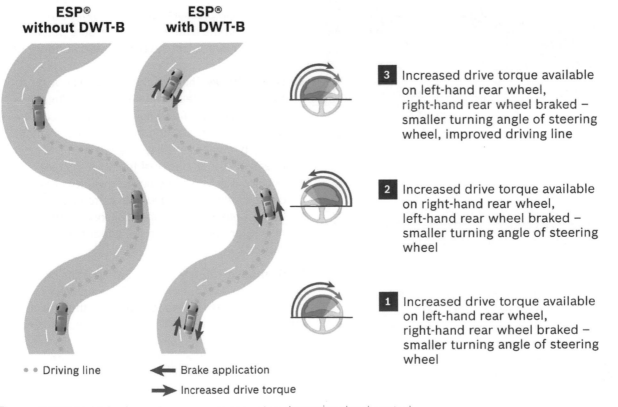

ESP® without DWT-B

ESP® with DWT-B

3 Increased drive torque available on left-hand rear wheel, right-hand rear wheel braked – smaller turning angle of steering wheel, improved driving line

2 Increased drive torque available on right-hand rear wheel, left-hand rear wheel braked – smaller turning angle of steering wheel

1 Increased drive torque available on left-hand rear wheel, right-hand rear wheel braked – smaller turning angle of steering wheel

• • Driving line

⬅ Brake application

➡ Increased drive torque

Figure 8.217 Vehicle dynamic management using dynamic wheel control

brakes to help control the vehicle and keep it in the direction intended by the driver.

Immediate reaction If the vehicle threatens to leave its track, ESP can apply individual braking forces at each wheel in order to correct vehicle movement. If, for instance, the vehicle threatens to oversteer in a curve, ESP applies the brakes at the front wheel on the outside of the curve and generates a reaction torque which stabilizes the vehicle. Furthermore, if necessary, ESP is able to influence engine torque and therefore the slip at the driven wheels. ESP reacts very quickly and thus neutralizes critical situations before they have a chance to materialize.

The ESP control system The complex ESP control system would be impossible without high-performance sensors and electronics. The system calculates the desired trajectory based on steering angle, while wheel speed sensors are measuring the rotational velocity of all wheels. A central element within this system is a yaw rate sensor designed to consistently monitor the vehicle's tendency to rotate around its vertical axis.

Figure 8.218 Yaw is the turning of a vehicle about a vertical axis

Summary ESP saves lives by improving vehicle stability in critical conditions. However, do remember it can't change the laws of physics, so beyond a certain point a crash will happen no matter how complex the system!

None the less, the computing power of ESP means that it is now possible to network active and passive safety systems so as to address other causes of crashes. For example, sensors may detect when a vehicle is following another car too closely and slow down the vehicle, straighten seat backs, and tighten seat belts to avoid or prepare for a crash.

> Using images and text, create a short presentation to show how a component or system works

8.4 Wheels and tyres

8.4.1 Tyre basics

Introduction The tyre performs four functions. These are:

1. **Support** The tyre supports the vehicle when it is stationary. However, when it is in motion the tyre must resist considerable load shifts during acceleration and braking. At times, a car tyre must carry over fifty times its own weight

2. **Traction and braking** The tyre must be able to transmit forces. These are the engine's power output as acceleration and the braking forces when the brakes are applied. How well these forces are transmitted depends on the quality of just a few square centimetres of tyre in contact with the ground

3. **Steering** The tyre should steer the vehicle accurately, regardless of road and weather conditions. The car's ability to keep a straight path depends on the tyre's ability to maintain its course. The tyre must absorb transverse forces without deviating from its trajectory. Each vehicle has a set inflation pressure for the tyres of each axle. By respecting the differences in pressure between the front and rear axles, the best driving accuracy can be obtained

4. **Road Shocks** The tyre absorbs road shocks to make life more comfortable for the driver and passengers. This also helps the vehicle components to last longer. The main characteristic of the tyre is its flexibility, especially in a vertical direction. The elasticity of the air in the tyre enables it to withstand successive deformations, caused by obstacles and uneven road surfaces. Correct pressures ensure a reasonable degree of comfort, and maintain the correct steering capacity.

Performance The tyre must continue at its best performance level for millions of revolutions of the wheel. Wear patterns depend on how the tyre is used, but especially on the quality of the ground contact. Tyre pressure plays a major role because it affects the size and shape of the contact area. It also affects the distribution of forces, to the different parts of the tyre, in contact with the ground. Several different materials are used in the construction of the tyre as shown.

Figure 8.219 Overview of tyre materials (Source: Continental)

Beads Steel bead core wires hold the tyre firmly on the rim and steel belt cords provide driving characteristics and improve mileage. Carcass fabric cords keep the tyre in shape and cap ply fabric cords improve high speed performance.

Figure 8.220 Steel and fabric reinforcing materials (Source: Continental)

Casing The tyre is flexible casing, which contains air. Tyres are manufactured from reinforced synthetic rubber. The tyre is made from an inner layer of fabric plies, which are wrapped around bead wires at the inner edges. The bead wires hold the tyre in position on the wheel rim. The fabric plies are coated with rubber, which is moulded to form the side walls and the tread of the tyre. Behind the tread is a reinforcing band, usually made of steel, rayon, or glass fibre. Modern tyres are mostly tubeless, so they also have a thin layer of rubber coating the inside to act as a seal.

Construction Beads clamp the radial tyre firmly against the wheel rim and can withstand very high forces. The tyre has supple rubber walls, which protect the tyre against the impacts (with kerbs, etc.) that might otherwise damage the carcass. There is also a hard rubber link between the tyre and the rim. Crown plies, consist of oblique overlapping layers of rubber

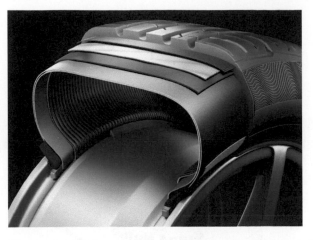

Figure 8.221 Tyre construction (Source: Continental)

reinforced with very thin, but very strong, steel wires. The overlap between these wires, and the carcass cables, forms a series of non-deformable triangles. This arrangement lends great rigidity to the tyre structure.

As the sidewalls of the radial tyre are very flexible, they stretch in proportion to the increase in force. The sidewall acts like a moving hinge between the wheel and the crown. This allows the crown to remain flat against the ground. The path of the tyre therefore remains constant, even when subject to lateral forces.

Figure 8.222 Tyre components: 1. Bead core, 2. Apex, 3. Inner liner, 4. textile carcass, 5. Bead reinforcement, 6 Flange cushion or rim strip, 7. Steel cord belt, 8. Cap ply, 9. Sidewall, 10. Tread (Source: Continental)

Markings There are various forms of tyre size markings and these differ to differentiate between tyre types. The size markings should be treated the same as a part number on a vehicle, so the driver should ensure that the tyres on their vehicle carry the precise markings indicated in the vehicle handbook or are an approved alternative fitment.

Most size markings indicate the dimensions, the type of structure and the speed capacity of the tyre.

8

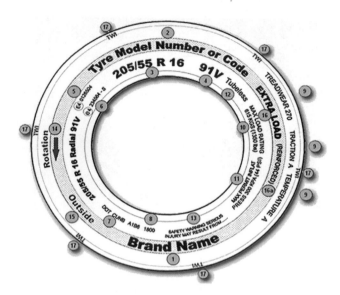

Figure 8.223 Tyre markings: 1. Tyre brand name, 2. Pattern or model code, 3. Tyre size, 4. Service description (Load index & Speed symbol), 5. ECE or EEC type approval number (C & U regulations stipulate all tyres used in the UK must carry an E number), 6. ECE noise approval, 7. North American recall code, 8. Date of manufacture code, 9. UTQG quality rating, 10. Maximum permitted load (USA), 11. Maximum permitted inflation pressure (USA), 12. Denotes tubeless construction, 13. Safety warning, 14. Direction of rotation (Directional or Composite tyres), 15. Rim fitment instruction (Asymmetrical or Composite tyres), 16. Denotes extra load tyre construction, 16a. Denotes reinforced tyre construction, 17. Position of tread wear indicators.

Load index The following table shows the weight that each index specification can carry:

Load Index	Load in kg	Load Index	Load in kg	Load Index	Load in kg
62	265	84	500	106	950
63	272	85	515	107	975
64	280	86	530	108	1000
65	290	87	545	109	1030
66	300	88	560	110	1060
67	307	89	580	111	1090
68	315	90	600	112	1120
69	325	91	615	113	1150
70	335	92	630	114	1180
71	345	93	650	115	1215
72	355	94	670	116	1250
73	365	95	690	117	1285
74	375	96	710	118	1320
75	387	97	730	119	1360
76	400	98	750	120	1400
77	412	99	775	121	1450

Load Index	Load in kg	Load Index	Load in kg	Load Index	Load in kg
78	425	100	800	122	1500
79	437	101	825	123	1550
80	450	102	850	124	1600
81	462	103	875	125	1650
82	475	104	900	126	1700
83	487	105	925		

Speed ratings The following table shows the speed ratings in mph and kph:

Speed Rating	Mile/ Hour	Kilometres/ Hour	Speed Rating	Miles/ Hour	Kilometres/ Hour
N	87	140	U	124	200
P	93	150	H	130	210
Q	99	160	V	149	240
R	106	170	Z	150+	240+
S	112	180	W	168	270
T	118	190	Y	186	300

Tyre labelling The tyre labelling regulation introduces labelling requirements regarding the display of information on the fuel efficiency, wet grip and external rolling noise of tyres. Its aim is to increase the safety and the environmental and economic efficiency of road transport by promoting fuel-efficient and safe tyres with low noise levels. This regulation allows end-users to make more informed choices when purchasing tyres by considering this information.

Tread The rubber compound of a tyre provides grip – not the tread. However, without the tread to disperse water, very little grip can be achieved. Formula 1, and other racing tyres do not have tread. These are known as 'slicks'. A version of these high-performance tyres has grooves to disperse water in wet conditions!

Cross-ply tyres are not now used on any mass produced modern car. However, the construction details are useful to show how tyre technology has developed. Several textile plies are laid across each other, running from bead to bead in alternate directions. The same number of plies is used on the crown and the sidewalls. There is no difference between the sidewalls and crown because each has the same plies.

Bias belted tyre Another tyre not now in common use is the bias belted tyre. This starts with two or more bias-plies to which stabilizer belts are bonded directly beneath the tread. This construction provides a smoother ride that is like the cross-ply tyre, while lessening rolling resistance because the belts increase tread stiffness.

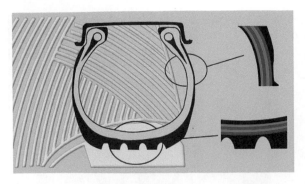

Figure 8.224 Plies embedded in layers of rubber

Functions The functions of security, comfort and economy must continue for the whole lifetime of the tyre. The tyre needs air to function correctly and be long lasting. Its inflation pressure must be checked regularly. Tread depth and general condition should also be checked regularly.

 Create a word cloud for one or more of the most important screens or blocks of text in this section

8.4.2 Tread patterns ❶❷

Introduction All tyre designs and tread patterns are, to some extent, a compromise between wet grip, dry grip and rolling resistance. There are four basic tread pattern concepts:

1. Symmetrical
2. Directional
3. Composite
4. Asymmetrical

Figure 8.225 Symmetrical (Source: Continental)

Figure 8.226 Directional (Source: Continental)

Figure 8.227 Composite (Source: Continental)

Figure 8.228 Asymmetrical (Source: Continental)

Symmetrical or multi-directional tyres have a pattern design that may be fitted either way round on the rim and to any position on the vehicle. The advantages are that they can be fitted on any position on the vehicle, without considering rotation and sidewall 'outside' marking, and they have good overall performance characteristics. The disadvantage is that aquaplaning and pass-by noise may be compromised.

Tread patterns There are several different directional tyre tread patterns. They have a pattern which is designed for one direction of rotation only. The pattern grooves generally follow the shape of an arrowhead. They offer very good protection against aquaplaning, have good directional stability and are attractive. However, there is an increased complexity at fitting and limitations when rotating after usage.

Figure 8.229 Water dispersal (Source: Continental)

Composite tyres are both directional and asymmetric and are for specialist applications. These tyres can provide a combination of advantages of both

8

asymmetric and directional tyres. They are therefore optimized for handling in both wet and dry conditions, offer good aquaplaning protection, have good directional stability and low noise emissions. They are however more complex to fit and rotation after usage is limited.

Asymmetrical pattern tyres have a tread pattern which differs between inner and outer shoulder. The outer shoulder has more rubber contact area to the road and the inside contact area is more open. They offer optimal adaptation to the road and driving conditions in both wet and dry conditions. They can be mounted at any position on the vehicle (if the 'outside' mark is correctly positioned). Pass-by noise is reduced and they have an improved wear potential. The asymmetrical design may have a slight compromise in aquaplaning vs. directional concept but this depends on the particular design.

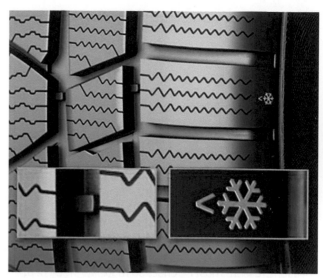

Figure 8.231 Once the tread is worn down to 4mm, the Winter Indicator becomes level with the surface of the tread pattern, indicating that the tire has reached its limit of winter suitability (Source: Continental)

Figure 8.230 ContiSportContact™ 5 (Source: Continental)

Winter tyres Specially formulated natural rubber compound of winter tyres ensures rubber flexibility at a wide range of low temperature contrary to summer tire tread compound. The winter tyre tread design offers its benefits especially on roads covered with snow or mud. The snow gets pressed into the wider and deeper tread grooves and by this it utilises the effect of shear forces on snow for additional grip. The wider grooves can also absorb more slush compared to summer tyres and therefore have more contact with the road surface

Summary This section has outlined the four main tread patterns and the winter (or M+S) tyre concept. All aspects of tyre operation are a compromise to

some extent, but a key point to remember is that you generally get what you pay for!

> Use the media search tools to look for pictures and videos relating to the subject in this section

8.4.3 Wheels and rims ❶ ❷

Introduction Together with the tyre, a road wheel must support the weight of the vehicle. It must also withstand side thrusts when cornering, and torsional forces when driving. Road wheels must be strong, but light weight. They must be cheap to produce, easy to clean, and simple to remove and refit.

The centre of this type of wheel is made by pressing a disc into a dish shape, to give it greater strength. The rim is a rolled section, which is circled and welded. The rim is normally welded to the flange of the centre disc. The centre disc has several slots under the rim. This is to allow ventilation for the brakes as well as the wheel itself.

The manufacture of this type of wheel makes it cheap to produce and strong. The bead of a tyre is made from wire, which cannot be stretched for fitting or removal. The wheel rim, therefore, must be designed to allow the tyre to be held in place, but also allow for easy removal.

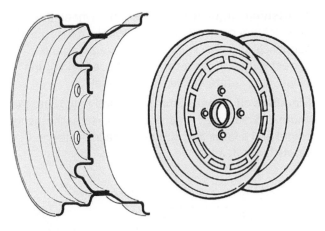

Figure 8.232 Standard wheel design

Fitting To facilitate fitting and removal a 'well-base' is manufactured into the rim. For tyre removal, one bead must be forced into the well. This then allows the other bead to be levered over the edge of the rim. The bead seats are made with a taper so that as the tyre is inflated the bead is forced up the taper by the air pressure. This locks the tyre on to the rim making a good seal. Most wheels in current use have safety humps. The rim base width will be different on one side compared to the other. The safety hump provides some handling control characteristics when a tyre is in a fully deflated condition as it ensures the bead doesn't totally dislodge into the well-base.

Figure 8.233 All car rims are well base. Early ones were intended for cross-ply, tubed type small section tyres which were usually less than 5" rim width and a low flange height

Figure 8.234 From left to right: Flat hump rim (FH), Contre pente rim, Double hump rim (H2)

Steel wheels are a very popular design. They are very strong and cheap to produce. Steel wheels are usually covered with plastic wheel trims. Trims are available in many different styles. To improve the appearance of the wheel trim some designs protrude above the wheel rim flange, i.e. their outer diameter is greater than the diameter of the wheel rim at the tip of the flange.

Wheel trims can abrade the tyre under load and, if not a tight fit rotationally, can rotate around the wheel rim and foul the tyre inflation valve. A severe case could lead to valve damage and air leakage.

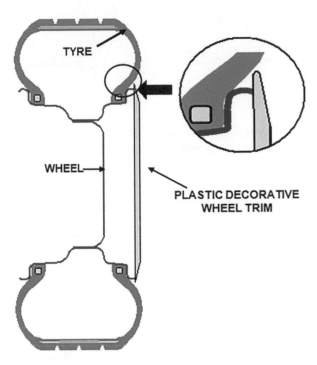

Figure 8.235 Wheel trim

Alloy wheels, or 'alloys', are good, attractive looking wheels. They tend to be fitted to higher specification vehicles. Many designs are used. They are light weight but can be difficult to clean. Wheels of this type are generally produced from aluminium alloy castings, which are then machine finished. Alloy wheels can be easily damaged by 'kerbing'!

Figure 8.236 Alloy wheels

Split rims Many commercial vehicles use split rims, either of a two, or three-piece construction. The tyre is held in place by what could be described as a very large circlip. Do NOT remove or fit tyres on this type of wheel unless you have received proper instruction.

8

423

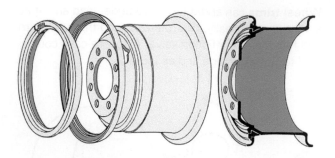

Figure 8.237 Do not work on split rims unless trained

Divided rims On a few specialist vehicles, the rims are divided into two halves which are bolted together. The nuts and bolts holding them together should be specially marked. Undoing them with the tyre inflated would be very dangerous.

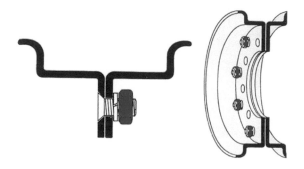

Figure 8.238 Do not work on divided rims unless trained

Wheel fixing Light vehicle road wheels are usually held in place by four or five nuts or bolts. The fixing holes in the wheels are stamped or machined to form a cone shaped seat. The wheel nut or bolt heads, fit into this seat. This ensures that the wheel fits in exactly the right position. In the case of the steel pressed wheels, it also strengthens the wheel centre around the stud holes.

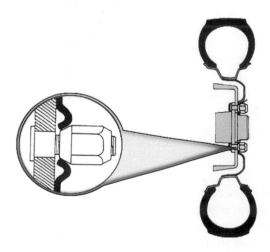

Figure 8.239 Wheel fixing

Tightening sequence When fitting a wheel, the nuts or bolts must be tightened evenly in a diagonal sequence. It is also vital that they are set to the correct torque. Ensure the cone shaped end of the wheel nuts is fitted towards the wheel.

Markings on the wheel rims are used to indicate the width and diameter. A code letter indicates the flange height and a letter and number code indicate the safety hump feature.

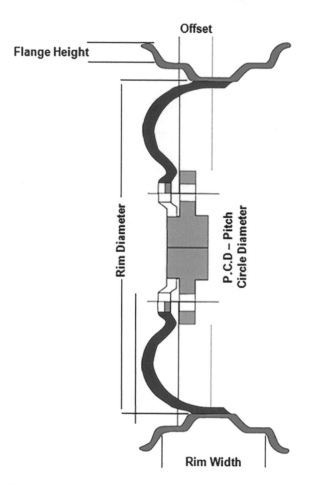

Figure 8.240 Rim sizes

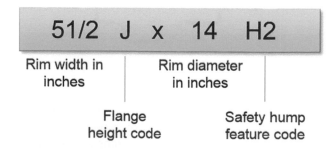

Figure 8.241 Diameter is the distance between bead seats, width is measured between the vertical faces of the flanges and the flange height is denoted by a code such as J 17.3mm, JK 18mm, K 19.6mm, L 21.6mm etc.

Rim offset The rim offset is measured in millimetres and is the distance between the centre of the rim and inner surface of the wheel disc (the hub contact face). This measurement can be positive or negative. The rim offset is normally specified by the car manufacturer and determines aspects such as wheel track. Different cars may have different offsets for the same tire size. Rims used for upsizing may have a smaller rim offset to increase the wheel track to ensure a fit into the wheelhouse.

On the front axle, it is extremely important to keep offset within the manufacturer's specs. The consequences of incorrect offset are: increased steering effort, increased steering wheel kick-back during acceleration, and increased load on wheel bearings and suspensions

On the rear axle, negative offset can increase wheel track and improve stability and handling, but will increase the loading on the vehicle's suspension.

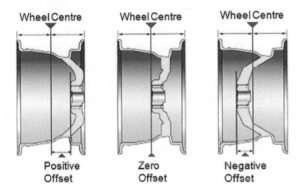

Figure 8.242 From left to right: positive, zero and negative offset

Pitch circle diameter (PCD) Always ensure that the correct wheel PCD dimensions are used and consider the tyre dimensions carefully. Increasing the wheel diameter will require adjustment to the tyres aspect ratio to ensure the overall rolling circumference remains the same. Always remember to check the rear overall clearance dimensions carefully in a loaded and unloaded condition. Try jacking each wheel on the car as far as it will go, check with the suspension at full height.

Valve The valve is to allow the tyre to be inflated with air under pressure, prevent air from escaping after inflation, and to allow the release of air for adjustment of pressure. The valve assembly is contained in a brass tube, which is bonded into a rubber sleeve and mounting section.

The valve core consists of a centre pin, which has metal and rubber disc valves. When the tyre is inflated, the centre pin is depressed, the disc valve moves away from the bottom of the seal tube and allows air to enter the tyre. To release air, or for

pressure checking, the centre pin is depressed. During normal operation, the disc valve is held onto its seat by a spring and by the pressure of air. If all the air needs to be released, the valve core assembly can be removed. The upper part of the valve tube is threaded to accept a valve cap. This prevents dirt and grit from entering and acts as a secondary seal.

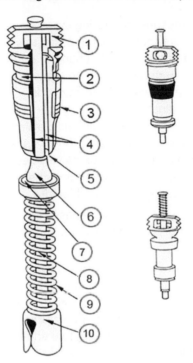

Figure 8.243 Details of the valve construction. 1. Strong bridge, 2. Sturdy swivel connection, 3. Wedge-fit, 4. Clear passage, 5. Knife edge seat, 6. Special bell-shaped guide, 7. Seat washer, 8. Full length plunger pin, 9. Special spring, 10. Spring cup

Tubeless valve The tubeless valve core is as described previously. However, the valve body must be made so that when fitted into the wheel, an airtight seal is formed. Wheel rims used for tubeless tyres must be sealed and airtight. Most wheels and tyres in use are of the tubeless design.

Figure 8.244 Various 'tubeless' valves

8

Make a simple sketch to show how one of the main components or systems in this section operates

8.4.4 Tyre pressure monitoring ❷

Introduction A tyre pressure monitoring system (TPMS) is a safety feature that continually monitors a vehicles' tyres and alerts the driver to changes in tyre pressure. The changes in pressure can be detected by either direct or indirect means. Both methods will, as a minimum, illuminate a warning light on the vehicle dashboard display and sound an audible alert when 25% deflation has occurred. Early TPMS were introduced as an option on high-end luxury vehicles as early as the 1980s, although it wasn't until the year 2000 that it was first fitted as a standard feature.

Pressure It is estimated that 50% of all passenger car tyres have the wrong inflation pressure. This has several serious consequences:

▶ If the inflation pressure is not adapted to the axle loads, tyre flexing increases, resulting in energy loss and therefore higher fuel consumption.

▶ Increased flexing leads to higher temperatures, which cause damage to the tyre structure and can even lead to tyre failure.

▶ Hidden tyre damage will not be eliminated by later adjustment of the tyre pressure.

▶ The influence on a tyre's life expectancy is disadvantageous.

▶ Driving safety is influenced because in case of wrong inflation, the risk of tyre damage increases significantly.

Direct TPMS means each wheel of the vehicle has a sensor fixed to monitor the changes in pressure from the tyre. They also measure temperature. Each sensor sends its signal to the receiver inside the vehicle using a wireless connection. In Europe, the transmission frequency is 433Mhz.

▶ If low pressure or a leak is detected (generally 25% less than normal operating pressure), the driver is alerted by the in-car system and generally the deflated tyre is identified

▶ Direct TPMS is very accurate measuring to 1 or 2 psi

▶ A puncture after parking is immediately identified

▶ Sensors send their signals approximately every 30 seconds, when parked they transmit every 20 to 30 minutes. At 25 km/h the sensor switches back on to transmit every 30 seconds

▶ Sensors have an approximate life of 5 years or 160,000 km.

Figure 8.245 This sensor measures pressure and temperature and transmits the data to a control unit

Indirect TPMS is generally fitted to a vehicle that has had fitted or can be fitted with run flat tyres. This is because it is difficult to see or feel deflation in this type of tyre. Indirect tyre pressure monitoring systems do not use pressure sensors to monitor tyre pressure, they work from the ABS or speed sensors on the vehicle. Indirect systems monitor tyre pressure by assessing the rotational speeds of each tyre, and work on the premise that an under-inflated tyre has a slightly different diameter than a fully inflated tyre. An algorithm is used to assess the differences in wheel speeds. The under-inflated tyre would therefore rotate at a different speed than the correctly inflated one, causing a tyre pressure warning. The deflated tyre is not identified so the driver must check all four.

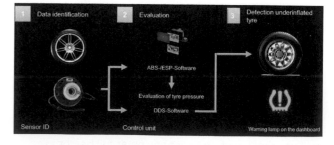

Figure 8.246 Indirect TPMS operation (Source: Continental)

Negative aspects of indirect TPMS:

▶ The system is not very accurate

▶ When tyres are re-inflated, the system needs to be re-calibrated

▶ When tyre positions are changed, the system needs to be re-calibrated

▶ When the tyres are replaced, the system needs to be re-calibrated

▶ The system can be re-calibrated by the driver without first ensuring that the pressure is correct in all tyres

▶ A puncture after parking is not immediately identified.

Tyre pressure monitoring and the law in Europe

The law is not currently retrospective, and does not apply to older vehicles. This law applies to passenger vehicles only, with no more than 7 seats.

▶ From November 2012, all new type vehicles in the M1 category (vehicles under 3.5 Tonnes with less than 8 seats) were required by law to have TPMS installed. This applies to the road wheels not the spare

▶ From November 2014, all new passenger vehicles had to have TPMS installed by the manufacturer.

Repairs When a tyre is replaced or a puncture repaired on a vehicle without TPMS, it is normal practice to replace the rubber valve. With a TPMS sensor it is also important to replace the rubber sealing grommet which deteriorates just like the rubber valve. At the same time as replacing the rubber grommet, it is good practice to replace other parts too.

▶ The Metal Sealing Washer (if fitted) often becomes distorted, replacing this ensures a good seal when re-assembling the sensor

▶ The Outer Collar is replaced because it becomes metal fatigued, and sometimes badly corroded. This metal fatigue can often make the outer securing collar crack and sometimes this will fall off completely

▶ The Core is replaced to create a good airtight seal in the sensor internally. Only Nickel plated cores should be used in Metal Valves

▶ The Cap is replaced with a new pressure cap. The rubber washer in the old cap deteriorates, replacing the cap helps keep the sensor airtight and free from debris, or fluids that might affect the operation of the core.

MOT TPMS is now part of the annual vehicle test (in the UK), and has applied to all newly registered cars from January 1st 2012. This means that a car with a faulty TPMS has failed the vehicle test since January 2015. Different European and other countries may interpret the EU legislation differently, so please refer to your own country legislation.

There are hundreds of thousands of cars on European roads with sensors removed because the cost of replacement is high. Rubber tyre valves are used to replace the sensor and the warning light disabled, however this practice was prohibited on new-type 2012 cars. Insurance implications will be the same as for disabling an airbag or ABS. Disabling a safety system on a vehicle may become an offence.

 Create an information wall to illustrate the features of a key component or system

8.4.5 Spare wheels and extended mobility ❶ ❷

Introduction At first view, it would seem obvious that a vehicle should carry a spare wheel. However, modern tyre technology and associated extended mobility solutions means that this is not necessarily the case. The information presented here is supplied by Continental Tyres but work by other manufacturers on similar systems has produced similar results. There are four main ways to deal with a roadside puncture:

1. Spare wheel (full size or space-saver)
2. ContiKit (or similar kit from other manufacturers)
3. ContiSeal (or other self-sealing system)
4. Self-supporting run-flat (SSR) systems

Most manufacturers prefer to eliminate the spare wheel to reduce weight, gain space and save costs. The choice of alternatives now allows this to happen. Other than a spare wheel, the three main options are outlined here.

Repair kit A repair kit is designed to be simple for the driver to use. It is compact and versatile whilst being cost effective. The pump is adequate for all car tyre sizes up to 225/55 R16 with an inflation pressure up to 3 bar. A hole up to 6mm in diameter can be sealed followed by a driving distance of about 200 km. Various sizes of the sealant bottles are available to suit the tyres fitted to the car. NOTE: the lifetime of the sealant is limited. Always check the latest data but a four-year life is typical. The system will work in the temperature range: –30°C to 70°C.

Figure 8.247 ContiComfortKit (Source: Continental)

8

The ContiSeal system shown here has an extra air-proof sealant layer in the tyre tread area. It can automatically seal punctures up to 5mm diameter in the tread area. There is no influence on ride comfort or rolling resistance and it can be mounted on standard rims. It can seal about 80% of all punctures and works on all tyre types and dimensions. It has no effect on vehicle dynamics and distance and speed are not affected due to the instant seal. It is, however, recommended that this system is used in conjunction with a tyre pressure monitoring system (TPMS).

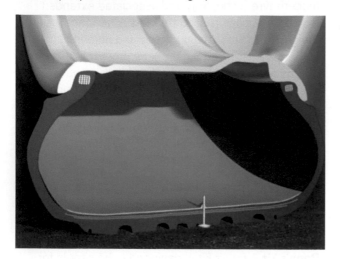

Figure 8.248 ContiSeal system (Source: Continental)

Run-flat tyres are designed to keep going in the event of full deflation. Stiff rubber inserts are used inside the tyre walls to carry the temporary additional load in the case of pressure loss. The system is compatible with current standard rims. They will keep working for up to 150 km at 80 km/h. There is an additional weight of about 20% associated with run-flat tyres. It is essential that a pressure warning system is used and ESP/DSC is also recommended.

Figure 8.249 Normal usage with full air pressure (Source: Continental)

Summary There are several different extended mobility systems (EMS) on the market, none of them offer a perfect solution as even a full-size spare has

only a 70% success rate! For this reason, conventional spares, mini-spares, sealant kits, sealant tyres and run-flat tires co-exist in the market.

Figure 8.250 Pressure loss (Source: Continental)

Figure 8.251 Mobility kit (Source: Continental)

Use a library or the web search tools to further examine the subject in this section

8.4.6 Wheel balancing ❶ ❷

Correctly balanced wheels are an important comfort and safety issue. An out of balance wheel produces vibration and a reduction in steering control. It will also result in abnormal tyre wear. However, the wheel and tyre are not always to blame. Worn steering joints, wheel bearings or driveshaft joints can also cause vibration.

Balance A wheel and tyre may be out of balance either statically or dynamically. Static balance relates to a stationary wheel. Dynamic balance relates to the conditions of a rotating wheel.

Figure 8.252 Wheel with an out of balance mass

Static balance A wheel and tyre that is in perfect static balance has the mass evenly distributed around its centre. When mounted on a free bearing and spun, it comes to rest in any position.

Static imbalance A simple example of static imbalance is a bike wheel. When spun freely, it will always come to rest with the valve at the bottom. The effect of static imbalance on a vehicle is to cause it to 'tramp' up and down. The effect becomes progressively worse at higher speeds. This puts a strain on the steering and suspension components.

Curing static imbalance To cure static imbalance a compensating mass, or masses, is placed on the wheel. One method is to place a large mass on the wheel flange. Another is to use two smaller masses as shown in Figure 8.253. These methods are used because it is not normally possible to put an extra mass on the wheel centreline.

Figure 8.253 Wheels that have been statically balanced

Dynamic balance The term 'dynamic' is used because the effect is only noticeable when the wheel is in motion. This is felt, when driving, as a 'steering wobble'. It can be dangerous if excessive and, at the least, it results in premature tyre wear.

Dynamic imbalance Dynamic imbalance is best explained by imagining a crank, like the pedals of a bike. If the weights are equal and at the same distance from the bearing centreline, the wheel will be statically balanced. However, when the

crank rotates a force will act on each weight in an outwards direction. This will result in a twisting force on the bearing, which is described as a 'couple'. The direction of the 'couple' reverses every half turn, resulting in a rocking movement of the vehicle steering. The force increases with speed.

Car wheel dynamic imbalance The wheel shown in Figure 8.254 has been statically balanced. Mass B has been added to compensate for the out of balance mass A. However, because the masses A and B are not in the same plane, a twist or couple will be set up. This will result in a dynamic imbalance.

Figure 8.254 Statically balanced but dynamically imbalanced

Curing dynamic imbalance The out-of-balance wheel in Figure 8.254, can be balanced in three ways. Adding a weight on the centreline is fine, but this can only be done on spoked wheels. Weight on one flange will only statically balance the wheel. Smaller weights on both flanges will result in a statically, and dynamically, balanced wheel.

Curing static and dynamic imbalance In reality, an out of balance mass is usually away from the wheel centreline. Static and dynamic imbalance occurs. To compensate for this, weights are added as indicated by a balancing machine. The weights may not be directly opposite the out of balance mass.

Figure 8.255 Weights A and B cure the imbalance due to M

Wheel balancing machine When a balancing machine is operated, the wheel is spun at high speed. An indication is given to the users of the required masses and their positions on the wheel. Do not use this type of equipment unless you have been trained.

Balance weights Small lead weights are used for rectifying out of balance wheels. The weights either clip onto the rim of steel wheels, or bond with adhesive to the rim of alloy wheels.

Summary Correct wheel and tyre balance is essential for the safe operation of the vehicle. Excessive tyre wear results from imbalance. A wheel must be balanced both statically and dynamically. Modern wheel balancers will achieve both these requirements quickly and easily.

Figure 8.256 Lead weights

 Select a routine from section 1.3 and follow the process to study a component or system

CHAPTER 9

Transmission systems

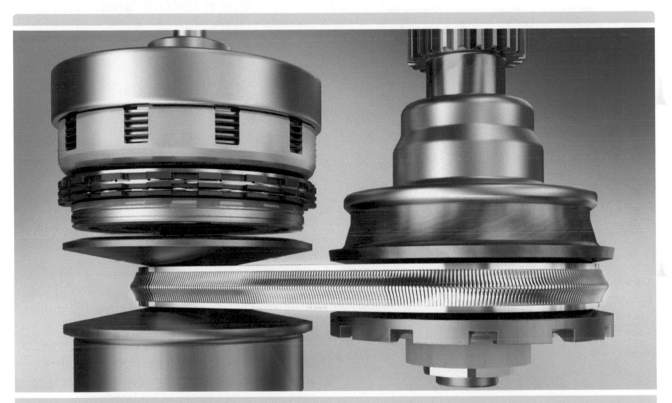

After successful completion of this chapter you will be able to show you have achieved these objectives:

- Understand how light vehicle clutch systems operate.
- Understand how light vehicle manual gearbox systems operate.
- Understand how light vehicle automatic gearbox systems operate.
- Understand how light vehicle driveline and final drive systems operate.
- Understand how to check, replace and test transmission and driveline units and components.
- Understand how to diagnose and rectify faults in light vehicle transmission and driveline systems.

DOI: 10.1201/9781003173236-9

9.1 Clutch

9.1.1 Purpose of the clutch ❶

A clutch is a device for disconnecting and connecting rotating shafts. In a vehicle with a manual gearbox, the driver depresses the clutch when changing gear, thus disconnecting the engine from the gearbox. It allows a temporary neutral position for gear changes and also a gradual way of taking up drive from rest.

Automatic transmission Cars with automatic transmission do not have a clutch as described here. Drive is transmitted from the flywheel to the automatic gearbox by a torque converter, sometimes called a fluid clutch.

Figure 9.1 Torque converter components

Gearbox For most light vehicles, a gearbox has five forward gears and one reverse gear. It is used to allow operation of the vehicle through a suitable range of

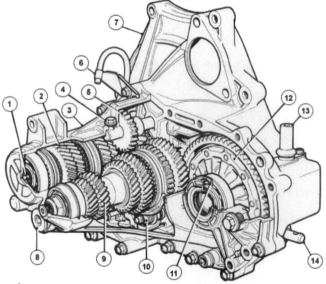

Figure 9.2 Gearbox components

speeds and torque. A manual gearbox needs a clutch to disconnect the engine crankshaft from the gearbox while changing gears. The driver changes gears by moving a gear lever, which is connected to the box by a mechanical linkage.

Clutch components Each of the following sections covers one or more typical clutch components. Some are more important than others. The driven plate and the pressure plate are the two main parts.

Figure 9.3 Driven plate and pressure plate

In conjunction with a sensor, the reluctor ring provides a signal for the ignition and fuel systems. It supplies information on engine speed and position. However, it is not part of the clutch!

The flywheel keeps the engine running smoothly between power strokes. It also acts as a surface against which the driven plate can press. A locking plate is used for security of the flywheel.

The driven plate is a friction material plate, which is clamped between the pressure plate and the flywheel. It is splined on to the gearbox input shaft. The small coil springs are to prevent the clutch snatching as drive is taken up.

This cover of the pressure plate is fixed to the flywheel with a ring of bolts. The fingers in the centre act as springs and levers to release the pressure. Drive is transmitted unless the fingers are pressed in towards the flywheel.

A release shaft transfers the movement of the cable to the release fork and bearing. The bearing pushes against the clutch fingers when the pedal is depressed to release the drive. A return spring is used so that, when the clutch pedal is not depressed, the bearing allows the clutch fingers to return outwards. A seal is fitted to keep out water and dirt.

The clutch cable makes a secure connection to the clutch pedal. Strong steel wire is used. Movement of the pedal is, through this, transferred to the release bearing. A few vehicles use hydraulics to operate the clutch.

A support is made for the ball end of the cable. Many different methods are used and this is just one example. The rubber pad prevents metal-to-metal contact. A retaining clip secures the end of the cable.

A general cover is used for the clutch assembly but it is also the way to secure the clutch and gearbox to the engine. Some front-wheel drive clutches are covered with a thin pressed steel plate.

 Create a word cloud for one or more of the most important screens or blocks of text in this section

9.1.2 Clutch mechanisms ❶

The driver operates the clutch by pushing down a pedal. This movement has to be transferred to the release mechanism. There are two main methods used. These are cable and hydraulic. The cable method is the most common. Developments are taking place and an electrically operated clutch will soon be readily available.

Cable A steel cable is used, which runs inside a plastic-coated steel tube. The cable 'outer' must be fixed at each end. The cable 'inner' transfers the movement. One problem with cable clutches is that movement of the engine, with respect to the vehicle body, can cause the length to change. This results in a judder when the clutch is used. This problem has been almost eliminated, however, by careful positioning and quality engine mountings.

This clutch cable works on a simple lever principle. The clutch pedal is the first lever. Movement is transferred from the pedal to the second lever, which is the release fork. The fork, in turn, moves the release bearing to operate the clutch.

Hydraulics A hydraulic mechanism involves two cylinders. These are termed the master and slave cylinders. The master cylinder is connected to the clutch pedal. The slave cylinder is connected to the release lever.

The clutch pedal moves the master cylinder piston. This pushes fluid through a pipe, which in turn forces a piston out of the slave cylinder. The movement ratio can be set by the cylinder diameters and the lever ratios.

The **electronic clutch** was developed for racing vehicles to improve their getaway performance. For production vehicles, a strategy has been developed to interpret the driver's intention. With greater

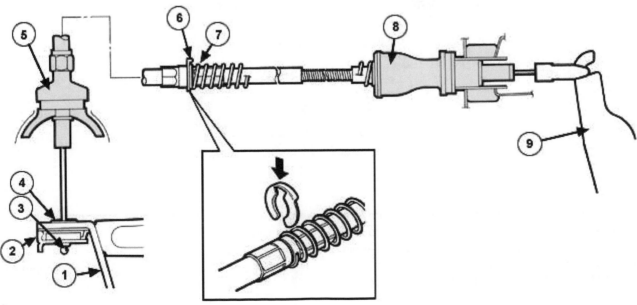

Figure 9.4 Clutch cable

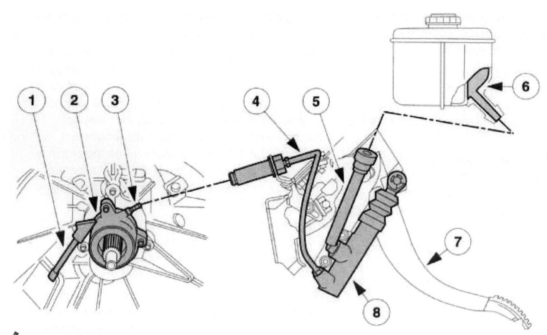

Figure 9.5 Figure 9 5 Clutch hydraulic components 1. Lever 2. Thrust bearing 3. Cylinder 4. Pipe 5. Pipe 6. Tank 7 Pedal 8. Master cylinder

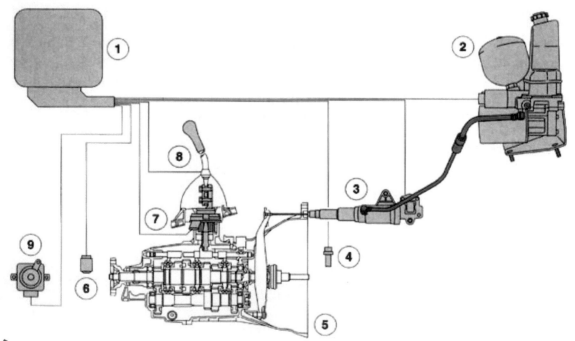

Figure 9.6 Electronic clutch 1. ECU 2. Pump and accumulator 3. Actuating cylinder 4. Sensor 5. Bellk housing 6. Speed sensor 7. Gearchange 8. Stick 9. Pedal sensor

throttle openings, the strategy changes to prevent abuse and drive line damage. Electrical control of the clutch release bearing position is by a solenoid actuator, which can be modulated by signals from the ECU. This allows the time to reach the ideal take off position to be reduced and the ability of the clutch to transmit torque to be improved.

Efficiency of the whole system can therefore be increased.

 Look back over the previous section and write out a list of the key bullet points from this section

9.1.3 Coil spring clutch ❶❷

Coil spring pressure plate assemblies use helical springs that are evenly spaced around the inside of the pressure plate cover. These springs exert pressure to hold the pressure plate against the flywheel.

Figure 9.9 Each lever has two pivot points

Figure 9.7 Coil spring clutch

Release levers During clutch disengagement, release levers release the holding force of the springs and the clutch disc no longer rotates with the pressure plate and flywheel. Usually, these pressure plates have three release levers. Each lever has two pivot points.

Figure 9.8 Disengagement levers

Pivot point One pivot point attaches the lever to a pedestal cast into the pressure plate. The other attaches the lever to a release yoke that is bolted to the cover. The levers pivot on the pedestals and release lever yokes. This moves the pressure plate through its engagement and disengagement operations.

Disengagement To disengage the clutch, the release bearing pushes the inner ends of the release levers toward the flywheel. The outer ends of the release levers move to pull the pressure plate away from the clutch disc. This action compresses the coil springs and disengages the clutch.

Engagement When the clutch is engaged, the release bearing moves and allows the springs to exert pressure. This holds the pressure plate against the clutch disc, which in turn forces the disc against the flywheel. The engine power is therefore transmitted to the gearbox through the clutch disc.

Figure 9.10 The disc is forced against the flywheel

Use a library or the web search tools to further examine the subject in this section

9.1.4 Diaphragm clutch ❶❷

Basic functions A clutch is a device for disconnecting and connecting rotating shafts. In a vehicle with a manual gearbox, the driver pushes down the clutch when changing gear to disconnect the engine from the gearbox. It also allows a temporary neutral position for, say, waiting at traffic lights and a gradual way of taking up drive from rest.

Figure 9.12 Driven plate and pressure plate together with bearings kit and an alignment tool (Source: http://haysclutches.com/)

Figure 9.11 Diaphragm clutch

Clutch location The exact location of the clutch varies with vehicle design. However, the clutch is always fitted between the engine and the transmission. With a few exceptions, the clutch and flywheel are bolted to the rear of the engine crankshaft.

Main parts The clutch is made of two main parts – a pressure plate and a driven plate. The driven plate, often termed the clutch disc, is fitted on the shaft which takes the drive into the gearbox.

Engagement When the clutch is engaged, the pressure plate presses the driven plate against the engine flywheel. This allows drive to be passed to the gearbox. Depressing the clutch moves the pressure plate away, which frees the driven plate.

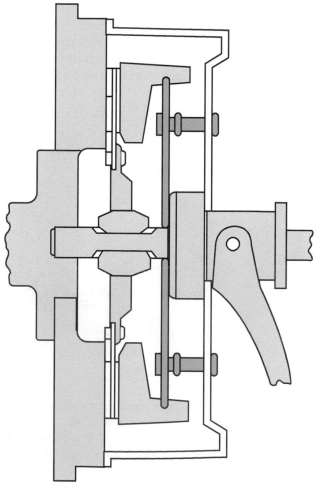

Figure 9.13 Clutch engaged

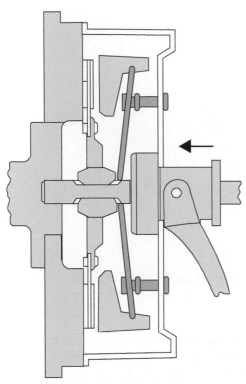

Figure 9.14 Clutch disengaged

Coil springs Earlier clutches, and some heavy-duty types, use coil springs instead of a diaphragm. However, the diaphragm clutch replaced the coil spring type because it has the following advantages:

▶ it is not affected by high speeds (coil springs can be thrown outwards)
▶ the low pedal force makes for easy operation
▶ it is light and compact
▶ the clamping force increases or at least remains constant as the friction lining on the plate wears.

Movement of the diaphragm clutch The online animation shows the movement of the diaphragm during clutch operation. The method of controlling the clutch is quite simple. The mechanism consists of either a cable or hydraulic system.

The **clutch shaft**, or gearbox input shaft, projects from the front of the gearbox. Most shafts have a smaller section or spigot which projects from its outer end. This rides in a spigot bearing in the engine crankshaft flange. The splined area of the shaft allows the clutch disc to move along the splines. When the clutch is engaged, the disc drives the gearbox input shaft through these splines.

Figure 9.16 Gearbox input shaft

The **clutch disc** is a steel plate covered with frictional material. It fits between the flywheel face and the pressure plate. In the centre of the disc is the hub, which is splined to fit over the splines of the input shaft. As the clutch is engaged, the disc is firmly

Figure 9.15 Coil spring clutch assembly

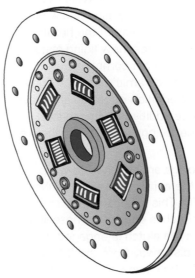

Figure 9.17 Clutch disc or driven plate

9

437

squeezed between the flywheel and pressure plate. Power from the engine is transmitted by the hub to the gearbox input shaft. The width of the hub prevents the disc from rocking on the shaft as it moves along the shaft.

Frictional facings The clutch disc has frictional material riveted or bonded on both sides. These frictional facings are either woven or moulded. Moulded facings are preferred because they can withstand high-pressure plate loading forces. Grooves are cut across the face of the friction facings to allow for smooth clutch action and increased cooling. The cuts also make a place for the facing dust to go as the clutch lining material wears.

Figure 9.19 These springs eliminate chatter

Figure 9.18 Friction material

Types of clutch disc There are two types of clutch discs – rigid and flexible. A rigid clutch disc is a solid circular disc fastened directly to a centre splined hub. The flexible clutch disc has torsional dampener springs that circle the centre hub.

Health hazards The frictional material wears as the clutch is engaged. At one time asbestos was in common use. Due to awareness of the health hazards resulting from asbestos, new lining materials have been developed. The most commonly used types are paper-based and ceramic materials. They are strengthened by the addition of cotton and brass particles and wire. These additives increase the torsional strength of the facings and prolong the life of the clutch.

Wave springs The facings are attached to wave springs, which cause the contact pressure on the facings to rise gradually. This is because the springs flatten out when the clutch is engaged. These springs eliminate chatter when the clutch is engaged. They also help to move the disc away from the flywheel when it is disengaged. The wave springs and facings are attached to the steel disc.

Figure 9.20 Solid and flexible discs

Shock absorbing The dampener is a shock-absorbing feature built into a flexible clutch disc. The primary purpose of the flexible disc is to absorb power impulses from the engine that would otherwise be transmitted directly to the gears in the transmission. A flexible clutch disc has torsion springs and friction discs between the plate and hub of the clutch.

Figure 9.21 Damping springs

When the clutch is engaged, the springs cushion the sudden loading by flexing and allowing some twist between the hub and the plate. When the loading is over, the springs release and the disc transmits power normally. The number, and tension, of these springs is determined by the amount of engine torque and the weight of the vehicle. Stop pins limit this torsional movement to a few millimetres.

Pressure plate assembly The pressure plate squeezes the clutch disc onto the flywheel when the clutch is engaged. It moves away from the disc when the clutch is disengaged. These actions allow the clutch disc to transmit, or not transmit, the engine's torque to the gearbox.

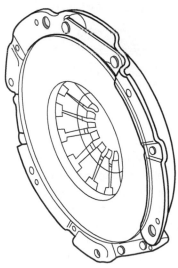

Figure 9.22 Details of the pressure plate

Spring loading A pressure plate is a large spring-loaded clamp, which is bolted to, and rotates with, the flywheel. The assembly includes a metal cover, heavy release springs and a metal pressure ring that

provides a friction surface for the clutch disc. It also includes a thrust ring or fingers for the release bearing and release levers.

Figure 9.23 Pressure plate

Release levers The release levers release the holding force of the springs when the clutch is disengaged. Some pressure plates are of a 'semi-centrifugal' design. They use centrifugal weights, which increase the clamping force on the thrust springs as engine speed increases.

Figure 9.24 Levers release the holding force

Diaphragm spring The diaphragm spring assembly is a cone-shaped diaphragm spring between the pressure plate and the cover. Its purpose is to clamp the pressure plate against the clutch disc. This spring is normally secured to the cover by rivets. When pressure is exerted on the centre of the spring, the outer diameter of the spring tends to straighten out. When pressure is released, the spring resumes its normal cone shape.

9

439

Figure 9.25 Cone-shaped diaphragm spring

Clutch release The centre portion of the spring is slit into a number of fingers that act as release levers. When the clutch is disengaged, these fingers are depressed by the release bearing. The diaphragm spring pivots over a fulcrum ring. This makes its outer rim move away from the flywheel. The retracting springs pull the pressure plate away from the clutch disc to disengage the clutch.

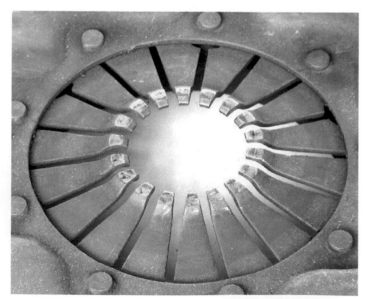

Figure 9.26 Fingers

As the clutch is engaged, the release bearing is moved away from the release fingers. As the spring pivots over the fulcrum ring, its outer rim forces the pressure plate tightly against the clutch disc. At this point, the clutch disc is clamped between the flywheel and pressure plate.

Clutch assembly The individual parts of a pressure plate assembly are contained in the cover. Most covers are vented to allow heat to escape and air to enter. Other covers are designed to provide a fan action to force air circulation around the clutch assembly. The effectiveness of the clutch is affected by heat. Therefore, by allowing the assembly to cool, it works better.

 Create an information wall to illustrate the features of a key component or system

9.1.5 Other types of clutch ❷

The simple definition of a clutch is something that engages or disengages drive. A number of different types of clutch are used for this purpose. Some of these are examined briefly over the following pages.

Automatic transmissions use a torque converter, or fluid flywheel, to couple the engine and the gearbox. The torque converter is a fluid coupling in which one rotating part causes transmission fluid to rotate. This imparts a rotation to another part, which is connected to the gearbox.

Figure 9.27 Automatic gearbox

Torque converter The coupling action of the torque converter, or fluid clutch, allows slippage for when the car is starting from rest. As the car gains speed, the slippage is reduced and, at cruising speeds, the driven member turns almost as fast as the driving member does. Some modern systems lock the two together at high speed to eliminate slip. An automatic gearbox usually contains epicyclic or planetary gears. Clutches and brake bands are used for engaging the desired gears.

Multi-plate clutches are used in specialist applications such as for very high performance vehicles. Some motorcycles and heavy commercials

Figure 9.28 Fluid clutch

also use clutches of this type. The principle is the same as a single plate clutch except that, with multiple plates, greater power can be transmitted.

Figure 9.29 Motor cycle clutch

Automatic gearbox clutch A common use of a multi-plate clutch is in an automatic gearbox. This is

Figure 9.31 High-performance racing clutch

because a number of clutches are needed to control the gears. As space is limited, multiple plates are used to allow all the power to be transmitted. Modern limited slip differentials also make use of the multi-plate clutch technique.

High-performance clutch Many high-performance clutch assemblies use multiple clutch discs. An intermediate plate is used in these assemblies to separate the clutch discs.

Operation When the clutch is engaged, the first clutch disc is held between the clutch pressure plate and intermediate plate, and the second clutch disc is held between the intermediate plate and the flywheel. When disengaged, the intermediate plate, flywheel, and pressure plate assembly rotate as a unit, while the clutch discs, which are not in contact with the plates, rotate freely within the assembly and do not transmit power to the transmission.

A clutch will continue to work for many miles of trouble-free motoring. However, a sensible driving technique and regular quick checks can help to avoid problems.

 Select a routine from section 1.3 and follow the process to study a component or system

 9

Figure 9.30 Clutches as part of an automatic gearbox

9.1.6 **Electronic clutch** 3

Introduction An electronic clutch (eClutch) saves fuel and makes driving easier. Traffic congestion is bad enough at the best of times, but it is worse without automatic transmission through the need for constant switching between clutch, throttle and brake. In stop-and-go traffic, the eClutch allows drivers using manual transmission to use first gear without using the clutch. They can simply use the brake and throttle pedal, just like in an automatic transmission, without accidentally stalling the engine.

Features There are a number of benefits and features that can be achieved with an electronically controlled clutch actuator. The Bosch eClutch system, for example, costs significantly less than a conventional automatic transmission and is thus an attractive alternative in the compact car segment, where price competition is tough. Sport mode settings are also available on some systems. In addition, the eClutch makes a coasting function possible, which saves fuel. Independently of the driver, the clutch decouples the engine from the transmission if the driver is no longer accelerating, and the engine is then stopped. The result is a real fuel saving in the region of 10%.

Figure 9.32 This electronically controlled clutch closes the gap between automatic and manual transmission (Source: Bosch Media)

Drive by wire This clutch system is described as drive-by-wire as there is no direct connection between the pedal and the clutch actuator. Under normal operating conditions, the actuator position will just follow the driver's clutch pedal position. However, on some systems it will intervene if the pedal is released too quickly. The operation of different types of actuator varies, but in general an electric motor is used that drives a segment of a wheel via a worm gear. The worm wheel presses the release pin forwards, thus preventing the flow of power through the clutch. The large tensioning forces to separate the clutch are balanced by a spring, which means that the electric motor only supplies the torque to overcome friction.

Figure 9.33 Clutch actuator sectioned view (Source: www.technolab.org)

Start-stop function The principle behind start-stop coasting is simple. The system detects the driver's easing of pressure on the throttle, decouples the engine from the transmission, and therefore prevents the engine from consuming fuel. Drivers can already manually simulate this effect by disengaging the clutch on a downhill stretch. In the future, the system will automatically assume this function, while stopping the engine at the same time. This is technically sophisticated but worthwhile, leading to a 10% reduction in fuel consumption.

Improved gear shifts As well as the stop-and-go feature and the possibility of saving fuel, the eClutch offers a number of other functions. For example, it can be used to support gear shifts, making them smoother. A special sensor detects the start of a gear shift and adjusts engine speed. The result is a smooth, easy gear change.

Hybrids If the powertrain is electrified, the electronically controlled clutch means that a combination of hybrid powertrain and manual transmission is possible. Up to now, it has only been possible to combine a hybrid powertrain with an automatic transmission since it is not possible to coordinate a combustion engine and an electrical powertrain using a standard manual transmission. In this respect, the eClutch offers two advantages – manual transmission is still possible in hybrid vehicles, and the price of entry-level hybrids can be reduced since a fully automatic transmission is no longer necessary.

Summary The cost of an eClutch system is a lot more than a clutch cable! However, the potential benefits will soon outweigh the extra initial expense. This is particularly the case as manufacturers have to meet ever-lower emissions targets.

Use a library or the web search tools to further examine the subject in this section

9.2 Manual gearbox

9.2.1 Gearbox operation ❶

A transmission system gearbox is required because the power of an engine consists of speed and torque. Torque is the twisting force of the engine's crankshaft and speed refers to its rate of rotation. The transmission can adjust the proportions of torque and speed that are delivered from the engine to the drive shafts. When torque is increased, speed decreases; and when speed is increased, the torque decreases. The transmission also reverses the drive and provides a neutral position when required.

Types of gear Helical gears are used for almost all modern gearboxes. They run more smoothly and are quieter in operation. Earlier 'sliding mesh' gearboxes used straight cut gears, as these were easier to manufacture. Helical gears do produce some sideways force when operating, but this is dealt with by using thrust bearings.

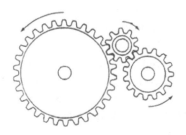

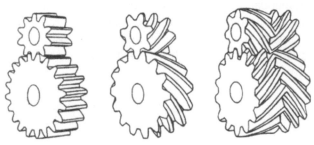

Figure 9.34 Straight cut and helical gears

Gearbox For most light vehicles, a gearbox has five forward gears and one reverse gear. It is used to allow operation of the vehicle through a suitable range of speeds and torque. A manual gearbox needs a clutch to disconnect the engine crankshaft from the gearbox while changing gears. The driver changes gears by moving a lever which is connected to the box by a mechanical linkage.

Power, speed and torque The gearbox converts the engine power by a system of gears, providing different ratios between the engine and the wheels. When the vehicle is moving off from rest, the gearbox is placed in first, or low, gear. This produces a high torque but low wheel speed. As the car speeds up, the next higher gear is selected. With each higher gear, the output turns faster but with less torque.

Figure 9.35 Pontiac six-speed gear selector

9

Top gears Fourth gear on most rear-wheel drive light vehicles is called direct drive because there is no gear reduction in the gearbox. In other words, the gear ratio is 1:1 The output of the gearbox turns at the same speed as the crankshaft. For front-wheel drive vehicles, the ratio can be 1:1 or slightly different. Most modern light vehicles now have a fifth gear. This can be thought of as a kind of overdrive because the output always turns faster than the engine crankshaft.

443

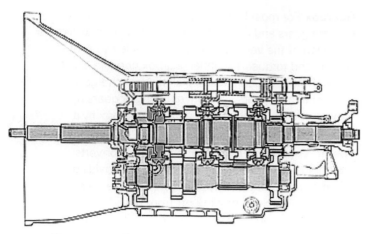

Figure 9.36 Fourth gear is often 'straight through'

Gearbox input Power travels in to the gearbox via the input shaft. A gear at the end of this shaft drives a gear on another shaft called the countershaft or layshaft. A number of gears of various sizes are mounted on the layshaft. These gears drive other gears on a third motion shaft also known as the output shaft.

was, therefore, a skill that took time to master! These have now been replaced by constant mesh gearboxes.

Constant mesh The modern gearbox still produces various gear ratios by engaging different combinations of gears. However, the gears are constantly in mesh. For reverse, an extra gear called an idler operates between the countershaft and the output shaft. It turns the output shaft in the opposite direction to the input shaft.

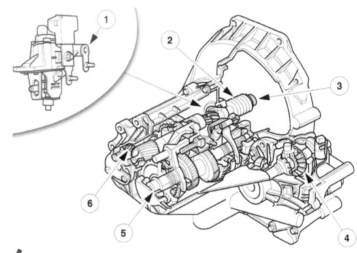

Figure 9.38 FWD gearbox (transaxle)

Note how in each case, with the exception of reverse, the gears do not move. This is why this type of gearbox has become known as constant mesh. In other words the gears are running in mesh with each other at all times.

In constant mesh boxes, dog clutches are used to select which gears will be locked to the output shaft. These clutches, which are moved by selector levers, incorporate synchromesh mechanisms.

A manual gearbox allows the driver to select the gear appropriate to the driving conditions. Low gears produce low speed but high torque; high gears produce higher speed but lower torque.

Figure 9.37 Sectioned view of a gearbox

Older vehicles used sliding-mesh gearboxes. With these gearboxes, the cogs moved in and out of contact with each other. Gear changing

Create a mind map to illustrate the features of a key component or system

9.2.2 Gear change mechanisms ❶ ❷

On all modern gearboxes, the selection of different ratios is achieved by locking gears to the mainshaft. A synchromesh and clutch mechanism does this when moved by a selector fork. The selector fork is moved by a rod, or rail, which in turn is moved by the external mechanism and the gearstick.

Single rail system To save space, some manufacturers use a single selector shaft. This means the shaft has to twist and move lengthways. The twisting allows a finger to contact with different selector forks. The lengthways movement pushes the synchronisers into position. All the selector forks are fitted on the same shaft.

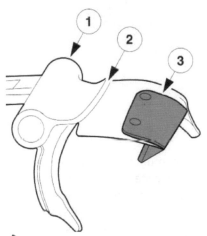

Figure 9.39 Gear shift or selector fork

Two-rail system On a two-shaft system, the main selector shaft often operates the first/second gear selector fork. An auxiliary shaft operates the third/fourth selector fork.

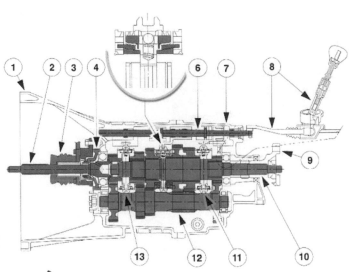

Figure 9.40 Double selector shaft

Three-rail system The three-rail, or three shaft system, is similar to the two shaft type. However, each shaft can be moved lengthways. In turn, the shafts will move the first/second, third/fourth or fifth/reverse forks.

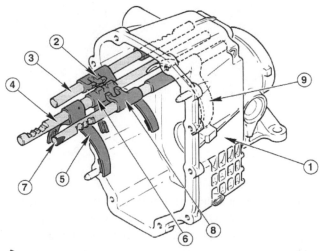

Figure 9.41 Triple selector shaft

External linkages A common external linkage is shown here. Movement of the shift lever is transferred to the gearbox by a shift rod. The rod will only move to select reverse gear when the lock sleeve is lifted. This prevents accidental selection of reverse gear.

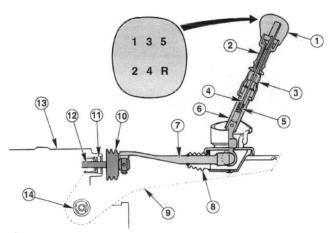

Figure 9.42 Rod-operated shift mechanism

Cable system A recent development is the cable shift mechanism. The advantage of this system is that the shift lever does not have to be fixed to the gearbox or in a set position. This allows designers more freedom.

Detent mechanism A detent mechanism is used to hold the selected gear in mesh. In most cases, this is just a simple ball and spring acting on the selector shaft(s). Figure 9.43 shows a gearbox with the detent mechanisms highlighted.

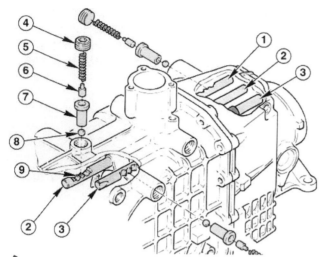

Figure 9.43 Ball and spring detent

Interlocks Gear selection interlocks are a vital part of a gearbox. They prevent more than one gear from being engaged at any one time. When any selector clutch is in mesh, the interlock will not allow the remaining selectors to change position. As the main selector shaft is turned by side-to-side movement of the gear stick, the gate restricts the movement. The locking plate, shown as number (15) in Figure 9.44, will only allow one shaft to be moved at a time. Because the gate restricts the movement, selection of more than one gear is prevented.

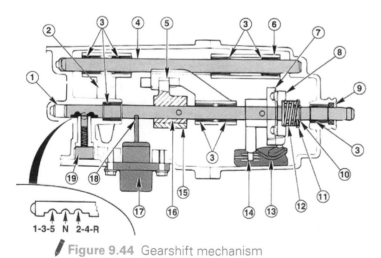

1-3-5 N 2-4-R

Figure 9.44 Gearshift mechanism

Figure 9.45 Gate and reverse gear lock

Sliding plunger interlock When three rails are used to select the gears, plungers or locking pins can be used. These lock the two remaining rails when one has moved. In the neutral position, each of the rails is free to move. When one rail (rod or shaft) has moved, the pins move into the locking notch, preventing the other rails from moving.

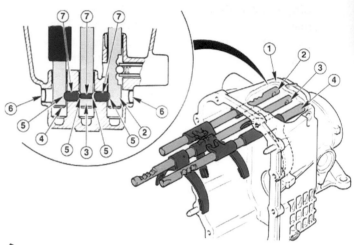

Figure 9.46 Plunger-type interlock

Summary Gear selection must be a simple process for the driver. In order to facilitate changing, a number of mechanical components are needed. The external shift mechanism must transfer movement to the internal components. The internal mechanism must only allow selection of one gear at a time by use of an interlock. A detent system helps to hold the selected gear in place.

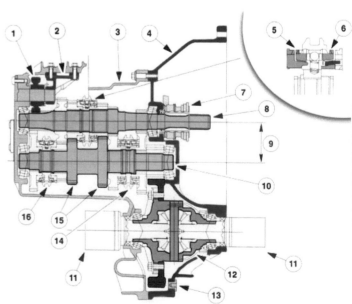

Figure 9.47 Design features of a transaxle

Make a simple sketch to show how one of the main components or systems in this section operates

9.2.3 Gears and components ❶ ❷

Introduction There is a wide range of gearboxes in use. However, although the internal components differ, the principles remain the same. The examples in this section are, therefore, useful for learning the way in which any gearbox works.

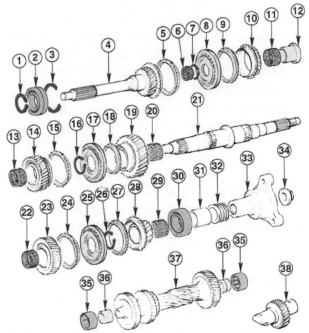

Figure 9.48 Gearbox components

Input shaft The input shaft transmits the torque from the clutch, via the countershaft, to the transmission output shaft. It runs inside a bearing at the front and has an internal bearing, which runs on the mainshaft, at the rear. The input shaft carries the countershaft driving gear and the synchroniser teeth and cone for fourth gear.

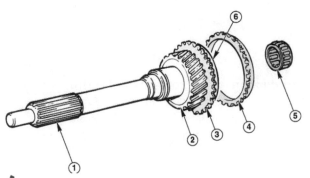

Figure 9.49 Details of the input shaft

Mainshaft or output shaft The mainshaft is mounted in the transmission housing at the rear and the input shaft at the front. This shaft carries all the main forward gears, the selectors and clutches. All the gears run on needle roller bearings. The gears run freely unless selected by one of the synchroniser clutches.

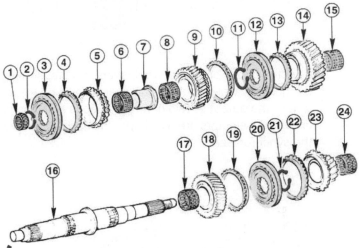

Figure 9.50 Details of the mainshaft

Countershaft The countershaft is sometimes called a layshaft. It is usually a solid shaft containing four or more gears. Drive is passed from here to the output shaft, in all gears except fourth. The countershaft runs in bearings, fitted in the transmission case, at the front and rear.

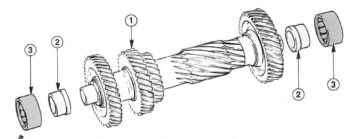

Figure 9.51 Details of the countershaft

Reverse idler gear An extra gear has to be engaged to reverse the direction of the drive. A low ratio is used for reverse – even lower than first gear in many cases. The reverse idler connects the reverse gear to the countershaft.

Selector mechanism An interlock is used on all gearboxes, to prevent more than one gear being

Remember! There is more support on the website that includes additional images and interactive features: www.tomdenton.org

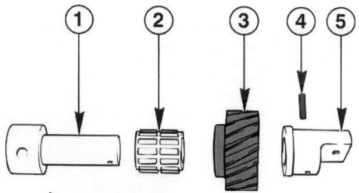

Figure 9.52 Details of the reverse gear idler

selected at any one time. If this were not prevented, the gearbox would lock, as the gears would be trying to turn the output at two different speeds at the same time. The selectors are 'U' shaped devices that move the synchronisers.

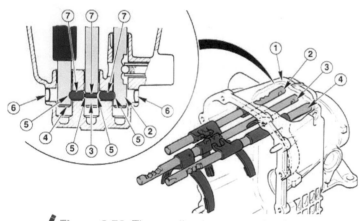

Figure 9.53 Three rail selector mechanism

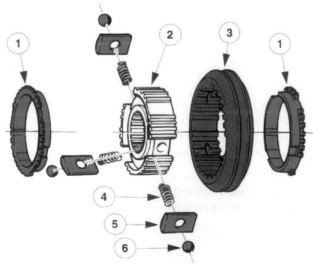

Figure 9.54 Details of the selector and synchroniser

Selector and synchroniser Most gearboxes have three synchronisers. Their task is to bring the chosen gear to the correct speed for easy selection. The unit consists of cone clutches to synchronise speed, and dog clutches to connect the drive.

Transmission fluid The transmission fluid must meet the following requirements:

▶ viscosity must be largely unaffected by temperature
▶ high ageing resistance (gearboxes are usually filled for life)
▶ minimal tendency to foaming
▶ compatibility with different sealing materials.

Only the specified transmission fluid should be used when topping up or filling after dismantling and reassembly. Bearing and tooth flank damage can occur if this is disregarded.

Overdrive On earlier vehicles, a four-speed gearbox was the norm. Further improvements in operation could be gained by fitting an overdrive. This was mounted on the output of the gearbox (RWD). In fourth gear, the drive ratio is usually 1:1. Overdrive would allow the output to rotate faster than the input, hence the name. Most gearboxes now incorporate a fifth gear, which is effectively an overdrive but does not form a separate unit.

Summary The transmission gearbox on all modern cars is a sophisticated component. However, the principle of operation does not change because it is based on simple gear ratios and clutch operation. Most current gearboxes are five-speed, constant mesh and use helical gears.

Using images and text, create a short presentation to show how a component or system works

9.2.4 Synchromesh mechanisms 2

Introduction A synchromesh mechanism is needed because the teeth of dog clutches clash if they meet at different speeds. Shown here is part of a synchroniser. The dog clutch and cone clutch are highlighted. A synchromesh system synchronises the speed of two shafts before the dog clutches are meshed – hence the name.

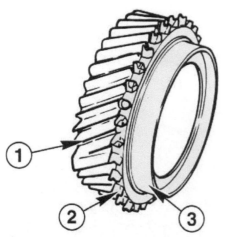

Figure 9.55 Part of a synchroniser

Synchromesh The system works like a friction-type cone clutch. The collar is in two parts and contains an outer toothed ring, which is spring loaded to sit centrally on the synchromesh hub. When the outer ring, or synchroniser sleeve, is made to move by the action of the selector mechanism, the cone clutch is also moved because of the blocker bars.

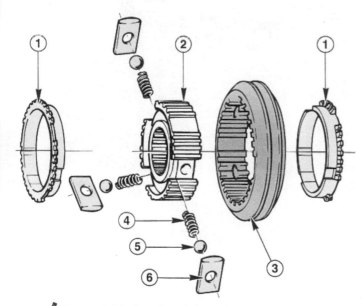

Figure 9.56 Synchroniser components

Neutral position In the neutral position, the shift ring and blocker bars are centralised. There is no connection between the shift ring and the gear wheel. The gear wheel can turn freely on the shaft.

Synchronising position When the shift fork is moved by the driver, the shift ring is slid towards the gear wheel. In the process, the shift ring carries three

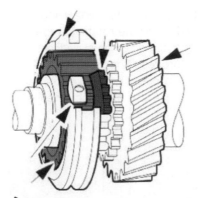

Figure 9.57 Synchroniser in neutral

blocker bars, which move the synchroniser ring axially and press it onto the friction surface (cone clutch) of the gear wheel. As long as there is a difference in speed, the shift ring cannot move any further. This is because the frictional force turns the synchroniser ring causing the tooth flanks to rest on the side of the synchroniser body.

Shift position Once the shift ring and gear are turning at the same speed, circumferential force no longer acts on them. The force still acting on the shift ring turns it until it slides onto the teeth of the gear wheel. The gear wheel is now locked to its shaft.

The online animation shows the three stages of engagement:

▶ neutral
▶ synchronising
▶ shift position.

The system ensures that shifting gear is easy and that damage does not occur to the teeth.

Reverse gear An extra shaft carries the reverse gear cog. Because reverse gear is selected with the car at rest and at low engine speed, some earlier gearboxes did not have a synchroniser on reverse. However, many modern boxes now include this facility.

9

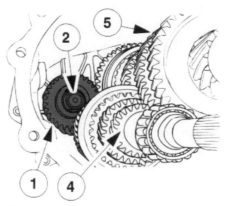

Figure 9.58 Reverse gear in position

Summary For two rotating shafts to mesh using a dog clutch, they should ideally be rotating at the same speed. Early motorists had to be skilled in achieving this through a process known as double-declutching. However, all modern gearboxes make life much easier for us by the use of synchromesh systems!

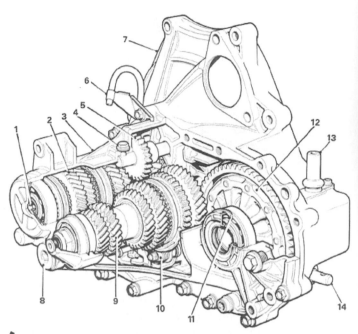

Figure 9.60 Internal transaxle features

Figure 9.59 Synchromesh components

 Use the media search tools to look for pictures and videos relating to the subject in this section

9.2.5 **Front- and rear-wheel drive** ❷

Introduction Rear-wheel drive cars usually have the engine mounted lengthways in the car. The gearbox is mounted on the back of the engine in the same direction. It passes the drive via a propshaft to the rear axle. Front-wheel drive cars usually have the engine mounted transversely (sideways). The gearbox fits on to the back of the engine but then straight gears pass the drive, via the final drive, to the driveshafts and wheels.

Front-wheel drive (FWD) Most front-wheel drive cars have a transmission system where the gearbox and final drive are combined. This is often described as a transaxle. The unit shown in Figure 9.61 is a five-speed box.

Example FWD gearbox The selector fork, synchroniser and the helical gears can be seen in the cutaway gearbox in Figure 9.62. The gears, as with all modern boxes, are in constant mesh. The correct lubricant is essential for these gearboxes. Damage will occur if the wrong type is used.

Figure 9.61 Cable change FWD gearbox (Ford)

FWD gearbox mountings Front-wheel drive transmission gearboxes are solid-mounted on the engine. They are secured to the vehicle body or chassis with rubber mountings. This reduces noise and vibration for the passengers.

Speedometer drive The drive for the speedometer is taken from the gearbox output shaft on most vehicles. This shaft rotates at a speed proportional to road speed. Some manufacturers still use speedometer cables but many now opt for speed sensors, which provide a signal for an electronic gauge.

with the shift lever acting directly. Gearboxes, where the shift lever acts via a linkage, are often described as indirect or remote operated.

Figure 9.62 Rubber mountings reduce vibration

Figure 9.63 Speed sensor and cable connection

Rear-wheel drive (RWD) Rear-wheel drive gearboxes do not usually contain final drive components. The exception of this may be on four-wheel drive vehicles. The gearbox casing attaches to a bell housing, which in turn bolts to the engine and covers the clutch. Larger vehicles used larger gearboxes because of the extra strength required. The operating principles are the same for all types.

Example RWD gearbox General Motors use the box in Figure 9.65 in the 'Chevy Crew Cab'. The gearbox is made by ZF. It is a six-speed manual transmission

Figure 9.64 Volvo transverse engine 4WD

Figure 9.65 Heavy-duty RWD gearbox

RWD gearbox mountings Rear-wheel drive transmission gearboxes are solid-mounted on the engine. They are secured to the vehicle body or chassis with rubber mountings. Usually a cross-member is fitted at the rear to support the weight. Rubber mountings reduce noise and vibration for the passengers.

Figure 9.66 Rubber mountings reduce noise

Reverse light switch Most reverse light switches are simple on/off types. The switch body is fitted in the side of the gearbox casing. The toggle of the switch is moved by a selector shaft or other component when reverse gear is engaged.

451

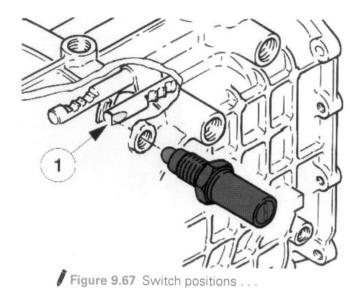

Figure 9.67 Switch positions . . .

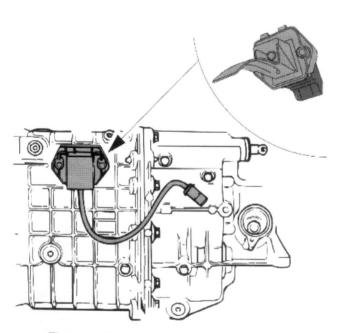

Figure 9.68 . . . vary on different cars

Summary The gearbox is the main transmission component. Transmission is a general term used to describe all the components required to transmit power from the engine to the wheels. The requirement is to convert the power from the relatively high velocity and low torque of the engine crankshaft, to the variable, usually lower speed and higher torque, needed at the wheels. FWD and RWD gearboxes may look different but their operating principles are the same. Automatic transmission is another story . . .

Make a simple sketch to show how one of the main components or systems in this section operates

9.3 Automatic transmission
9.3.1 Torque converter ❶ ❷

Introduction An automatic gearbox contains special devices that automatically provide various gear ratios, as they are needed. Most automatic gearboxes have three or four forward gears and one reverse gear. Instead of a gearstick, the driver moves a lever called a selector. Most automatic gearboxes now have selector positions for Park, Neutral, Reverse, Drive, 2 and 1 (or 3, 2 and 1 in some cases). The fluid flywheel, or torque converter, is the component that makes automatic operation possible.

Safety circuit The engine will only start if the selector is in either the park or neutral position. In park, the drive shaft is locked so that the drive wheels cannot move. It is also now common, when the engine is running, to only be able to move the selector out of park if you are pressing the brake pedal. This is a very good safety feature as it prevents sudden, uncontrolled movement of the vehicle.

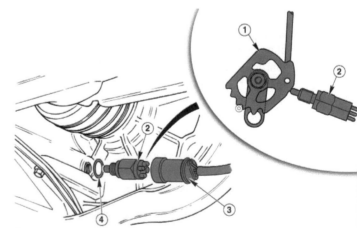

Figure 9.69 Starter circuit with inhibitor switch

Automatic gearbox For ordinary driving, the driver moves the selector to the 'Drive' position. The transmission starts out in the lowest gear and automatically shifts into higher gears as the car picks up speed. The driver can use the lower positions of the gearbox for going up or down steep hills or driving through mud or snow. When in position 3, 2, or 1, the gearbox will not change above the lowest gear specified. Figure 9.70 shows a modern automatic gearbox used on rear-wheel drive vehicles.

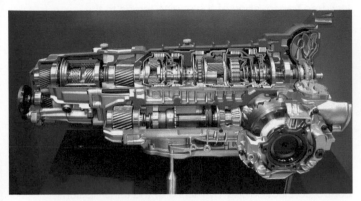

Figure 9.70 Modern auto-box

Figure 9.71 Torque converter

A **fluid flywheel** consists of an impeller and turbine, which are immersed in oil. They transmit drive from the engine to the gearbox. The engine driven impeller faces the turbine, which is connected to the gearbox. Each of the parts, which are bowl-shaped, contains a number of vanes. They are both a little like half of a hollowed out orange facing each other. When the engine is running at idle speed oil is flung from the impeller into the turbine, but not with enough force to turn the turbine.

Energy As engine speed increases so does the energy of the oil. This increasing force begins to move the turbine and hence the vehicle. The oil gives up its energy to the turbine and then recirculates into the impeller at the centre starting the cycle over again. As the vehicle accelerates the difference in speed between the impeller and turbine reduces until the slip is about 2%.

Fluid flywheel development A problem, however, with a basic fluid flywheel is that it is slow to react when the vehicle is moving off from rest. This can be improved by fitting a reactor or stator between the impeller and turbine. We now know this device as a torque converter. All modern cars fitted with automatic transmission use a torque converter.

The **torque converter** delivers power from the engine to the gearbox like a basic fluid flywheel, but also increases the torque when the car begins to move. Similar to a fluid flywheel, the torque converter resembles a large doughnut sliced in half. One half, called the pump impeller, is bolted to the drive plate or flywheel. The other half, called the turbine, is connected to the gearbox input shaft. Each half is lined with vanes or blades. The pump and the turbine face each other in a case filled

with oil. A bladed wheel called a stator is fitted between them.

Impeller The engine causes the pump (impeller) to rotate and throw oil against the vanes of the turbine. The force of the oil makes the turbine rotate and send power to the transmission. After striking the turbine vanes, the oil passes through the stator and returns to the pump. When the pump reaches a specific rate of rotation, a reaction between the oil and the stator increases the torque. In a fluid flywheel, oil returning to the impeller tends to slow it down. In a torque converter, the stator or reactor diverts the oil towards the centre of the impeller for extra thrust.

Turbine When the engine is running slowly, the oil may not have enough force to rotate the turbine. However, when the driver presses the accelerator pedal, the engine runs faster and so does the impeller. The action of the impeller increases the force of the oil. This force gradually becomes strong enough to rotate the turbine and moves the vehicle. Torque converters can double the applied torque when moving off from rest. As engine speed increases, the torque multiplication tapers off until at cruising speed there is no increase in torque. The reactor or stator then freewheels on its one-way clutch at the same speed as the turbine.

Converter housing and impeller The converter housing is bolted to the crankshaft and driven directly. It is welded to the impeller and filled with automatic transmission fluid. The impeller:

▶ is welded to the converter housing
▶ has blades arranged radially
▶ turns at the same speed as the engine
▶ conveys fluid to the turbine blades and, as a result, produces a radial force at the turbine.

9

Figure 9.72 Impeller

Turbine The turbine is splined to, and drives, the transmission input shaft. It has blades arranged in a curved pattern, which allows fluid to flow inwards due to the reduced centrifugal force compared with the impeller. The fluid is then passed to the stator.

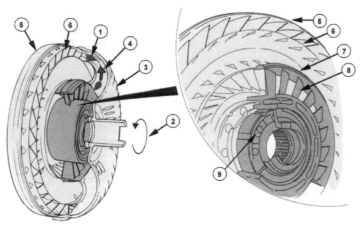

Figure 9.73 Torque converter

The purpose of the stator is to deflect the stream of fluid into the impeller until the coupling speed ratio is reached. The stator and one-way clutch assembly is located between the impeller and the turbine. It is splined on the stator support, which is locked to the fluid pump housing and hence to the transmission housing. The stator has blades arranged in a curved pattern. It locks, counter to the normal direction of rotation of the engine, and runs freely in the normal direction of rotation of the engine. The purpose is to boost torque, through ram pressure, up to the coupling point. It is exposed to flow from the rear until a turbine to impeller speed ratio of 85% is reached. The stator now rotates with the converter.

The fluid flywheel action, of a torque converter or fluid flywheel, reduces efficiency because the pump tends to rotate faster than the turbine. In other words some slip will occur. This is usually about 2%. To improve efficiency, many transmissions now include a lock-up facility. When the pump reaches a specific rate of rotation, the pump and turbine are locked together, allowing them to rotate as one.

The converter lock-up clutch allows slip-free and hence loss-free transmission of the engine torque to the automatic transmission. When engaged, it creates a frictional connection between the converter housing and the turbine. It consists of a clutch pressure plate with a friction lining and a torsional vibration damper, to damp the crankshaft torsional vibrations. It is connected positively to the turbine and is exposed to fluid pressure from one side for clutch disengagement and engagement. A modulating valve is often used to allow controlled pressure build-up and reduction. This is to ensure smooth opening and closing. The valve is controlled electronically by means of the transmission ECU.

Summary The purpose of the torque converter and lock-up clutch, can be summarised as follows:

▶ to transmit engine torque
▶ to boost torque, particularly when pulling away
▶ to bypass the torque converter to increase the efficiency of the automatic transmission.

> **?** Create a mind map to illustrate the features of a key component or system

9.3.2 Automatic transmission operation ❶❷

Introduction The main parts of the automatic transmission system are:

▶ the torque converter with converter lock-up clutch
▶ the fluid pump with stator support
▶ the planetary gear train with clutches and brakes
▶ the intermediate gear stage
▶ the final drive assembly (if FWD)
▶ the control unit or valve body assembly.

Epicyclic gearbox operation Epicyclic gears are a special set of gears that are part of most automatic gearboxes. In their basic form they consist of three main elements:

▶ a sun gear, located in the centre
▶ the carrier that holds two, three, or four planet gears, which mesh with the sun gear and revolve around it
▶ an internal gear or annulus, which is a ring with internal teeth – it surrounds the planet gears and meshes with them.

Planetary gears Any part of a set of planetary gears can be held stationary or locked to one of the others.

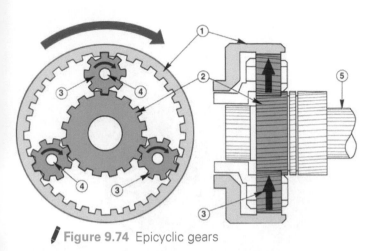

Figure 9.74 Epicyclic gears

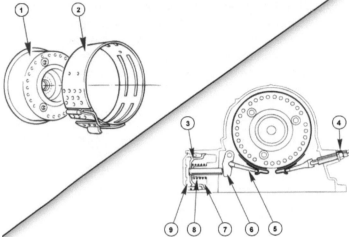

Figure 9.75 Brake bands lock different components

This will produce different gear ratios. Most automatic gearboxes have two sets of planetary gears that are arranged in line. This provides the necessary number of gear ratios.

Automatic transmission system As the gear selector is moved into different positions, the power flow through the gearbox changes. Refer to the series of images and animations online.

Power flow The power flows shown here are a representation of what occurs in an auto-box. Note in particular that only the top half is shown! In other words, the complete picture would include a reflection of what is represented here. On this online animation, click each button in turn to show how the power flow changes.

Valves and brake bands The appropriate elements in the gear train are held stationary by a system of hydraulically operated brake bands and clutches. These are worked by a series of hydraulically operated valves, usually in the lower part of the gearbox.

Oil pressure to operate the clutches and brake bands is supplied by a pump. The supply for this is the oil in the sump of the gearbox. Three forward gears and one reverse gear are achieved from two sets of epicyclic gears. Unless the driver moves the gear selector to operate the valves, automatic gear changes are made depending on just two factors:

1 Throttle opening – a cable is connected from the throttle to the gearbox.
2 Road speed – when the vehicle reaches a set speed a governor allows pump pressure to take over from the throttle.

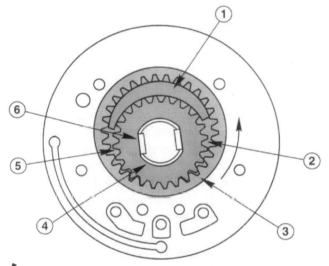

Figure 9.76 Oil pump and governor

Kick down The cable from the throttle also allows a facility known as 'kick down'. This allows the driver to change down a gear, such as for overtaking, by pressing the throttle all the way down.

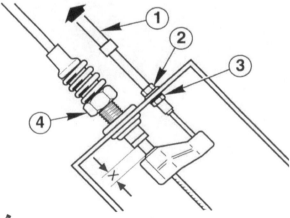

Figure 9.77 Cable position

9

Standard gear systems Many automatic transaxle gearboxes use gears the same as in manual boxes. The changing of ratios is similar to the manual operation except that hydraulic clutches and valves are used. Figure 9.78 is an example of this system.

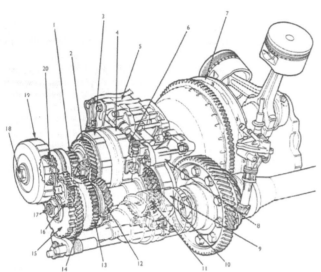

Figure 9.78 Automatic gearbox

The nine-speed automatic transmission was designed for passenger cars with front-transverse engine configurations. Model ranges cover 200 to 480 Nm. Four individual gear sets are nested within the transmission, with six shifting elements and nine gear ratios. The 9HP also features interlocking dog clutches for power shifting; hydraulically operated constant-

Figure 9.80 Gearsets producing 9-speeds (Source: ZF)

mesh elements reduce overall transmission length and optimize efficiency. With nine speeds, the 9HP improves fuel economy and achieves small gear ratios. RPMs are lowered by approximately 700 at a speed of 75 mph (120 km/h), reducing noise. A torque converter is the 9HP's standard starting element, and a multilevel torsion damper system minimizes hydraulic losses.

Key features of this transmission are:

▶ Wide transmission ratio spread with small ratio steps
▶ Interlocking dog clutches
▶ Torque converter with excellent vibration and oscillation isolation for optimal comfort during drive-off and shifting
▶ Proven ZF control systems technology to provide excellent shift quality.

The design makes the most of the available space and creates the best possible conditions for versatile front transverse applications.

Summary Cars fitted with modern automatic transmission systems are a pleasure to drive. Traditionally auto-box cars used more fuel than those with manual transmission. However, the difference is now very small. The main reason for this is the ability to lock the converter and, therefore, eliminate slip.

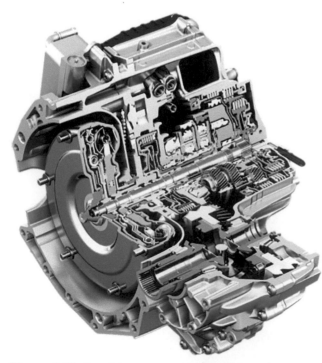

Figure 9.79 Front transverse engine, 9-speed transmission (Source: ZF)

 Create an information wall to illustrate the features of a key component or system

9.3.3 Electronic and hydraulic control ❸

Introduction The main aim of electronically controlled automatic transmission (ECAT) is to improve on conventional automatic transmission in the following ways:

- gear changes should be smoother and quieter
- improved performance
- reduced fuel consumption
- reduction of characteristic changes over system life
- increased reliability.

Gear changes and lock-up of the torque converter are caused by hydraulic pressure – but under electronic control.

Figure 9.81 Porsche Carrera auto-gearbox

In an **ECAT system**, electrically controlled solenoid valves can influence this hydraulic pressure. Most ECAT systems now have a transmission ECU that is in communication with the engine control system. Control of gearshift and torque converter lockup is determined by the ECU. With an ECAT system, the actual point of gearshift is determined from pre-programmed memory within the ECU. Data from other sensors is also taken into consideration. Actual gearshifts are initiated by changes in hydraulic pressure, which is controlled by solenoid valves.

The two **main control functions** of this system are hydraulic pressure and engine torque. A temporary reduction in engine torque during gear shifting allows smooth operation. This is because the peaks of gearbox output torque, which cause the characteristic surge during gear changes on conventional automatics, are suppressed. Smooth gearshifts are possible because of the control functions and, due to the learning ability of some ECUs, the characteristics remain throughout the life of the system.

The ability to **lock-up** the torque converter has been used for some time, even on vehicles with more conventional automatic transmission. This gives better fuel economy, quietness and improved driveability. Lock-up is carried out using a hydraulic valve, which can be operated gradually to produce a smooth transition. The timing of lock-up is determined from ECU memory in terms of the vehicle speed and acceleration.

Hydraulic systems Oil pressure is generated by the oil pump. It is then regulated and controlled by the main regulator valves. The same automatic transmission fluid (ATF) is used throughout the gearbox. The control valves are operated by hydraulic pressure, which can be a function of road speed, or by electrical solenoids controlled by the ECU.

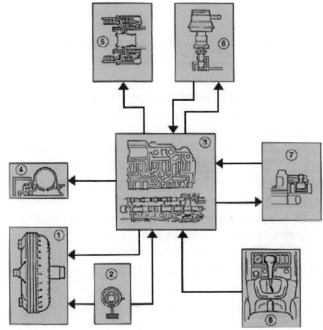

Figure 9.82 Hydraulic components and connections

Electronic control valve block Figure 9.83 shows a modern electronically controlled valve block. The

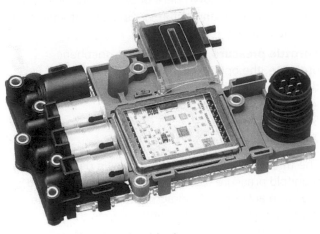

Figure 9.83 Bosch valve block

9

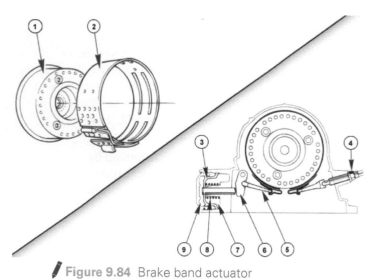

Figure 9.84 Brake band actuator

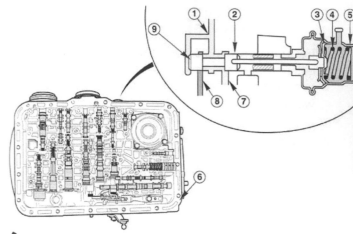

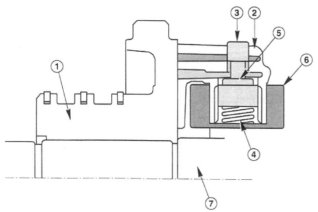

Figure 9.85 Throttle pressure valve

ECU is built in to the system. A connection to other ECUs is made via a CAN (controller area network) connection.

Brake band movement Oil pressure is used to actuate clutches and break bands. Figure 9.84 shows an actuator that tightens a brake band onto a rotating 'drum'. The drum, which in some cases is the outer side of the gear set annulus, is stopped by the action of the brake band. This changes the ratio of the epicyclic gears.

Hydraulic functions Hydraulic components in automatic transmission systems can be split into three groups as shown here. These are:

A Components that receive ATF independent of vehicle operation and selector position.

B Components with control and monitoring functions during gearshifts and while driving.

C Actuators to make the gearshifts while driving.

The hydraulic circuits can appear quite complex; however, they can be read like a circuit diagram. A full circuit is not shown here because they differ so much between manufacturers. Specific information is essential for any repair work.

Throttle pressure In addition to atmospheric pressure, the operation of an automatic box is determined by three other pressures:

▶ main line pressure
▶ governor pressure
▶ throttle pressure.

The throttle pressure in the gearbox is determined by manifold pressure (vacuum if you prefer). Manifold pressure acts on a diaphragm, which in turn moves the throttle pressure valve and controls throttle hydraulic pressure inside the transmission.

Governor Most governors on earlier boxes were centrifugal types as shown in Figure 9.86. As is common when showing automatic transmission, only half of the component is shown. The complete assembly can be imagined as a reflection about the horizontal centreline. When shaft (7) rotates, centrifugal force acts on weight (6). This acts on the spring and, in the control range, determines the governor pressure due to action of valve (3).

Figure 9.86 Centrifugal governor

There is a wide range of electronic and hydraulic control systems. All types, however, serve to operate brake bands and/or clutches. Electronically controlled system operation is determined by ECU programming. Other systems work by sensing a combination of road speed (governor pressure) and throttle (throttle pressure). The transmission (main line pressure) is then used to operate clutches and bands, which are under the control of valves.

Transmission control and GPS Compared to manual transmissions, modern automatic transmissions enhance driving comfort and save on fuel, because they independently determine the point at which the engine runs most efficiently. In series-produced vehicles, there are now up to nine gears to choose from. Modern transmissions are equipped with a great deal of digital intelligence so that they are always capable of identifying the engine's ideal operating point. The control unit is a high-tech miniature computer that enables the complex operation of different types of automatic transmissions. Navigation systems know the lay of the land and can transmit this data to the automatic transmission, which, in turn, can shift into neutral and use the residual momentum if city limits are coming up after a long bend, for example. This can provide additional fuel savings.

Figure 9.88 Ford CTX

Figure 9.87 Transmission ECU (Source: Bosch Media)

 Select a routine from section 1.3 and follow the process to study a component or system

9.3.4 Constantly variable transmission ❸

Introduction Shown here is the Ford CTX (Constantly Variable Transaxle) transmission. This type of automatic transmission uses a pair of cone-shaped pulleys connected by a metal belt.

Drive belt The key to this system is the high-friction drive belt. The belt, made from high-performance steel, transmits drive by thrust rather than tension. The ratio of the rotations, or the gear ratio, is determined by how far the belt rides from the centres of the pulleys. The transmission can produce an unlimited number of ratios. As the car changes speed, the ratio is continuously adjusted.

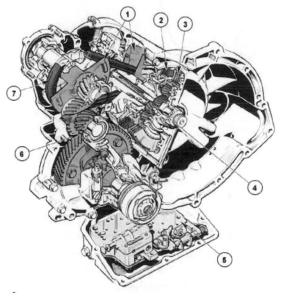

Figure 9.89 The belt transmits drive by thrust

Efficiency Cars with this system are said to use fuel more efficiently than cars with set gear ratios. Within the gearbox, hydraulic control is used to move the pulleys and hence change the drive ratio. An epicyclic gear set is used to provide a reverse gear as well as a fixed ratio.

Transmission components Shown here are the main components of the CTX automatic transmission. Although the gear ratios are achieved through different means, the overall operation of the transmission is similar to other automatic boxes.

Reverse and forward clutch To achieve forward and reverse, a standard epicyclic gear set is used. The drive is taken from this by operating one of two clutches. These are multi-plate clutches and are operated hydraulically.

459

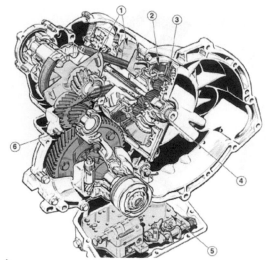

Figure 9.90 CTX components

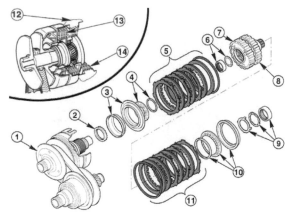

Figure 9.91 Clutches

Cone pulleys and drive belt The drive belt transmits torque from the primary cone pulley to the secondary cone pulley unit. The belt is V-shaped – it consists of 450 steel elements held together by 10 steel strips. The tension of the belt is determined by the current ratio and the torque to be transmitted.

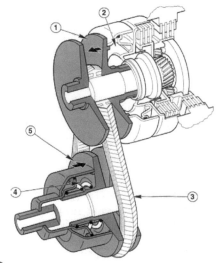

Figure 9.92 Cone pulley unit

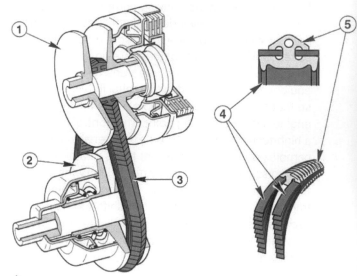

Figure 9.93 Drive belt

Two **control cables** are used on this system – one from the selector lever and the other from the throttle. The position of the accelerator cable is transmitted to the shift cable through the throttle cable and cam plate. Correct adjustment is vital to ensure drive ratios are correctly produced.

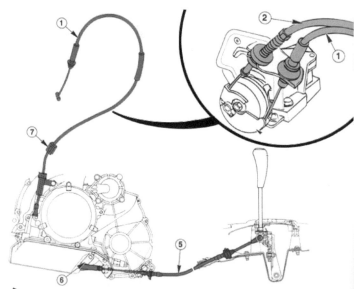

Figure 9.94 Throttle and selector controls

Shown here is a representation of the power flow when reverse, neutral or drive is selected. The clutches are the parts that change power flow. From the output shaft, drive is transmitted to the

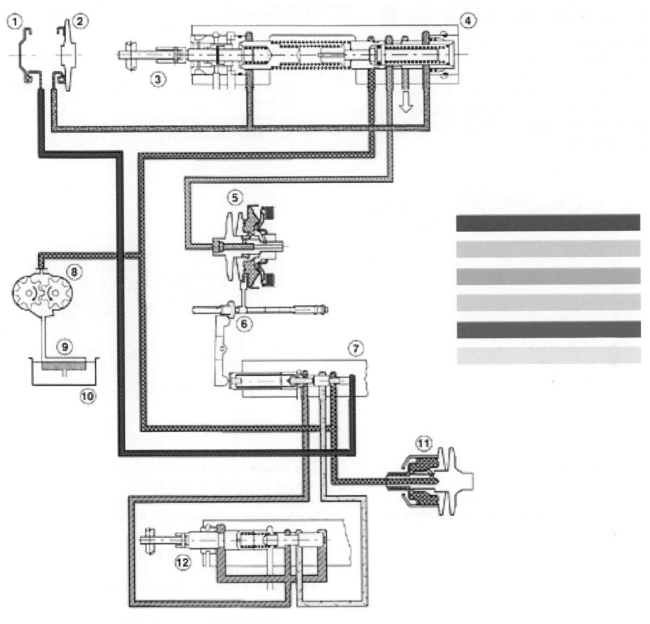

Figure 9.95 Hydraulic system components

final drive reduction gear, differential and front axle driveshafts.

The range of ratios varies from about 2.5:1 to about 15:1. The ratio is set by moving the two halves of the pulleys. The pulleys are moved by hydraulic pressure in the primary and secondary cylinders. The changes in the two cylinders take place at the same time so that the required length of the drive belt remains constant.

Hydraulic control Figure 9.95 shows the components of the hydraulic system. Clutches are controlled by hydraulic valves. Different pressures are used to move the pulleys. The term 'pitot' refers to sensing by means of a special nozzle in a rotating chamber. The pressure produced is proportional to the speed of rotation. This speed is either engine speed or road speed.

Summary The CTX transmission is an innovative design. It feels unusual to drive at first but the user soon becomes accustomed to it.

 Use the media search tools to look for pictures and videos relating to the subject in this section

9

9.3.5 Transaxle transmission ❸

Introduction The description that follows is based on the Ford AG4 (Automatic Gearbox, 4-speed), a fully automatic, electronically controlled transmission with four forward gears. Diagnostics are carried out, using dedicated test equipment, through the data bank link connector (DLC) in the passenger compartment.

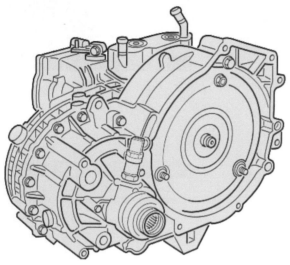

Figure 9.96 AG4 gearbox

Transmission control system Shift operations are controlled by the transmission module using fuzzy logic. The shift timing is variable within consumption and power shift characteristics depending on:

▶ individual driving style
▶ the current driving situation
▶ the current road resistance.

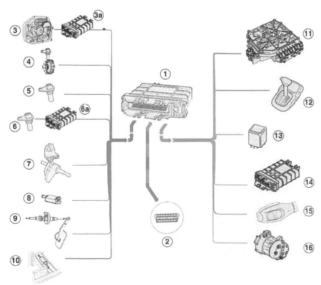

Figure 9.97 Control components

Main Assemblies The AG4 consists of the following main assemblies:

▶ a torque converter with converter lock-up clutch
▶ a fluid pump with stator support
▶ Ravigneaux planetary gear train with clutches and brakes
▶ an intermediate gear stage
▶ the final drive assembly.

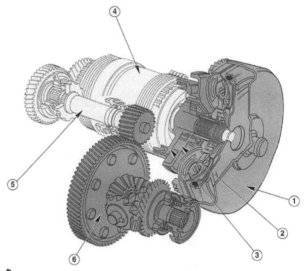

Figure 9.98 Main parts of the transmission`

The **torque converter** housing is bolted to the crankshaft and driven directly by it. The housing is welded to the impeller. The converter is filled with automatic transmission fluid. The blades are arranged radially and turn at the same speed as the engine. This conveys fluid to the turbine blades and as a result produces a radial force at the turbine. The turbine, which has blades arranged in a curved pattern, is splined to the transmission input shaft. The curved pattern allows fluid to flow inwards, due to the reduced centrifugal force compared with the impeller, and pass to the stator.

Stator with roller one-way clutch The stator is located between the impeller and the turbine. It is splined on the stator support, which is locked to the fluid pump housing and hence to the transmission housing. The stator has blades arranged in a curved pattern. It runs freely in the normal direction of rotation of the engine, but locks in the other direction. Its purpose is to boost torque (through ram pressure) up to the coupling point.

Converter lock-up clutch The torque converter lock-up clutch allows slip-free, and hence loss-free, transmission of the engine torque to the automatic transmission. When engaged, it creates a frictional connection between the converter housing and the turbine. It consists of a clutch pressure plate with a friction lining and a torsional vibration damper to

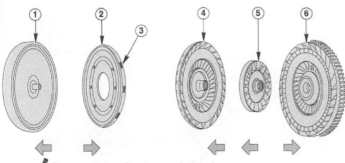

Figure 9.99 Stator and clutch

damp the crankshaft torsional vibrations. The clutch is connected positively to the turbine. It is exposed to fluid pressure from one side for clutch disengagement and engagement and, as a result, is pressed against the converter housing or moved clear of it. A modulating valve allows controlled pressure build-up and reduction to ensure smooth opening and closing. The clutch is controlled electronically.

Fluid pump and stator support The fluid pump is a crescent gear pump. It is driven directly from the crankshaft through two engaging pins on the converter housing. It supplies the hydraulic system with working pressure. The stator support is bolted to the fluid pump, which is splined to the stator. The fluid pump housing is bolted to the transmission housing.

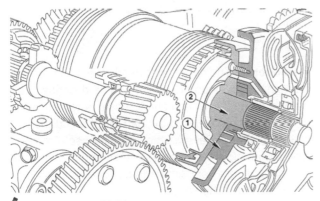

Figure 9.100 Fluid pressure pump

The **planetary gear train** is a Ravigneaux planetary gear set. The long planet gears of the planetary gear set are stepped to achieve the optimum spacing between third and fourth gears. The different transmission ratios are selected by actuating and releasing the individual clutches and brakes. The drive can pass through the small sun wheel, the large sun wheel or the planet carrier. Drive always passes through the annulus.

The **clutch unit** consists of three multi-plate clutches, two multi-plate brakes and a roller one-way clutch. The appropriate clutches or brakes are supplied with pressure by the transmission module according to the selected drive range and shift programme. The required components are driven, locked or allowed

to rotate freely to select the different transmission ratios. The clutch unit acts on the components of the planetary gear train.

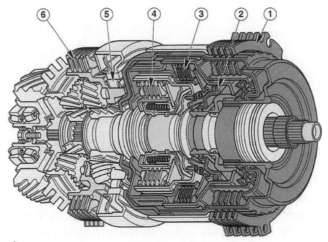

Figure 9.101 Multi-plate clutches

Power flow In first gear, drive passes through the first–third gear clutch to the small sun wheel. The planet carrier first gear one-way clutch is locked. The short planet gears drive the long planet gears, which then drive the annulus. The large sun wheel is allowed to rotate freely. Drive passes via the reverse gear clutch to the large sun wheel. In reverse gear, the planet carrier is locked by means of the reverse gear brake. The large sun wheel drives the long planet gears; these, in turn, drive the annulus. This gives one ratio in the opposite direction to the normal direction of rotation of the engine. The long planet gears also mesh in the short planet gears, which rotate freely with the small sun wheel.

Control unit (valve body assembly) The control unit (valve body assembly) accommodates the solenoid valves for hydraulic control of the drive ranges. They are actuated by the transmission module. A conductor foil connects the solenoid valves to each other and, via a multiplug, to the transmission module loom.

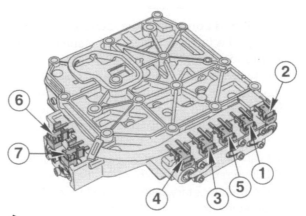

Figure 9.102 Control unit with valves

9

463

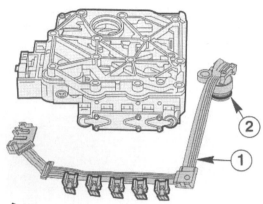

Figure 9.103 Connector foil removed

Parking pawl The parking gear is located on the intermediate shaft. The engaging plunger is actuated by means of an engaging lever, which is connected securely to the shift shaft. When the manual selector lever is moved to position 'P', the engaging plunger moves the parking pawl against the force of a spring and presses it into the teeth of the parking gear. This immobilises the transmission. When the manual selector lever is moved out of position 'P', the engaging plunger releases the parking pawl.

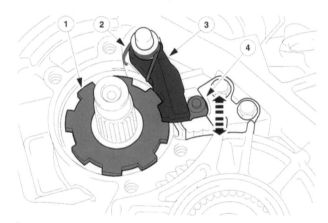

Figure 9.104 Locking mechanism for parking

Fluid circuit and control On the low-pressure side of the pump, the fluid passes from the sump via the fluid filter to the pump. On the high-pressure side, the fluid is conveyed to the control unit (valve body assembly) at working pressure and passed from there by the solenoid valves to the corresponding clutches and brakes. The transmission module switches the clutches and brakes by means of the solenoid valves in the control unit according to the selected drive range. The power then flows via the clutches to the planetary gear set, while the brakes and the one-way clutch lock the corresponding components.

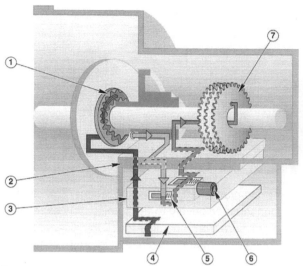

Figure 9.105 Fluid circuit

Summary This type of transaxle gearbox has been in use now for a number of years. It is typical of many and it highlights some of the techniques used by different systems.

 Make a simple sketch to show how one of the main components or systems in this section operates

9.3.6 Direct shift gearbox ❸

Introduction The direct shift gearbox (DSG) is an interesting development as it could be described as a manual gearbox that can change gear automatically. In fact it can be operated by 'paddles' behind the steering wheel, a lever in the centre console or in a fully automatic mode. The gear train and synchronising components are very similar to a normal manual change gearbox however.

Basic principles The direct shift gearbox is made of two transmission units that are independent of each other. Each transmission unit is constructed in the same way as a manual gearbox. Each transmission unit is connected by a multi-plate clutch. Both multi-plate clutches are the wet type and work in oil. They are regulated, opened and closed by a mechatronics system.

Gear selection 1st, 3rd, 5th and reverse gears are selected via multi-plate clutch 1; 2nd, 4th and 6th gears are selected via multi-plate clutch 2. One transmission unit is always in gear and the other transmission unit has the next gear selected ready for the next change, but with its clutch still in the open position.

Torque input Torque is transmitted from the crankshaft to a dual mass flywheel. The splines of the flywheel, on the input hub of the double clutch, transmit the torque to the drive plate of the multi-plate clutch. This is joined to the outer plate carrier of clutch 1 with the main hub of the multi-plate clutch. The outer plate carrier of clutch 2 is also positively joined to the main hub.

Multi-plate clutches Torque is transmitted into the relevant clutch through the outer plate carrier. When the clutch closes, the torque is transmitted further into the inner plate carrier and then into the relevant gearbox input shaft. One multi-plate clutch is always engaged.

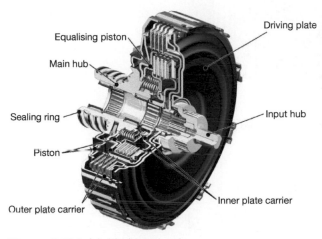

Figure 9.106 Multi-plate clutch

Multi-plate clutch 1 is the outer clutch and transmits torque into input shaft 1 for the 1st, 3rd, 5th and reverse gears. To close the clutch, oil is forced into the pressure chamber. Plunger 1 is therefore pushed along its axis and the plates of clutch 1 are pressed together. Torque is then transmitted via the plates of the inner plate carrier to input shaft 1. When the clutch

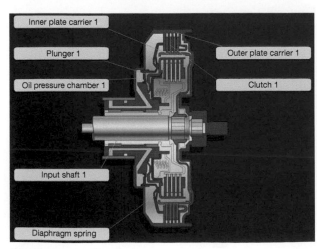

Figure 9.107 Clutch 1

opens, a diaphragm spring pushes plunger 1 back into its start position.

Multi-plate clutch 2 is the inner clutch and transmits torque into input shaft 2 for 2nd, 4th and 6th gears. As with clutch 1, oil is forced into the pressure chamber so that plunger 2 then joins the drive via the plates to input shaft 2. The coil springs press plunger 2 back to its start position when the clutch is opened.

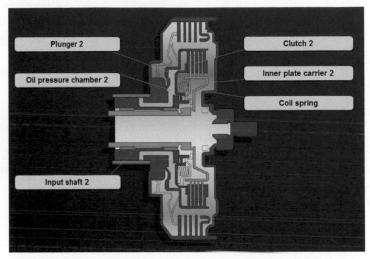

Figure 9.108 Clutch 2

Input shaft 2 is shown in relation to the installation position of input shaft 1. It is hollow and is joined via splines to multi-plate clutch 2. The helical gear wheels for 6th, 4th and 2nd gears can be found on input shaft 2. For 6th, 4th and 2nd gears, a common gear wheel is used. A pulse wheel is used to measure the speed of input shaft 2. The sender is adjacent to the gear wheel for 2nd gear.

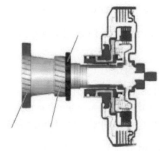

Figure 9.109 Output shaft (top), input shaft (bottom)

465

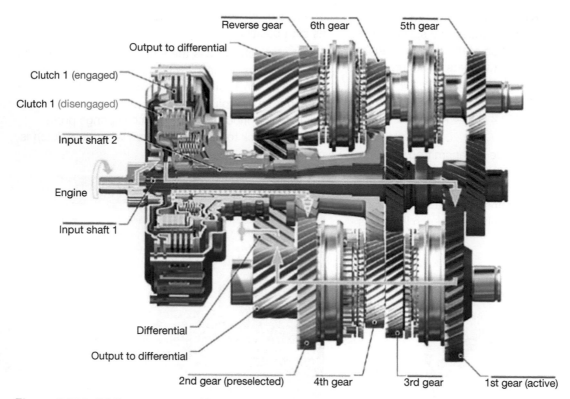

Figure 9.110 DSG components (Source: Audi Media)

Input shaft 1 rotates inside shaft 2 and is joined to multi-plate clutch 1 via splines. Located on input shaft 1 are the helical gear wheels for 5th gear, the common gear wheel for 1st and reverse gears and the gear wheel for 3rd gear. A second pulse wheel is used to measure the speed of input shaft 1. The sender is between the gear wheels for 1st/reverse gear and 3rd gear.

Output shafts In line with the two input shafts, the direct shift gearbox also has two output shafts. Because the gear wheels for 1st and reverse gear are the same, and 4th and 6th gear are on the input shafts, it was possible to reduce the length of the gearbox.

Output shaft 1 Located on output shaft 1 are:

▶ the three-fold synchronised selector gears for 1st, 2nd and 3rd gears
▶ the single synchronised selector gear for the 4th gear
▶ the output shaft gear for meshing into the differential.

The output shaft meshes into the final drive gear wheel of the differential.

Output shaft 2 Located on output shaft 2 are:

▶ the pulse wheel for gearbox output speed
▶ the selector gears for 5th, 6th and reverse gears and
▶ the output shaft gear for meshing into the differential.

Both output shafts transmit the torque further into the differential via their output shaft gears.

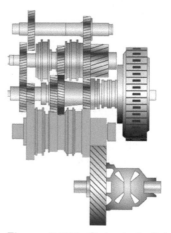

Figure 9.111 Output shaft 1

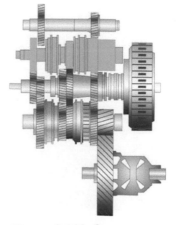

Figure 9.112 Output shaft 2

The **reverse shaft** changes the direction of rotation of output shaft 2 and therefore the direction of rotation of the final drive in the differential. It engages in the common gear wheel for 1st gear and reverse gear on input shaft 1 and the selector gear for reverse gear on output shaft 2.

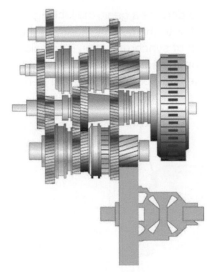

Figure 9.114 Final drive and differential

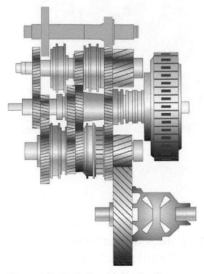

Figure 9.113 Reverse shaft

Differential Both output shafts transmit the torque to the input shaft of the differential. The differential transmits the torque via the drive shafts to the road wheels.

Parking lock A parking brake is integrated in the differential to secure the vehicle in the parked position and to prevent the vehicle from creeping forwards or backwards unintentionally when the handbrake is not applied. Engagement of the locking pawl is by mechanical means via a cable between the selector lever and the parking brake lever on the gearbox.

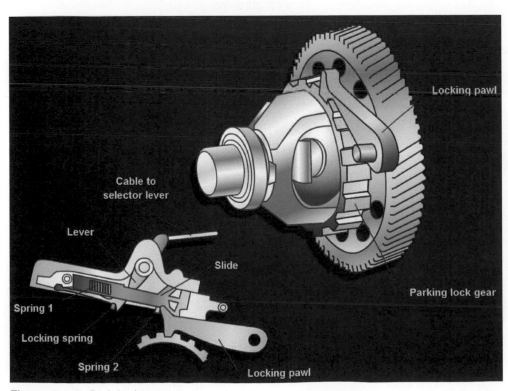

Figure 9.115 Park lock

Synchronisation 1 To engage a gear, the locking collar must be pushed onto the selector teeth of the selector gear. The task of synchronisation is to balance the speed between the engaging gear wheels and the locking collar. Molybdenum-coated brass synchro-rings form the main part of synchronisation.

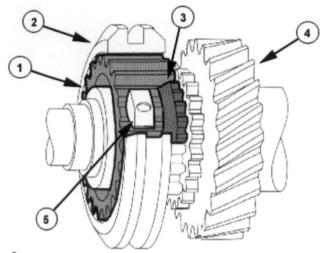

Figure 9.116 Synchromesh operation

Synchronisation 2 1st, 2nd and 3rd gears are equipped with three-fold synchronisation. Compared with a simple cone system, a considerably larger friction area is provided. Synchronisation efficiency is increased as there is a greater surface area to transfer heat. This three-fold synchronisation consists of:

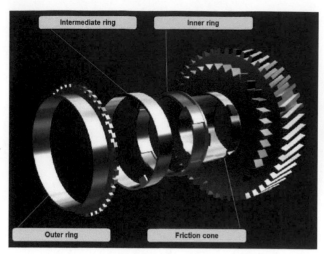

Figure 9.117 Synchro

- ▶ an outer ring (synchro-ring)
- ▶ an intermediate ring
- ▶ an inner ring (2nd synchro-ring)
- ▶ a friction cone on the selector gear/gear wheel.

Synchronisation 3 Balancing of the large speed differences between different selector gears is faster in the low gears and less effort is required to engage the gears. 4th, 5th and 6th gears are equipped with a simple cone system. The speed differences here are not as great as when gears are selected. The balancing of speed is therefore faster. Less effort is required for synchronisation. Reverse gear is equipped with dual cone synchronisation which consists of:

- ▶ a synchro-ring
- ▶ a friction cone on the selector gear/gear wheel.

Direct shift gearbox

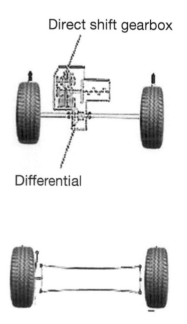

Differential

Bevel box

Haldex coupling

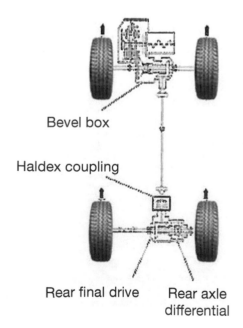

Rear final drive   Rear axle differential

Figure 9.118 4WD

Torque transmission in the vehicle The engine torque is transmitted via the dual mass flywheel to the direct shift gearbox. On front-wheel drive vehicles, the drive shafts transmit the torque to the front road wheels. On four-wheel drive vehicles, the torque is also transmitted to the rear axle via a bevel box. A propshaft transmits the torque to a haldex coupling. Integrated in this rear final drive is a differential for the rear axle.

Transmission route through gears The torque in the gearbox is transmitted either via the outer clutch 1 or the inner clutch 2. Each clutch drives an input shaft. Input shaft 1 (inner) is driven by clutch 1 and input shaft 2 (outer) is driven by clutch 2. Power is transmitted further to the differential via:

▶ output shaft 1 for 1st, 2nd, 3rd and 4th gears
▶ output shaft 2 for 5th, 6th and reverse gears.

Mechatronics 1 The mechatronics are housed in the gearbox, surrounded by oil. They comprise of an electronic control unit and an electro-hydraulic control unit. The mechatronics form the central control unit in the gearbox. Housed in this compact unit are 12 sensors. Only two sensors are located outside the mechatronics system.

Mechatronics 2 The mechatronic control unit uses

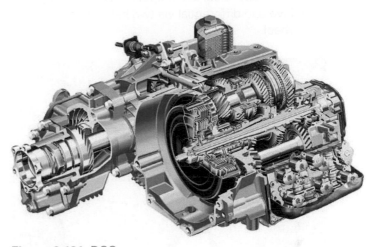

Figure 9.121 DSG

hydraulics to control or regulate eight gear actuators via six pressure valves. It also controls the pressure and flow of cooling oil from both clutches. The mechatronics control unit learns and remembers (adapts) the position of the clutches, the positions of the gear actuators when a gear is engaged, and the main pressure.

Summary The DSG is clearly an innovative development in automotive transmission. It will be interesting to see how it develops.

A version used by Audi is shown in Figure 9.122.

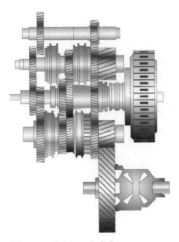

Figure 9.119 DSG

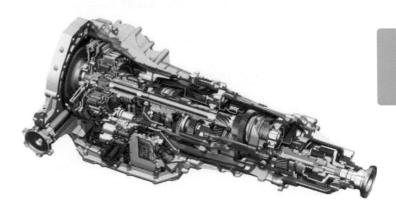

Figure 9.122 Audi DSG

Figure 9.120 Electronic control

 Make a simple sketch to show how one of the main components or systems in this section operates

9

9.4 Driveline

9.4.1 Propshafts ❶ ❷

Introduction Propshafts, with universal joints, are used on rear- or four-wheel drive vehicles. They transmit drive from the gearbox output to the final drive in the rear axle. Drive then continues through the final drive and differential via two half shafts to each rear wheel.

Figure 9.123 Propshaft

Mainshaft A hollow steel tube is used for the mainshaft. This is lightweight, but will still transfer considerable turning forces. It will also resist bending forces.

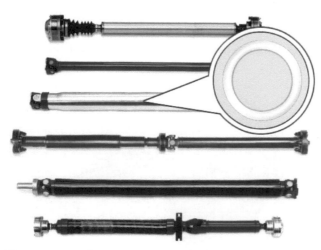

Figure 9.124 Section of a propshaft

Universal joints (UJs) allow for the movement of the rear axle with the suspension, while the gearbox remains fixed. Two joints are used on most systems and must always be aligned correctly.

Because of the angle through which the drive is turned, a variation in speed results. This is caused

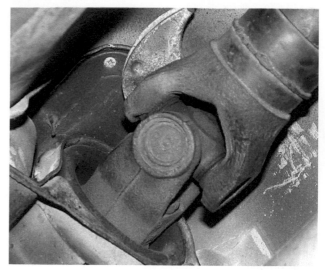

Figure 9.125 Details of a UJ

because two arms of the UJ rotate in one plane and two in another. The cross of the UJ, therefore, has to change position twice on each revolution. However, this problem can be overcome by making sure the two UJs are aligned correctly.

Universal joint alignment If the two UJs on a propshaft are aligned correctly, the variation in speed caused by the first can be cancelled out by the second. However, the angles through which the shaft works must be equal. The main body of the propshaft will run with variable velocity but the output drive will be constant.

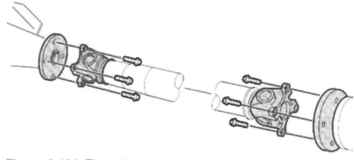

Figure 9.126 These joints are aligned correctly

UJ bearings The simplest and most common type of UJ consists of a four-point cross, which is sometimes called a spider. Four needle roller bearings are fitted, one on each arm of the cross. Two bearings are held in the driver yoke and two in the driven yoke.

UJ Developments Several types of UJ have been used on vehicles. These developed from the simple 'Hooke' type joint, to the later cross type, often known as a Hardy Spicer. Rubber joints are also used on some vehicles.

Figure 9.127 Details of a universal joint

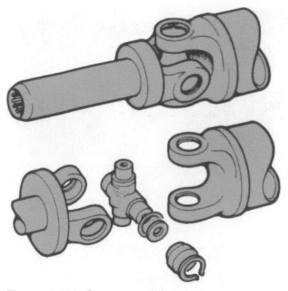

Figure 9.130 Cross-type joint

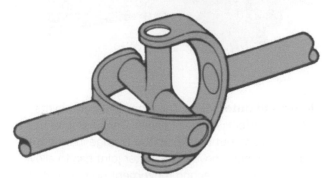

Figure 9.128 Hooke-type joint

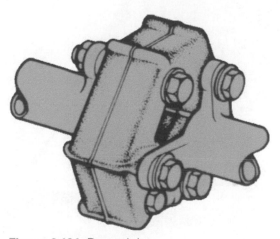

Figure 9.131 Donut joint

As the suspension moves up and down, the length of the driveline changes slightly. As the rear wheels hit a bump, the axle moves upwards. This tends to shorten the driveline. The splined sliding joint allows for this movement.

A **sliding joint** allows for axial movement. However, it will also transfer the rotational drive. Internal splines are used on the propshaft so that the external surface is smooth. This allows an oil seal to be fitted in to the gearbox output casing.

Centre bearings When long propshafts are used, there is a danger of vibration. This is because the weight of the propshaft can cause it to sag slightly and therefore 'whip' (like a skipping rope) as it rotates. Most centre bearings are standard ball bearings mounted in rubber.

Summary Propshafts are used on rear- or four-wheel drive vehicles. They transmit drive from the gearbox output to the rear axle. Most propshafts

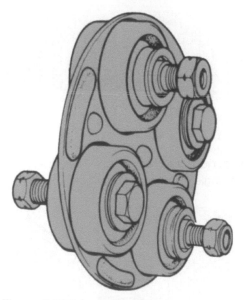

Figure 9.129 Layrub joint

Rubber couplings The donut coupling has the advantage that it is flexible and absorbs torsional shocks. It will also tend to reduce vibrations caused by other joints. Its other main advantage is that it allows some axial (back and forth) movement.

9

Figure 9.132 A splined joint connects to the gearbox

Figure 9.133 This bearing prevents propshaft whip

contain two universal joints (UJs). A single joint produces rotational velocity variations, but this can be cancelled out if the second joint is aligned correctly. Centre bearings are used to prevent vibration due to propshaft whip.

 Create a mind map to illustrate the features of a key component or system

Remember! There is more support on the website that includes additional images and interactive features: www.tomdenton.org

9.4.2 Driveshafts **1 2**

Driveshafts with constant velocity joints transmit drive from the output of the final drive and differential to each front wheel. They must also allow for suspension and steering movements.

Constant velocity (CV) joint A CV joint is a universal joint; however, it is constructed so that the output rotational speed is the same as the input speed. The speed rotation remains constant even as the suspension and steering move the joint.

Figure 9.134 Outer CV joint

Inner and outer joints The inner and outer joints have to perform different tasks. The inner joint has to plunge in and out to take up the change in length as the suspension moves. The outer joint has to allow suspension and steering movement up to about 45°. A solid steel shaft transmits the drive.

Figure 9.135 Inner CV joint

CV joint operation When a normal UJ operates, the operating angle of the cross changes. This is

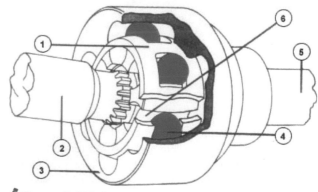

Figure 9.136 Details of a CV joint

what causes the speed variations. A CV joint spider (or cross) operates in one plane because the balls or rollers are free to move in slots. The cross bisects the driving and driven planes.

The **rubber boot or gaiter** is to keep out the dirt and water and keep in the lubricant. Usually a graphite or molybdenum grease is used but check the manufacturer's specifications to be sure.

There are a number of different types of constant velocity joint. The most common is the Rzeppa (pronounced reh-ZEP-ah). The inner joint must allow for axial movement due to changes in length as the suspension moves.

Rzeppa joint The Rzeppa joint is one of the most common. It has six steel balls held in a cage between an inner and outer race inside the joint housing. Each ball rides in its own track on the inner and outer races. The tracks are manufactured into an arch shape so that the balls stay in the midpoint at all times, ensuring that the angle of the drive is bisected. This joint is used on the outer end of a driveshaft. It will handle steering angles of up to 45°.

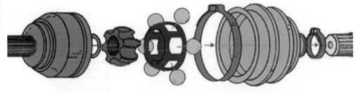

Figure 9.137 CV joint – Rzeppa type

Cross groove joint The cross groove CV joint is like a compact version of the Rzeppa joint. However, unlike the Rzeppa joint, the cross groove type can plunge up to about 52mm (2 inches). It is more compact but the operating angle is limited to about 22°. It can be used where space is at a premium.

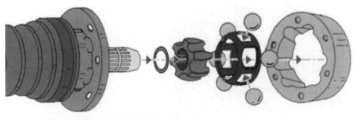

Figure 9.138 CV joint – cross groove type

Double offset joint The double offset joint is a further variation of the Rzeppa joint. The main difference is that the outer race has long straight tracks. This allows a plunge (axial movement) of up to 55mm (2.1 inches) and a steering angle of up to 24°.

Figure 9.139 CV joint – double offset type

Tripod joint The tripod joint is different from other CV joints. A component called a spider splits the drive angle. The arms of the spider give it the tripod name. Each arm of the spider has needle roller bearings and a roller ball. The roller balls work in grooves in the housing. This joint is suitable for inner or outer positions.

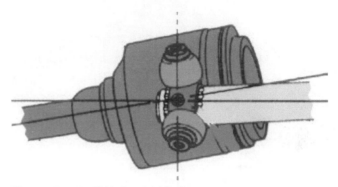

Figure 9.140 CV joint – tripod type

Summary Driveshafts with CV joints are used on front-wheel drive vehicles. They transmit drive from the differential to each front wheel. They must also allow for suspension and steering movements. Inner joints must 'plunge' to allow for changes in length of the shaft. Several types of CV joint are used. All types work on the principle of bisecting the drive angle to produce a constant velocity output.

> Make a simple sketch to show how one of the main components or systems in this section operates

9.4.3 Rear-wheel drive bearings ❶ ❷

Types of bearing There are two main types of bearing used in rear wheel hubs. These are ball bearings and roller, or tapered roller, bearings.

Rear-wheel bearings Axle shafts transmit drive from the differential to the rear-wheel hubs. An axle shaft has to withstand:

▶ torsional stress due to driving and braking forces
▶ shear and bending stress due to the weight of the vehicle
▶ tensile and compressive stress due to cornering forces.

Figure 9.141 Ball bearing

Figure 9.142 Roller bearing

A number of bearing layouts are used, depending on application, to handle these stresses.

Semi floating Shown here is a typical axle mounting used on many rear-wheel drive cars. A single bearing is used, which is mounted in the axle casing. With this design, the axle shaft has to withstand all the operating forces. The shaft is therefore strengthened and designed to do this. An oil seal is incorporated because oil from the final drive can work its way along the shaft. The seal prevents the brakes being contaminated.

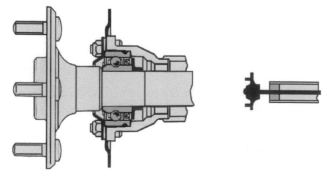

Figure 9.143 Wheel bearing – semi floating

The **three-quarter floating** bearing shown in Figure 9.144 reduces the main shear stresses on the axle shaft but the other stresses remain. The bearing is mounted on the outside of the axle tube. An oil seal

is included to prevent the brake linings from being contaminated.

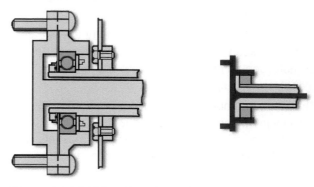

Figure 9.144 Wheel bearing – three-quarter floating

Fully floating systems are generally used on heavy or off-road vehicles. This is because the stresses on these applications are greater. Two widely spaced bearings are used, which take all the loads, other than torque, off the axle shaft. Bolts or studs are used to connect the shaft to the wheel hub. When these are removed, the shaft can be taken out without jacking up the vehicle.

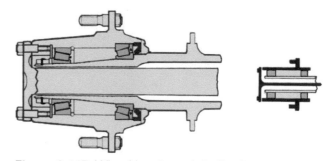

Figure 9.145 Wheel bearing – fully floating

Front-wheel bearings Front hubs on rear-wheel drive cars consist of two bearings. These are either ball

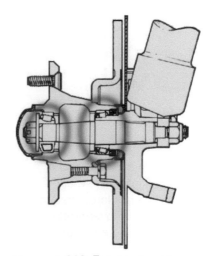

Figure 9.146 Front hub with tapered roller bearings

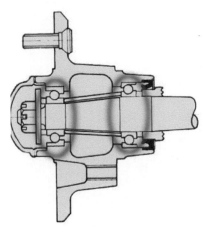

Figure 9.147 Front hub with ball bearings

or tapered roller types. The roller types are generally used on earlier vehicles. They have to be adjusted by tightening the hub nut and then backing it off by about half a turn. The more modern hub bearings, known as contact-type ball races, do not need adjusting. This is because the hub nut tightens against a rigid spacer. This nut must always be set at a torque specified by the manufacturer.

Summary The most common systems for rear-wheel drive cars are semi-floating rear bearings at the rear, and twin ball bearings at the front. The front bearings are designed to withstand side forces as well as vertical loads.

Figure 9.148 Rear hub

Figure 9.149 Front hub

Look back over the previous section and write out a list of the key bullet points from this section

9.4.4 Front-wheel drive bearings ❶ ❷

Introduction Wheel bearings must allow smooth rotation of the wheel but also be able to withstand high stresses such as when cornering. Front-wheel drive arrangements must also allow the drive to be transmitted via the driveshafts.

Front bearings The front hub works as an attachment for the suspension and steering as well as supporting the bearings. It supports the weight of the vehicle at the front, when still or moving. Ball or roller bearings are used for most vehicles, with specially shaped tracks. This is so the bearings can stand side loads when cornering. The bearings support the driveshaft as well as the hub.

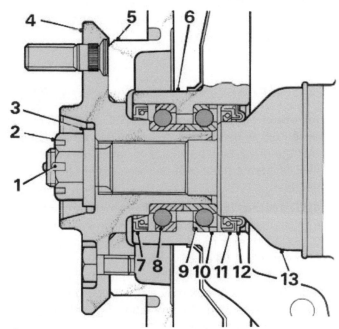

Figure 9.150 Front hub and bearings

Rear bearings The stub axle, which is solid-mounted to the suspension arm, fits in the centre of two bearings. The axle supports the weight of the vehicle at the rear, when still or moving. Ball bearings are used for most vehicles, with specially shaped tracks for the balls. This is so the bearings can stand side loads when cornering. A spacer is used to ensure the correct distance between, and pressure on, the two bearings.

9

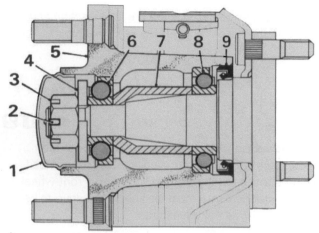

Figure 9.151 Rear hub and bearings

Summary The hub and bearing arrangement on the front of a front-wheel drive car, must bear weight, withstand driving forces and support the driveshaft. The rear hub and bearings must support the vehicle and withstand side forces.

 Look back over the previous section and write out a list of the key bullet points from this section

9.4.5 Four-wheel drive systems ❸

Introduction Four-wheel drive (4WD) systems can be described as part-time or full-time. Part-time means that the driver has the choice to select the drive. All 4WD systems must include some type of transfer gearbox.

4WD system layout The main components of a four-wheel drive system are shown in Figure 9.152. Each axle must be fitted with a differential. A transfer box takes drive from the output of the normal gearbox and distributes it to the front and rear. The transfer box may also include gears to allow the selection of a low ratio. High ratio is a straight through drive.

Part-time 4WD A 4WD system, when described as part-time, means that the driver selects four-wheel drive only when the vehicle needs more traction. When the need no longer exists, the driver reverts to the normal two-wheel drive. This keeps driveline friction, and therefore the wear rate, to a minimum.

Full-time 4WD A 4WD system, when described as full-time, means that the drive is engaged all the time. The driver may still be able to select a low range setting. To prevent 'wind up', which

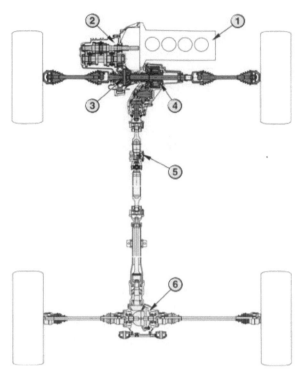

Figure 9.152 The main components of a four-wheel drive system

Figure 9.153 Selection control

would occur when the front and rear axles rotate at different speeds, a centre differential or viscous drive is used.

All-wheel drive (AWD) An all-wheel drive system automatically transfers drive to the axle with better traction. It is designed for normal road use. A low ratio option is not available. The system is described as part-time if the driver can select either front- or all-wheel drive. It is described as full-time if selection is not possible. The drive, on full-time systems, is passed to the rear via a viscous coupling. When the front wheels spin, the viscous coupling locks and transfers drive to the rear.

Figure 9.154 Volvo S60 AWD vehicle

Transfer box The transfer box of a part-time 4WD system usually allows the driver to choose from four options: Neutral, 2WD High, 4WD High and 4WD Low. A typical system will have the transfer box attached to the normal rear-wheel drive gearbox, in place of the extension housing. A two-speed transfer box is shown in Figures 9.155, 9.156, 9.157 and 9.158.

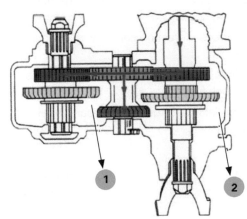

Figure 9.155 Neutral 1-Low ratio gear selector, 2-High ratio gear selector

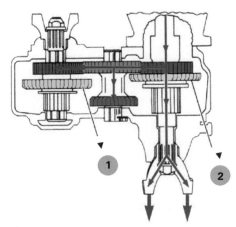

Figure 9.156 Two wheel drive high 1-Low ratio gear selector, 2-High ratio gear selector

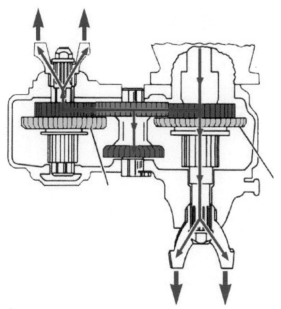

Figure 9.157 Four wheel drive high 1-Low ratio gear selector, 2-High ratio gear selector

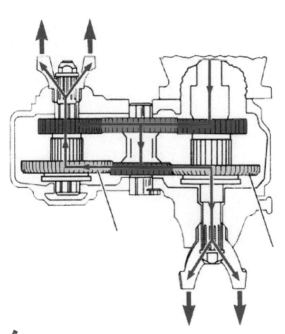

Figure 9.158 Four wheel drive low 1-Low ratio gear selector, 2-High ratio gear selector

Centre Differential A differential allows its two outputs to be driven at different speeds. This is normally important for the drive axle of a vehicle. When a vehicle is cornering, the outer wheels travel faster than the inner wheels. On 4WD systems, it is possible for, say, the front axle to rotate faster than the rear axle. This could produce driveline 'wind up' of the transmission. Centre differentials are designed to allow for this. On modern vehicles, they often consist of planetary type gears.

9

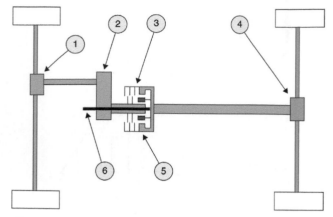

Figure 9.159 A viscous coupling transfers torque when axle speeds differ: 1-Front differential, 2-Drive chain transfer box, 3-Viscous coupling, 4-Rear differential, 5-Planetary gear train, 6-Transmission gearbox output shaft

Viscous coupling A viscous coupling is designed to transmit drive when the axle speeds differ. This occurs because the difference in speed of the two axles increases the friction in the coupling. This results in greater torque transmission, which in turn reduces the speed difference. As the speed difference reduces, less torque is transmitted. In this way, the torque is 'shared' proportionally between the two axles.

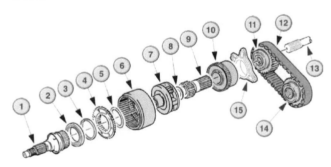

Figure 9.160 Transfer box using planetary gears and a drive chain: 1-Output shaft, 2-Ball bearing, 3-Spacer, 4-Drive plate, 5-Spacer, 6-Annulus, 7-Planet carrier and planet gears, 8-Thrust washer, 9-Sun wheel shaft, 10-Viscomatic locking unit, 11-Driving sprocket and bearing, 12-Drive chain, 13-Transmission gearbox output shaft, 14-Driven sprocket and bearing, 15-Bracket

Chain drive A 'silent' drive chain is used on many newer vehicles to pass the drive to the auxiliary output shaft. The chain takes up less space than gears. It is designed to last the life of the vehicle and adjustment is not normally possible. The steel chain is similar in design to timing gear chains only it is wider and stronger.

Active driveline is a system used by Land Rover. The key to the system is the synchroniser, which disconnects the drive to the rear wheels when not

needed but can re-engage the drive within 300ms. This reduces friction losses and therefore improves economy. With this system, a rear differential gear is not needed and instead drive to each wheel is via multi-plate clutches that are under electronic control. When in 2WD the clutches are released to reduce drag.

In off-road conditions, these clutches allow power to be diverted to the wheel with most traction or the axle can be fully locked. During normal road use, the system can push more power to an individual wheel. This enhances the cornering response, and is known as torque biasing. All settings can be adjusted by the driver. There are several active driveline functions available with the system shown here.

▶ When 4WD is not required, the system automatically selects 2WD

▶ All major driveline components downstream of the power transfer unit (PTU) come to rest in 2WD

▶ 4WD reconnects within 300ms when required

▶ 4WD is connected when stationary, providing increased traction when the vehicle pulls away

▶ The system disconnects at 35km/h (22mph) under normal driving conditions

▶ Provides increased traction capacity in low grip conditions

▶ The rear drive unit (RDU) twin clutch provides a cross-axle lock function

▶ The system delivers torque to the rear wheel with most traction.

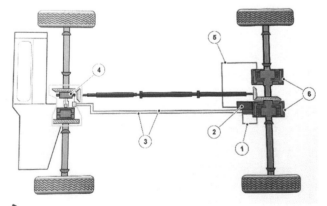

Figure 9.161 Active driveline control. 1. Hydraulic clutch line (left), 2. Valve block, 3. PTU synchroniser hydraulic feed and return, 4. PTU synchroniser, 5. Hydraulic clutch line (right), 6. RDU clutches (Source: Land Rover)

Summary Four-wheel drive systems use a combination of propshafts and driveshafts, together with viscous couplings and transfer boxes. A number of variations are possible. These are described as either full-time or part-time.

Use the media search tools to look for pictures and videos relating to the subject in this section

9.5 Final drive and differential

9.5.1 Final drive ❶ ❷

Introduction Because of the speed at which an engine runs, and in order to produce enough torque at the road wheels, a fixed gear reduction is required. This is known as the final drive. It consists of just two gears. The final drive is fitted after the output of the gearbox on front-wheel drive vehicles. It is fitted in the rear axle after the propshaft on rear-wheel drive vehicles.

The ratio is normally between about 2:1 and about 4:1. In other words, at 4:1, when the gearbox output is turning at 4000 rpm, the wheels will turn at 1000 rev/min.

Rear-wheel drive The final drive gears turn the drive through 90° on rear-wheel drive vehicles. Four-wheel drive vehicles will also have this arrangement as part of the rear axle.

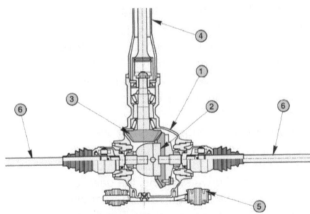

Figure 9.162 Rear axle final drive gears: 1-Rear axle housing, 2-Differential, 3-Crown wheel and pinion final drive gears, 4-Extension tube, 5-Mounting, 6-Driveshafts

Front-wheel drive Most cars now have a transverse engine which drives the front wheels. The power of the engine therefore does not have to be carried through a right angle to the drive wheels. The final drive contains ordinary reducing gears rather than bevel gears.

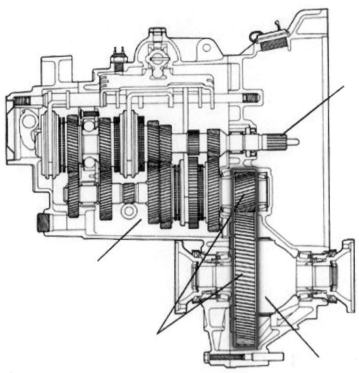

Figure 9.163 Transaxle final drive gears 1-Layshaft, 2-Final drive gears, 3-Differential 4-Input shaft

Bevel gears The crown wheel and pinion are types of bevel gears because they mesh at right angles to each other. They carry power through a right angle to the drive wheels. The crown wheel is driven by the pinion, which receives power from the propeller shaft.

Reduced speed and increased torque Final drive gears reduce the speed from the propeller shaft and increase the torque. The reduction in the final drive multiplies any reduction that has already taken place in the transmission.

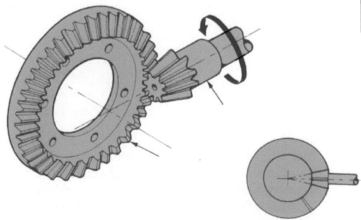

Figure 9.164 Bevel gears change ratio and drive angle

9

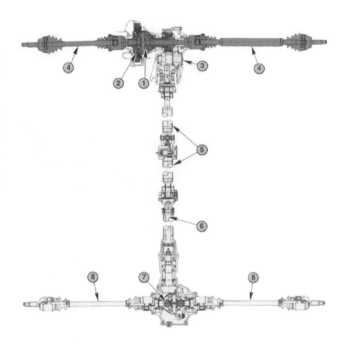

Figure 9.167 Lubrication is important

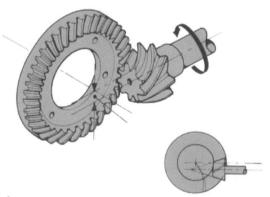

Figure 9.165 The main components of a four-wheel drive system: 1-Engine, 2-Transmission, 3-Front axle differential and final drive, 4-Transfer box with longitudinal differential, 5-Two-piece driveshaft, 6-Rear final drive and differential

Hypoid gear The crown wheel gear of a rear-wheel drive system is usually a hypoid type, which is named after the way the teeth are cut. As well as quiet operation, this allows the pinion to be set lower than the crown wheel centre, thus saving space in the vehicle because a smaller transmission tunnel can be used.

Figure 9.166 The design allows a lower propshaft to be used

Hypoid gear oil Because the teeth of hypoid gears cause 'extreme pressure' on the lubrication oil, a special type is used. This oil may be described as 'Hypoid Gear Oil' or 'EP', which stands for extreme pressure. As usual, refer to manufacturers' recommendations when topping up or changing oil.

The complete rear axle assembly consists of other components as well as the final drive gears. The other main components are the differential, the halfshafts and bearings. Components that make up a solid axle are shown here. Some rear-wheel drive and four-wheel drive vehicles have a split axle. On these types, the final drive is mounted to the chassis and driveshafts are used to connect to the wheels.

The front-wheel drive axle, where a transaxle system is used, always consists of the final drive and two driveshafts. The gearbox, final drive and one driveshaft are shown here. The final drive gears provide the same reduction as those used on rear-wheel drives, but do not need to turn the drive through 90°.

Four-wheel drive The general layout of a four-wheel drive system is shown in Figure 9.165. A representation of how torque is distributed is also shown. The variation in torque is achieved by differential action. This is examined in some detail later.

Summary To produce enough torque at the road wheels, a fixed gear reduction is required. This is known as the final drive. It consists of just two gears. On rear-wheel drive systems, the gears are bevelled to turn the drive through 90°. On front-wheel drive systems, this is not necessary. The drive ratio is similar for front- or rear-wheel drive cars.

 Use the media search tools to look for pictures and videos relating to the subject in this section

9.5.2 Differential operation ❶ ❷

Introduction The differential is a set of gears that divides the torque evenly between the two drive wheels. The differential allows one wheel to rotate faster than the other. As a car goes around a corner, the outside driven wheel travels further than the inside one. The outside wheel must therefore rotate faster than the inside one to cover the greater distance in the same time. Tyre scrub and poor handling would be the result if a fixed axle were used.

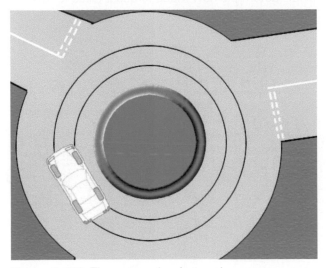

Figure 9.168 The outer wheels travel a greater distance

The differential consists of sets of bevel gears, and pinions within a cage, attached to the large final drive gear. The bevel gears can be described as sun and planet gears. The sun gears provide the drive to the wheels via halfshafts or driveshafts. The planet gears either rotate with the sun gears or rotate around them, depending on whether the car is cornering or not.

The small pinion brings the drive from the gearbox to the larger final drive gear. A fixed gear reduction is produced by the crown wheel and pinion. On rear-wheel drive cars, bevel gears are used to turn the drive through 90°.

The bearings support the differential casing, which is in turn bolted to the final drive gear. The casing transmits the drive from the final drive gear to the planet gear pinion shaft.

The planet gears are pushed round by their shaft. The sun gear pinions, which are splined to the drive shafts, take their drive from the planet gears. The sun gears always rotate at the same speed as the road wheels.

The planet shaft is secured in the differential casing so that it pushes the planet gears. If the sun gears, which are attached to the road wheels via the driveshafts, are moving at the same speed, the planet gears do not spin on their shaft. However, when the vehicle is cornering, the sun gears need to move at different

speeds. In this case, the planet gears spin on the shaft to make up for the different wheel speeds.

When the vehicle is travelling in a straight line, the bevel pinions (planet gears) turn with the sun gears but do not rotate on their shaft. This occurs because the two sun gears attached to the driveshafts are revolving at the same speed.

When the vehicle is cornering, the bevel pinions (planet gears) roll round the sun gears and rotate on their shaft. This rotation is what allows the outer wheel to turn faster than the inner.

A standard differential can be described as a torque equaliser. This is because the same torque is provided to each wheel, even if they are revolving at different speeds. At greater speeds, more power is applied to the wheel, so the torque remains the same.

Extreme example One further way to understand the differential action is to consider the extreme situation. This is when the corner is so sharp, the inner wheel does not move at all! Now of course this is impossible, but it can be simulated by jacking up one wheel of the car. All the drive is transferred to the free wheel. The planets roll around the stationary sun wheel but drive the free wheel because they are rotating on their shaft.

The example given in the last paragraph highlights the one problem with a differential. If one of the driven wheels is stuck in the mud, all the drive is transferred to that wheel and it normally spins. Of course, in this case, drive to the wheel on the hard ground would be more useful. The solution to this problem is the limited slip differential.

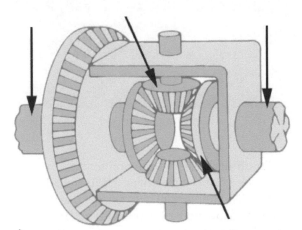

Figure 9.169 All the drive is transferred to the free wheel

Summary As a car goes around a bend, the outside driven wheel travels further than the inside one. The

481

outside wheel must therefore rotate faster to cover the greater distance in the same time. The differential allows this difference in speed.

 Using images and text, create a short presentation to show how a component or system works

9.5.3 Limited slip differentials ❸

Introduction Some higher performance vehicles use a limited slip differential (LSD). Clutch plates, or similar, are connected to the two output shafts and can, therefore, control the amount of slip. This can be used to counteract the effect of one wheel losing traction when high power is applied.

Standard differential A standard differential always applies the same amount of torque to each wheel. Two factors determine how much torque can be applied to a wheel. In dry conditions, when there is plenty of traction, the amount of torque applied to the wheels is limited by the engine and gearing. When the conditions are slippery, such as on ice, the torque is limited by the available grip.

Figure 9.170 This differential is sometimes described as an open type

Limited slip differential The solution to the problems of the normal differential, is the limited slip differential (LSD). Limited slip differentials use various mechanisms to allow normal differential action when going around turns. However, when a wheel slips, they allow more torque to be transferred to the non-slipping wheel.

The clutch-type LSD The clutch-type LSD is the most common. It is the same as a standard differential, except that it also has a spring pack and a multi-plate

clutch. The spring pack pushes the sun gears against the clutch plates, which are attached to the cage. Both sun gears spin with the cage, when both wheels are moving at the same speed, and the clutches have little or no effect. However, the clutch plates try to prevent either wheel from spinning faster than the other. The stiffness of the springs and the friction of the clutch plates determine how much torque it takes to make it slip.

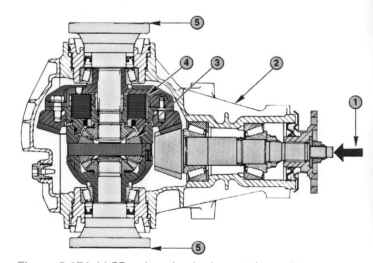

Figure 9.171 LLSD using clutch plates: 1-Input drive from propshaft, 2-Housing, 3-Differential, 4-Clucth plates, 5-Output drive flanges (Source: Ford Motor Company)

Slippery surface If one drive wheel is on a slippery surface and the other one has good traction, drive can be transmitted to this wheel. The torque supplied to the wheel not on the slippery surface is equal to the amount of torque it takes to overpower the clutches. The result is that the car will move, but not with all the available power.

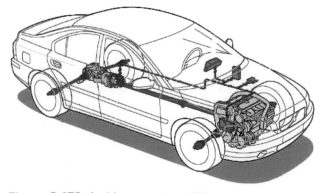

Figure 9.172 4×4 layout using LSDs

Viscous coupling The viscous coupling is often found in all-wheel drive vehicles. It is commonly used to link the back wheels to the front wheels so that when one set of wheels starts to slip, torque will be transferred to the other set. The viscous coupling has two sets of plates inside a sealed housing that is filled with a thick fluid. One set of plates is connected to each output shaft. Under normal conditions, both sets of plates and the viscous fluid spin at the same speed. However, when one set of wheels spins faster, there will be a difference in speed between the two sets of plates.

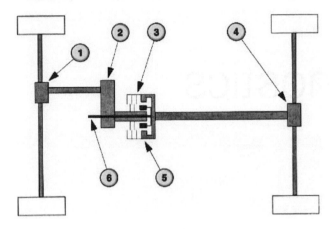

Figure 9.173 Viscous-type LSD

The viscous fluid between the plates tries to catch up with the faster discs, dragging the slower discs along. This transfers more torque to the slower wheels. When a vehicle is cornering, the difference in speed between the wheels is not as large as when one wheel is slipping. The faster the plates

spin, relative to each other, the more torque the coupling transfers. This effect can be demonstrated by spinning an egg! Spin the egg and then stop it. Let go, and it will start to spin again as the viscous fluid inside is still spinning and drags the shell around with it.

Electronic control Conventional limited slip differentials cannot be designed for optimum performance because of the effect on the vehicle when cornering and on the steering. These issues prompted the development of electronic control. The slip limiting action is controlled by a multi-disc clutch as discussed previously. The pressure on the clutch plates is controlled by hydraulic pressure, which in turn is controlled by a solenoid valve under the influence of an ECU. It is able, if required, to fully lock the axle. Data is provided to the ECU from standard ABS-type wheel sensors.

Summary The two main types of limited slip differentials are the plate type and the viscous coupling type. A speed difference between wheels or axles must overcome plate friction on the clutch type. The viscous type works because the friction between plates increases as the speed difference increases.

> Use the media search tools to look for pictures and videos relating to the subject in this section

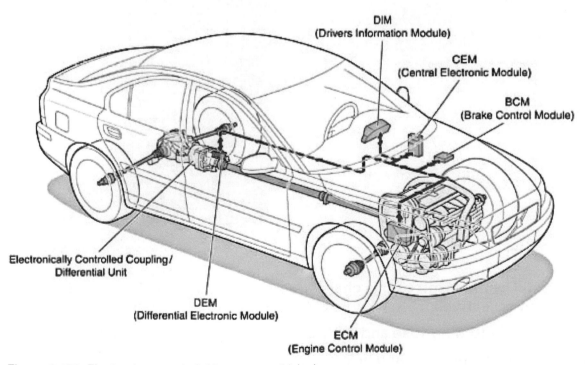

Figure 9.174 Electronic control of drive system (Volvo)

CHAPTER 10

Advanced diagnostics

After successful completion of this chapter you will be able to show you have achieved these objectives:

1 Understand how to diagnose and rectify faults in light vehicle systems

2 Understand the purpose of and how to use a range of test equipment

3 Understand OBD and fault codes

DOI: 10.1201/9781003173236-10

10.1 Diagnostic techniques

10.1.1 Introduction ❶ ❷ ❸

Logic Diagnostics or faultfinding is a fundamental part of an automotive technician's work. The subject of diagnostics does not relate to individual areas of the vehicle. If your knowledge of a vehicle system is at a suitable level, the same logical process is used for diagnosing the fault, whatever the system.

Figure 10.1 Diagnostics

Terminology 1 The terminology included in the following tables is provided to ensure we are talking the same language. These tables are provided as a simple reference source.

Symptom	The effect of a fault noticed by the driver, user or technician
Fault	The cause of a symptom/problem
Root cause	The cause of the fault
Diagnostics	The process of tracing a fault by means of its symptoms, applying knowledge and analysing test results
Knowledge	The understanding of a system that is required to diagnose faults
Logical procedure	A step by step method used to ensure nothing is missed
Concern, cause, correction	A reminder of the process starting from what the driver reports, to the correction of the problem
Report	A standard format for the presentation of results

Terminology 2 General terminology

System	A collection of components that carry out a function
Efficiency	This is a simple measure of any system. It can be scientific for example if the power out of a system is less than the power put in, its percentage efficiency can be determined (P-out/P-in x 100%). This could for example, be given as say 80%. In a less scientific example, a vehicle using more fuel than normal is said to be inefficient
Noise	Emanations of a sound from a system that is either simply unwanted or is not the normal sound that should be produced
Active	Any system that is in operation all the time (steering for example)
Passive	A system that waits for an event before it is activated (an air bag is a good example)
Short circuit	An electrical conductor is touching something that it should not be (usually another conductor or the chassis)
Open circuit	A circuit that is broken (a switched off switch is an open circuit)
High resistance	In relation to electricity, this is part of a circuit that has become more difficult for the electricity to get through. In a mechanical system a partially blocked pipe would have a resistance to the flow of fluid
Worn	This works better with further additions such as: 'Worn to excess', 'worn out of tolerance' or even 'worn, but within tolerance'
Quote	Exact information on the price of a part or service. A quotation may often be considered to be legally binding
Estimate	A statement of the anticipated cost of a job (e.g. a service or repairs). An estimate is normally an approximation, and is not legally binding
Bad	Not good – and also not descriptive enough really…
Dodgy, knackered or @#%&*.	Words often used to describe a system or component, but they mean nothing. Get used to describing things so that misunderstandings are eliminated

Information Information and data relating to vehicles are available for carrying out many forms of diagnostic work. The data may come as a book, online or on CD/DVD. This information is vital and will ensure that you find the fault – particularly if you have developed the diagnostic skills to go with it. The general type of information available is as follows:

▶ engine diagnostics, testing and tuning
▶ servicing, repairs and times
▶ fuel and ignition systems
▶ auto electrics data
▶ component location
▶ body repairs, tracking and tyres
▶ diagnostic procedures
▶ recall information

10

485

Figure 10.2 Data source

Where to stop? This is one of the most difficult skills to learn. It is also one of the most important. The secret is twofold:

1 know your own limitations – it is not possible to be good at everything;
2 leave circuits alone where you could cause more damage or even injury – for example air bag circuits.

Often with the best of intentions, a person new to diagnostics will not only fail to find the fault but introduce more faults into the system in the process. I would suggest you learn your own strengths and weaknesses; you may be confident and good at dealing with mechanical system problems but less so when electronics is involved. Of course you may be just the opposite of this.

Remember that diagnostic skill is in two parts – the knowledge of the system and the ability to apply diagnostics. If you do not yet fully understand a system – leave it alone until you do.

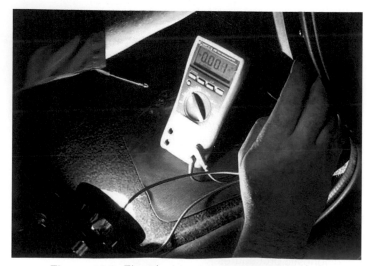

Figure 10.3 Electrical testing

 Select a routine from section 1.3 and follow the process to study a component or system.

10.1.2 Equipment overview 🄴 ❶ ❷

Introduction Diagnostic techniques are very much linked to the use of test equipment. In other words, you must be able to interpret the results of tests. In most cases this involves comparing the result of a test to the reading given in a data source. By way of an introduction, the following table lists some of the basic words and descriptions relating to tools and equipment.

Table 10.1 Tools and equipment

Hand tools	Spanners and hammers and screwdrivers and all the other basic bits!
Special tools	A collective term for items not held as part of a normal tool kit. Or items required for just one specific job
Test equipment	In general, this means measuring equipment. Most tests involve measuring something and comparing the result of that measurement to data. The devices can range from a simple ruler to an engine analyser
Dedicated test equipment	Some equipment will only test one specific type of system. The large manufacturers supply equipment dedicated to their vehicles. For example, a diagnostic device which plugs in to a certain type of fuel injection ECU
Accuracy	Careful and exact, free from mistakes or errors and adhering closely to a standard
Calibration	Checking the accuracy of a measuring instrument
Serial port	A connection to an electronic control unit, a diagnostic tester or computer for example. Serial means the information is passed in a 'digital' string like pushing black and white balls through a pipe in a certain order
Code reader or scanner	This device reads the 'black and white balls' mentioned above or the on-off electrical signals, and converts them into language we can understand
Combined diagnostic and information system	This equipment has an operating system, and can be used to carry out tests on vehicle systems, they also contain electronic data. Test sequences guided by software can also be carried out, recordings made and reports generated
Oscilloscope	The main part of a 'scope' is the display, which is a screen. A scope is a voltmeter but instead of readings in numbers it shows the voltage levels by a trace or mark on the screen (like a graph). The marks on the screen can move and change very fast allowing us to see the way voltages change

 Use the media search tools to look for pictures and videos relating to the subject in this section.

10.1.3 Diagnostic process ❷ ❸

Six-stage process 1 A key checklist – the six stages of fault diagnosis – is given in this list:

1 Verify: Is there actually a problem, can you confirm the symptoms?
2 Collect: Get further information about the problem, by observation and research
3 Evaluate: Stop and think about the evidence
4 Test: Carry out further tests in a logical sequence
5 Rectify: Fix the problem
6 Check: Make sure all systems now work correctly

Six-Stage Process 2 Here is a very simple example to illustrate the diagnostic process. The reported fault is excessive use of engine oil:

1 Question the motorist to find out how much oil is being used (is it excessive?) check service history.
2 Examine the vehicle for oil leaks, and blue smoke from the exhaust. Are there any service bulletins?
3 If leaks are found, the engine could still be burning oil but leaks would be a likely cause.
4 A compression test, if the results were acceptable, would indicate a leak to be the most likely fault. Cleaning down the engine and then run will show a leak more easily.
5 Change a gasket or seal, etc.
6 Run through an inspection of the vehicle systems particularly associated with the engine. Double check the fault has been rectified and that you have not caused any further problems.

Figure 10.4 Engine

Six-stage process 3 The six-stage diagnostic process will be used extensively to illustrate how a logical process can be applied to any situation.

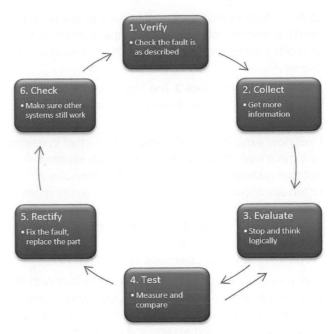

Figure 10.5 Six-stage diagnostic process in flowchart form

The art of diagnostics 1 The knowledge needed for accurate diagnostics is in two parts:

1 understanding of the system in which the problem exists
2 having the ability to apply a logical diagnostic routine

The knowledge requirement and use of diagnostic skills can be illustrated with a very simple example:

After connecting a hose pipe and turning on the tap, no water comes out of the end. Your knowledge of this system tells you that water should come out providing the tap is on, because the pressure from a tap pushes water through the pipe, and so on. This is where your diagnostic skills become essential. The following shows the required stages.

The art of diagnostics 2

1 Confirm that no water is coming out by looking down the end of the pipe.
2 Check if water comes out of the other taps, or did it come out of this tap before you connected the hose?
3 Consider what this information tells you; for example, if the answer is 'Yes' the hose must be blocked or kinked
4 Walk the length of the pipe looking for a kink
5 Straighten out the hose
6 Check that water now comes out and that no other problems have been created.

Much simplified, but the procedure you have just followed made the hose work and it is also guaranteed to find a fault in any system. It is easy to see how it works in connection with a hose pipe and I'm sure anybody could have found that fault (well most people anyway).

The art of diagnostics 3 The higher skill is to be able to apply the same logical routine to more complex situations.

I will now explain each of these steps further in relation to a more realistic automotive workshop situation – not that getting the hose to work is not important! Often electrical faults are considered to be the most difficult to diagnose – but this is not true. I will use a vehicle cooling system fault as an example. Remember that the diagnostic procedure can be applied to any problem, mechanical, electrical or even medical.

However, let's assume that the reported fault with the vehicle is overheating. As is quite common in many workshop situations that's all the information we have to start with. The next screen will explain the stages in more detail.

Diagnostics example

1 Quick examination for obvious problems such as leaks, broken drive belts or lack of coolant. Run the vehicle and confirm that the fault exists. It could be the temperature gauge for example.

2 Is the driver available to give more information? For example, does the engine overheat all the time or just when working hard? Check records, if available, of previous work done to the vehicle.

3 Consider what you now know. Does this allow you to narrow down what the cause of the fault could be? For example, if the vehicle overheats all the time and it had recently had a new cylinder head gasket fitted, would you be suspicious about this? Don't let two and two make five, but do let it act as a pointer. Remember that in the science of logical diagnostics, two and two always makes four. However, until you know this for certain then play the best odds to narrow down the fault.

4 The further tests carried out would now be directed by your thinking at stage three. You don't yet know if the fault is a leaking head gasket, the thermostat stuck closed or some other problem. Playing the odds, a cooling system pressure test would probably be the next test. If the pressure increases when the engine is running then it is likely to be a head gasket or similar problem. If no pressure increase is noted, then move on to the next test and so on. After each test go back to stage 3 and evaluate what you know, not what you don't know.

5 Let's assume the problem was a thermostat stuck closed – replace it and top up the coolant, etc.

6 Check that the system is now working. Also check that you have not caused any further problems such as leaks or loose wires.

Summary This example is simplified a little, but like the hose pipe problem it is the sequence that matters, particularly the 'stop and think' at stage 3. It is often possible to go directly to the cause of the fault at this stage, providing that you have an adequate knowledge of how the system works.

Concern, cause, correction 1 The 3 C's, as concern, cause and correction are sometimes described, is another reminder that following a process for automotive repairs and diagnostics is essential. It is in a way a simplified version of our six-stage process as shown in this table:

Six-stage process	CCC
Verify	Concern
Collect	Cause
Evaluate	
Test	
Rectify	
Check	Correction

Concern, cause, correction 2 This table is a further example where extra suggestions have been added as a reminder of how important it is to collect further information. It is also recommended that this information and process is included on the job sheet so the customer is kept informed. Most customer complaints come about because of poor work or poor communication – this may be acceptable in some poor quality establishments but not in any that you and I are involved in – be professional and you will be treated like one (lecture over, sorry).

Process outline	Example situation	Notes
Customer Concern:	Battery seems to be discharged and will sometimes not start the car. It seems to be worse when the headlights are used	This should set you thinking that the cause is probably a faulty battery, a charging system fault, a parasitic discharge or a starter motor problem (the symptoms would suggest a charging fault is most likely but keep an open mind)
Vehicle service history information:	Car is five years old, has done 95,000 miles but has a good service history. A new battery was fitted one year ago and the cam belt was replaced two years ago	Battery probably ok and drive belt adjustment likely to be correct (still suspicious of a charging fault)

Process outline	Example situation	Notes
Related technical service bulletins:	New camshaft drive belt should be fitted every 50,000 miles	Not connected but it would be good to recommend that the belt was changed at this time
Diagnostic procedures performed:	Battery voltage and discharge test – ok Drive belt tension – ok (but a bit worn) Alternator charging voltage – 13V Checked charging circuit for volt drop – ok	14V is the expected charging voltage on most systems
Cause:	Alternator not producing correct voltage	An auto-electrician may be able to repair the alternator but for warranty reasons a new or reconditioned one is often best (particularly at this mileage)
Correction:	Reconditioned alternator and new drive belt fitted and checked – charging now ok at 14V	Note how by thinking about this process we had almost diagnosed the problem before doing any tests, also note that following this process will make us confident that we have carried out the correct repair, first time. The customer will appreciate this – and will come back again

Summary So, while the concern, cause, correction sequence is quite simple, it is very effective as a means of communication as well as a diagnosis and repair process.

Root cause analysis 1 The phrase 'root cause analysis' (RCA) is used to describe a range of problem solving methods aimed at identifying the root causes of problems or events. I have included this short section

because it helps to reinforce the importance of keeping an open mind when diagnosing faults, and again, stresses the need to work in a logical and structured way. The root cause of a problem is not always obvious.

Root cause analysis 2 Let's assume the symptom was that one rear light on a car did not work. Using the six-stage process, a connector block was replaced as it had an open circuit fault. The light now works ok but what was missed was that a small leak from the rear screen washer pipe dripped on the connector when the washer was operated. This was the root cause.

Root cause analysis 3 The practice of RCA is based, quite rightly, on the belief that problems are best solved by attempting to address, correct or eliminate the root causes, as opposed to just addressing the faults causing observable symptoms. Dealing with root causes ensures that problems will not reoccur. RCA is best considered to be an iterative process because complete prevention of recurrence by one corrective action is not always realistic.

Root cause analysis 4 The list shown here is a much simplified representation of a failure-based RCA process. Note that the key steps are numbers 3 and 4. This is because they direct the corrective action at the truo root cause of the problem.

1 Define the problem
2 Gather data and evidence
3 Identify the causes and root causes
4 Identify corrective action(s)
5 Implement the root cause correction(s)
6 Ensure effectiveness.

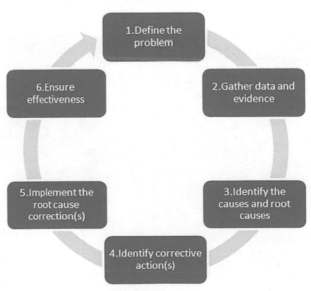

Figure 10.7 RCA process

As an observant reader, you will also note that these steps are very similar to our six-stage fault finding process.

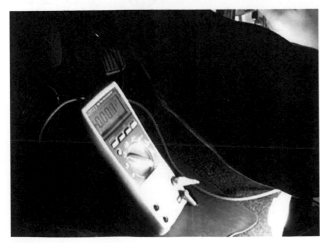

Figure 10.6 Electrical testing

Summary I have introduced the six-stage process of diagnostics, not so that it should always be used as a checklist but to illustrate how important it is to follow a process. Much more detail will be given later, in particular about stages 3 and 4. The purpose of this set process is to ensure that 'we' work in a set, logical way.

Figure 10.8 'Logic is the beginning of wisdom, not the end' (Spock to Valeris, Star Trek IV)

 Create a mind map to illustrate the features of a key component or system

10.1.4 Mechanical diagnostics ❸

Check the obvious first Start all 'hands on' diagnostic routines with 'hand and eye checks'. In other words look over the vehicle for obvious faults. For example,

Figure 10.9 Mechanical systems

if automatic transmission fluid is leaking on to the floor then put this right before carrying out complicated stall tests. Here are some further suggestions that will at some point save you a lot of time.

▶ If the engine is blowing blue smoke out of the exhaust – consider the worth of tracing the cause of a tapping noise in the engine

▶ When an engine will not start – check that there is fuel in the tank

Noise, vibration and harshness Noise, vibration and harshness (NVH) concerns have become more important as drivers have become more sensitive to these issues. Drivers have higher expectations of comfort levels. Noise, vibration and harshness issues are more noticeable due to reduced engine noise and better insulation in general. The main areas of the vehicle that produce NVH are:

▶ tyres
▶ engine accessories
▶ suspension
▶ driveline

It is necessary to isolate the NVH into its specific area(s) to allow more detailed diagnosis. A road test as outlined later is often the best method but there are specialised tools that are sometimes used for NVH.

Figure 10.10 1914 Ford Model T (Source: Ford Media)

Noise, vibration and harshness 2 The five most common sources of non-axle noise are exhaust, tyres, roof racks, trim and mouldings, and transmission. Ensure that none of the following conditions is the cause of the noise before proceeding with a driveline strip down and diagnosis.

1 In certain conditions, the pitch of the exhaust may sound like gear noise or under other conditions like a wheel bearing rumble.

2 Tyres can produce a high pitched tread whine or roar, similar to gear noise. This is particularly the case for non-standard tyres.

3 Trim and mouldings can cause whistling or whining noises.

4 Clunk may occur when the throttle is applied or released due to backlash somewhere in the driveline.

5 Bearing rumble sounds like marbles being tumbled.

Noise conditions Noise is very difficult to describe. However, the following are useful terms and are accompanied by suggestions as to when they are most likely to occur:

▶ Gear noise is typically a howling or whining due to gear damage or incorrect bearing preload. It can occur at various speeds and driving conditions, or it can be continuous.

▶ 'Chuckle' is a rattling noise that sounds like a stick held against the spokes of a spinning bicycle wheel. It usually occurs while decelerating.

▶ Knock is very similar to chuckle though it may be louder and occurs on acceleration or deceleration.

Check and rule out tyres, exhaust and trim items before any disassembly to diagnose and correct gear noise. Remember to also check the spare wheel is correctly stowed along with any other gear!

Figure 10.11 New Tyre being fitted (Source: Bosch media)

Vibration conditions 1 Clicking, popping or grinding noises may be noticeable at low speeds and be caused by the following:

▶ inner or outer CV joints worn (often due to lack of lubrication so check for split gaiters);
▶ loose drive shaft;
▶ another component contacting a drive shaft;

▶ damaged or incorrectly installed wheel bearing, brake or suspension component.

Vibration conditions 2 The following may cause vibration at normal road speeds:

▶ out-of-balance wheels;
▶ out-of-round tyres.

The following may cause shudder or vibration during acceleration:

▶ damaged power train/drive train mounts;
▶ excessively worn or damaged out-board or in-board CV joints.

Figure 10.12 Wheel Balancer (Source: Bosch media)

Road test route A vehicle will produce a certain amount of noise. Some noise is acceptable and may be audible at certain speeds or under various driving conditions such as on a new road. One of the ratings that tyres are assessed for at the point of sale to the motorist is noise measured in decibels.

Carry out a thorough visual inspection of the vehicle before carrying out the road test. Keep in mind anything that is unusual. A key point is to not repair or adjust anything until the road test is carried out. Of course this does not apply if the condition could be dangerous or the vehicle will not start.

Establish a route that will be used for all diagnostic road tests. This allows you to get to know what is normal and what is not. The roads selected should have sections that are reasonably smooth, level and

10

free of undulations as well as lesser quality sections needed to diagnose faults that only occur under particular conditions. A road that allows driving over a range of speeds is best. Gravel, dirt or bumpy roads are unsuitable because of the additional noise they produce.

Road test Road test the vehicle and define the condition by reproducing it several times during the road test. During the road test recreate the following conditions.

1 **Normal driving speeds** of 20 to 80 km/h (15 to 50 mph) with light acceleration, a moaning noise may be heard and possibly a vibration is felt in the front floor pan. It may get worse at a certain engine speed or load.

2 **Acceleration/deceleration** with slow acceleration and deceleration, a shake is sometimes noticed through the steering wheel seats, front floor pan, front door trim panels, etc.

3 **High speed** a vibration may be felt in the front floor pan or seats with no visible shake, but with an accompanying sound or rumble, buzz, hum, drone or booming noise. Coast with the clutch pedal down or gear lever in neutral and engine idling. If vibration is still evident, it may be related to wheels, tyres, front brake discs, wheel hubs or wheel bearings.

4 **Engine rpm** sensitive a vibration may be felt whenever the engine reaches a particular speed. It may disappear in neutral coasts. Operating the engine at the problem speed while the vehicle is stationary can duplicate the vibration. It can be caused by any component, from the accessory drive belt to the clutch or torque converter, which turns at engine speed when the vehicle is stopped.

5 **Noise and vibration** while turning clicking, popping or grinding noises may be due to the following: damaged CV joint; loose front wheel half shaft joint boot clamps; another component contacting the half shaft; worn, damaged or incorrectly installed wheel bearing; damaged power train/drive train mounts.

After the road test After a road test, it is often useful to do a similar test on a hoist or lift. When carrying out a 'shake and vibration' diagnosis or 'engine accessory vibration' diagnosis on a lift, observe the following precautions:

▶ If only one drive wheel is allowed to rotate, speed must be limited to 55 km/h (35 mph) indicated on the speedometer. This is because the actual wheel speed will be twice that indicated on the speedometer.

▶ The suspension should not be allowed to hang free. If a CV joint were run at a high angle, extra vibration as well as damage to the seals and joints could occur.

Summary A test on the lift may produce different vibrations and noises than a road test because of the effect of the lift. It is not unusual to find a vibration on the lift that was not noticed during the road test. If the condition found on the road can be duplicated on the lift, carrying out experiments on the lift may save a great deal of time.

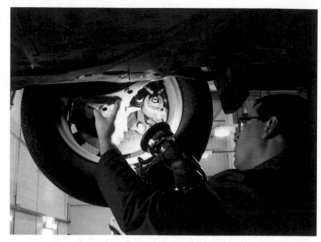
Figure 10.13 Checking suspension

Engine noises How do you tell a constant tapping from a rattle? Worse still, how do you describe a noise in a book? I'll do my best. Try this table as a non-definitive guide to the source or cause of engine or engine ancillary noises:

Noise description	Possible source
Tap	Valve clearances out of adjustment, cam followers or cam lobes worn
Rattle	A loose component, broken piston ring or component
Light knock	Small end bearings worn, cam or cam follower
Deep knock or thud	Big end bearings worn
Rumble	Main bearings worn
Slap	Worn pistons or bores
Vibration	Loose or out of balance components
Clatter	Broken rocker shaft or broken piston rings
Hiss	Leak from inlet or exhaust manifolds or connections
Roar	Air intake noise, air filter missing, exhaust blowing or a seized viscous fan drive
Clunk	Loose flywheel, worm thrust bearings or a loose front pulley/damper
Whine	Power steering pump or alternator bearing

Noise description	Possible source
Shriek	Dry bearing in an ancillary component
Squeal	Slipping drive belt

Sources of engine noise The table shown here is a further guide to engine noise. Possible causes are listed together with the necessary repair or further diagnosis action as appropriate.

Sources of engine noise	Possible cause	Required action
Misfiring/ backfiring	Fuel in tank has wrong octane/cetane number, or is wrong type of fuel	Determine which type of fuel was last put in the tank
	Ignition system faulty	Check the ignition system
	Engine temperature too high	Check the engine cooling system
	Carbon deposits in the combustion chamber start to glow and cause misfiring	Remove the carbon deposits by using fuel additives and driving the vehicle carefully
	Timing incorrect, which causes misfiring in the intake/exhaust system.	Check the timing.
Valve train faulty	Valve clearance too large due to faulty bucket tappets or incorrect adjustment of valve clearance.	Adjust valve clearance if possible and renew faulty bucket tappets – check cam condition
	Valve timing incorrectly adjusted valves and pistons are touching	Check the valve timing and adjust if necessary
	Timing belt broken or damaged	Check timing belt and check pistons and valves for damage – renew any faulty parts
Engine components faulty	Pistons	Disassemble the engine and check components.
	Piston rings	
	Cylinder head gasket	
	Big-end and/or main bearing journals.	
Ancillary components	Engine components or ancillary components loose or broken	Check that all components are secure, tighten/ adjust as required. Renew if broken

> **?** Create a mind map to illustrate the features of a key component or system.

10.1.5 Electrical diagnostics ❸

Check the obvious first. Start all 'hands on diagnostic' routines with 'hand and eye checks'. In other words

look over the vehicle for obvious faults. For example, if the battery terminals are loose or corroded then put this right before carrying out complicated voltage readings. Here are some further suggestions that will at some point save you a lot of time:

▶ A misfire may be caused by a loose plug lead – it is easier to look for this than interpret the ignition waveforms on a scope.

▶ If the ABS warning light stays on – look to see if the wheel speed sensor(s) are covered in mud or oil.

Figure 10.14 Electrical system

Test lights and analog meters – Warning A test lamp is ideal for tracing faults in, say, a lighting circuit because it will cause a current to flow, which tests out high resistance connections. However, it is this same property that will damage delicate electronic circuits – so don't use it for any circuit that contains an ECU.

A digital multimeter is ideal for all forms of testing, most have an internal resistance in excess of 10M, which means that the current they draw is almost insignificant. An LED test lamp or a logic probe is also acceptable.

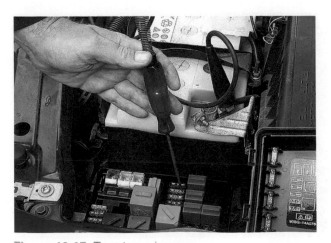

Figure 10.15 Test lamp in use

Generic electrical testing procedure The following procedure is very generic but with little adaptation can be applied to any electrical system. Refer to manufacturer's recommendations if in any doubt.

Volt drop testing Volt drop is a term used to describe the difference between two points in a circuit. In this way we can talk about a voltage drop across a battery (normally about 12.6V) or the voltage drop across a closed switch (ideally 0V but may be 0.1 or 0.2V).

The first secret to volt drop testing is to remember that the sum of all volt drops around a circuit always add up to the supply. The second secret is to ensure the circuit is switched on and operating – or at least the circuit should be 'trying to operate'.

In this picture this means that if the circuit is operating correctly, V1 + V2 + V3 = Vs. When electrical testing therefore, and if the battery voltage is measured at say 12V, a reading of less than 12V at V2 would indicate a volt drop between the terminals of V1 and/or V3. Likewise the correct operation of the switch, that is, it closes and makes a good connection, would be confirmed by a very low reading on V1.

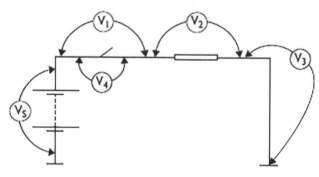

Figure 10.16 Volt drop testing

Bad earth or bad ground What is often described as a 'bad earth' (when what is meant is a high resistance to earth), could equally be determined by the reading

Figure 10 17 Headlight Connections

on V3. To further narrow the cause of a volt drop down, simply measure across a smaller area. The voltmeter V4 for example, would only assess the condition of the switch contacts.

Testing for short circuits to earth This fault will normally blow a fuse – or burn out the wiring completely. To trace a short circuit is very different to looking for a high resistance connection or an open circuit. The volt drop testing above will trace an open circuit or a high resistance connection. My preferred method of tracing a short, after looking for the obvious signs of trapped wires, is to connect a bulb or test lamp across the blown fuse and switch on the circuit. The bulb will light because on one side it is connected to the supply for the fuse and on the other side it is connected to earth via the short circuit fault. Now disconnect small sections of the circuit one at a time until the test lamp goes out. This will indicate the particular circuit section that has shorted out.

Figure 10.18 Short circuit testing

On and off load tests On load means that a circuit is drawing a current; off load means it is not. One example where this may be an issue is when testing a starter circuit. Battery voltage may be 12V (well, 12.6V) off load, but may be as low as 9V when on load (cranking a cold engine perhaps).

A second example is the supply voltage to the positive terminal of an ignition coil via a high resistance connection (corroded switch terminal for example). With the ignition on and the vehicle not running, the reading will almost certainly be battery voltage because the ignition ECU switches off the primary circuit and no volt drop will show up. However, if the circuit were switched on (with a fused jumper lead if necessary) a lower reading would result showing up the fault.

Black box technique The technique outlined here is known as 'black box faultfinding'. This is an excellent technique and can be applied to many vehicle systems from engine management and ABS to cruise control and instrumentation.

As most systems now revolve around an ECU, the ECU is considered to be a 'black box', in other words we know what it should do but the exact details of how it does it are less important.

Treating the ECU as a 'black box' allows us to ignore its complexity. The theory is that if all the sensors and associated wiring to the 'black box' are OK, all the output actuators and their wiring are OK and the supply/earth (ground) connections are OK, then the fault must be the 'black box'. Most ECUs are very reliable however and it is far more likely that the fault will be found in the inputs or outputs.

Sensors and Actuators Normal faultfinding or testing techniques can be applied to the sensors and actuators. For example, if an ABS system uses four inductive type wheel speed sensors, then an easy test is to measure their resistance. Even if the correct value were not known, it would be very unlikely for all four to be wrong at the same time so a comparison can be made. If the same resistance reading is obtained on the end of the sensor wires at the ECU then almost all of the 'inputs' have been tested with just a few ohmmeter readings.

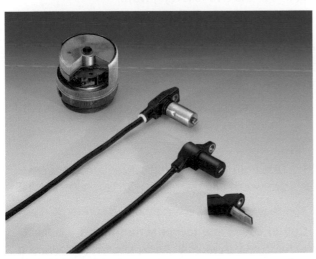

Figure 10.19 Wheel speed sensors (Source: Bosch Media)

Problems! The same technique will often work with 'outputs'. If the resistance of all the operating windings in say a hydraulic modulator were the same, then it would be reasonable to assume the figure was correct. Sometimes however, it is

almost an advantage not to know the manufacturers recommended readings. If the 'book' says the value should be between 800 and 900Ω, what do you do when your ohmmeter reads 905Ω? Answers on a postcard please…

Finally, don't forget that no matter how complex the electronics in an ECU, they will not work without a good power supply and an earth.

Figure 10.20 Hydraulic Modulator (Source: Bosch Media)

Sensor to ECU method This technique is simple but very useful. The picture here shows a resistance test being carried out on a component. Ω_1 is a direct measure of its resistance whereas Ω_2 includes the condition of the circuit. If the second reading is the same as the first then the circuit must be in good order.

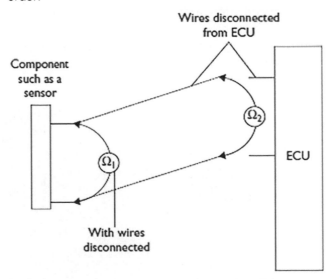

Figure 10.21 Ohmmeter testing. Warning: The circuit supply must always be off when carrying out ohmmeter tests.

Flight recorder tests 1 It is said that the best place to sit in an aeroplane is on the black box flight recorder.

Personally, I would prefer to be in 'first class'! Also, apart from the black box usually being painted bright orange so it can be found after a crash – my reason for mentioning it is to consider how the flight recorder principle can be applied to automotive diagnostics.

Most digital oscilloscopes have flight record facilities. This means that they will save the signal from any probe connection in memory for later play back. The time duration will vary depending on the available memory and the sample speed, but this is a very useful feature.

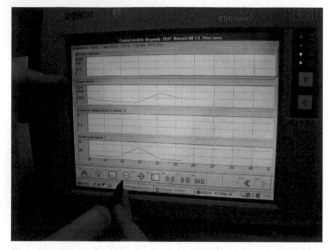

Figure 10.22 Recorded Data.

Flight Recorder Tests 2 As an example, consider an engine with an intermittent misfire that only occurs under load. If a connection is made to the suspected component (coil HT output for example), and the vehicle road tested, the waveforms produced can be examined afterwards. Many engine (and other system), ECUs have built in flight recorders in the form of self-diagnostic circuits. If a wire breaks loose causing a misfire but then reconnects the faulty circuit will be 'remembered' by the ECU.

Figure 10.23 Electric window switches

Faultfinding by luck – Or is it logic? Actually, what this section considers is the benefit of playing the odds which, while sometimes you get lucky, is still a logical process. If four electric windows stopped working at the same time, it would be very unlikely that all four motors had burned out. On the other hand if just one electric window stopped working, then it may be reasonable to suspect the motor. It is this type of reasoning that is necessary when fault finding. However, be warned it is theoretically possible for four motors to apparently burn out all at the same time.

Playing the odds 1 Using this 'playing the odds' technique can save time when tracing a fault in a vehicle system. For example, if both stop lights do not work and everything else on the vehicle was OK, I would suspect the switch (stages one to three of the six-stage process). At this stage though, the fault could be anywhere – even two or three blown bulbs. Nonetheless a quick test at the switch with a voltmeter would prove the point. Now, let's assume the switch is OK and it produces an output when the brake pedal is pushed down. Testing the length of wire from the front to the back of the vehicle further illustrates how 'luck' comes into play.

Playing the odds 2 Figure 10.24 represents the main supply wire from the brake switch to the point where the wire 'divides' to each individual stop light (the odds say the fault must be in this wire). For the purpose of this illustration we will assume the open circuit is just before point 'I'. The procedure continues in one of the two following ways:

One

▶ Guess that the fault is in the first half and test at point F.
▶ We were wrong. Guess that the fault is in the first half of the second half and test at point I.
▶ We were right. Check at H and we have the fault . . . On test number THREE

Two

▶ Test from A to K in a logical sequence of tests.
▶ We would find the fault ... On test number NINE

You may choose which method you prefer.

Colour codes and terminal numbers 1 It is useful to become familiar with a few key wire colours and terminal numbers when diagnosing electrical faults. As seems to be the case for any standardisation a number of colour code systems are in operation.

A system used by a number of manufacturers is based broadly on the information in the given table. After some practice with the use of colour codes the job of the technician is made a lot easier when fault finding an electrical circuit.

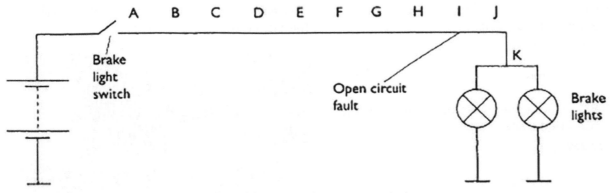

Figure 10.24 Faultfinding by playing the odds – sometimes you get lucky

Colour	Symbol	Destination/Use
Red	Rt	Main battery feed
White/Black	Ws/Sw	Headlight switch to dip switch
White	Ws	Headlight main beam
Yellow	Ge	Headlight dip beam
Grey	Gr	Side light main feed
Grey/Black	Gr/Sw	Left hand side lights
Grey/Red	Gr/Rt	Right hand side lights
Black/Yellow	Sw/Ge	Fuel injection
Black/Green	Sw/Gn	Ignition controlled supply
Black/White/Green	Sw/Ws/Gn	Indicator switch
Black/White	Sw/Ws	Left side indicators
Black/Green	Sw/Gn	Right side indicators
Light Green	LGn	Coil negative
Brown	Br	Earth
Brown/White	Br/Ws	Earth connections
Pink/White	KW	Ballast resistor wire
Black	Sw	Reverse
Black/Red	Sw/Rt	Stop lights
Green/Black	Gn/Sw	Rear Fog light

Colour codes and terminals 2 A system now in use almost universally is the terminal designation system in accordance with DIN 72 552. This system is to enable easy and correct connections to be made on the vehicle, particularly in after sales repairs. Note that the designations are not to identify individual wires but are to define the terminals of a device. Listed here are some of the most popular numbers.

1	Ignition coil negative
4	Ignition coil high tension
15	Switched positive (ignition switch output)
30	Input from battery positive
31	Earth connection
49	Input to flasher unit

49a	Output from flasher unit
50	Starter control (solenoid terminal)
53	Wiper motor input
54	Stop lamps
55	Fog lamps
56	Headlamps
56a	Main beam
56b	Dip beam
58L	Left side lights
58R	Right side lights
61	Charge warning light
85	Relay winding out
86	Relay winding input
87	Relay contact input (change over relay)
87a	Relay contact output (break)
87b	Relay contact output (make)
L	Left side indicators
R	Right side indicators
C	Indicator warning light (vehicle)

Colour codes and terminals 3 Ford Motor Company, and many others, use a circuit numbering and wire identification system. Used worldwide, it is known as Function, System-Connection (FSC). The system was developed to assist in vehicle development and production processes. It is also useful to help the technician with fault finding. Many of the function codes are based on the DIN system. Note that earth wires are now black. The system works as shown here:

31S-AC3A || 1.5 BK/RD

Function:
  31 = ground/earth
  S = additionally switched circuit
System:
  AC = headlamp levelling

497

Connection:

    3 = switch connection
    A = branch

Size:

    1.5 = 1.5mm²

Colour:

    BK = Black (determined by function 31)
    RD = Red stripe

Colour codes

Code	Colour
BK	Black
BN	Brown
BU	Blue
GN	Green
GY	Grey
LG	Light-Green
OG	Orange
PK	Pink
RD	Red
SR	Silver
VT	Violet
WH	White
YE	Yellow

Ford system codes It should be noted that the colour codes and terminal designations given in this section are for illustration only.

Letter	Main system	Examples
D	Distribution systems	DE = earth
A	Actuated systems	AK = wiper/washer
B	Basic systems	BA = charging BB = starting
C	Control systems	CE = power steering
G	Gauge systems	GA = level/pressure/temperature
H	Heated systems	HC = heated seats
L	Lighting systems	LE = headlights
M	Miscellaneous systems	MA = air bags
P	Powertrain control systems	PA = engine control
W	Indicator systems ('indications' not turn signals)	WC = bulb failure
X	Temporary for future features	XS = could mean too much?

Back probing connectors If you are testing for a supply (for example) at an ECU, then use the probes of your digital meter with care. Connecting to the back of the terminals will not damage the connecting surfaces providing excessive force is not applied. Sometimes, a pin clamped in the test lead's crocodile/alligator clip is ideal for connecting 'through' the

insulation of a wire without having to disconnect it. This picture shows this technique.

Figure 10.25 Test the voltage by back probing a connector with care

Summary The key to electrical diagnostics, as with all other systems, is to work methodically and logically. And finally, remember to take care making test connections so that you do not cause more faults as you carry out tests!

 Make a simple sketch to show how one of the main components or systems in this section operates.

10.1.6 Real world diagnostics ❸

Introduction This section will look at carrying out diagnostics in the real world. It is not intended to be the definitive answer, rather I hope it will make you think about the subject and how you go about it, as well as customer expectation and how to manage their experience.

'Just plug it in and the computer will tell you what's wrong!'

What do you know? When faced with a fault on a vehicle we have to make decisions about what to do first, and then what to do next! Let's take a simple example where the brake lights are not working. It is very easy to start thinking about all the things it could be, but at this stage we need to find out some facts. In other words, build up the information from what we know, NOT from what we don't know. To do this, observe or measure something, consider what the observation means or compare the measurement to data. Then, based on your conclusions so far, take the next step.

Figure 10.26 This process can be thought of as: Sense, think, act

Example In our simple example, after having observed that none of the brake lights work, we would probably measure the fuse. I used the word measure on purpose because observing a fuse is often not good enough. Switch on the ignition and check both sides of the fuse with a test lamp for example. Assuming that this is ok we now have some facts to build upon. We would probably decide to check the supply into, and out of the brake light switch next. Based on this result we could then decide the next step and so on. Each time building up what we know about the fault.

Training There are many excellent training and CPD courses available to help you practice and develop your diagnostic skills. You can learn a lot from online content, magazine articles and textbooks. An experienced teacher can be valuable, but the absence of one does not stop the learning process and self-learning can help you get the best experience from a taught environment.

Figure 10.27 PicoScope connected to view waveforms (Source: PicoScope Media)

The customer Let's start with the most important point about our customers: without them we would not have a job. That said, some can be challenging to work with! The answer to this is to be very clear about what you will be doing and not doing as part of a diagnostic process. Also, be willing to explain things to help them understand. One key skill to develop is how to get the appropriate information from a customer. For example, if a driver says: '*My car won't start*', this could mean several things, but three in particular:

▶ The starter is not operating
▶ The starter is operating but not properly
▶ The starter is operating properly but the engine will not run

Cost In this example we would need to narrow it down by talking to them. Another really important customer-skill is to be very clear about what you will spend and what you will cost, and at what point you will stop and consult them further.

Relationships A third customer relationship skill is to understand that their lack of knowledge means they can come to the wrong conclusion over something, or worse, there are some who try and take advantage of a situation. How many times have you heard something like this? '*Ever since you did X to my car Y has stopped working.*' To help counter some of these issues, a set diagnostic routine or sequence can be very useful, along with records of what job was performed 'last time' and when!

Figure 10.28 ArtiPad scanner in use

How much does diagnostics cost? Multimeters, oscilloscopes, scanners and lots of other equipment are needed in our armoury to diagnose faults. This

Figure 10.29 Voltage reading at the fuse/relay distribution box

10

equipment costs serious money and it is important to remember this when pricing diagnostic work. We also need to generate our wages and a suitable profit margin, whether employed by an organisation, or self-employed. Equipment and training should be thought of as an investment, not a cost, so to get the return on this investment, it is essential to charge appropriately for diagnostic work and to take into account any calibrations, software updates and insurance costs to maintain your kit.

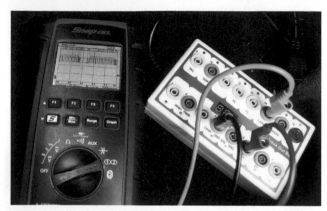

Figure 10.30 CAN high and low on a scope using a breakout box connected to the diagnostic socket

Figure 10.31 Test leads and adapters: at the time of writing, even this simple set of leads cost over £50

How much should you charge? Prices change and vary depending on geographic area, reputation, volume of demand, and over time. The figures stated here are just an outline guide, but make sure you charge what you are worth. If a customer is not willing to pay for proper diagnostic work, then they may not be the type of customer you want.

One thing is easy to state as a given, and that is diagnostic charges should be fixed NOT time related (don't diagnose at an hourly rate). This is because the more money you invest in equipment, the more experience and skill you develop through attending and paying for courses, for example, the quicker you will become at finding the fault. A time related charge would mean you earn less, the better at the job you get! Let's assume you currently have an hourly rate of X. We can compare this with a two tier method of charging.

Example method A useful way of pricing diagnostics is to have two tiers:

▶ **Tier one: 1.25 times X (for example):** For this you will carry out non-invasive diagnostic work that does not involve removing any components other than maybe engine covers or similar. It also does not involve complex testing using specialist equipment. At the end of this process you will be able to state the fault OR will be able to explain to the customer why the next tier (level) is needed. Also explain that this may involve removing some components that have a risk of breakage. The diagnostic sequence in Table 10.2 below is an example of a tier one process. Most faults will be traced at tier 1, but repairs will cost extra.

▶ **Tier two: 2 times X:** At this level, and with the customer's permission to proceed, we can start digging much deeper. This will involve carrying out more invasive measurements that may involve some dismantling work. Even at this stage, the cost of the actual repair may not be included, unless it is relatively simple to carry out. At the end of this process you would guarantee to either have fixed the fault or be able to state exactly what is needed.

Diagnostic sequence Every set of symptoms you come across will be different and so the diagnostic routine will need to adapt to find the fault. However, it is useful to have a generic sequence in mind. Doing this means you cover all the steps but also cover your back if needed! Further, there could be opportunities for generating more work. Let's take an example where all we know at this stage is that the malfunction indicator lamp (MIL) is on. Remember, this is a guide to make you consider different sequences and methods, it is not a definitive routine. It follows the six-stages of diagnosis we examined earlier, with an extra one at the start (prepare) and at the end (report). I have estimated the time each step will take, but again this is a rough guide only.

Figure 10.32 Malfunction indicator lamp (MIL) or check engine warning light

Routine The routine here totals to about 50 minutes, but in many cases, it will be less. Your experience will tell you when certain tests, like the battery charging and starting check, could be missed out, if you know of and normally service the vehicle for example.
It is good practice to keep a note of what you do. Sometimes it may take longer, but on average you should spend less than an hour on tier 1.

Table 10.2 Diagnostic sequence (tier 1)

Steps	Comments	Time (minutes)
Prepare Check fluids	This is to make sure it is safe to operate. In some cases, it may be necessary to run the engine at high speed, so better to be safe than sorry. It may also start to give you some information, if the oil is low and very dirty it may indicate lack of servicing, or an absence of fuel or brake fluid may indicate a problem.	2
Carry out battery, charging and starting test	Not essential but if the tests you carry out require the engine to be cranked over several times, a starter motor on its last legs could fail altogether – better to know before you start. Once again, before we even begin looking in any detail, this stage does give us more information. I use a device for this test that allows you to print out the results.	5
Connect battery saver	Any test routine will involve using up the battery to some extent. A suitable battery saver will ensure that it remains at a suitable level.	1
Verify the symptoms	In this case it is easy, the warning light is on. It is still worth running and listening to the engine at idle, low and high speeds because there may be other symptoms that the customer did not notice like excess smoke, or a misfire.	2
Collect more information	Is there a service history or has any other work done? Access manufacturer's or other form of data.	5
Carry out a full scan for DTCs	Most scanners will allow you to save and/or print this file. Doing this regardless of the symptoms gives you more detailed information about the fault but it also acts as a record in case the: 'Ever since you…' comment is used! Let's assume an EGR diagnostic trouble code (DTC) is shown, which cause the MIL to be illuminated. Also present is a braking system fault.	5
Hand and eye (and ear!) checks	Look and listen for anything obvious, loose wires, loose connections, oil leaks, noises etc., particularly in areas related to the DTC.	2
Evaluate	Stop and think to decide what specific tests should now be done.	3

Steps	Comments	Time (minutes)
Test Check live data	In this case we would probably look at the signal or voltages to and from the EGR sensor. Compare any readings to manufacturer's date and look for anomalies: Does something look like an apple when it should be an orange?!	5
Carry out cylinder balance check	This can be done if appropriate to the symptoms, but also as a generic process to learn more and allow you to provide detailed information about the condition of the engine/vehicle to the customer. For example, if there is a serious loss of compression on one cylinder, will they really want you to chase down a smaller EGR fault? Some diagnostic technicians choose to carry out this test in the preparation stage. It can even be done using a DLC breakout box and by measuring voltage drop on a scope from inside the vehicle. However, the ideal is to measure starter current using an amp clamp and a scope. Compare the current on each cylinder for balance. If the average current is about three times the battery Ah capacity, then the compression is probably ok. (Note, there is an excellent free phone app from Exide that gives useful data about what battery should be on a vehicle)	5
Guided by any DTCs and the live data, carry out appropriate tests on sensors, actuators and wiring.	This can be done with a multimeter set to measure voltage. If very detailed checks with a multichannel oscilloscope are necessary, it may be appropriate to stop at this stage and report to the customer that you need to go to tier two.	5
Rectify	A sensor may need to be fitted or an actuator removed and cleaned. If a simple wiring fault you may be able to fix the fault but remember the diagnostic routine is just that, it does not cover the cost of repairs. If appropriate, clear the DTC.	0
Check	Physically check all areas of the vehicle where you have worked (and more) but also carry out another full scan for DTCs. This then shows either that you have rectified the fault or not caused any more compared to the first scan.	5
Report	Let the customer know what has been done, what has or has not been fixed and what your recommendations are, for example the brake system fault may need to be followed up depending on its priority.	5

10

Figure 10.33 Scanner health report showing two DTCs

Select a routine from section 1.3 and follow the process to study a component or system.

10.1.7 Pass-through overview 3

Introduction A pass-through device is used in conjunction with a computer to reprogram vehicle control modules through the OBD-II/CANbus port. It is sometimes necessary to reprogram ECUs to regulate and repair vehicles equipped with OBD systems, which do not conform with pollution emission values. Alternatively, these ECUs, or others, may require updating to improve other functions or to recognise new components that have been fitted.

Each manufacturer has their own methods, but SAE International standardised the J-2534 universal requirements in 2004. This required all manufacturers of vehicles sold in the USA and Europe to accept powertrain reprogramming through specific universal parameters. In the USA and Europe vehicle manufacturers must therefore provide ECU reprogramming functionality to all workshops, whether independent or franchised.

Figure 10.34 Battery, charging and starting tester (Source: Bosch Media)

Figure 10.36 Pass through II device (Source: Snap-on)

Standards Because of the existence of pass-through, vehicle manufacturers have had to ensure that the reprogramming software applications (APIs) are compatible with standardised J2534 vehicle communication interfaces (VCIs). Because of this, independent workshops can access OEM applications by subscribing to the appropriate websites. By downloading the software to a PC, and connecting it to the vehicle with a J2534 VCI, you have the same level of access to the vehicle as a main dealer.

In summary, to reprogram a vehicle ECU you need:

1 Computer equipped with a Windows operating system
2 J2534 vehicle communication interface (VCI)
3 OEM application programming interface (API)
4 Knowledge of how to use the software!

Figure 10.35 GYS Battery Charger/Saver connected to a vehicle

Summary Diagnostics is not the easiest part of our job, but it can be most satisfying. It can also generate a good income if carried out correctly and charged for accordingly. It is also an opportunity to generate more work if communication with the customer is good or as a service to other businesses not strong in this area.

'The computer does not tell you exactly what is wrong when you plug it in, it is just the first step in a process!'

of the cable between the J2534 device and the vehicle is 5 metres. If the vehicle manufacturer doesn't use DLC, necessary information for connection must be provided.

Figure 10.39 1962 data link connector (DLC)

Reprogramming an ECU using J2534 is done from a PC, preferably a laptop computer, with a Windows operating system. Each vehicle manufacturer has their own software application (API) used for analysing and programming their vehicles. The application will have complete information on the ECUs that are supported by it. The application also includes a user interface where choices can be made, depending on the ECU, and what action to perform.

Summary The intention is that every J2534 tool should be capable of communicating with all protocols supported by the standard. The connection and initialisation process starts by information being sent to the hardware tool, about which protocol is being used. Thereafter it is up to the hardware tool to manage the connection to the vehicle with the desired protocol. The PC application will send messages in the earlier determined protocol format to the hardware tool, which buffers the messages and transmits them in the order they were received.

 Use a library or the web search tools to further examine the subject in this section.

10.2 Diagnostic tools and methods

10.2.1 Multimeters ❷ ❸

Introduction There are lots of different options or settings available when using a multimeter, but the three most common measurements are: voltage (volts), resistance (ohms) and current (amps).

Voltage To measure voltage the meter is connected in parallel with the circuit. Most voltage

Figure 10.37 Bosch pass-through device (Source: Bosch Media)

Hardware The J2534 hardware works like a bridge between the vehicle ECU and the PC. This pass-through device translates messages sent from the PC into messages with the same protocol being used in the vehicle ECU. J2534 supports a range of protocols.

The connection between the PC and the J2534 hardware can be chosen by the manufacturer of the device, for example, RS-232, USB or a wireless interface. The vehicle manufacturer's programming application is not dependent on the hardware connection. Therefore, any device can be used for programming any vehicle regardless of the manufacturer.

The connection between the J2534 hardware and the vehicle should be the SAE J1962 connector, more commonly called the OBD connector or the data link connector (DLC). The maximum recommended length

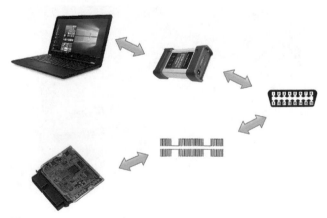

Figure 10.38 Pass-through principle and stages

measurements on a vehicle are DC. Remember to set the range of the meter (some are auto-ranging) and if in doubt, start with a higher range and work downwards.

Figure 10.40 Voltage supply to a fuse box

Resistance To measure resistance the meter must be connected across (in parallel with) the component or circuit under test. However, the circuit must be switched off or isolated. If not, the meter will be damaged. Likewise, because an ohmmeter causes a current to flow, there are some circuits such as Hall effect sensors, that can be damaged by the meter.

Figure 10.41 Checking a simple resistor

Current can be measured in two ways:

1 Connecting the meter in series with the circuit (in other words break the circuit and reconnect it through the meter)

2 Using an inductive amp clamp around the wire, which is a safer way to measure, but less accurate at low values.

Figure 10.42 Inductive ammeter clamp on a high voltage cable (measuring the current drawn by the EV cabin heater)

What is an open circuit voltage? This internal resistance of a meter can affect the reading it gives on some circuits. It is recommended that this should be a minimum of 10MΩ, which ensures accuracy because the meter only draws a very tiny (almost insignificant) current. This stops the meter loading the circuit and giving an inaccurate reading, and it prevents damage to sensitive circuits (in an ECU for example).

However, the very tiny current draw of a good multimeter can also be a problem. A supply voltage of say 12V, can be shown on a meter when testing a circuit, but does not prove the integrity of the supply. This is because a meter with a 10MΩ internal resistance connected to a 12V supply, will only cause a current of 1.2µA (I=V/R) – that's 1.2 millionths of an amp, which will not cause any noticeable voltage loss even if there is an unwanted resistance of a few thousand ohms in the supply circuit. A test lamp can be connected in parallel with the meter to load the circuit (make more current flow), but should be used

Figure 10.43 Ghost voltage caused by shaking the red lead

carefully so you don't damage sensitive electronic switching circuits that may be present.

How do you know zero is zero? Voltmeters can display a 'ghost' voltage rather than zero when the leads are open circuit. In other words, if checking the voltage at an earth/chassis connection we would expect a 0V reading. However, the meter will also display zero before it is connected, so how do we know the reading is correct, when it is connected? The answer is to shake the multimeter leads, a 'ghost' voltage will fluctuate, a real voltage will not!

Voltage drop testing example Shown here is a slightly simplified lighting circuit. The symptom on the vehicle is that the LH side high beam headlight is dimmer than the RH side. The voltmeter negative lead is connected to a good earth, on the battery negative terminal (position marked 0). The ignition and the lights are switched on and the following voltage readings are obtained when the meter positive lead is connected to the numbered locations:

1 12.62V
2 12.59V
3 12.58V
4 12.52V
5 4.33V
6 0.06V (while shaking the leads!)

What is the fault? What were the reasons for testing voltage at points 1 and 2?

Answer: The fault is a high resistance in the earth wire between 5 and 6. Testing battery voltage (1) is a reference for all other tests, testing the voltage at 2 means you can compare it with the voltage at 3.

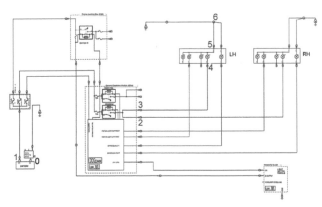

Figure 10.44 Simplified lighting circuit from a Ford vehicle (Source: Ford Motor Company)

 Look back over the previous section and write out a list of the key bullet points from this section.

10.2.2 Oscilloscopes ❷❸

Introduction An oscilloscope (often shortened to scope) is an instrument used to display amplitude and period of a signal, as well as its shape. Amplitude is its height, and period is the time over which it repeats. A scope draws a two-dimensional graph. Normally, voltage is displayed on the vertical, or Y-axis and time on the horizontal, or X-axis.

Settings The settings for these can be changed and are usually described as volts per division (volts/div) and time per division (time/div). The graphs can be moved up and down on the screen with a control known as Y-shift. This means that the zero voltage position can, for example, be set to the middle of the screen. This means that the blue line in this image shows a DC voltage of 0V. The red line represents a DC voltage of about 12V because the volts per division (each small vertical square) in this case, is set at 5V. The green trace shows an AC signal that, assuming the volts/div is still set at 5V, has a peak-to-peak voltage of 20V.

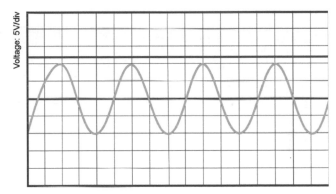

Figure 10.45 Oscilloscope screen

Information Much more information is shown compared to other instruments such as multimeters or frequency meters. For example, when using a scope, you can determine the amplitude and shape, how much noise is present and the frequency of a signal.

The frequency of a signal can be worked out from the time per division settings. In the image show, the time/div (each small horizontal square), is set to 25ms. This means that the sinewave repeats every 100ms (four divisions). Because frequency is measured in cycles per second (hertz or Hz), we simply need to divide 1s (1000ms) by the time period, 100ms in this case, which means that this signal has a frequency of 10Hz.

As well as directly displaying electric signals, scopes can measure non-electrical values if appropriate transducers are used. Transducers change one kind of variable into another. For example, a pressure transducer (often described as a sensor), produces a voltage that is proportional to pressure.

Time base The speed at which the trace moves across the screen is known as the time base, which can be adjusted either automatically or manually. The start of the trace moving across the screen is known as the trigger. This can be internal, such that it flies back and re-starts every two seconds or whatever, or it can be external, so it starts every time a fuel injector operates for example. The voltage from the item under test can either be amplified or attenuated (reduced), much like changing the scale on a voltmeter.

Recording Now almost all automotive oscilloscopes are digital and use a computer screen to display signals. This also allows values such as the voltage and time base scales, and frequency to be shown on the screen. It also means the waveforms can be saved and shared. waveforms gathered over a period of time can also be saved so that they can be replayed after, for example, a road test.

High voltages and safety Some scopes such as the Snap-on Verdict shown below have a CAT rating (see the section on multimeters for more information), so it can be used directly with higher voltages. However, in most cases when high voltages are measured, special attenuating leads are used. These reduce the voltage by, say, a factor of ten so a 300V signal would be attenuated to 30V before entering the instrument. The scale is then set and adjusted accordingly.

AC-DC In most cases a scope is set so as to measure DC voltage, even though it may appear to be showing an AC signal! This is because at any point in time it is actually measuring DC – it is just that this DC voltage varies quickly. However, for some measurements it is useful to display and measure just the AC component of a signal. This is known as AC coupling. The voltage across the battery on a car is normally about 14V when the engine is running, and the alternator is charging the battery. However, because the alternator rectifies AC into DC and the voltage regulator controls voltage, the voltage across the battery is 14V but with very small variations. The figure below shows this variation (ripple) to be about 0.2V. If the display

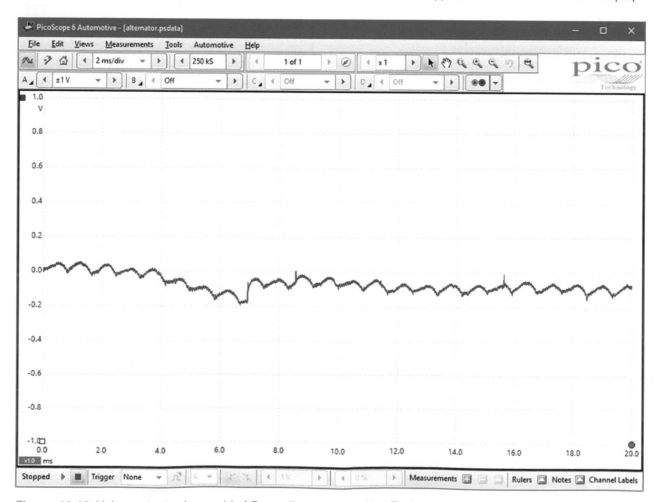

Figure 10.46 Voltage ripple show with AC coupling (Source: Pico Technology Ltd., www.picoauto.com)

was set to show the DC components, then the scale would not be sensitive enough to show details of the 0.2V ripple.

Channels The number of traces that can be displayed on an oscilloscope at the same time is known as the number of channels. Most scopes designed for automotive use will have a minimum of two channels and more often four. This is so that different waveforms (traces, patterns, signals etc.) can be displayed at the same time, using different voltage scales if necessary, so that comparisons can be made. Figure 10.49 shows a fuel injector signal compared to the fuel pressure. Figure 10.50 shows three signals being compared: injector voltage, secondary waveform and intake manifold pressure.

Figure 10.47 Pico 8-channel scope (Source: Pico Technology Ltd., www.picoauto.com)

Making connections Modern vehicle wiring harnesses and associated plugs and sockets are very well sealed to keep out water (well most are!).

Figure 10.48 Adapter leads to make a secure connection to a lambda sensor

This means that it is sometimes difficult to make a connection with scope leads. It is very important not to damage connections when taking a reading, and unless absolutely essential, the insulation on a wire should not be pierced.

Non-invasive measurement An invasive measurement is when dismantling is required to connect and measure something. It can also mean that the act of measuring affects the reading you get. Non-invasive measurement is a great way to get a first impression of a vehicle without removing any major items which introduce a risk of damage. A good example of this is to carry out a relative compression test. This is done by measuring either starter current or battery voltage while the engine is cranking. The voltage and time scales are set so that the current flow or voltage drop during compression of each cylinder can be captured and compared.

Wiggly lines The phrase 'wiggly lines' is an affectionate term for waveforms or scope traces! On multichannel scopes comparing one waveform with another is a very useful diagnostic technique. In this image for example, the injector signal is being compared with a fuel pressure signal. In this case a small drop in pressure is shown as the injector fires.

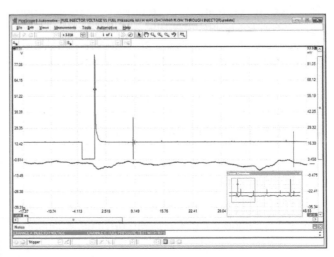

Figure 10.49 Fuel injector voltage vs. fuel pressure (Source: Pico Technology Ltd., www.picoauto.com)

Comparing signals In Figure 10.50, the injector waveform, secondary waveform (the high voltage that creates the spark), and intake manifold pressure are being compared. This is a great way to check for correct operation of the electrical components, but also by comparing each cylinder, it can also tell you a lot about engine condition generally and valve condition in particular.

10

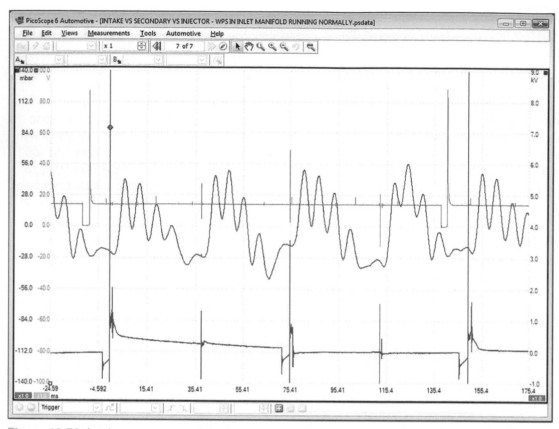

Figure 10.50 Intake pressure vs. injector voltage vs. secondary waveform (Source: Pico Technology Ltd., www. picoauto.com)

PicoScope There are a range of PicoScope automotive diagnostic kits. They integrate with software on a PC to measure and test virtually all of the electrical and electronic components and circuits in any modern vehicle. Common systems to measure include:

▶ Ignition (primary and secondary)
▶ Injectors and fuel pumps
▶ Batteries, alternators and starter motors
▶ Lambda, airflow, ABS and MAP sensors
▶ Electronic throttle control
▶ CAN bus, LIN bus and FlexRay

A two or four-channel PicoScope is recommended for general workshop use. On these devices, a separate ground connection is used for each channel and the instrument is protected up to 200V.

Software The software version at the time of writing is version 6 and version 7 is in beta. Features such as auto setup, tutorials and guided tests are available to get new users up to speed. Advanced features such as math channels, waveform buffers, advanced triggers and reference waveforms ensure

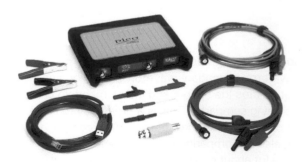

Figure 10.51 PicoScope 2-channel starter kit (Source: Pico Technology Ltd., www.picoauto.com)

Figure 10.52 PicoScope 4-channel master kit (Source: Pico Technology Ltd., www.picoauto.com)

the experienced user will not run out of power. The regular software updates include new features and new tests, and are free for the life of the product.

With sampling rates of up to 400 million samples per second, it is possible to capture complex automotive waveforms, and then zoom in on areas of interest. Being PC-based these waveforms can then be saved for future reference, printed or emailed, and are also easily shared within the PicoScope forum for peer analysis.

PicoScope features The PicoScope is a favourite device for all complex work because of the advanced features previously mentioned. Key benefits include:

▶ The software is intuitive and regularly updated
▶ Guided testing and auto-setup mean new users learn quickly
▶ A database of known good waveforms is available for making comparisons
▶ Automotive options range from basic kits (at a price point to get everyone started), through to master kits that contain everything needed for even the most complex diagnostic work
▶ Maths channel options for advanced users.

 Use the media search tools to look for pictures and videos relating to the subject in this section.

10.2.3 CAT Ratings ❸

Introduction Meters and their leads have category ratings that give the voltage levels up to which they are safe to use. CAT ratings can be a little confusing but there is one simple rule of thumb: Select a multimeter rated to the highest category in which it could possibly be used. In other words, err on the side of safety. The table below lists some of the different ratings.

Table 10.3 CAT ratings

Category	Working voltage (voltage withstand)	Peak impulse (transient voltage withstand)	Test source impedance
CAT I	600V	2500V	30Ω
CAT I	1000V	4000V	30Ω
CAT II	600V	4000V	12Ω
CAT II	1000V	6000V	12Ω
CAT III	600V	6000V	2Ω
CAT III	1000V	8000V	2Ω
CAT IV	600V	8000V	2Ω

Voltages The voltages listed in Table 10.3 are those that the meter will withstand without damage or risk to the user. A test procedure (known as IEC 1010) is used, and takes three main criteria into account:

▶ steady-state working voltage
▶ peak impulse transient voltage
▶ source impedance

These three criteria together will tell you a multimeter's true '*voltage withstand*' values. However, this is confusing because it can look as if some 600V meters offer more protection than 1000V ones. Impedance is the total opposition to current flow in an AC circuit (in a DC circuit it is described as resistance).

Impedance Within a category, a higher working voltage is always associated with a higher transient voltage. For example, a CAT III 600V meter is tested with 6000V transients while a CAT III 1000V meter is tested with 8000V transients. This indicates that they are different, and that the second meter clearly has a higher rating. However, the 6000V transient CAT III 600V meter and the 6000V transient CAT II 1000V meter are not the same even though the transient voltages are. This is because the source impedance has to be considered.

Figure 10.53 Cat III 1000V and CAT IV 600V meter

Ohm's Law (I = V/R) shows that the 2W test source for CAT III will have six times the current of the 12W test source for CAT II. The CAT III 600V meter therefore offers better transient protection, compared to the CAT II 1000V meter, even though in this case the voltage rating appears to be lower. The combination of working voltage and category determines the total 'voltage withstand' rating of a multimeter (or any other test instrument), including the very important 'transient voltage withstand' rating.

For working on vehicle high voltage systems, you should choose a calibrated CAT III or CAT IV meter AND leads.

Figure 10.54 Cat III 1000V and CAT IV 600V leads

 Using images and text, create a short presentation to show how a component or system works

10.2.4 Other test equipment 2 3

Insulation testing An insulation tester does exactly as its name suggests. This test is particularly used for electric and hybrid vehicles. The device shown here

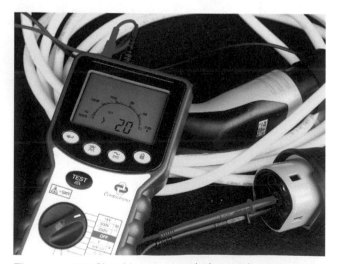

Figure 10.55 Checking the insulation resistance between conductors in an EV charging lead (in this case the reading is greater than 20GΩ)

is known as a Megger and uses 1000V to test the resistance of the insulation on a wire or component. A reading in excess of 10 megaohms is typical. The high voltage is used because it puts the insulation under pressure and will show up faults that would not be apparent if you used an ordinary ohmmeter.

Figure 10.56 Megger multimeter and insulation tester

Engine analysers Some form of engine analyser was an essential tool for fault finding modern vehicle engine systems. The latest machines are now based around a personal computer. This allows more facilities that can be added to by simply changing the software.

Whilst engine analysers are designed to work specifically with the motor vehicle, it is worth remembering that the machine consists basically of three parts.

▶ Multimeter
▶ Gas analyser
▶ Oscilloscope

However, separate systems such as the Pico Automotive kit will now do as many tests as the engine analyser, currently with the exception of exhaust emissions.

four gasses. The Greek symbol lambda (λ) is used to represent the ideal air fuel ratio (AFR) of 14.7:1 by mass. In other words, just the right amount of air to burn up all the fuel. Table 10.4 lists gas, lambda and AFR readings for a closed loop lambda control system, before (or without) and after the catalytic converter. These are for a modern engine in excellent condition and are a guide only – always check current data for the vehicle you are working on.

Table 10.4 Exhaust examples

Reading:	CO%	HC ppm	CO_2%	O_2%	Lambda(l)	AFR
Before catalyst	0.5	100	14.7	0.7	1.0	14.7
After catalyst	0.1	12	15.3	0.1	1.0	14.7

The composition of exhaust gas is now a critical measurement, and hence a certain degree of accuracy is required. To this end the infrared measurement technique has become the most suitable for CO, CO_2 and HC. Each individual gas absorbs infrared radiation at a specific rate. Oxygen is measured by electro-chemical means in much the same way as the on vehicle lambda sensor. Accurate measurement of exhaust gas is not only required for annual tests, but is essential to ensure an engine is correctly tuned.

Exhaust analyser There are lots of different pieces of equipment that can be used to measure exhaust emissions. Some fixed and some mobile and battery operated.

The mobile equipment shown here as uses Bluetooth connectivity and lithium-ion batteries. It will test petrol/gasoline emissions and diesel.

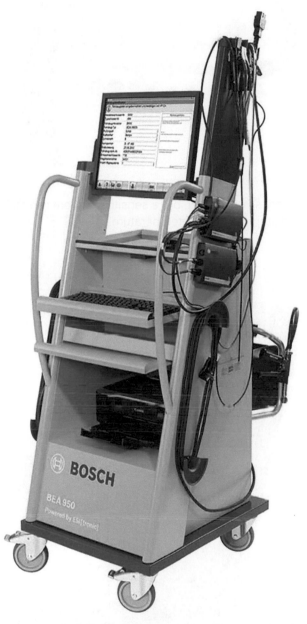

Figure 10.57 Bosch engine analysers (Source: Bosch Media)

Exhaust gas measurement It has now become standard to measure four of the main exhaust gases namely:

► Carbon monoxide (CO)
► Carbon dioxide (CO_2)
► Hydrocarbons (HC)
► Oxygen (O_2)

On many analysers, lambda value and the air fuel ratio are calculated and displayed in addition to the

Figure 10.58 Mobile exhaust gas measuring equipment (Source: Bosch media)

10

Figure 10.59 Fixed (workshop based) exhaust gas measuring equipment (Source: Bosch media)

Figure 10.60 OBDII breakout box

a very small current, but a sensitive camera can still register a change of temperature.

Breakout box A breakout box is a simple piece of electrical test equipment used to support diagnostics by providing easy access to test signals. The type shown here connects to a vehicle diagnostic link connector (DLC) and because it has a plug and corresponding socket, other test equipment can be used at the same time. The terminals on the box can then be used to, for example, connect an oscilloscope to monitor CAN signals.

Thermal camera A thermal camera can be used for a number of things. If, for example, a car battery discharges overnight, it may be due to a parasitic current draw. Taking a thermal image of the vehicle fuse box can help to show where the problem is because current flow causes heat, not much if it is

Figure 10.61 Flir thermal camera for fitting to an iPhone

The image shown here is of a heated rear window on a car as it was first switched on. In this case it shows all the elements to be intact.

Figure 10.62 Thermal image of a rear screen

Figure 10.63 COP ignition probe (Source: Pico Technology Ltd., www.picoauto.com)

Coil on plug (COP) diagnostics Connecting a normal probe to a COP ignition system is very difficult. It is sometimes possible to back probe the primary circuit but not always. A signal probe is the easiest and fastest non-intrusive way to check coil-on-plug ignition coils and spark plugs. The probe shown will display a scope pattern of the secondary, faster than scoping the primary. It is simply held on top of the ignition coil. A typical known good waveform is shown here.

Sensor simulator A sensor simulator is used to replace a suspected faulty sensor for test purposes. By simulating the sensor output you can check the signals generated are being sent and received correctly by an ECU. Simulation measurements available on this device include frequency, voltage and O_2 sensor signals. Other modes include simulation of crankshaft and camshaft inductive and Hall effect

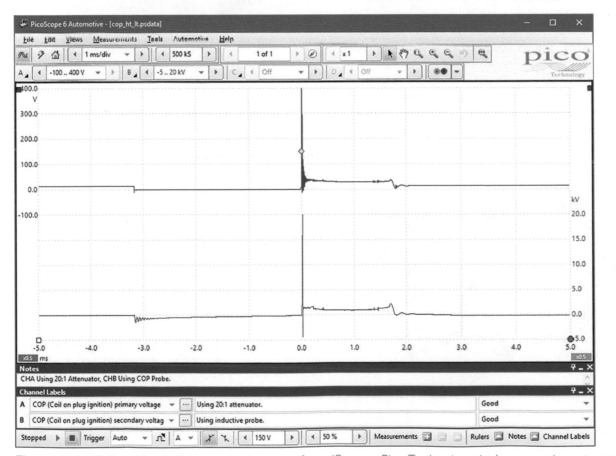

Figure 10.64 Coil-on-plug ignition secondary waveform (Source: Pico Technology Ltd., www.picoauto.com)

513

sensors. In these modes it is possible to adjust the number of active teeth and the number of missing teeth. The simulator is useful for circuit tests. This is because the device can simulate the voltages and you can check the live data readings on a scan tool to confirm the wiring. Also, simulating an O_2 sensor signal, for example, can be used to check ECU response and operation.

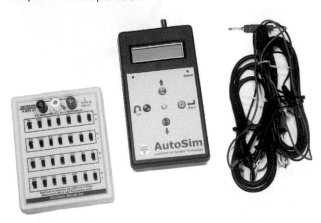

Figure 10.65 A decade box that allows resistance values to be selected and an AutoSim sensor simulator

Actuator driver Automotive actuators work by being supplied appropriate electrical signals. There is the simple on or off signal as supplied to most petrol fuel injectors, or a more complex set of signals to operate a stepper motor. The device shown here can supply a pulse width modified (PWM) signal at a current of a few amps if necessary. PWM signals are used for many different actuators but a typical example would be a throttle control valve. When connected in place of the normal supply from the ECU, it will prove if the actuator is working or not.

Figure 10.66 The unit will produce pulse width modified (PWM) output to drive an actuator such as a throttle controller

Use the media search tools to look for pictures and videos relating to the subject in this section

10.2.5 **Pressure analysis**

Introduction PicoScope automotive scopes are top of the range for waveform analysis, and with the addition of a pressure transducer, now allow detailed examination of pressures at all levels. The Pico Technology pressure transducer simply converts pressure values to voltage which are then relayed to the scope, allowing pressure to be displayed against time. This device will actually replace vacuum, coolant, compression (petrol and diesel), oil pressure, turbo boost and fuel pressure gauges, and more.

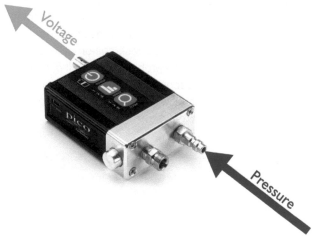

Figure 10.67 WPS500X Pressure Transducer (Source: Pico Technology Ltd., www.picoauto.com)

The compression test is an example where using a transducer, when coupled to a PicoScope, will reveal far more than the maximum compression pressure. Connecting the device to a compression hose will allow for a compression measurement to be taken identical to a typical compression tester, but with the results displayed on the scope screen rather than reading from a mechanical gauge. Setting up PicoScope to measure compression is straight forward using the Guided Tests built into the PicoScope software.

Advantages To appreciate the advantages of this testing method, we need to understand the waveform acquired during a typical compression test. Using a suitable adaptor, the transducer is connected after removing a spark plug or glow plug. The injectors should also be disconnected to prevent fuel being delivered.

The waveform shown reveals the compression peaks at 170 psi as would a typical compression

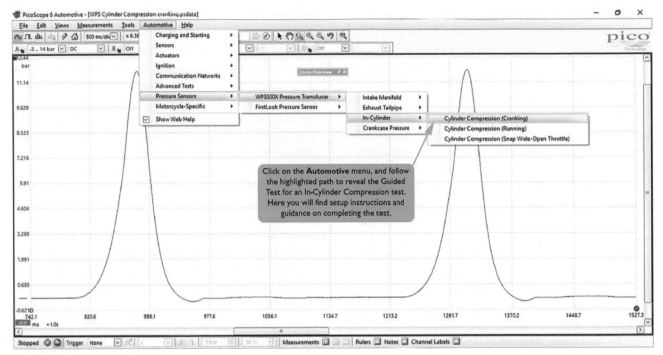

Figure 10.68 Setting up the PicoScope

tester. However, we can now see repeated even compression peaks (towers) as the crankshaft rotates, and more importantly events taking place between compressions that would not be visible with a standard compression tester.

Compression events Using PicoScope we can equally divide the distance between compression events to reveal the position of the crankshaft (degrees of rotation) using our rotation markers. If we know the position of the crankshaft, we can identify each of the four stroke cycles between compressions.

Look closely at the base of each compression tower, you can see the expansion pocket dropping below the zero psi rule (line) indicating the cylinder pressure momentarily dropped to negative (below atmospheric pressure). This indicates both intake and exhaust

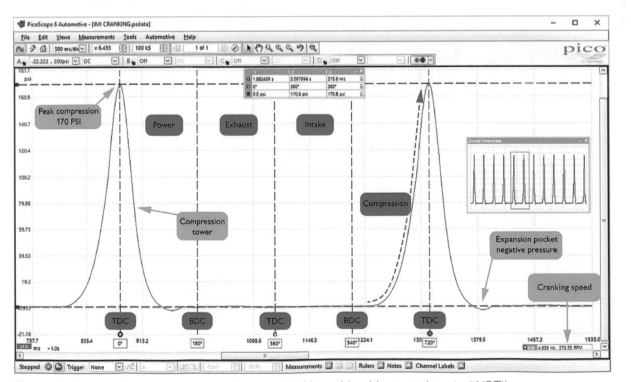

Figure 10.69 Good compression waveform – (Cranking with wide open throttle (WOT))

valves remain closed with adequate sealing as the piston descends the cylinder towards the end of the power stroke. The power stroke is referred to here as the expansion stroke as there is no combustion (the integrity of the piston compression rings, and cylinder face can also be confirmed via the expansion pocket).

Using the time-rulers, we can also measure the time it takes (frequency) for the crankshaft to rotate 360 degrees and multiply this value by 60 to reveal the cranking speed (278 RPM).

Compression towers As with a conventional gauge, we can confirm peak pressure to be correct for the engine under test. However, by using the pressure transducer, we can also see uniform compression towers confirming cylinder efficiency, not only as the pressure builds but also as it naturally decays during the expansion stroke. For the compression towers to be symmetrical, the mechanical integrity of the piston/cylinder and valve gear must be efficient.

Expansion pockets The presence of the expansion pocket confirms our cylinder can hold a vacuum and must, therefore, be airtight (valve seat and piston ring integrity OK). We can also note that there is sufficient cranking speed, adequate intake and exhaust flow so

achieving the correct peak pressure (no restrictions), and the repeatability of peak compression for every completion of the four stroke cycle is good.

Pressure peak In this image, which shows an engine with a fault, we can see an additional pressure peak during the exhaust event that should not be present. As the piston rises from Bottom Dead Centre (BDC) of the expansion stroke, the exhaust valve will open to release the cylinder pressure out to the atmosphere (through atmospheric pressure) via the exhaust system. The waveform indicates a pressure increase to 125 psi approximately during the exhaust stroke because the exhaust valve is not opening! In effect what we have here is another compression stroke during the 4 stroke cycle: Compression-Expansion-Compression-Intake. A typical compression tester cannot detect this condition and would read a normal compression value.

Looking a little deeper at the exhaust stroke we can see what looks to be another compression tower as a result of the exhaust valve remaining closed, only this time the tower is no longer symmetrical.

In this scenario we have approximately 125 psi present inside our cylinder when the intake valve opens, abruptly releasing this pressure into the

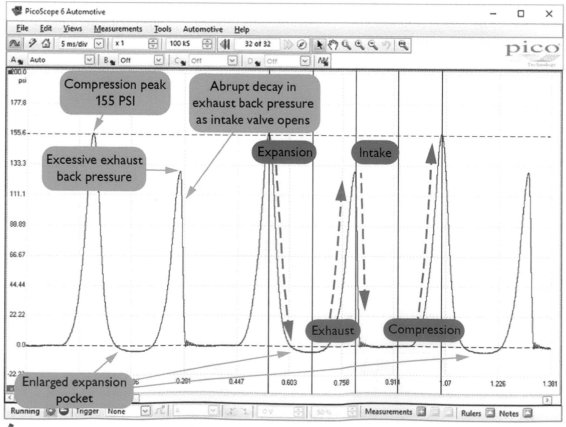

Figure 10.70 Faulty compression waveform

intake manifold, hence the rapid drop in pressure and asymmetric tower. Such an event would manifest itself as a popping sound via the intake manifold.

Accurate diagnosis The previous example highlighted just a small number of the advantages when viewing pressure against time given the accuracy and responsiveness of the transducer. The possibilities for accurate diagnosis increase further with a running engine. It is extremely important, in any in-cylinder pressure analysis, to disconnect the relevant fuel injector to prevent bore wash, oil contamination, and catalyst damage. In the case of a diesel engine, combustion would take place; potentially resulting in injury, as well as damaging the compression hose and transducer.

With fuel removed from the cylinder in question, we are left with a simple air pump system where the air is drawn in via the intake, compressed, decompressed and then released via the exhaust (the four-stroke cycle). Each stage of the four-stroke cycle reveals information about the efficiency of the cylinder and the timings of each stroke relevant to the degrees of crankshaft rotation.

Dynamic in-depth analysis of the four-stroke cycle for each cylinder is now possible. Specific errors, such as timing chain elongation, worn camshaft lobes, poor alignment (or installation) of camshafts, broken and insecure rockers, compressed hydraulic lifters, or camshaft lobes that have spun independently of

the camshaft, can all be determined. All of these conditions will affect the independent valve timing of each cylinder.

Example of a good cylinder waveform This image shows a known good cylinder waveform from a petrol engine, running at idle speed. Like all diagnostic techniques, we need to know what the waveform from a good cylinder looks like before we begin to formulate theories about the capture we have taken from the engine under diagnosis.

The waveforms below have additional features in comparison to the in-cylinder pressure waveform from a cranking engine. It is now possible to detect and measure the intake and exhaust events based on the formation of the expansion and intake pockets. As a general rule of thumb, the expansion and intake pockets should be equal in depth, and measure the typical negative pressure found in the intake manifold at idle speed (approximately −650 to −750 mbar).

In addition to the depth of the expansion pockets, their duration relates directly to the opening and closing of both the exhaust and intake valves. This time can be measured against crankshaft rotation by using the rotation rulers found in the PicoScope software. The compression towers should be symmetrical and uniform in structure, with an equal build and decay in pressure (remember there is no combustion taking place inside the cylinder).

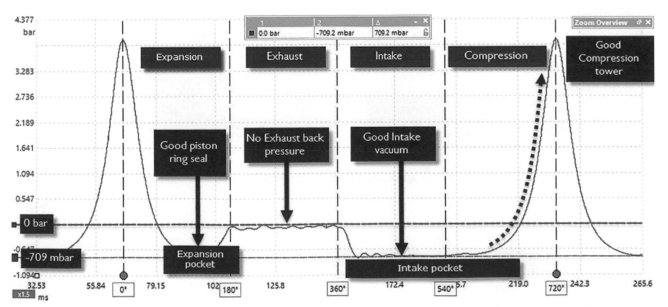

Figure 10.71 Good example of the in-cylinder waveform revealing the four stroke cycle at idle (Petrol)

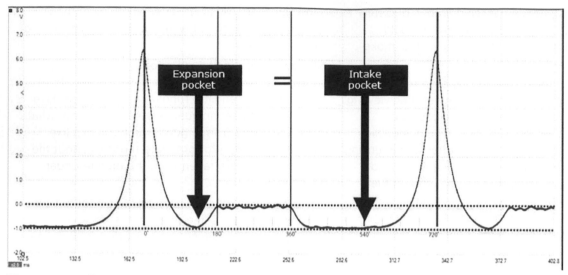

Figure 10.72 Expected results when evaluating valve timing and duration (Petrol)

Valve overlap activity Once familiar with the features of a good waveform it is now possible to consider what effects certain faults will have. For example, valve overlap activity can be monitored and measured for correct timing and duration in relation to crankshaft rotation. This is done during the transition of the piston at top dead centre (TDC) on the exhaust stroke, through to the commencement of the intake stroke. Valve timing and lift errors can be identified if the valve overlap events do not take place at the correct rotational position of the crankshaft.

The waveform in shown here confirms a retarded exhaust valve open event based on the oversized expansion pocket, as well as a momentarily retarded

intake valve open event with minimal duration. Here we have a classic valve timing error on both camshafts, along with insufficient valve lift and duration of the intake valve.

Restrictions present in the exhaust system can also be rapidly confirmed, with the presence of backpressure measured inside the cylinder, as the piston moves upwards during the exhaust stroke. Here we can identify diesel particulate filter (DPF) and catalyst restrictions without intrusion to the exhaust system (exhaust valve open should result in cylinder pressure equalising to atmospheric pressure). The waveform shown here indicates over 600 mbar of backpressure at idle speed. This would only increase

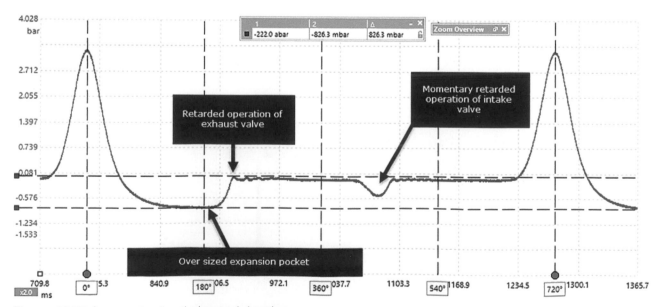

Figure 10.73 Incorrect valve timing and duration

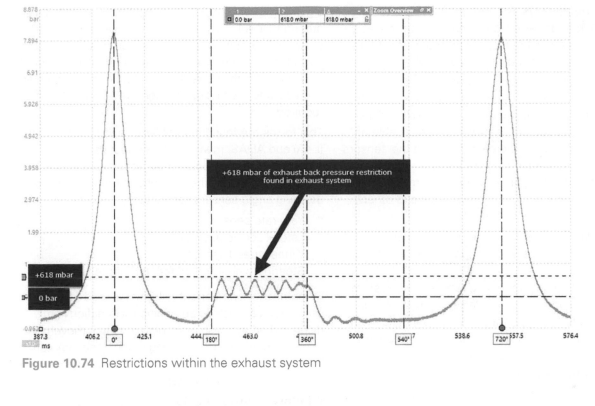

1	2	Δ
0.0 bar	618.0 mbar	618.0 mbar

+618 mbar of exhaust back pressure restriction found in exhaust system

+618 mbar

0 bar

Figure 10.74 Restrictions within the exhaust system

with engine speed and will result in power loss, and as a consequence, damage to relevant components.

Summary Often, in-cylinder waveforms require close analysis where errors are not immediately apparent. The Pico pressure transducer has an ultra-fast response time of 100ms, and when combined with the zoom and scaling features of scope, even the smallest anomalies can be revealed.

These features allow us to measure variations in individual cylinder valve timings, as a result of valve clearance errors, camshaft lobe wear, or valve lifter and rocker failure. Connecting the pressure transducer to a PicoScope opens lots of new diagnostic possibilities, for example visualising in-cylinder events under various engine operating conditions. However, the pressure transducer can be connected to the intake manifold, exhaust tailpipe, lambda sensor aperture or crankcase via the dipstick tube, to name just a few accessible locations. We are just scratching the surface of this tool's potential, and it is now as essential for many engine (and other system) diagnostic procedures as using a current clamp to determine cylinder balance. More details about the highly-recommended PicoScope and the associated pressure transducer can be found here: http://www.picoauto.com

 Select a routine from section 1.3 and follow the process to study a component or system

10.2.6 Calibrating components ❸

Introduction As vehicles increase the amount of information that they gather for the motorist, the method of that collection, usually through sensors, may need calibration. Some sensors are remarkable in that they need little intervention, such as tyre pressure sensors, but others may need calibration after a repair.

Batteries Due to the complexity of systems that require continuous power, there are often battery connection and disconnection procedures found in the manufacturer's data to ensure functionality is not lost following replacement.

Figure 10.75 Battery (Source: Bosch Media)

Tyre pressure sensors The tyre pressure monitoring system (TPMS) has become so common place that it has become part of the inspection routine for MOT testing. There are two main types of TPMS

Direct TPMS uses sensors within the tyre that monitors pressure levels, warning the motorist via the dash and sometimes also with an audible alert. This system gives an accurate reading, can be capable of an accurate read out on the dashboard but the sensors are generally mounted in the tyre valve, and they can be expensive to replace. Some systems require motorist input via a dash within the cabin and others reset themselves after several rotations.

Indirect TPMS is a much simpler system that uses the ABS wheel sensors to pick up that one wheel that is revolving at a different speed (an underinflated tyre will have a slightly smaller rolling radius). This system is less accurate and can give a false alarm. If all four tyres are underinflated by the same amount (albeit an unlikely event) then it cannot detect any difference and would not warn the motorist.

Steering angle sensors There are several sensors, usually packaged together in a single unit sited within the steering column. They are common systems that require calibration:

▶ Electronic stability control (ESC), usually part of the ABS module
▶ Electric power steering / variable effort power steering
▶ Advanced driver assistance systems (ADAS)

Analog sensors rely on voltage differences to determine information about angle and turn direction, digital sensors employ an LED that accurately measures the angle of the steering input. Relaying this information to the ESC module, which in turn uses an algorithm to ensure the steering is accurate from the steering wheel to the wheels. An advantage of ESC systems is that it can determine from the information collected if the driver is in control of the vehicle or not, assisting with the correction of steering for the driver and applying the brakes through the ABS module if required.

This forms the foundation of lane keep assist (LKA) and ADAS, meaning the correct calibration of all equipment associated with these systems is important to ensure the safe operation of the vehicle. Steering angle sensors can include

▶ Yaw rate sensor
▶ Lateral acceleration sensor
▶ Wheel speed sensor
▶ Steering wheel angle/steering torque sensor

Symptoms of component failure or a need for calibration can include an engine malfunction light displayed on the dash or as play or erratic movement in the steering wheel. Steering wheel feedback can be more common after a mechanical change has been made to the steering system.

Window/sunroof regulators If a vehicle has automatic open and close functionality in the operation of the windows, usually instructed by a long press of the window activation button (a short press allows a small amount of travel) then the vehicle needs to 'learn' when to stop the motor when the desired position has been reached. This calibration is usually required after there has been an interruption to power supply, such as a change of battery. The system is normally reset using a prescribed routine and is found in the manufacturer's information.

Wiper motors If a vehicle has wiper blades that work in opposite directions, they usually have one lead

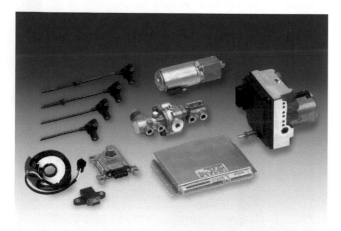

Figure 10.76 Steering angle sensor (bottom left) as part of the stability control system (Source: Bosch Media)

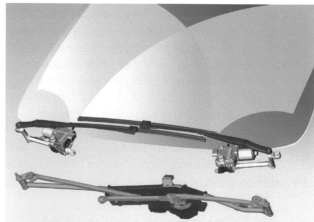

Figure 10.77 Twin wiper motors (Source: Bosch Media)

motor and one which follows it, rather than the more traditional method of one motor operating a linkage that pulls the wipers in the same direction. If the lead motor is changed, it often needs to be coded and the follower calibrated, although the performing each process in practice is by the same means, through a diagnostic tablet.

 Look back over the previous section and write out a list of the key bullet points from this section

10.3 OBD

10.3.1 On-board diagnostics ❷❸

On-board diagnostics (OBD) is a generic term referring to a vehicle's self-diagnostic and reporting system. The amount of diagnostic information available via OBD has varied considerably since its introduction in the early 1980s. Early versions of OBD would simply illuminate a malfunction indicator light (MIL) if a problem was detected, but did not provide any information about the problem. The current versions are OBD2 and in Europe EOBD2. These two standards are quite similar. Modern OBD systems use a standardized digital communications port to provide real-time data in addition to a standardized series of diagnostic trouble codes (DTCs).

Compatibility All 1996 and newer vehicles are OBD2 compatible. However, the amount of OBD2 parameters obtained will depend on the specific OBD2 protocol of the vehicle. Under the original OBD2 specification, up to 36 parameters were available. Newer OBD2 vehicles that support the CAN-BUS protocol can have up to 100 generic parameters. This includes trouble codes in systems such as ABS, transmission, and airbags. Most scanners now also access information via CAN as well as the OBD information.

Diagnostic trouble codes (DTCs) or fault codes, are stored by an on-board computer diagnostic system. These codes are stored when, for example, a sensor in the car produces a reading that is outside its expected range. DTCs identify a specific problem area and are a guide as to where a fault might be occurring within the vehicle. Parts or components should not be replaced with reference only to a DTC. No matter what some customers may think, the computer does not tell us exactly what is wrong! For example, if a DTC reports a sensor fault, replacement of the sensor is unlikely to resolve the underlying problem. The fault

is more likely to be caused by the systems that the sensor is monitoring, but can also be caused by the wiring to the sensor.

DTCs may also be triggered by faults earlier in the operating process. For example, a dirty MAF sensor might cause the car to overcompensate its fuel-trim adjustments. As a result, an oxygen sensor fault may be set as well as a MAF sensor code. The figure below shows how to interpret OBD2 codes.

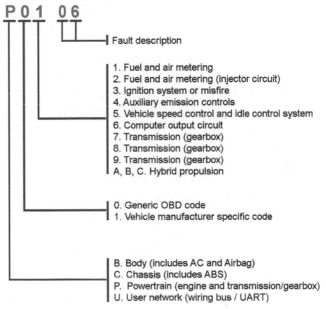

Figure 10.78 On board diagnostic trouble codes

Example codes The following are example DTCs:

▶ P0106 Manifold Absolute Pressure/Barometric Pressure Circuit Range/Performance Problem
▶ P0991 Transmission Fluid Pressure Sensor/Switch E Circuit Intermittent
▶ C1231 Speed Wheel Sensor Rear Center Circuit Open (DTCs use USA spellings)
▶ B1317 Battery Voltage High
▶ U2004 Audio Steering Wheel Control Unit is Not Responding

Summary OBD is a very powerful tool to help us diagnose faults on a vehicle. However, diagnostic skill is still required to interpret the codes correctly. All but the most basic DIY scanners will translate the codes and present the text so no need to refer to a list. Remember, a DTC is a guide, not an answer!

 Use a library or the web search tools to further examine the subject in this section

10

10.3.2 Scanners ❷ ❸

Introduction In this section we have outlined just three different scanners:

► VCDS
► AtriPad
► Snap-on

Many more are available. Interestingly, it is suggested by those using these tools that often three different ones are needed to ensure coverage of all vehicles. This is why many garages are choosing to specialise in one area; Ford, Peugeot or VAG for example.

VCDS This professional system is designed specifically for VW, Audi, Seat and Skoda passenger cars and remains current due to regular software updates. Like all scanners, it has a vehicle communication interface (VCI) that plugs into the diagnostic socket on the car. It can then be connected by USB cable or wirelessly to a Windows computer running the VCDS software.

Facilities The facilities available include scanning for faults as well as reading and logging live data, resetting service interval warnings, and much more. It costs a fraction of the price of a top range system but of course the vehicle coverage is restricted.

Figure 10.80 VCDS vehicle connection interface

Priority numbers Using the VCDS we scanned for faults and found the tyre pressure warning as shown. The fault priority is 3, note that this shows in a different way on the ArtiPad live data examples. The DTC priority numbers are shown here:

Number	Meaning
0	Undefined by manufacturer.
1	The fault has a strong influence on drivability, immediate stop is required.
2	The fault requires an immediate service appointment.

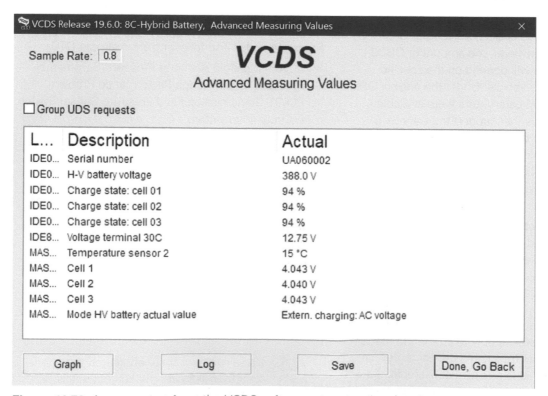

Figure 10.79 A screenshot from the VCDS software showing live data from the high voltage battery on a PHEV

Number	Meaning
3	The fault doesn't require an immediate service appointment, but it should be corrected with the next service appointment.
4	The fault recommends an action to be taken, otherwise drivability might be affected.
5	The fault has no influence on drivability.
6	The fault has a long term influence on drivability.
7	The fault has an influence on the comfort functions, but doesn't influence the car's drivability.
8	General Note

Figure 10.82 TopDon ArtiPad1 main unit, the VCI is plugged into the diagnostic socket

ArtiPad The ArtiPad outlined here is just one example of a comprehensive professional level scanner. However, my experience, and that of others, shows it to be a very powerful tool for the price. This diagnostic tool is based on the Android operating system. It will show fault codes from a very wide range of vehicles and associated live data. However, it is also capable of ECU programming for Mercedes-Benz, BMW, VW, Audi and Ford. The device itself is very well packaged and robust, as is the software, which is updated regularly. The battery has an excellent life and the screen is clear and easy to read.

Feedback I have included this device because it has received excellent feedback from technicians working in the trade as being a high quality device at a very competitive price. It also allows some pass-through operations. It allows quick and complete diagnoses, providing the user operates it correctly of course, and can interpret the information presented. With the

associated Bluetooth VCI device, this diagnostic tool is capable of ECU programming for Benz, BMW, VW, Audi and more. This diagnostic scanner is suitable for over 2000 European-based and newer OBD2 vehicles.

Example use I recently used this device on a PHEV and as an example of its capabilities, I could log the hybrid battery voltage and current while driving the car. This can then be replayed and graphed. The interface is easy to use and very responsive. Adapters are available but in its basic form it does not work as a scope or test meter.

Snap-on Verdict The Verdict diagnostic and information system includes a wireless display, scanner and meter/scope. The advantage of this is that the information

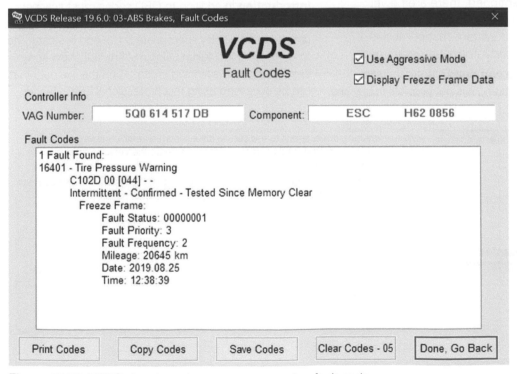

Figure 10.81 VCDS showing a tyre pressure warning fault code

Figure 10.83 TopDon Atripad1 main unit and VCI

Figure 10.84 Snap-on Verdict display unit, scanner and multimeter/oscilloscope

is mirrored to the main unit, which in turn can be connected to a large display screen, or used on its own. It runs the software in Windows. You can view the readings and control the scanner and scope remotely from the main display unit. This is very convenient, for example, if you want to compare a scope waveform with a known good example. A repair information system is built into the software package. The Verdict is a good few years old now and newer more comprehensive models are available, but it is still a good example – and I like it!

The wireless scanner module connects to the display tablet that is operated using a stylus. One keyless adapter covers OBD2 compliant vehicles. It will connect to dozens of systems on most vehicle makes, including European, Asian and USA vehicle coverage.

Channels The wireless meter has 2-channels for component testing. The high-speed oscilloscope and digital graphing meter have CATIII/CATIV certification, so it is safe for use on hybrid and electric vehicles. It can be used with the Verdict display, or as a standalone tool. As mentioned, there are newer systems available from Snap-on, but this tool is a good example of an integrated system that combines a scanner, meter, scope and information system.

How many scanners do you need? Notwithstanding claims by some manufacturers, it is not possible to buy one

scanner that will cover the entire range of vehicles. All will read E/OBD2 data but when it comes to digging deeper, every device has some gaps in its capabilities, for example when coding a new component that has been fitted, or resetting some service intervals. Most of the professional level scanners cover a significant range but the consensus, out in the real world, is that you need three different devices to be certain of covering everything!

Summary and which is the best scanner? Hard to say, but probably the one that covers most of the range of vehicles you work on. But also, and perhaps more importantly, the one you are familiar with and are comfortable using.

> Use a library or the web search tools to further examine the subject in this section

10.3.3 Live data ❷ ❸

Introduction In addition to OBD codes most scanners now communicate with all the electronic modules on the data bus and display appropriate DTCs and live data. The following images, taken using my ArtiPad, show some examples of the scanning process and the live data that can be accessed using this professional level scanner.

	After connecting the scanner to the car via its associated vehicle communication interface (VCI), it reads the VIN and automatically selects the vehicle. After this the menu shown here is displayed. The two menu items of interest for now, are the options: 02 Read DTC, and 08 Read Data Stream (live data).
	After selecting option 02 on the touch screen the scanner takes a minute or so to cycle through all connected modules and display the state as 'No Fault Code' or 'Find Fault Code'. The number indicates how many codes are stored.

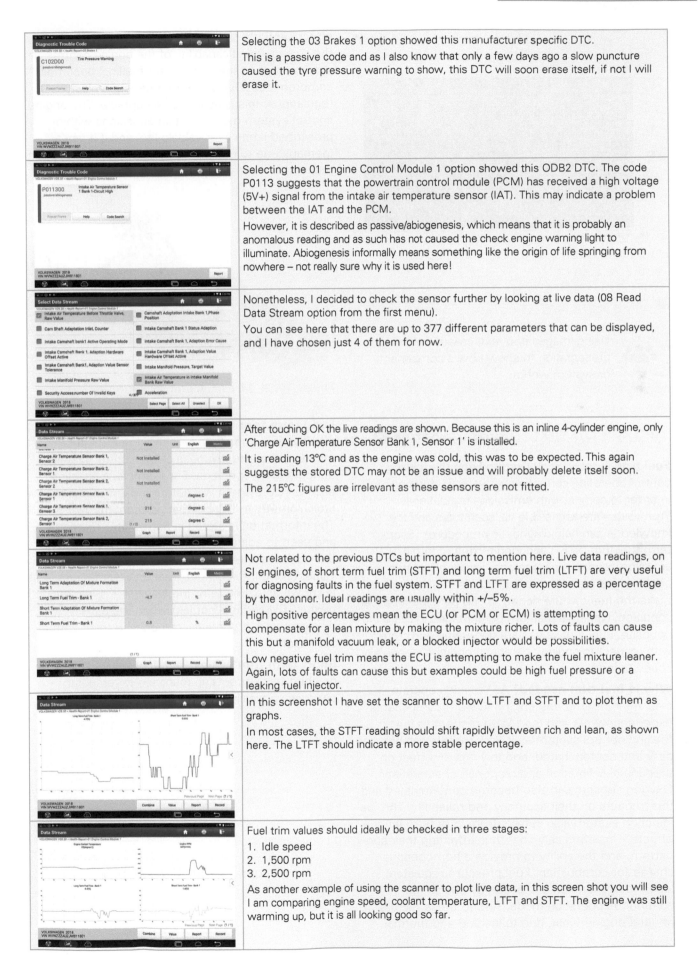

Selecting the 03 Brakes 1 option showed this manufacturer specific DTC.

This is a passive code and as I also know that only a few days ago a slow puncture caused the tyre pressure warning to show, this DTC will soon erase itself, if not I will erase it.

Selecting the 01 Engine Control Module 1 option showed this ODB2 DTC. The code P0113 suggests that the powertrain control module (PCM) has received a high voltage (5V+) signal from the intake air temperature sensor (IAT). This may indicate a problem between the IAT and the PCM.

However, it is described as passive/abiogenesis, which means that it is probably an anomalous reading and as such has not caused the check engine warning light to illuminate. Abiogenesis informally means something like the origin of life springing from nowhere – not really sure why it is used here!

Nonetheless, I decided to check the sensor further by looking at live data (08 Read Data Stream option from the first menu).

You can see here that there are up to 377 different parameters that can be displayed, and I have chosen just 4 of them for now.

After touching OK the live readings are shown. Because this is an inline 4-cylinder engine, only 'Charge Air Temperature Sensor Bank 1, Sensor 1' is installed.

It is reading 13°C and as the engine was cold, this was to be expected. This again suggests the stored DTC may not be an issue and will probably delete itself soon.

The 215°C figures are irrelevant as these sensors are not fitted.

Not related to the previous DTCs but important to mention here. Live data readings, on SI engines, of short term fuel trim (STFT) and long term fuel trim (LTFT) are very useful for diagnosing faults in the fuel system. STFT and LTFT are expressed as a percentage by the scanner. Ideal readings are usually within +/–5%.

High positive percentages mean the ECU (or PCM or ECM) is attempting to compensate for a lean mixture by making the mixture richer. Lots of faults can cause this but a manifold vacuum leak, or a blocked injector would be possibilities.

Low negative fuel trim means the ECU is attempting to make the fuel mixture leaner. Again, lots of faults can cause this but examples could be high fuel pressure or a leaking fuel injector.

In this screenshot I have set the scanner to show LTFT and STFT and to plot them as graphs.

In most cases, the STFT reading should shift rapidly between rich and lean, as shown here. The LTFT should indicate a more stable percentage.

Fuel trim values should ideally be checked in three stages:

1. Idle speed
2. 1,500 rpm
3. 2,500 rpm

As another example of using the scanner to plot live data, in this screen shot you will see I am comparing engine speed, coolant temperature, LTFT and STFT. The engine was still warming up, but it is all looking good so far.

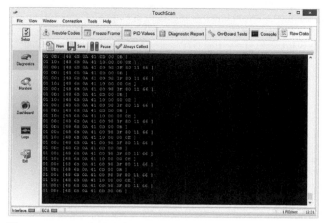

Figure 10.85 Raw data transmission to the TouchScan scan tool

 Using images and text, create a short presentation to show how a component or system works

10.3.4 ECU fuel trim diagnostics ❸

Fuel trim is technology used to keep the fuelling control operating correctly over vehicle lifetime, in order to comply with emissions regulations. Trim values are accessible via scan tools and this knowledge can help diagnostic procedures. Short-term fuel trim (STFT) and long-term fuel trim (LTFT) are expressed as a percentage, Positive fuel trim percentages indicate that the ECU is attempting to make the fuel mixture richer, to compensate for a lean condition. Negative fuel trim percentages indicate the opposite. Fuel trim values are generated and stored with respect to engine operating condition – generally speed or load.

Why is fuel trim needed? As vehicles operate over their lifetime, the engine system components experience wear and tear. With respect to the fuelling control system – air flow meters can become dirty and contaminated, and this has an effect on their initial calibration and response characteristic. Oxygen sensors can also become contaminated and this impacts on their accuracy and reliability. The fuel injection system also suffers from long term fatigue effects – injectors can clog, thus affecting their spray pattern, and the ability to provide a homogenous charge for combustion. Fuel pressure regulators lose their calibration and accuracy due to the continuous operation mode. Add to these factors, general engine wear, due to loading and thermal

effects and it's clear that over a long period of time, the accuracy and capability of the engine control system, with respect to correct fuelling, becomes compromised. Due to ever tightening emissions regulations, this is no longer acceptable. The engine control system must maintain emissions within prescribed limits over vehicle life, and if it can't (perhaps due to component failure) then the system must inform the driver.

Figure 10.86 Manifold/port fuel injector

Long term trim values are the result of an adaptive learning strategy within the ECU, this monitors the effectiveness of the control system during engine operation. For example, a gasoline engine ECU system will determine deviations from stoichiometry in the exhaust gas content over time. The information is stored in non-volatile memory and used to adjust or offset the fuelling value from the original stored fuel map value to the required value, taking into account wear and tear factors. This is known as a long term trim value – it is learned by the ECU and it is not lost when the battery is disconnected, it generally can only be reset with a scan tool.

Short term trim values are offset values based on short term effects, in response to temporary changes. Typically the short term trim values are a

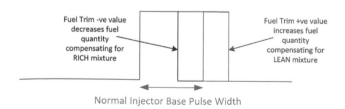

LONG + SHORT TERM FUEL TRIM

Figure 10.87 Fuel trim is effected by modifying the basic injector demand, for the engine operating condition

result of the closed loop lambda control system; the main part of this loop being the pre-catalyst oxygen sensor. Short and long term fuel trim correction values work together to maintain the correct, required lambda value. The short term trim will provide immediate correction, and if this correction is needed to maintain the required value on an on-going basis, then the long term fuel trim will learn this and provide a permanent offset to the basic fuelling map. Note though, this process is monitored and maintained within calibrated limits that correlate with tailpipe emissions. If the long term trim value becomes greater than a certain, pre-set limit, a trigger in the diagnostic system will inform the driver of an emissions related problem.

Fuel trim diagnostics A useful aspect of fuel trim values is the potential to use them as part of a diagnostic procedure. Many fault code readers, including low cost, generic devices, can provide fuel trim values. If you know and understand what these values are, and how they are generated, then you can use them to help you understand the nature and root cause of many fuel systems related faults that you may come across.

During normal engine operation, the ECU records long (LTFT) and short (STFT) term fuel trim values as ECU labels. STFT normally cycles at the same frequency as the oxygen sensor switches, in effect it reflects the operation of the closed loop lambda control. LTFT is normally quite stable, and any adjustment of the value is made over a long period of time, so is less noticeable. Remember that LTFT is held in non-volatile memory (also known as KAM – keep alive memory), but STFT is dynamic, thus constantly changes during run-time, but starting at zero, both values re-set when DTCs are cleared.

Diagnosing faults The cycle-to cycle, and cylinder to cylinder variations of an engine cause loss of efficiency and roughness. The ECU is therefore continually adjusting spark timing and fuelling to compensate. As injectors wear, the delivered quantity and spray pattern are compromised, and this can cause cylinder specific fuelling faults that can be identified very efficiently using fuel trim values. The long and short term fuel trim value will reflect any variations in injector delivery or combustion efficiency for a specific cylinder, and can help diagnose faults like leaking blocked or dirty injectors.

Trim values should generally not exceed 10% either way. Positive means the ECU is trying to compensate for a weak mixture, it does this by extending the basic injector pulse width. This could indicate faults caused by air leaks in the flow system, manifold or vacuum pipework. In addition, MAF (mass airflow meter) or oxygen sensor issues. Lean mixture problems are the most common failure mode. Negative trims mean that the engine is running rich, and the base pulse width is reduced to compensate. This could indicate problems with EGR, MAF, fuel pressure regulator, leaking injectors or oxygen sensor problems. Individual cylinder contributions can also be evaluated. Disconnecting the injectors one-by-one and monitoring fuel trim values allows the

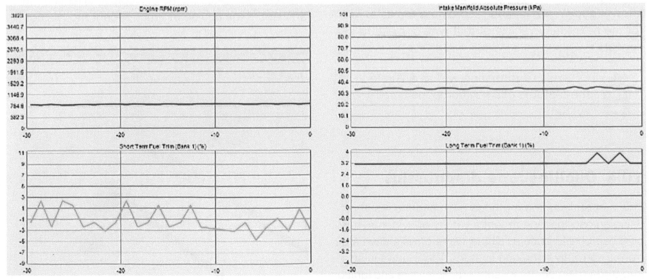

Figure 10.88 Typical fuel trim values at idle, as shown on a standard scan tool, connected to the vehicle OBD connector. Note that STFT is negative and follows closed loop lambda operation. LTFT is much more stable over time, and is a positive value. Both are within acceptable limits

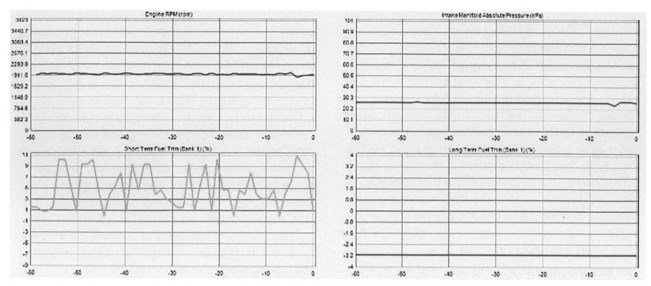

Figure 10.89 Fuel trim – high idle speed. STFT is now positive, LTFT has switched negative, this shows how both work together to achieve the required target of Lambda 1.

technician, to compare each cylinder with the others. Thus individual misfire or injector problems can be identified clearly.

Summary Different ECUs respond differently to a given fault or diagnostic stimulus, in particular with respect to how LTFT and STFT work together. This is part of the ECU calibration strategy, so there are no hard and fast rules, but if you understand what LT and ST fuel trims are, and what they do, then you can observe what you see and make an informed judgement. There is no doubt that being familiar with the concept will help your diagnostics of fuelling related problems. If you have time, it's always worth making some observational measurements of a known 'good' vehicle, and record/note the data. Ready for the next faulty vehicle of that type or make to appear!

 Create an information wall to illustrate the features of a key component or system

10.4 Oscilloscope diagnostics

This part outlines the methods used and the results of using an oscilloscope to test a variety of systems. It will be a useful reference as all the waveforms shown are from a correctly operating system. The module is split into three main sections: sensors, actuators and ignition.

The waveforms are all available from PicoScope: www.picoauto.com/library **if you have access or**

alternatively in the IMI eLearning or from: www.tomdenton.org

A good way to learn what a waveform looks like is to draw it. We recommend that you make a rough sketch and then once you have researched the answer, make a copy of that in the spaces provided in this book.

You will find downloadable resources or a link to other sources here:

https://www.tomdenton.org

10.4.1 Sensors ❸

ABS speed sensor waveform The ABS wheel speed sensors have become increasingly smaller and more efficient in the course of time. Recent models not only

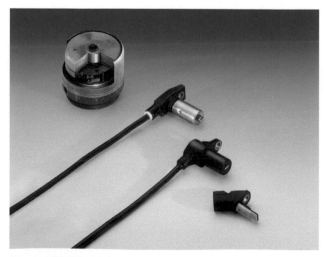

Figure 10.90 ABS wheel speed sensors (Source: Bosch Press)

measure the speed and direction of wheel rotation but can be integrated into the wheel bearing as well.

The Anti-lock Braking System (ABS) relies upon information coming in from the sensors to determine what action should be taken. If, under heavy braking, the ABS electronic control unit (ECU) loses a signal from one of the road wheels, it assumes that the wheel has locked and releases that brake momentarily until it sees the signal return. It is therefore imperative that the sensors are capable of providing a signal to the ABS ECU. If the signal produced from one wheel sensor is at a lower frequency that the others the ECU may also react.

The operation of an ABS sensor is similar to that of a crank angle sensor. A small inductive pick-up is affected by the movement of a toothed wheel, which moves in close proximity. The movement of the wheel next to the sensor, results in a 'sine wave'. The sensor, recognisable by its two electrical connections (some may have a coaxial braided outer shield), will produce an output that can be monitored and measured on the oscilloscope.

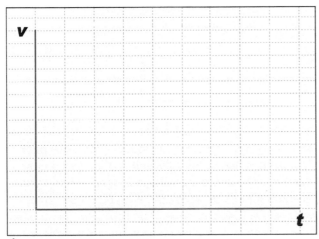

Figure 10.91 ABS speed sensor waveform

Air flow meter – air vane waveform The vane type air flow meter is a simple potentiometer that produces a voltage output that is proportional to the position of a vane. The vane in turn positions itself in a position proportional to the amount of air flowing.

The voltage output from the internal track of the air flow meter should be linear to flap movement; this can be measured on an oscilloscope and should look similar to the example shown online.

The waveform should show approximately 1.0V when the engine is at idle, this voltage will rise as the engine is accelerated and will produce an initial peak. This peak is due to the natural inertia of the air vane and drops momentarily before the voltage is seen to

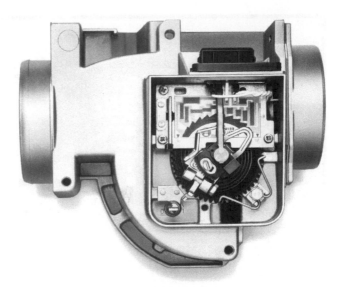

Figure 10.92 Vane or flap type air flow sensor (Source: Bosch)

rise again to a peak of approximately 4.0 to 4.5V. This voltage will however depend on how hard the engine is accelerated, so a lower voltage is not necessarily a fault within the air flow meter. On deceleration the voltage will drop sharply as the wiper arm, in contact with the carbon track, returns back to the idle position.

This voltage may in some cases 'dip' below the initial voltage before returning to idle voltage. A gradual drop will be seen on an engine fitted with an idle speed control valve as this will slowly return the engine back to base idle as an anti-stall characteristic.

A time base of approximately 2 seconds plus is used, this enables the movement to be shown on one screen, from idle, through acceleration and back to idle again. The waveform should be clean with no 'drop-out' in the voltage, as this indicates a lack of electrical continuity. This is common on an AFM with a dirty or faulty carbon track. The problem will show as a 'flat spot' or hesitation when the vehicle is driven, this is a typical problem on vehicles with high mileage that have spent the majority of their working life with the throttle in one position.

The 'hash' on the waveform is due to the vacuum change from the induction pulses as the engine is running.

Air flow meter – hot wire waveform Figure 10.94 shows a micro mechanic mass airflow sensor from Bosch. This type has been in use since 1996. As air flows over the hot wire it cools it down and this produces the output signal. The sensor measures air mass because the air temperature is taken into account due to its cooling effect on the wire.

10

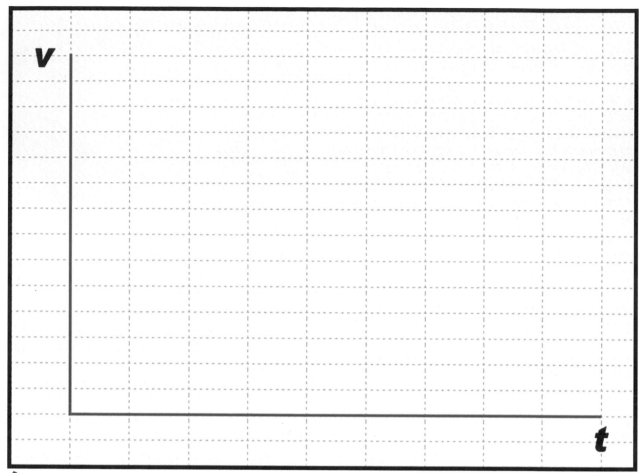

Figure 10.93 Air vane output voltage waveform

Figure 10.94 Hot wire air mass meter (Source: Bosch Press)

The voltage output should be linear to airflow. This can be measured on an oscilloscope and should look similar to the example shown online. The waveform should show approximately 1.0V when the engine is at idle. This voltage will rise as the engine is accelerated and air volume is increased producing an initial peak. This peak is due to the initial influx of air and drops momentarily before the voltage is seen to rise again to another peak of approximately 4.0 to 4.5V. This voltage will however depend on how hard the engine is accelerated; a lower voltage is not necessarily a fault within the meter.

On deceleration the voltage will drop sharply as the throttle butterfly closes, reducing the airflow, and the engine returns back to idle. The final voltage will drop gradually on an engine fitted with idle speed control valve as this will slowly return the engine back to base idle as an anti-stall characteristic. This function normally only effects the engine speed from around 1200 rpm back to the idle setting.

A time base of approximately 2 seconds plus is used because this allows the output voltage on one screen, from idle, through acceleration and back to idle again. The 'hash' on the waveform is due to airflow changes caused by the induction pulses as the engine is running.

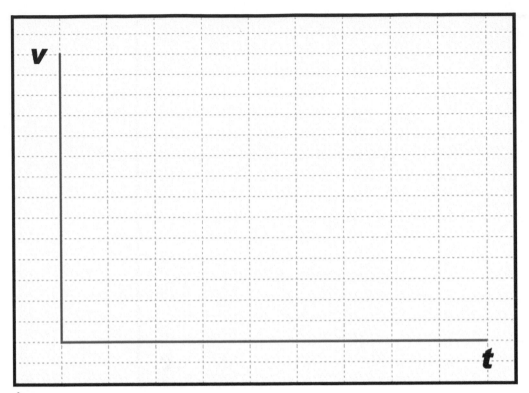

/ **Figure 10.95** Air mass meter waveform

Inductive crankshaft and camshaft sensor waveform The inductive type crank and cam sensors work in the same way. A single tooth, or toothed wheel, induces a voltage into a winding in the sensor. The cam sensor provides engine position information as well as which cylinder is on which stroke. The crank sensor provides engine speed. It also provides engine position in many cases by use of a 'missing' tooth.

Figure 10.96 Crank sensor in position near the engine flywheel

In this online waveform we can evaluate the output voltage from the crank sensor. The voltage will differ between manufacturers and it also increases with engine speed. The waveform will be an alternating voltage signal.

The gap in the picture is due to the 'missing tooth' in the flywheel or reluctor and is used as a reference for the ECU to determine the engine's position. Some systems use two reference points per revolution.

The camshaft sensor is sometimes referred to as the cylinder identification (CID) sensor or a 'phase' sensor and is used as a reference to time sequential fuel injection.

This particular type of sensor generates its own signal and therefore does not require a voltage supply to power it. It is recognisable by its two electrical connections, with the occasional addition of a coaxial shielding wire.

The voltage produced by the camshaft sensor will be determined by several factors, these being the engine's speed, the proximity of the metal rotor to the pick-up and the strength of the magnetic field offered by the sensor. The ECU needs to see the signal when the engine is started for its reference; if absent it can alter the point at which the fuel is injected. The driver of the vehicle may not be aware that the vehicle has a problem if the CID sensor fails, as the drivability may not be affected. However, the MIL should illuminate.

The characteristics of a good inductive camshaft sensor waveform is a sine wave that increases in magnitude as the engine speed is increased and usually provides one signal per 720° of crankshaft rotation (360° of camshaft rotation). The voltage will

10

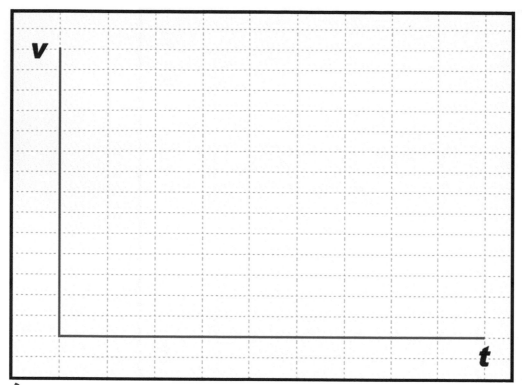

Figure 10.97 Camshaft sensor output signal waveform

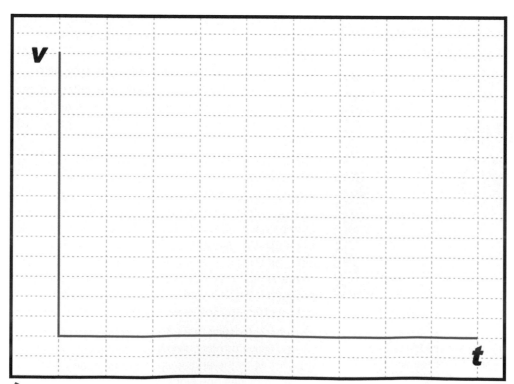

Figure 10.98 Crankshaft sensor output signal waveform

be approximately 0.5V peak to peak while the engine is cranking, rising to around 2.5V peak to peak at idle as seen in the example shown.

Coolant temperature sensor waveform Most coolant temperature sensors are NTC thermistors; their resistance decreases as temperature increases. This can be measured on most systems as a reducing voltage signal.

The coolant temperature sensor (CTS) will usually be a two wire device with a voltage supply of approximately 5V.

The resistance change will therefore alter the voltage seen at the sensor and can be monitored for any discrepancies across its operational range. By selecting a time scale of 500 seconds and connecting the oscilloscope to the sensor, the output voltage can be monitored. Start the engine and in the majority of cases the voltage will start in the region of 3 to 4V and fall gradually. The voltage will depend on the temperature of the engine.

The rate of voltage change is usually linear with no sudden changes to the voltage, if the sensor displays a fault at a certain temperature, it will show up in this test.

Figure 10.99 Temperature sensor

Hall effect distributor pick-up waveform Hall sensors are now used in a number of ways. The ignition distributor is very common but they are also used by ABS for monitoring wheel speed and as transmission speed sensors, for example.

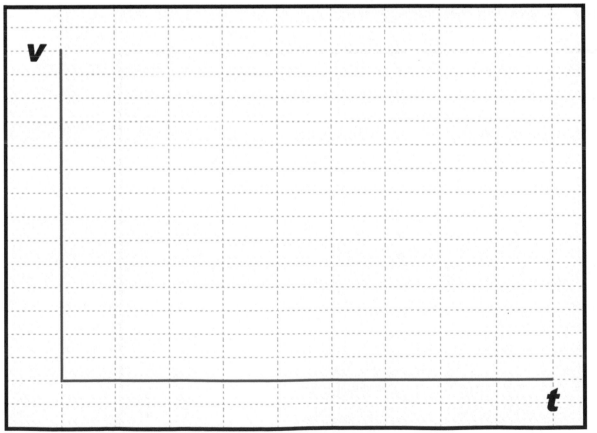

Figure 10.100 Decreasing voltage waveform

This form of trigger device is a simple digital 'on / off switch' which produces a square wave output that is recognised and processed by the ignition control module or engine management ECU.

Figure 10.101 Distributors usually contain a Hall effect or inductive pulse generator (Source: Bosch Press)

The trigger has a rotating metal disc with openings that pass between an electromagnet and the semiconductor (Hall chip). This action produces a square wave that is used by the ECU or amplifier.

The sensor will usually have three connections which are: a stabilised supply voltage, an earth and the output signal. The square wave when monitored on an oscilloscope may vary in amplitude; this is not usually

a problem as it is the frequency that is important, not the height of the voltage. However, in most cases the amplitude/voltage will remain constant.

Inductive distributor pick-up waveform This particular type of pick-up generates its own signal and therefore does not require a voltage supply to power it. The pick-up is used as a signal to trigger the ignition amplifier or an ECU. The sensor normally has two connections. If a third connection is used it is normally a screen to reduce interference.

As a metal rotor spins, a magnetic field is altered which induces an AC voltage from the pick-up. This type

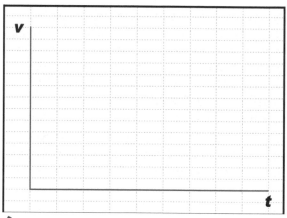

Figure 10.103 Inductive pick-up output signal waveform

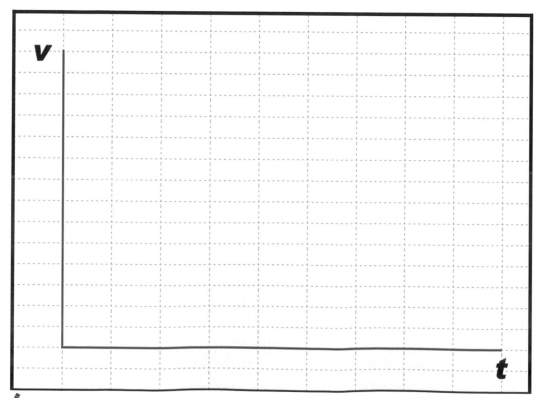

Figure 10.102 Hall output waveform

of pick-up could be described as a small alternator because the output voltage rises as the metal rotor approaches the winding, sharply dropping through zero volts as the two components are aligned and producing a voltage in the opposite direction as the rotor passes. The waveform is similar to a sine wave, however, the design of the components are such that a more rapid switching is evident.

The voltage produced by the pick-up will be determined by three main factors:

▶ Engine speed – the voltage produced will rise from as low as 2 to 3V when cranking, to over 50V, at higher engine speeds

▶ The proximity of the metal rotor to the pick-up winding – an average air gap will be in the order of 0.2 to 0.6mm (8 to 14 thou), a larger air gap will reduce the strength of the magnetic field seen by the winding and the output voltage will be reduced

▶ The strength of the magnetic field offered by the magnet – the strength of this magnetic field determines the effect it has as it 'cuts' through the windings and the output voltage will be reduced accordingly.

A difference between the positive and the negative voltages may also be apparent as the negative side of the sine wave is sometimes attenuated (reduced) when connected to the amplifier circuit, but will produce perfect AC when disconnected and tested under cranking conditions.

Knock sensor waveform The optimal point at which the spark ignites the air/fuel mixture is just before knocking occurs. However, if the timing is set to this value, under certain conditions knock (detonation) will occur. This can cause serious engine damage as well as increasing emissions and reducing efficiency. A knock sensor is used by some engine management systems. The sensor is a small piezo-electrical

Figure 10.104 Knock sensor

crystal that, when coupled with the ECU, can identify when knock occurs and retard the ignition timing accordingly.

The frequency of knocking is approximately 15 kHz. As the response of the sensor is very fast an appropriate time scale must be set, in the case of the example waveform 0–500ms and a 0–5V volt scale. The best way to test a knock sensor is to remove the knock sensor from the engine and to tap it with a small spanner; the resultant waveform should be similar to the example shown.

Note: When refitting the sensor tighten to the correct torque setting as over tightening can damage the sensor and/or cause it to produce incorrect signals.

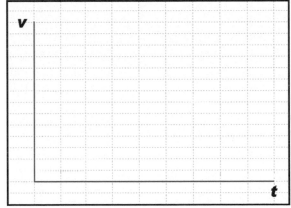

Figure 10.105 Knock sensor output signal waveform

Oxygen sensor (Titania) waveform The lambda sensor, also referred to as the oxygen sensor, plays a very important role in the control of exhaust emissions on a catalyst equipped vehicle.

The main lambda sensor is fitted into the exhaust pipe before the catalytic converter. The sensor will have four electrical connections. It reacts to the oxygen content in the exhaust system and will produce an oscillating voltage between 0.5V volt (lean) to 4.0V or above (rich) when running correctly. A second sensor to monitor the catalyst performance may be fitted downstream of the converter.

Titania sensors, unlike Zirconia sensors, require a voltage supply as they do not generate their own voltage. A vehicle equipped with a lambda sensor is said to have 'closed loop', this means that after the fuel has been burnt during the combustion process, the sensor will analyse the emissions and adjust the engine's fuelling accordingly.

Titania sensors have a heater element to assist the sensor reaching its optimum operating temperature. The sensor when working correctly will switch approximately once per second (1 Hz) but will only start to switch when at normal operating temperature. This

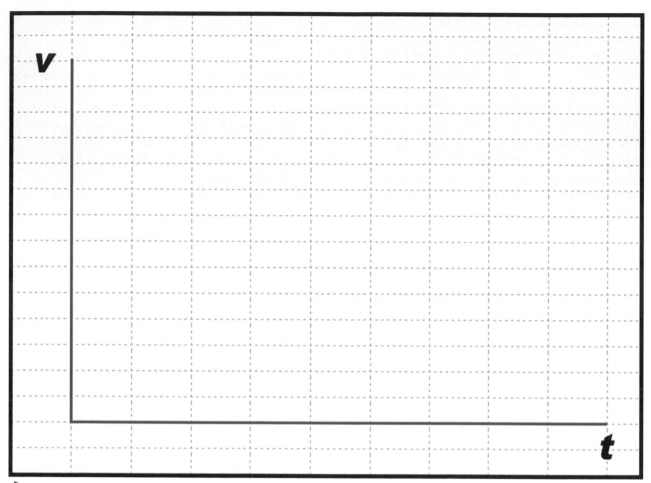

Figure 10.106 Titania lambda sensor output waveform

switching can be seen on the oscilloscope, and the waveform should look similar to the one in the example.

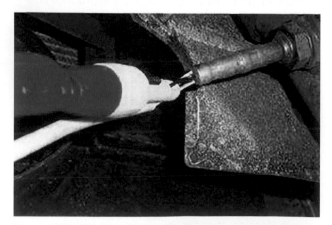

Figure 10.107 Titania knock sensor

Oxygen sensor (Zirconia) waveform The lambda sensor is also referred to as the oxygen sensor or a heated exhaust gas oxygen (HEGO) sensor and plays a very important role in control of exhaust emissions on a catalytic equipped vehicle. The lambda sensor is fitted into the exhaust pipe before the catalytic converter. A second sensor to monitor the catalyst performance may be fitted downstream of the converter.

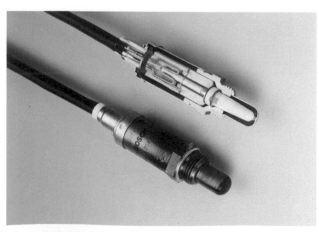

Figure 10.108 Zirconia type oxygen sensor

The sensor will have varying electrical connections and may have up to four wires; it reacts to the oxygen content in the exhaust system and will produce a small voltage depending on the Air/Fuel mixture seen at the time. The voltage range seen will, in most cases, vary between 0.2 and 0.8V. The 0.2V indicates a lean mixture and a voltage of 0.8V shows a richer mixture.

Lambda sensors can have a heater element to assist the sensor reaching its optimum operating temperature. Zirconia sensors when working correctly will switch approximately once per second (1 Hz) and will only start to switch when at normal operating temperature. This switching can be seen on the oscilloscope, and the waveform should look similar to the one in the example waveform.

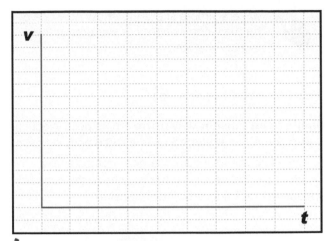

Figure 10.110 Throttle pot output voltage signal

A good throttle potentiometer should show a small voltage at the throttle closed position, gradually rising in voltage as the throttle is opened and returning back to its initial voltage as the throttle is closed. Although many throttle position sensor voltages will be manufacturer specific, many are non-adjustable and the voltage will be in the region of 0.5 to 1.0V at idle rising to 4.0V (or more) with a fully opened throttle. For the full operational range, a time scale around 2 seconds is used.

Road speed sensor (Hall effect) To measure the output of this sensor, jack up the driven wheels of the vehicle and place on axle stands on firm level ground. Run the engine in gear and then probe each of the three connections (+, − and signal).

As the road speed is increased the frequency of the switching should be seen to increase. This change can also be measured on a multimeter with frequency capabilities. The sensor will be located on either the speedometer drive output from the gearbox or to the rear of the speedometer head if a cable is used. The signal is used by the engine ECU and if appropriate, the transmission ECU.

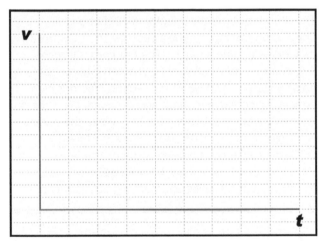

Figure 10.109 Zirconia oxygen sensor output waveform

Throttle position potentiometer waveform This sensor or potentiometer is able to indicate to the ECU the exact amount of throttle opening due to its linear output.

The majority of modern management systems use this type of sensor. It is located on the throttle butterfly spindle. The 'throttle pot' is a three-wire device having a 5 volt supply (usually), an earth connection and a variable output from the centre pin. As the output is critical to the vehicle's performance, any 'blind spots' within the internal carbon track's swept area, will cause 'flat spots' and 'hesitations'. This lack of continuity can be seen on an oscilloscope.

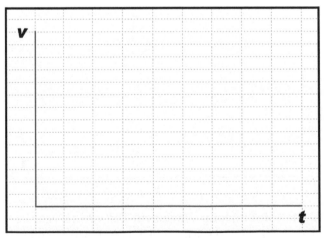

Figure 10.111 Hall effect road speed sensor waveform

Make a simple sketch to show how one of the main components or systems in this section operates

10.4.2 Actuators 3

Single point injector waveform Single point injection is also sometimes referred to as throttle body injection.

A single injector is used (on larger engines two injectors can be used) in what may have the outward appearance to be a carburettor housing.

Figure 10.112 Throttle body with a single injector

The resultant waveform from the single point system shows an initial injection period followed by multi-pulsing of the injector in the remainder of the trace. This 'current limiting' section of the waveform is called the supplementary duration and is the part of the injection trace that expands to increase fuel quantity.

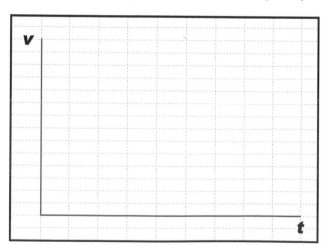

Figure 10.113 Signal point injector waveform

Multi-point injector waveform The injector is an electromechanical device which is fed by a 12 volt supply. The voltage will only be present when the engine is cranking or running because it is controlled

by a relay that operates only when a speed signal is available from the engine. Early systems had this feature built into the relay; most modern systems control the relay from the ECU.

Figure 10.114 Multipoint injectors on the rail

The length of time the injector is held open will depend on the input signals seen by the ECU from its various engine sensors. The held open time or 'injector duration' will vary to compensate for cold engine starting and warm-up periods. The duration time will also expand under acceleration. The injector will have a constant voltage supply while the engine is running and the earth path will be switched via the ECU, the result can be seen in the example waveform. When the earth is removed, a voltage is induced into the injector and a spike approaching 60V is recorded.

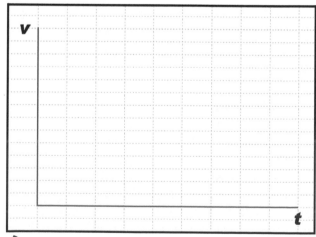

Figure 10.115 Multipoint injector waveform

The height of the spike will vary from vehicle to vehicle. If the value is approximately 35V, it is because a Zener diode is used in the ECU to clamp the voltage. Make sure the top of the spike is squared off, indicating the Zener dumped the remainder of the spike. If it is not squared, that indicates the spike is not strong enough to make the Zener fully dump, meaning there is a problem with a weak injector winding. If a Zener diode is not used in the computer, the spike from a good injector will be 60V or more.

Multi-point injection may be either sequential or simultaneous. A simultaneous system will fire all 4 injectors at the same time with each cylinder receiving two injection pulses per cycle (720° crankshaft rotation). A sequential system will receive just one injection pulse per cycle, this is timed to coincide with the opening of the inlet valve.

Monitoring the injector waveform using both voltage and amperage, allows display of the 'correct' time that the injector is physically open. The current waveform (the one starting on the zero line) shows that the waveform is 'split' into two defined areas.

The first part of the current waveform is responsible for the electromagnetic force lifting the pintle; in this example the time taken is approximately 1.5ms; this is often referred to as the solenoid reaction time. The remaining 2ms is the actual time the injector is fully open. This, when taken as a comparison against the injector voltage duration, is different to the 3.5ms shown. The secret is to make sure you compare like with like!

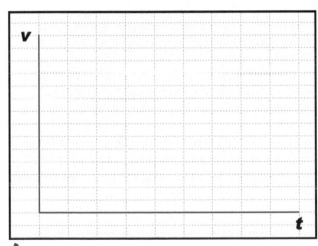

Figure 10.116 Current waveform

Bosch common rail diesel injector waveform

Common rail diesel systems are becoming more common!

It can be clearly seen from the example waveform that there are two distinctive points of injection, the first being the 'pre injection' phase, with the second pulse being the 'main' injection phase.

As the throttle is opened, and the engine is accelerated, the 'main' injection pulse expands in a similar way to a petrol injector. As the throttle is released, the 'main' injection pulse disappears until such time as the engine returns to just above idle.

Under certain engine conditions a third phase may be seen, this is called the 'post injection' phase and is predominantly concerned with controlling the exhaust emissions.

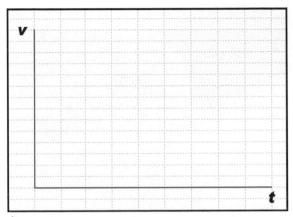

Figure 10.117 Injector voltage waveform

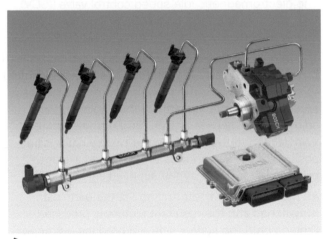

Figure 10.118 Common rail diesel pump, rail, injectors and ECU (Source: Bosch Press)

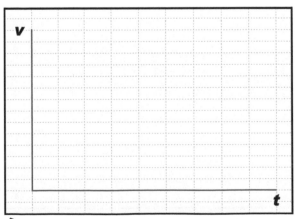

Figure 10.119 CR injector (current) waveform showing pre and main injection pulses

10

Electromagnetic idle speed control valve waveform
This device contains a winding, plunger and spring.
When energized the port opens and when not it closes.

Figure 10.120 Electromagnetic idle speed control valve

The electromagnetic idle speed control valve (ISCV) will have two electrical connections; usually a voltage supply at battery voltage and a switched earth.

The rate at which the device is switched is determined by the ECU to maintain a prerequisite speed according to its programming. The valve will form an air by-pass around the throttle butterfly. If the engine has an adjustable air by-pass and an ISCV, it may require a specific routine to balance the two air paths. The position of the valve tends to take up an average position determined by the supplied signal.

As the example waveform shows, the earth path is switched and the resultant picture is produced. Probing onto the supply side will produce a straight line at system voltage. When the earth circuit is monitored a 'saw tooth' waveform will be seen.

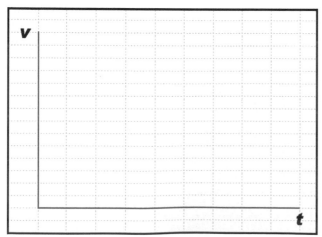

Figure 10.121 Signal produced by an electromagnetic idle speed control valve

Rotary idle speed control valve waveform The rotary idle speed control valve (ISCV) will have 2 or 3 electrical connections, with a voltage supply at battery voltage and either a single or a double switched earth path. The device is like a motor that only ever rotates about half a turn in each direction!

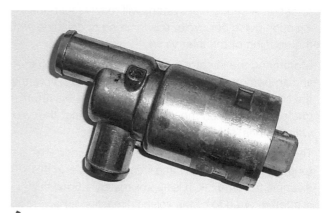

Figure 10.122 Rotary actuator

The rate at which the earth path is switched is determined by the ECU to maintain a prerequisite idle speed according to its programming.

The valve will form an air bypass past the throttle butterfly, to form a controlled air bleed within the induction tract. The rotary valve will have the choice of either single or twin earth paths, the single being pulled one way electrically and returned to its closed position via a spring; the double switched earth system will switch the valve in both directions. This can be monitored on a dual trace oscilloscope. As the example waveform shows the earth path is switched and the resultant picture is produced. The idle control device takes up a position determined by the on/off ratio (duty cycle) of the supplied signal.

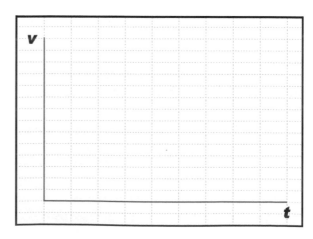

Figure 10.123 Signal supplied to a rotary idle control valve

Probing onto the supply side will produce a straight line at system voltage and when the earth circuit is monitored a square wave will be seen. The frequency can also be measured as can the on/off ratio.

Figure 10.124 Stepper motor and throttle potentiometer on a throttle body

Stepper motor waveform Stepper motors are used to control the idle speed when an idle speed control

valve is not employed. The stepper may control an 'air bypass' circuit by having 4 or 5 connections back to the ECU. The earth's enable the control unit to move the motor in a series of 'steps' as the contacts are earthed to ground. These devices may also be used to control the position of control flaps, for example, as part of a heating and ventilation system.

The individual earth paths can be checked using the oscilloscope. The waveforms should be similar on each path. Variations to the example shown here may be seen between different systems.

 Make a simple sketch to show how one of the main components or systems in this section operates

10.4.3 Ignition system ❸

Ignition primary waveform The ignition primary waveform is a measurement of the voltage on the negative side of the ignition coil. The earth path of the coil can produce over 350V. Different types of

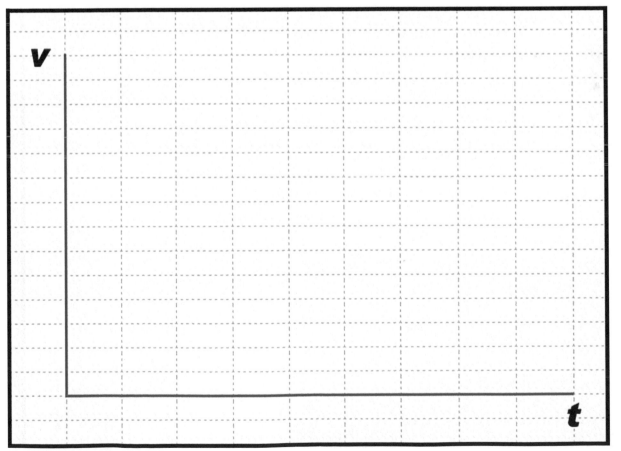

Figure 10.125 Stepper motor signal waveform

ignition coils produce slightly different traces but the fundamental parts of the trace and principles are the same.

Figure 10.126 Direct ignition coils in position

In the waveform shown, the horizontal voltage line at the centre of the oscilloscope is at fairly constant voltage of approximately 40V, which then drops sharply into what is referred to as the coil oscillation. The length of the horizontal voltage line is the 'spark duration' or 'burn time', which in this particular case is about 1ms. The coil oscillation period should display a minimum of 4 to 5 peaks (both upper and lower). A loss of peaks would indicate a coil problem.

There is no current in the coil's primary circuit until the dwell period. This starts when the coil is earthed and the voltage drops to zero. The dwell period is controlled by the ignition amplifier or ECU and the length of the dwell is determined by the time it takes to build up to about 8A. When this predetermined current has been reached, the amplifier stops increasing the primary current and it is maintained until the earth is removed from the coil. This is the precise moment of ignition.

The vertical line at the centre of the trace is in excess of 200V, this is called the 'induced voltage'. The induced voltage is produced by magnetic inductance. At the point of ignition, the coil's earth circuit is removed and the magnetic flux collapses across the coil's windings. This induces a voltage between 150 and 350V. The coil's high tension output will be proportional to this induced voltage. The height of the induced voltage is sometimes referred to as the primary peak volts.

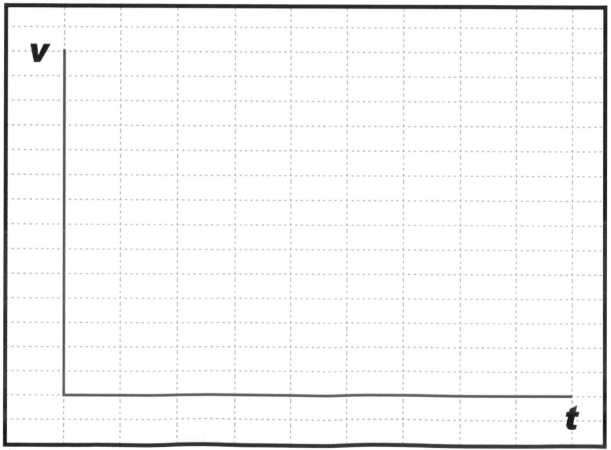

Figure 10.127 Primary ignition voltage trace waveform

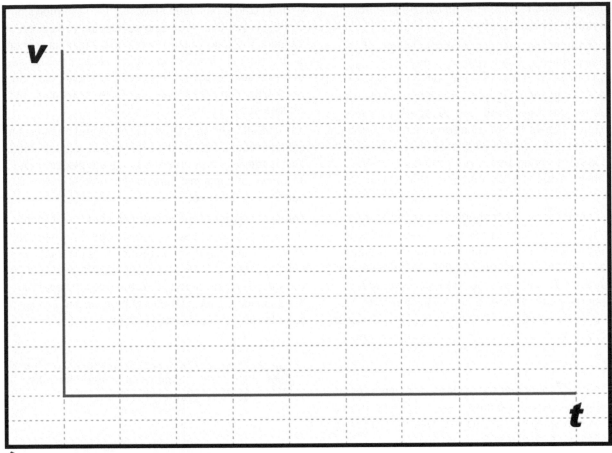

Figure 10.128 Ignition secondary trace waveform

From the example current waveform, the limiting circuit can be seen in operation. The current switches on as the dwell period starts and rises until the required value is achieved (usually about 8A). At this point the current is maintained until it is released at the point of ignition.

The dwell will expand as the engine revs are increased to maintain a constant coil saturation time. This gives rise to the term 'constant energy'. The coil saturation time can be measured and this will remain the same regardless of engine speed. The example shows a charge time of about 3.5ms.

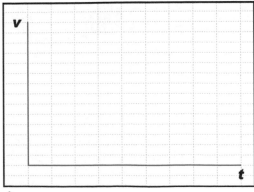

Figure 10.129 Primary ignition current trace waveform

Ignition secondary waveform The ignition secondary waveform is a measurement of the HT output voltage from the ignition coil. Some coils can produce over 50,000V. Different types of ignition coils produce slightly different traces but the fundamental parts of the trace and principles are the same.

Figure 10.130 Spark plugs (Source: Bosch Press)

The ignition secondary picture shown in the example waveform is from an engine fitted with electronic ignition. In this case, the waveform has been taken from the main coil lead (king lead). Suitable connection

543

methods mean that similar traces can be seen for other types of ignition system.

The secondary waveform shows the length of time that the HT is flowing across the spark plug electrode after its initial voltage, which is required to initially jump the plug gap. This time is referred to as either the 'burn time' or the 'spark duration'. In the trace shown it can be seen that the horizontal voltage line in the centre of the oscilloscope is at fairly constant voltage of approximately 3 or 4kV, which then drops sharply into the 'coil oscillation' period.

The coil oscillation period should display a minimum of 4 or 5 of peaks (both upper and lower). A loss of peaks indicates that the coil may be faulty. The period between the coil oscillation and the next 'drop down' is when the coil is at rest and there is no voltage in the secondary circuit. The 'drop down' is referred to as the 'polarity peak', and produces a small oscillation in the opposite direction to the plug firing voltage. This is due to the initial switching on of the coil's primary current.

The plug firing voltage is the voltage required to jump and bridge the gap at the plug's electrode, commonly known as the 'plug kV'. In this example the plug firing voltage is about 12 or 13kV.

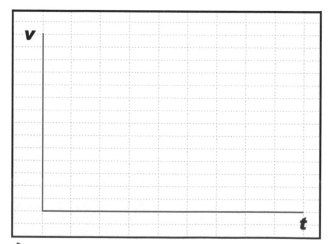

Figure 10.131 Distributorless ignition

When the plug kVs are recorded on a DIS or coil per cylinder ignition system, the voltage seen on the waveform should be in the 'upright position'. If the trace is inverted it would suggest that either the wrong polarity has been selected from the menu or in the case of DIS, the inappropriate lead has been chosen. The plug voltage, while the engine is running, is continuously fluctuating and the display will be seen to move up and down. The maximum voltage at the spark plug, can be seen as the 'Ch A: Maximum (kV)' reading at the bottom of the screen.

It is a useful test to snap the throttle and observe the voltage requirements when the engine is under load. This is the only time that the plugs are placed under any strain and is a fair assessment of how they will perform on the road.

The second part of the waveform after the vertical line is known as the spark line voltage. This second voltage is the voltage required to keep the plug running after its initial spark to jump the gap. This voltage will be proportional to the resistance within the secondary circuit. The length of the line can be seen to run for approximately 2ms.

 Make a simple sketch to show how one of the main components or systems in this section operates

10.4.4 Other components 🔞

Alternator waveform Checking the ripple voltage produced by an alternator is a very good way of assessing its condition.

Figure 10.132 Alternator

The example waveform illustrates the rectified output from the alternator. The output shown is correct and that there is no fault within the phase windings or the diodes (rectifier pack).

The three phases from the alternator have been rectified to dc from its original ac and the waveform shows that the three phases are all functioning.

If the alternator is suffering from a diode fault, long downward 'tails' appear from the trace at regular intervals and 33% of the total current output will be lost. A fault within one of the three phases will show a similar picture to the one illustrated but is three or four times the height, with the base to peak voltage in excess of 1V.

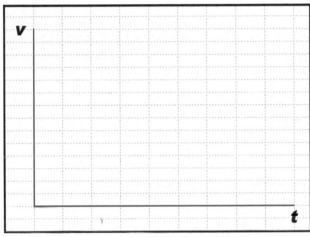

Figure 10.133 Alternator ripple voltage waveform

The voltage scale at the side of the oscilloscope is not representative of the charging voltage, but is used to show the upper and lower limits of the ripple. The 'amplitude' (voltage/height) of the waveform will vary under different conditions. A fully charged battery will show a 'flatter' picture, while a discharged battery will show an exaggerated amplitude until the battery is charged. Variations in the average voltage of the waveform are due to the action of the voltage regulator.

Relative compression petrol waveform Measuring the current drawn by the starter motor is useful to determine starter condition but it is also useful as an indicator of engine condition.

The purpose of this particular waveform is therefore to measure the current required to crank the engine and to evaluate the relative compressions. The amperage required to crank the engine depends on many factors, such as: the capacity of the engine, the number of cylinders,

the viscosity of the oil, the condition of the starter motor, the condition of the starter's wiring circuit and the compressions in the cylinders. To evaluate the compressions therefore, it is essential that the battery is charged, and the starter and associated circuit are in good condition.

The current for a typical 4 cylinder petrol/gasoline engine is in the region of 100 to 200 A.

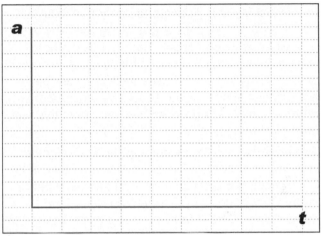

Figure 10.134 Spark ignition engine cranking amps waveform

In the waveform shown, the initial peak of current (approx. 400A) is the current required to overcome the initial friction and inertia to rotate the engine. Once the engine is rotating, the current will drop. It is also worth mentioning the small step before the initial peak, which is being caused by the switching of the starter solenoid. The compressions can be compared against each other by monitoring the current required to push

Figure 10.135 OBD socket – pin 6 is CAN-high and pin 14 is CAN-low

10

each cylinder up on its compression stroke. The better the compression the higher the current demand and vice versa. It is therefore important that the current draw on each cylinder is equal.

CAN-H and CAN-L waveform Controller area network (CAN) is a protocol used to send information around a vehicle on data bus. It is made up of voltage pulses that represent ones and zeros, in other words, binary signals. The data is applied to two wires known as CAN-high and CAN-low.

In this display, it is possible to verify that data is being continuously exchanged along the CAN bus. It is also possible to check that the peak to peak voltage levels are correct and that a signal is present on both CAN lines. CAN uses a differential signal, and the signal on one line should be a coincident mirror image (the signals should line up) of the data on the other line.

The usual reason for examining the CAN signals is where a CAN fault has been indicated by OBD, or to check the CAN connection to a suspected faulty CAN node. The vehicle manufacturers' manual should be referred to for precise waveform parameters.

The signal shown is captured on a fast timebase and allows the individual state changes to be viewed. This enables the mirror image nature of the signals, and the coincidence of the edges to be verified.

Summary Scope diagnostics is now an essential skill for the technician to develop. As with all diagnostic techniques that use test equipment, it is necessary for the user to know how:

1 the vehicle system operates
2 to connect the equipment
3 readings should be interpreted.

Remember that an oscilloscope is really just a voltmeter or ammeter but it draws a picture of the readings over a set period of time. Learn what good waveforms look like and then you will be able to make good judgements about what is wrong when they are not so good!

 Make a simple sketch to show how one of the main components or systems in this section operates

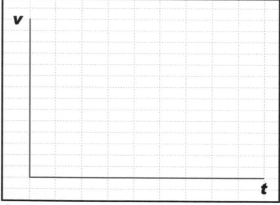

Figure 10.136 CAN high signal waveform

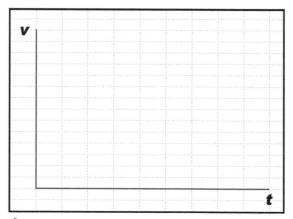

Figure 10.137 CAN low signal waveform

A trending future: Electric and hybrid vehicles and ADAS

DOI: 10.1201/9781003173236-11

11.1 Electric and hybrid vehicles overview

11.1.1 Introduction

Although it is not mandatory, it is recommended that learners in EV and Hybrid vehicle technology should have 3 GCSEs or Scottish Standard Grade/Intermediate in Maths, English and a Science based subject. Experience in the automotive industry is strongly recommended, and this content was optional learning at the time of writing. Due to the government's desire to increase uptake of this technology, it is expected to become part of the qualification in due course.

This unit is intended to provide **knowledge** that is required to work safely around vehicles that have high voltage systems and to understand their dangers. This information is not intended to allow work on high voltage systems, and this work should not be undertaken until suitably qualified.

11.1.2 Electric vehicle market ❶❷

The EV market has increased year on year, and the number of battery technology native drivers (meaning that this technology has been easily available to them from their first car) also rises. Toyota and Honda first entered the hybrid market in the 1990s, with Toyota dominating the market sector with the Prius. Nissan brought a cost-effective hatchback to the market with the Leaf in 2010, a market which now enjoys an offering from almost every manufacturer thus bringing increased choice to the consumer. 'Range anxiety' is the term applied to a fear that the vehicle could not complete a journey, and this can be a barrier to uptake of this technology. This problem, however, is becoming less apparent as the charging network increases together with motorist knowledge.

The government's intention to disallow new internal combustion engine (ICE) car sales from 2030 (correct at time of press) has been met with mixed reaction. There is speculation that this will age the UK car parc while others have embraced the target. In addition, the 2020 pandemic may bring changes in employment frameworks that might impact how cars are used. The pandemic resulted in a greater reliance on the private motor car particularly with concerns for safety using public travel. As more city centres create restrictions and/or financial penalties for using ICE cars in areas of high congestion, this may also impact uptake of electric car ownership, particularly small and cheap quadricycle models and/or shared ownership or rental solutions.

Figure 11.1 Citroën AMI - electric quadricycle (Source: Citroën Media)

11.1.3 The EV experience ❶❷

Introduction Living with an electric car is not that different to ownership of an ICE, any motorist with an appropriate license can purchase one and drive away with no special training. The standout difference is how the car receives power to run. An ICE vehicle has an advantage of a short refuelling time, the fuel is transferred from one place to another quickly and pumps are easily found owing to the large number of refilling stations. ICE vehicles are not as impacted in changes of environmental conditions, whereas an electric car can use more energy when the weather is extreme. A motorist used to the convenience of fuelling an ICE may find the transition to electric more difficult than an electric car native.

The electric car can be charged at home, once the necessary equipment is safely fitted, and uptake for charging availability at the workplace is increasing. Rapid charging can dump a lot of energy into a battery in a much shorter time in a more similar fashion to the ICE but this can have a detrimental effect on the battery if this is the primary method of charging. The advantage of charging the car at home and work not only removes the inconvenient visit to the fuel station entirely, but it also allows the development of bi-directional charging, the CHAdeMO protocol enables the use of a vehicle as temporary energy storage device, allowing it to charge during the day (whilst renewable energy is easy to make) and then when connected to the home, used to power the home using a home battery system, reducing the demand on the grid.

Driving the EV EVs deliver power faster than the ICE counterpart, giving swift acceleration and there are no gears to slow the progress. EVs are quite heavy, but that weight is mainly made up by the batteries, which can be placed to give the car a low centre of gravity and a planted driving experience. With no engine noise, the compartment is quiet and without vibration, in fact, in can be difficult to tell if the car is on without looking at the dash! Due to their quiet nature, EVs create an artificial sound when driving to alert pedestrians of their presence.

The maintenance and repair of an EV requires specialist equipment and qualified technicians, although you do not need a special driving licence to drive one. Roles such as valeting currently do not require formal qualifications to work with cars with high voltage systems, although it is an excellent idea to have a knowledge of the technology and how it is different to an ICE vehicle.

Maintenance and repairs of hybrid technologies need both the requirements of the EV and the ICE technology that are all packaged into the same engine bay as either a pure EV or standalone ICE – potentially increasing labour times and expense to maintain two systems that are tightly fitted into the same space!

For the motorist, maintenance of the pure EV is different as the vehicle does not require the fluids and filters of an ICE car. It is important to note that they do still require maintenance, and some of those maintenance tasks can be of a different technical level to that of an ICE. EVs still share a lot of components with their ICE cousins: steering, suspension and infotainment will still look remarkably familiar in their functionality, whereas brakes and air conditioning may be integrated into the high voltage systems. Some high voltage batteries have sophisticated cooling systems that are now adopting filtration.

11.1.4 History ❶ ❷

It may be surprising to learn that EVs were first invented in the early 1800s! Popular for replacing the horse and cart, particularly for doorstop deliveries, uptake in the UK of EVs was booming in the late 1800s to the early 1900s. Many of these vehicles were developed and built in the UK and shipped all over the world. World War 2 interrupted production but their popularity returned with around 40,000 of them in service in the 1970s. Although termed 'milk floats' the service sold anything that could be useful for a doorstep delivery including petfood, other dairy products and compost. One milkman was even convicted of selling cannabis to his elderly neighbours in 2009! The 2020 pandemic saw an unprecedented surge in demand for doorstep delivery and it is interesting to note there is a return to electric vehicles; the regular short run and stop use of the vehicle makes this technology an excellent choice once again.

Figure 11.2 supermarket delivery vehicle (source Waitrose & Partners Media)

Table 11.1 Key stages of EV development

Beginning	First Age	Boom and Bust	Second Age	Third Age
The earliest electric vehicles were invented in Scotland and the USA	Electric vehicles enter the market and start to find broad appeal	EVs reach historical peaks of production but are then displaced by petrol-engine cars	High oil prices and pollution created a new interest in electric vehicles	Public and private sectors now commit to vehicle electrification
1801–1850	1851–1990	1901–1950	1951–2000	2001–present
1832–39 Robert Anderson of Scotland built the first prototype electric carriage	1888 German engineer Andreas Flocken built the first four-wheeled electric car	1908 The petrol-powered Ford Model T was introduced to the market	1966 US Congress introduced legislation recommending EVs as a way of reducing air pollution	2008 Oil prices reached record highs
1834 Thomas Davenport of the USA invented the first direct current electrical motor in a car that operates on a circular electrified track	1897 The first commercial EVs entered the New York City taxi fleet. The Pope Manufacturing Company became the first large-scale EV manufacturer in the USA	1909 William Taft was the first US President to buy an automobile, a Baker Electric	1973 The OPEC oil embargo caused high oil prices, long delays at fuel stations and therefore renewed interest in EVs	2010 The Nissan LEAF was launched

11

Table 11.1 (Continued)

Beginning	First Age	Boom and Bust	Second Age	Third Age
	1899 The 'La Jamais Contente' (The Never Happy!), built in France, became the first electric vehicle to travel over 100 km/h.	1912 The electric starter motor was invented by Charles Kettering. This made it easier to drive petrol cars because hand-cranking was not now necessary	1976 The French government launched the 'PREDIT', which was a programme accelerating EV research and development	2011 The world's largest electric car sharing service, Autolib, was launched in Paris with a targeted stock of 3000 vehicles
	1900 Electricity-powered cars were the best-selling road vehicle in the USA with about 28% of the market	1912 The global stock of EVs reached around 30,000	1996 To comply with California's Zero Emission Vehicle (ZEV) requirements of 1990, GM produced the EV1 electric car	2011 The global stock of EVs reached around 50,000
		1930 By 1935, the number of EVs dropped almost to zero and the ICE vehicles dominated because of cheap petrol	1997 In Japan, Toyota began sales of the Prius, the world's first commercial hybrid car. 18,000 were sold in the first year	2011 The French government fleet consortium committed to purchase 50,000 EVs over four years
		1947 Oil rationing in Japan led carmaker Tama to release a 4.5 hp electric car. It used a 40 V lead–acid battery		2011 Nissan LEAF won the European Car of the Year award
				2012 The Chevrolet Volt PHEV outsold half the car models on the US market
				2012 The global stock of EVs reached around 180,000
				2014 Tesla Model S, Euro NCAP 5-star safety rating, autopilot-equipped, available all-wheel drive dual motor with 0–60 mph in as little as 2.8 seconds and a range of up to 330 miles
				2015 Car manufacturers were caught cheating emission regulations making EVs more prominent in people's minds as perhaps the best way to reduce consumption and emissions1

Beginning	First Age	Boom and Bust	Second Age	Third Age
				2015 The global stock of EVs reached around 700,000 and continues to grow (22,000 in the UK and 275,000 in the USA)

(Primary Source: Global EV outlook)

 Select a routine from section 1.13 and follow the process to study a component or system

11.2 Electric vehicle systems and components

11.2.1 Types of electric vehicles ❶❷

Electric Vehicles (EVs) usually refers to any vehicle that is powered, in part or in full, by a battery that can be directly plugged into the mains. The term "EV" can be generally used to cover all types of electric vehicle.

Pure-Electric Vehicles are electric vehicles powered only by a battery. They are sometimes referred to as a Battery Electric Vehicle (BEV), this is commonly being dropped in favour of just EV as a term.

Extended-Range Electric Vehicles (E-REVs) are like an EV, they generally have a smaller battery and a small ICE engine, used like a generator to put power into the battery whilst the vehicle is running. They are becoming less popular now and often seen as a progression step between the technologies.

Plug-In Hybrid Electric Vehicles (PHEVs) have an internal combustion engine (ICE) but also a battery range more than 10 miles (the average motorist completes less than 20 miles a day). Once the battery is depleted, the vehicle reverts to the full hybrid capability of utilizing both battery and ICE power.

Hybrid Electric Vehicles (HEVs) have a battery that assists the ICE, increasing its efficiency, but it is not possible to charge this battery externally in the same way as a PHEV or EV.

Micro Hybrid Vehicles normally employ a stop-start system and regenerative braking which charges the 12V battery. They do not have a high voltage battery, nor does the vehicle use the battery to drive the car.

Alternative Fuel Source Vehicles can refer to any vehicle that does not run on either petrol or diesel. This can include EVs! When used in the context of EV technology, alternative fuel is commonly used for hydrogen cell vehicles, vehicles that use methanol or other emerging fuel cell technology.

Figure 11.3 Nissan LEAF – pure EV (Source: Nissan Media)

Table 11.2 Summary of EVs and HEVs and their alternative names

Electric Vehicle/Car (EV), Electrically Chargeable Vehicle/Car	Generic terms for a vehicle powered, in part or in full, by a battery that can be plugged into the mains
Pure-EV, Pure-Electric Car, Vehicle, All Electric, Battery Electric Vehicle (BEV), Fully Electric	A vehicle powered only by a battery charged from mains electricity. Currently, typical pure-electric cars have a range of about 100 miles
Plug-In Hybrid Electric Vehicle (PHEV), Plug-In Hybrid Vehicle (PHV)	A vehicle with a plug-in battery and an internal combustion engine (ICE). Typical PHEVs will have a pure-electric range of 10–30 miles. After the pure-electric range is used up, the vehicle reverts to the benefits of full hybrid capability
Extended-Range Electric Vehicle (E-REV), Range-Extended Electric Vehicle (RE-EV)	A vehicle powered by a battery with an ICE-powered generator on board. E-REVs are like pure-EVs but with a shorter battery range of around 50 miles. Range is extended by an on-board generator providing additional miles of mobility. With an E-REV the vehicle is still always electrically driven and is known as a series hybrid (more on this later)

11

Hybrid Electric Vehicles (HEV), Full/Normal/Parallel/Standard hybrid	A hybrid vehicle is powered by a battery and/or an ICE. The power source is selected automatically by the vehicle, depending on speed, engine load and battery charge. This battery cannot be plugged in, so charge is maintained by regenerative braking supplemented by ICE-generated power
Mild Hybrid	A mild hybrid vehicle cannot be plugged in or driven solely on battery power. However, it does harvest power during regenerative braking and uses this during acceleration (current F1 cars are a type of mild hybrid
Micro Hybrid	A micro hybrid normally employs a stop-start system and regenerative braking which charges the 12 V battery
Stop-start Hybrid	A stop-start system shuts off the engine when the vehicle is stationary. An enhanced starter motor is used to support the increased number of engine starts
Alternative Fuel Vehicle (AFV)	Any vehicle that is not solely powered by traditional fuels (i.e. petrol or diesel) is referred to as alternative fuel
Internal Combustion Engine (ICE)	Petrol or diesel engine, as well as those adapted to operate on alternative fuels
Electric quadricycles	This is a four-wheeled vehicle that is categorized and tested in a similar way to a moped or three-wheeled motorcycle
Electric motorcycles	Battery only, so some full electric drive motorcycles may have a limited range. However, the Volt 220, which takes its name from its range of 220 kms, and will do up to 60 mph according to the manufacturer. Other manufacturers are also developing, the Harley Davidson shown below has a 146-mile range.

11.2.2 Identifying electric and hybrid vehicles ❶❷

Identifying electric and hybrid technology can sometimes be tricky if unfamiliar. Manufacturers are aware of this, and they wish to demonstrate the distinction of this new technology and so will brand a vehicle to distinguish it from the ICE counterparts. The government also announced in June 2020 that zero emission vehicles would be further identified by a green flash on the registration plate. The purpose of the registration plate flash is to make for easy identification, this could be important for areas that only allow zero emissions vehicles entry, and it is also hoped it may incentivise zero emission car ownership.

Six areas to check when identifying an EV, hybrid or alternative fuel vehicle

- Construction
- Badging
- Model
- Procedure to follow if initial identification is not possible
- Cable colouring
- Registration plate (post 2020)

Examples of where high voltage technology is available

- Cars that use this technology
  - EVs (or BEVs)
  - HEVs
  - PHEVs
  - Micro hybrid (cars that use start/stop technology)

Figure 11.4 manufacturer's badge (source Toyota Media)

- Fuel cell vehicle/Alternative fuel source vehicle
- Two-wheel vehicles (currently only full electric or ICE are offered)
- Commercial vehicles
- Passenger transport
- Mechanical handling equipment (MHE)
- Plant

Figure 11.5 Electric motorcycle (Source: Harley-Davidson Media)

11.2.3 High voltage components ❶❷

High voltage components are designed by manufacturers to be easily identifiable and high voltage cables are brightly coloured with an orange outer casing and a yellow inner casing. High voltage cables also require greater insulation, so will appear more heavy duty than the cabling used in lower voltage systems.

Manufacturers also use stickers, labels and tags to identify components that may not be easily brightly coloured or could be discoloured over time. It is important to be able to identify EV components in order to work with technology safely; in many cases, manufacturer's information will be required to assist with this task. Names of components can differ slightly between brands, but generally the main components are:

▶ high voltage battery
▶ motor
▶ high voltage cables and connectors
▶ relays (switching components)
▶ control units (power electronics)
▶ charging points
▶ isolators (safety devices)
▶ interlock devices (safety devices)
▶ inverter
▶ DC to DC converter
▶ battery management controller
▶ ignition key/key on control switch
▶ driver display panel/interface

Figure 11.6 Under bonnet components of a pure EV (source Media)

There are additions to these systems, such as braking, steering and even air conditioning that integrate into the high voltage system and are covered in later qualifications. The high voltage battery is not easily accessed, and most high voltage systems will still carry a 12V battery, one just like an ICE would use to run ancillary components and to activate the critical relays that bring the high voltage battery online and allow for it to be charged. A flat 12V battery will disable the high voltage system and will therefore not allow the vehicle to operate.

11.2.4 Motors and batteries overview ❶❷

Drive motors The electronically commutated motor (ECM) is, in effect, half way between an AC and a DC motor. The rotor contains permanent magnets and hence no slip rings. It is sometimes known as a brushless motor. The rotor operates a sensor, which provides feedback to the control and power electronics. This control system produces a rotating field, the frequency of which determines motor speed.

Brushless These motors are also described as brushless DC motors (BLDC) and they are effectively AC motors because the current through it alternates. However, because the supply frequency is variable, has to be derived from DC and its speed/torque characteristics are similar to a brushed DC motor, it is called a DC motor. Two typical motors are shown here, one is integrated with the engine flywheel and the other a separate unit. Both are DC brushless motors and are water cooled.

Figure 11.7 Bosch integrated motor generator (IMG) also called integrated motor assist (IMA) by some manufacturers

Figure 11.8 Separate motor unit showing the coolant connections on the side and the three main electrical connections on top

Lithium-ion batteries This technology is becoming the battery technology of choice, but it still has plenty of potential to offer. Today's batteries have an energy density of up to 140 Wh/kg or more in some cases, but have the potential to go as high as 280 Wh/kg. Much research in cell optimization is taking place to create a battery with a higher energy density and increased range. Lithium-ion technology is currently considered the safest.

Figure 11.9 Battery packs are several hundred volts

Operation The Li-ion battery works as follows. A negative pole (anode) and a positive pole (cathode) are part of the individual cells of a lithium-ion battery together with the electrolyte and a separator. The anode is a graphite structure and the cathode is layered metal oxide. Lithium-ions are deposited between these layers. When the battery is charging, the lithium-ions move from the anode to the cathode and take on electrons. The number of ions therefore determines the energy density. When the battery is discharging, the lithium-ions release the electrons to the anode and move back to the cathode.

11.2.5 Charging ❶ ❷

Charging, standards, and infrastructure
Most electric cars will be charged at home, but national infrastructures are developing. There are, however, competing organizations and commercial companies, so it is necessary to register with a few different organizations to access their charging points. Many businesses now also provide charging stations for staff and visitors. Some are pay in advance, some are pay as you go, and others require a monthly subscription. Many apps and websites are available for locating charge points. An example of this can be found at https://www.zap-map.com.

Plugs and connections There are different plugs, types of supply and methods of connection; it is important to use the correct plug as it is difficult to fit a plug into an incorrect socket due to the shapes of the available plugs. Some power supplies are 'tethered' meaning that the cable is fixed to the supply and is plugged into the car. Other supplies are simply a socket that require the motorist to select their own cable and connect themselves, and there is a lock facility to ensure that the cable cannot be removed by passers by. Once connected, before electricity begins to flow, a 'handshake' is completed to determine the situation is safe to operate the relays and begin charging. A supply can either be DC or AC. Most vehicles will provide status updates, such as unexpectedly becoming unplugged or reaching a full charge, and inform the motorist via an app.

Fuel stations The uptake of this technology is now prompting large fuel sellers to incorporate charging facilities into their existing business model and deploying through their established infrastructure, making the availability of charging along main routes more accessible.

Safety note: Although the rechargeable electric vehicles and equipment can be recharged from a domestic wall socket, a charging station has additional current or connection sensing mechanisms to disconnect the power when the EV is not charging. There are two main types of safety sensor:

▶ Current sensors monitoring the power consumed and only maintain the connection if the demand is within a predetermined range.

▶ Additional sensor wires, providing a feedback signal that requires special power plug fittings.

Wet weather It is safe to charge in wet weather. When you plug in the charge lead, the connection to the supply is not made until the plug is completely

Figure 11.10 Charging point on the roadside (Source: Citroën Media)

in position. Circuit breaker devices are also used for additional safety. Clearly some common sense is necessary, but EV charging is very safe.

Domestic charge points: It is strongly recommended that home charging sockets and wiring are installed and approved by a qualified electrician. A home charge point with its own dedicated circuit is the best way of charging an EV safely. For rapid charging, special equipment and an upgraded electrical supply would be required and is therefore unlikely to be installed at home, where most consumers will charge overnight.

Figure 11.11 A tethered domestic charging point (source EDF Media)

How long it takes to charge an EV depends on the type of vehicle, how discharged the battery is and the type of charge point used. Typically, pure-electric cars using standard charging will take between six and eight hours to charge fully and can be 'opportunity charged' whenever possible to keep the battery topped up.

Rapid charging Pure-EVs capable of using rapid charge points could be fully charged in around 30 minutes and can be topped up in around 20 minutes, depending on the type of charge point and available power. PHEVs take approximately

two hours to charge from a standard electricity supply. E-REVs take approximately four hours to charge from a standard electricity supply. PHEVs and E-REVs require less time to charge because their batteries are smaller.

Table 11.3 Estimated charging times

Charging time for 100-km range	Power supply	Power	Voltage	Max. current
6–8 hours	Single phase	3.3 kW	230 V AC	16 A
3–4 hours	Single phase	7.4 kW	230 V AC	32 A
2–3 hours	Three phase	10 kW	400 V AC	16 A
1–2 hours	Three phase	22 kW	400 V AC	32 A
20–30 minutes	Three phase	43 kW	400 V AC	63 A
20–30 minutes	Direct current	50 kW	400–500 V DC	100–125 A
10 minutes	Direct current	120 kW	300–500 V DC	300–350 A

Cost The cost of charging an EV depends on the size of the battery and how much charge is left in the battery before charging. As a guide, charging an electric car from flat to full will cost from as little as £1 to £4. This is for a typical pure-electric car with a 24 kWh battery that will offer around 100-mile range. This results in an average cost of a few pence per mile.

Overnight If you charge overnight, you may be able to take advantage of cheaper electricity rates when there is surplus energy. The cost of charging from public points will vary; many will offer free electricity in the short term. It is also possible to register with supply companies who concentrate on energy from renewable sources.

 Select a routine from section 1.13 and follow the process to study a component or system

11.3 Safety

11.3.1 Risks of working with EVs ❶ ❷

EVs introduce hazards into the workplace in addition to those normally associated with the repair and maintenance of vehicles, roadside recovery and other vehicle-related activities. These include:

▶ the presence of high-voltage components and cabling capable of delivering a fatal electric shock
▶ the storage of electrical energy with the potential to cause explosion or fire

- components that may retain a dangerous voltage even when a vehicle is switched off
- electric motors or the vehicle itself that may move unexpectedly due to magnetic forces within the motors
- manual handling risks associated with battery replacement
- the potential for the release of explosive gases and harmful liquids if batteries are damaged or incorrectly modified
- the possibility of people being unaware of vehicles moving, because when electrically driven they are silent in operation
- the potential for the electrical systems on the vehicle to affect medical devices such as pacemakers and insulin controllers

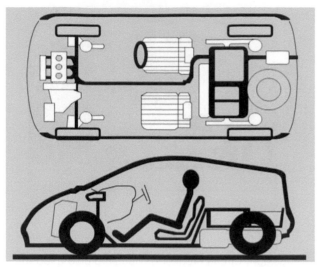

Figure 11.13 Motor and power pack locations on a typical hybrid: 1, power pack; 2, integrated motor

the high-voltage system. This will prevent the risk of electric shock or short circuit of the high-voltage system.

Risks The following table lists some identified risks involved with working on ALL vehicles. The table is by no means exhaustive but serves as a good guide.

Table 11.4 Risks and their reduction

Identified risk	Reducing the risk
Electric shock 1	Voltages and the potential for electric shock when working on an EV mean a high-risk level – see section 2.2 for more details.
Electric shock 2	Ignition HT is the most likely place to suffer a shock when working on an ICE vehicle; up to 40,000 V is quite normal. Use insulated tools if it is necessary to work on HT circuits with the engine running. Note that high voltages are also present on circuits containing windings due to back emf as they are switched off; a few hundred volts is common. Mains-supplied power tools and their leads should be in good condition and using an earth leakage trip is highly recommended. Only work on HEV and EVs if training in the high-voltage systems.
Battery acid	Sulphuric acid is corrosive, so always use good personal protective equipment. In this case, overalls and if necessary, rubber gloves. A rubber apron is ideal, as are goggles if working with batteries a lot.
Raising or lifting vehicles	Apply brakes and/or chock the wheels and when raising a vehicle on a jack or drive-on lift. Only jack under substantial chassis and suspension structures. Use axle stands in case the jack fails.

Figure 11.12 Orange high-voltage cables are easy to spot (Source: Renault Media)

Power unit Most of the high-voltage components are combined in a power unit. This is often located behind the rear seats or under the luggage compartment floor (or the whole floor in a Tesla). The unit is a metal box that is completely closed with bolts. A battery module switch, if used, may be located under a small secure cover on the power unit. The electric motor is located between the engine and the transmission or as part of the transmission on a hybrid or on a pure-EV; it is the main driving component. A few vehicles use wheel motors too. High-voltage wires are always brightly coloured: an orange outer casing with a yellow inner casing.

NOTE: Always follow the manufacturer's instructions – it is not possible to outline all the variations here.

Orange or brightly coloured wires The electrical energy is conducted to or from the motor by thick orange wires on most vehicles. If these wires must be disconnected, SWITCH OFF or DE-ENERGISE

Running engines	Do not wear loose clothing; good overalls are ideal. Keep the keys in your possession when working on an engine to prevent others starting it. Take extra care if working near running drive belts.
Exhaust gases	Suitable extraction must be used if the engine is running indoors. Remember it is not just the carbon monoxide that might make you ill or even kill you; other exhaust components could cause asthma or even cancer.
Moving loads	Only lift what is comfortable for you; ask for help if necessary and/or use lifting equipment. As a general guide, do not lift on your own if it feels too heavy!
Short circuits	Use a jump lead with an in-line fuse to prevent damage due to a short when testing. Disconnect the battery (earth lead off first and back on last) if any danger of a short exists. A very high current can flow from a vehicle battery, it will burn you as well as the vehicle.
Fire	Do not smoke when working on a vehicle. Fuel leaks must be attended to immediately. Remember the triangle of fire: Heat–Fuel–Oxygen. Don't let the three sides come together.
Skin problems	Use a good barrier cream and/or latex gloves. Wash skin and clothes regularly.

11.3.2 IMI TechSafe™ **1** **2**

The IMI TechSafe™ professional registration scheme is designed to ensure complex automotive technologies are repaired safely and that technicians work safely – particularly in the UK, but it is just as effective internationally.

Figure 11.14 Compact hatchback (Source: Citroën Media)

To be added to the register, a technician must: successfully complete a specified qualification (e.g. Electric/Hybrid Vehicles Level 2, 3 or 4), join the IMI Professional Register, and complete specified annual CPD to ensure current competency is maintained.

Figure 11.15 IMI TechSafe process

For EVs this will fully meet the requirements that anyone working on high voltages must be competent (Electricity at Work Regulations 1989). ADAS, and other areas will be covered in a similar way. Technology safe, means technician safe, means customer safe.

Figure 11.16 IMI TechSafe logo

 Select a routine from section 1.13 and follow the process to study a component or system

11.4 Advanced driver assistance systems overview

11.4.1 Introduction **3**

Advanced driver assistance systems (ADAS) are, as the name suggests, designed to help the driver. This improves safety because most road accidents occur due to human error. Automated systems help to minimize human error, which has been proven to reduce road fatalities. These are also the enabling technologies for full automated driving.

Example As an interesting example to show the benefits of ADAS,

> … there was an incident December 2016 when a Tesla saved a driver by suddenly activating the collision warning, and the autopilot turned on the emergency brake. It was not until after this manoeuvre that a car in front of the driver flipped and landed in the path of the oncoming Tesla. Apparently, the sensors could tell that a threat had developed two cars ahead of the Tesla, which is why the driver did not see it. In other words, an accident was prevented because the radar was able to detect a situation beyond the driver's field of vision.
>
> (Meyer-Hermann, Brenner et al. 2018)

11

Inputs ADAS relies on inputs from several sources such as LiDAR, RADAR, cameras and vehicle CAN data. Safety features are designed to alert the driver to potential problems, or to avoid collisions by implementing safeguards. In some cases, this means taking control of the vehicle.

Figure 11.17 Sensing activity

ADAS features can, for example, switch on lights, provide adaptive cruise control and collision avoidance, incorporate traffic warnings, alert the driver to other vehicles and dangers, warn if lane departure is detected or even initiate automated lane guidance. Cameras are used to see in blind spots. Many modern vehicles now have systems such as electronic stability control, anti-lock brakes, lane departure warning, adaptive cruise control and traction control. All these systems can be affected by mechanical alignment. For this reason, correct repairs, adjustments and servicing are essential.

Figure 11.18 Automatic parking (Source Park4U)

Some example advanced driver assistance systems are listed here:

- Adaptive cruise control
- Adaptive light control
- Anti-lock braking system

- Automatic parking
- Blind spot monitor
- Collision avoidance
- Collision warning
- Driver drowsiness detection
- Electric vehicle warning sounds
- Emergency driver assistant
- Glare-free high beam and pixel light
- Lane change assistance
- Lane departure warning
- Navigation system with traffic information
- Night vision
- Parking sensor
- Pedestrian protection
- Rain sensor
- Surround view
- Traffic sign recognition
- Tyre pressure monitoring

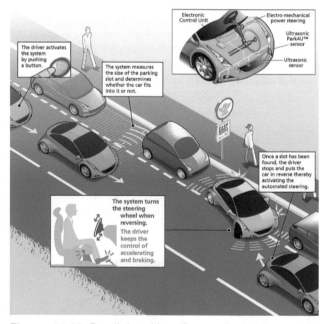

Figure 11.19 Parallel parking (Source Park4U)

Figure 11.20 Camera used for night vision (Source: First Sensor AG)

Create a mind map to illustrate the features of a key component or system.

11.4.2 Systems 3

Adaptive cruise control Conventional cruise control is not always practical on many European roads. This is because the speed of the general traffic varies constantly, and traffic is often very heavy. The driver must take over from a standard cruise control system on many occasions to speed up or slow down. Adaptive cruise control (ACC) can automatically adjust the vehicle speed to the current traffic situation. The system has three main features:

▶ Maintain a speed as set by the driver
▶ Adapt this speed and maintain a safe distance from the vehicles in front

Figure 11.21 Adaptive cruise control

▶ Provide a warning if there is a risk of collision.

The main **extra components**, compared to standard cruise control, are the headway sensor and the steering angle sensor; the first of these is clearly the most important. Information on steering angle is used to further enhance the data from the headway sensor by allowing greater discrimination between hazards and spurious signals. Two types of headway sensor are in use: RADAR and LiDAR. Both contain transmitter and receiver units. The RADAR system uses microwave signals of up to 80 GHz, and the reflection time of these gives the distance to the object in front. LiDAR uses a laser diode to produce infrared light signals, the reflections of which are detected by a photodiode.

Sensors These two types of sensors have advantages and disadvantages. The RADAR system is not affected by rain and fog, but the LiDAR can be more

Figure 11.22 Lidar sensor

selective by recognizing the standard reflectors on the rear of the vehicle in front. Radar can produce strong reflections from bridges, trees, posts and other normal roadside items. It can also suffer loss of signal return due to multipath reflections. Under ideal weather conditions, the LiDAR system appears to be the best, but it becomes very unreliable when the weather changes. A beam divergence of about 2.5° vertically and horizontally has been found to be the most suitable whatever headway sensor is used. An important consideration is that signals from other vehicles fitted with this system must not produce erroneous results. Figure 11.23 shows a typical headway sensor and control electronics.

Figure 11.23 Headway sensor is fitted at the front of a vehicle (Source: Bosch Media)

Operation Fundamentally, the operation of an adaptive cruise system is the same as a conventional system, except when a signal from the headway sensor detects an obstruction. In this case, the vehicle speed is decreased. If the optimum stopping distance cannot be achieved by just backing off the throttle, a warning is supplied to the driver. Later systems also take control of the vehicle transmission and brakes.

11

Figure 11.24 Forward sensing

Obstacle avoidance radar This system, sometimes called collision avoidance radar, can be looked at in two ways. First, an aid to reversing, which gives the driver some indication as to how much space is behind the car. Second, it can be used as a vision enhancement system. The principle of radar as a reversing aid is illustrated. This technique is, in effect, a range-finding system. The output can be audio or visual, the latter being perhaps most appropriate, as the driver is likely to be looking backwards. The audible signal is a 'pip-pip-pip' type sound, the repetition frequency of which increases as the car comes nearer to the obstruction and becomes almost continuous as impact is imminent. Many systems now also make the noise come from the appropriate speaker(s) to indicate direction.

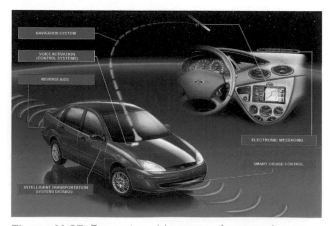

Figure 11.25 Reversing aid as part of a control system (Source: Ford)

Reversing aid A reverse sensing system is a reverse only parking aid system that uses sensors mounted in the rear bumper. Parking aid systems feature both front and rear sensors. Low-cost, high-performance ultrasonic range sensors are fitted to the vehicle. Generally, four sensors are used to form a detection zone as wide as the vehicle. A microprocessor

monitors the sensors and emits audible beeps during slow reverse parking to help the driver reverse or park the vehicle. The technique is relatively simple as the level of discrimination required is low and the system only has to operate over short distances.

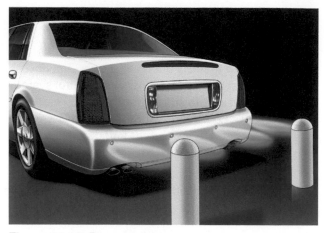

Figure 11.26 Rear sensors

Figure 11.27 Audi ultrasonic rear sensor

Rear radar Drivers are taught to assess surrounding traffic, before changing lanes, by checking their rear-view and side mirrors and looking over each shoulder. However, the area alongside and just behind the vehicle is a constant source of danger and often the cause of serious accidents. Drivers are not able to see into this area using either the rear view or side mirrors, but it is big enough for a vehicle to be missed by a cursory glance.

Lane change To help minimize this risk, a lane-changing assistant receives the information it needs from a mid-range radar sensor for rear-end applications. This means drivers are effectively looking over their shoulders all the time, because it reliably and accurately recognises other road users in their vehicle's blind spot.

A typical installation is to have two sensors in the rear bumper, one on the left, one on the right. These two rear sensors monitor the area alongside and behind the car. Powerful control software collates the sensor information to produce a complete picture of all traffic in the area behind the vehicle. Whenever another vehicle approaches at speed from behind or is already present in the blind spot, a signal such as a warning light in the side mirror alerts the driver to the hazard. Should the driver still activate the turn signal with the intention of changing lanes, the lane-changing assistant issues an additional acoustic and/or haptic warning.

Figure 11.28 Sensors monitor all traffic in the area behind the vehicle

Cross-traffic alert system The rear radar system can do more than just assist with lane-changing. These sensors also form part of a cross-traffic alert system, which supports drivers reversing out of perpendicular

Figure 11.29 Mid-range radar (MRR) sensor (Source: Bosch Media)

parking spaces when their rear view is obstructed. Able to recognise cars, cyclists and pedestrians crossing behind the reversing vehicle at up to 50m, the system alerts the driver to the imminent danger of collision by issuing an audible or visible signal.

Ranges Short range radar (SRR) works up to 30m, medium range radar (MRR) up to about 100m and long range radar (LRR) up to around 200m. The beam angle is narrower at longer ranges.

 Select a routine from section 1.3 and follow the process to study a component or system.

11.4.3 Radar ❸

Introduction Shown here is a block diagram to demonstrate the principle of a radar system. A frequency of 79GHz with a bandwidth of 4GHz is generally used and this has a resolution of 0.1m. 77GHz systems are still also in common use.

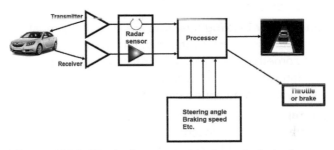

Figure 11.30 Block diagram animation of obstacle avoidance radar

Operation The operation of a basic radar system is as follows: a radio transmitter generates radio wave pulses, which are then radiated from an antenna. A target, such as another vehicle, scatters a small portion of the radio energy back to a receiving antenna. This weak signal is amplified and displayed on a screen. To determine its position, the distance (range) and bearing must be measured. Because radio waves travel at the speed of light (3 x 10⁸ m/s), the range may be found by measuring the time taken for a radio wave to travel from transmitter to obstacle, and back to the receiver.

Example If the range were 150m, the time for the round trip would be:

$$t = \frac{2d}{c}$$

where t = time, d = distance to object, and c = speed of light.

In this example it is 1 microsecond.

$$t = \frac{2 \times 150}{3 \times 10^8} = 1\mu S$$

Summary If the measured time for the round trip is 1 microsecond, in the above example, then the distance must be 150m. Relative closing speed can be calculated from the current vehicle speed. The type of display or output that may be used on a motor vehicle will vary from an audible warning to a warning light or series of lights or more likely a display screen.

 Make a simple sketch to show how one of the main components or systems in this section operates.

11.4.4 Camera ❸

Introduction Bosch has developed a stereo video camera such that an emergency braking system can function based solely on camera data. Normally, this would require a radar sensor or a combination of radar and video sensors.

Emergency braking systems are among the most effective assistance systems in the car. It is estimated that something like 70% of all rear-end collisions, resulting in personal injury, could be avoided if all vehicles were equipped with them.

Example Land Rover uses a stereo video camera together with the Bosch emergency braking system as standard in its Discovery Sport. When the camera recognizes another vehicle ahead in the lane as an obstruction, the emergency braking system prepares for action. If the driver does not react, then the system initiates maximum braking.

Figure 11.31 Land Rover using the camera system (Source: Bosch Media)

Other driver assistance functions can also be based on the stereo video camera. One such function is road-sign recognition, which keeps the driver informed about the current speed limit. Another is a lane-departure warning. This vibrates the steering wheel to warn drivers before they unintentionally drift out of lane.

Figure 11.32 Stereo video camera for ADAS (Source: Bosch Media)

Coverage With its light-sensitive lenses and video sensors, the camera covers a 50° horizontal field of vision and can take measurements in 3D at over 50m. Thanks to these spatial measurements, the video signal alone provides enough data to calculate, for example, the distance to vehicles ahead. Its pair of highly sensitive video sensors are equipped with colour recognition and CMOS (complementary metal oxide semiconductor) technology. They have a resolution of 1,280 by 960 pixels and can also process high-contrast images.

Figure 11.33 The camera recognizes another vehicle ahead in the same lane (Source: Bosch Media)

 Use the media search tools to look for pictures and videos relating to the subject in this section.

11.5 Alternative fuels ❶❷❸

Introduction The use of an alternative fuel can lessen dependence upon oil and reduce greenhouse gas emissions. There are a number of alternative fuels and each of these is outlined briefly in this section.

Ethanol Ethanol is an alcohol-based fuel made by fermenting and distilling starch crops, such as corn. It can also be made from plants such as trees and grasses. E10 is a blend of 10% ethanol and 90% petrol/gasoline. Almost all manufacturers approve the use of E10 in their vehicles. E85 is a blend of 85% ethanol and 15% petrol/gasoline and can be used in flexible fuel vehicles (FFVs). FFVs are specially designed to run on petrol/gasoline, E85, or any mixture of the two. These vehicles are offered by several manufacturers. There is no noticeable difference in vehicle performance when E85 is used. However, FFVs operating on E85 usually experience a 20–30% drop in miles per gallon due to ethanol's lower energy content.

Figure 11.34 E85 Vehicle

Advantages and disadvantages Here are some advantages and this screen lists some advantages and disadvantages of this alternative fuel:

Figure 11.35 E85 logo

Advantages:
▶ Lower emissions of air pollutants
▶ More resistant to engine knock
▶ Added vehicle cost is very small

Disadvantages:
▶ Can only be used in flex-fuel vehicles
▶ Lower energy content, resulting in fewer miles per gallon
▶ Limited availability

Biodiesel Biodiesel is a form of diesel fuel manufactured from vegetable oils, animal fats or recycled restaurant oils. It is safe, biodegradable and produces less air pollutants than petroleum-based diesel.

Biodiesel can be used in its pure form (B100) or blended with petroleum diesel. Common blends include B2 (2% biodiesel), B5, and B20. B2 and B5 can be used safely in most diesel engines. However, most vehicle manufacturers do not recommend using blends greater than B5, and engine damage caused by higher blends is not covered by some manufacturer warranties.

Figure 11.36 Biodiesel engine

Advantages and disadvantages Here are some advantages and this screen lists some advantages and disadvantages of this alternative fuel:

Figure 11.37 Biodiesel logo

11

Advantages:

▶ Can be used in most diesel engines, especially newer ones

▶ Less air pollutants (other than NOx) and less greenhouse gases

▶ Biodegradable

▶ Non-toxic

▶ Safer to handle

Disadvantages:

▶ Use of blends above B5 may not yet be approved by manufacturers

▶ Lower fuel economy and power (10% lower for B100, 2% for B20)

▶ More nitrogen oxide emissions

▶ B100 generally not suitable for use in low temperatures

▶ Concerns about B100's impact on engine durability

Natural gas Natural gas is a fossil fuel made up mostly of methane. It is one of the cleanest burning alternative fuels. It can be used in the form of compressed natural gas (CNG) or liquefied natural gas (LNG) to fuel cars and trucks.

Dedicated natural gas vehicles are designed to run on natural gas only, while dual-fuel or bi-fuel vehicles can also run on petrol/gasoline or diesel. Dual-fuel vehicles take advantage of the wide-spread availability of conventional fuels but use a cleaner, more economical alternative when natural gas is available. Natural gas is stored in high-pressure fuel tanks so dual-fuel vehicles require two separate fuelling systems, which takes up extra space.

Natural gas vehicles are not produced commercially in large numbers. However, conventional vehicles can be retrofitted for CNG.

Advantages and disadvantages Here are some advantages and this screen lists some advantages and disadvantages of this alternative fuel:

Figure 11.38 CNG logo

Advantages:

▶ 60–90% less smog-producing pollutants

▶ 30–40% less greenhouse gas emissions

▶ Less expensive than petroleum fuels

Disadvantages:

▶ Limited vehicle availability

▶ Less readily available

▶ Fewer miles on a tank of fuel

Propane or liquefied petroleum gas (LPG) Propane or liquefied petroleum gas is a clean-burning fossil fuel that can be used to power internal combustion engines. LPG-fuelled vehicles produce fewer toxic and smog-forming air pollutants.

Petrol/gasoline and diesel vehicles can be retrofitted to run on LPG in addition to conventional fuel. The LPG is stored in high-pressure fuel tanks, so separate fuel systems are needed in vehicles powered by both LPG and a conventional fuel.

Figure 11.39 Propane Tank (Source: http://www.rasoenterprises.com)

Advantages and disadvantages Here are some advantages and this screen lists some advantages and disadvantages of this alternative fuel:

Figure 11.40 LPG logo

Advantages:

▶ Fewer toxic and smog-forming air pollutants
▶ Less expensive than petrol/gasoline

Disadvantages:

▶ No new passenger cars or trucks commercially available but vehicles can be retrofitted for LPG
▶ Less readily available than conventional fuels
▶ Fewer miles on a tank of fuel

Hydrogen Hydrogen (H_2) can be produced from fossil fuels (such as coal), nuclear power, or renewable resources, such as hydropower. Fuel cell vehicles powered by pure hydrogen emit no harmful air pollutants. Hydrogen is being aggressively explored as a fuel for passenger vehicles. It can be used in fuel cells to power electric motors or burned in internal combustion engines. It is an environmentally friendly fuel that has the potential to dramatically reduce dependence on oil, but several significant challenges must be overcome before it can be widely used.

Figure 11.42 Hydrogen logo

Advantages:

▶ Can be produced from several sources, reducing dependence on petroleum
▶ No air pollutants or greenhouse gases when used in fuel cells
▶ It produces only NOx when burned in internal combustion engines (no CO or HC is produced)

Disadvantages:

▶ Expensive to produce and is only available at a few locations
▶ Fuel cell vehicles are currently too expensive for most consumers
▶ Hydrogen has a lower energy density than conventional petroleum fuels. For this reason it is difficult to store enough hydrogen on a vehicle to travel more than 200 miles

Summary This section has given an overview of some alternative fuels. All of them offer some significant advantages either commercially, environmentally or both. There are also some disadvantages not least of which is that the cost of production is high. This is likely to change however as the use of alternative fuels becomes more widespread.

Figure 11.41 Fuel cell vehicle (Source: Honda Media)

Advantages and disadvantages Here are some advantages and this screen lists some advantages and disadvantages of this alternative fuel:

 Select a routine from section 1.3 and follow the process to study a component or system.

Index

Page numbers in *italics* denote an illustration, **bold** indicates a table